Classical and Modern Potential Theory and Applications

NATO ASI Series

Advanced Science Institutes Series

A Series presenting the results of activities sponsored by the NATO Science Committee, which aims at the dissemination of advanced scientific and technological knowledge, with a view to strengthening links between scientific communities.

The Series is published by an international board of publishers in conjunction with the NATO Scientific Affairs Division

A Life Sciences **B Physics**	Plenum Publishing Corporation London and New York
C Mathematical and Physical Sciences **D Behavioural and Social Sciences** **E Applied Sciences**	Kluwer Academic Publishers Dordrecht, Boston and London
F Computer and Systems Sciences **G Ecological Sciences** **H Cell Biology** **I Global Environmental Change**	Springer-Verlag Berlin, Heidelberg, New York, London, Paris and Tokyo

NATO-PCO-DATA BASE

The electronic index to the NATO ASI Series provides full bibliographical references (with keywords and/or abstracts) to more than 30000 contributions from international scientists published in all sections of the NATO ASI Series.
Access to the NATO-PCO-DATA BASE is possible in two ways:

– via online FILE 128 (NATO-PCO-DATA BASE) hosted by ESRIN, Via Galileo Galilei, I-00044 Frascati, Italy.

– via CD-ROM "NATO-PCO-DATA BASE" with user-friendly retrieval software in English, French and German (© WTV GmbH and DATAWARE Technologies Inc. 1989).

The CD-ROM can be ordered through any member of the Board of Publishers or through NATO-PCO, Overijse, Belgium.

Series C: Mathematical and Physical Sciences - Vol. 430

Classical and Modern Potential Theory and Applications

edited by

K. GowriSankaran

Department of Mathematics and Statistics,
McGill University,
Montreal, Quebec, Canada

J. Bliedtner

Department of Mathematics,
J.W. Goethe University,
Frankfurt, Germany

D. Feyel

Department of Mathematics,
University Evry-Essone,
Evry, France

M. Goldstein

Department of Mathematics,
Arizona State University,
Tempe, Arizona, U.S.A.

W. K. Hayman

Department of Mathematics,
University of York,
Heslington, York, U.K.

and

I. Netuka

Faculty of Mathematics and Physics,
Charles University,
Praha, Republic of Czechia

Kluwer Academic Publishers

Dordrecht / Boston / London

Published in cooperation with NATO Scientific Affairs Division

Proceedings of the NATO Advanced Research Workshop on
Classical and Modern Potential Theory and Applications
Chateau de Bonas, France
July 25–31, 1993

A C.I.P. Catalogue record for this book is available from the Library of Congress.

ISBN 0-7923-2803-5

Published by Kluwer Academic Publishers,
P.O. Box 17, 3300 AA Dordrecht, The Netherlands.

Kluwer Academic Publishers incorporates the publishing programmes of
D. Reidel, Martinus Nijhoff, Dr W. Junk and MTP Press.

Sold and distributed in the U.S.A. and Canada
by Kluwer Academic Publishers,
101 Philip Drive, Norwell, MA 02061, U.S.A.

In all other countries, sold and distributed
by Kluwer Academic Publishers Group,
P.O. Box 322, 3300 AH Dordrecht, The Netherlands.

Printed on acid-free paper

Printed in the Netherlands

Contents

PREFACE

A NATO Advanced Research Workshop on Classical and Modern Potential Theory and Applications was held at the Chateau de Bonas, France, during the last week of July 1993. The workshop was organized by the Co-Directors M. Goldstein (Arizona) and K. GowriSankaran (Montreal). The other members of the organizing committee were J. Bliedtner (Frankfurt), D. Feyel (Paris), W.K. Hayman (York, England) and I. Netuka (Praha). The objective of the workshop was to bring together the researchers at the forefront of the aspects of the Potential Theory for a meaningful dialogue and for positive interaction amongst the mathematicians practising different aspects of the theory and its applications. Fifty one mathematicians participated in the workshop. The workshop covered a fair representation of the classical aspects of the theory covering topics such as approximations, radial behaviour, value distributions of meromorphic functions and the modern Potential theory including axiomatic developments, probabilistic theories, studies on infinite dimensional Wiener spaces, solutions of powers of Laplacian and other second order partial differential equations.

There were keynote addresses delivered by D. Armitage (Belfast), N. Bouleau (Paris), A. Eremenko (Purdue), S.J. Gardiner (Dublin), W. Hansen (Bielefeld), W. Hengartner (Laval U., Quebec), K. Janssen (Dusseldorf), T. Murai (Nagoya), A. de la Pradelle (Paris) and J.M. Wu (Urbana). There were thirty six other invited talks of one half hour duration each. The lectures covered quite a range of topics evenly distributed between the classical and the modern aspects of the subject. We want to thank all the speakers for their extremely interesting and stimulating talks. We also want to thank all the participants for such interesting scientific discussions and for the excellent research atmosphere that prevailed during the entire week of the workshop. A lot of credit goes to the staff at the Chateau who looked so well after the participants and their various needs giving us all the opportunity to engage in scientific activities. Our sincere thanks are due to them.

This volume consists of the Proceedings of the workshop. They include the contributions based on talks given at the workshop as well as a some of the research which culminated from the discussions that took place during that time. All the contributions were put through the normal refereeing procedure and only the papers accepted by the referees appear in this volume. The volume also includes the outcome of an afternoon session on the discussion of state of the art problems in the subject. Our thanks are due to Mr. Xun Zhang (Montreal) for his help in editing this part of the Proceedings.

This workshop was financed mostly by the Scientific Affairs Division of NATO. Additional support was provided by the Ministère de la Defense, Gouvernment de France. We express our sincere thanks to all of them for the generous support.

Frau C. Belz (Frankfurt) helped us in the preparation of the abstracts and the other material before the workshop and our thanks are due to her. We want to express our very special gratitude to the administrative staff in the Department of Mathematics and Statistics of McGill University, and in particular to Ms. Valerie McConnell, Elaine Swan and Elaine Tremblay who did so much work concerning the workshop and later with the editorial work of the Proceedings. Also Ms. Heather MacAuliffe and Elaine Swan retyped some of the manuscripts and reformatted many other manuscripts and our thanks are due to them for tremendously efficient work in such a short time framework. We would like to thank Ms. Luna Zaru and Mr. Xun Zhang for proofreading. Last but not the least, we would like to express our thanks to the Kluwer Academic Publishers, Dordrecht, The Netherlands, and in particular to Mrs. Odet Kiers-Pols for such efficient and helpful attention.

K. GowriSankaran
and
J. Bliedtner
D. Feyel
M. Goldstein
W.K. Hayman
I. Netuka

LIST OF PARTICIPANTS

David R. Adams, Dept. of Mathematics, University of Kentucky, Lexington, KY, 40506, USA (dave@ms.uky.edu).

Ömer Akin, Dept. of Mathematics, Faculty of Science, University of Ankara, 06100 Tandogan, Ankara, Turkey.

Ivan Andronov, Dept. of Mathematical Physics, State University of St. Petersburg, Russia (andronov@ceremab.greco-prog.fr).

David Armitage, Dept. of Mathematics, The Queen's University of Belfast, Belfast BT7 1NN, Northern Ireland, UK (pmg0398@vax2.queens-belfast.ac.uk)

Line Baribeau, Département de Mathématiques et Statistiques, Université Laval, Québec, Canada, G1K 7P4 (lbarib@mat.ulaval.ca).

Jürgen Bauer, Dept. of Mathematics, Johann Wolfgang Goethe-Universität, 600054 Frankfurt am Main, Germany.

Alexander Bendikov, Mathematisches Institut, Universität Erlangen-Nürnberg, Bismarckstrasse 1½, 91054 Erlangen, Germany.

Christian Berg, Kobenhavns Univ. Mat. Inst., Universitetsparken 5, DK-2100 Kobenhavn, Denmark (berg@math.ku.dk).

Emile Bertin, Mathematisch Istituut, Rijksuniversiteit, Utrecht, The Netherlands (bertin@math.ruu.nl).

Lucien Beznea, Mathematisch-Geographische Fakultät, Katholische Universität Eichstätt, Ostenstrasse 26-28, 85071 Eichstätt, Germany.

Marco Biroli, Dipartimento di Mathematica, Politecnico di Milano, Piazza Leonardo da Vinci 32, 20133 Milano, Italy.

Jürgen Bliedtner, Fachbereich Mathematik, Johann Wolfgang Goethe-Universität, 60054 Frankfurt am Main, Germany (jBli@mathematik.uni-frankfurt.400.de)

Nicu Boboc, Faculty of Mathematics, University of Bucarest, Str. Academiei 14, Ro-70109 Bucharest, Romania.

Nicolas Bouleau, Directeur du Lab de Math. Appliquées, U.A. CNRS 1502, Ecole Nationale des Ponts et Chausées, La Courtine, 93167 Noisy Le Grand, France (nb@clio.enpc.fr)

Bernhard Burgeth, Mathematisches Institut, Universität Erlangen-Nürnberg, Bismarckstrasse $1\frac{1}{2}$, 91054 Erlangen, Germany (leutwil@mi.uni-erlangen.de).

Ana Bela Cruzeiro, Centro de Mat e Aplicacoes Fundamentais, Ac Prof Gama Pinto, 1699 Lisboa, Portugal (cruzeiro@ptifm.2.ifm.pt).

Markus Denzer, Fachbereich Mathematik, Johann Wolfgang Goethe-Universität, 60054 Frankfurt am Main, Germany

Alexandre Eremenko, Department of Mathematics, Purdue University, West Lafayette, IN 47907, USA (eremenko@math.purdue.edu).

Sirkka-Liisa Eriksson-Bique, Department of Mathematics, University of Joensuu, P.O. Box 111, SF-80101 Joenssu, Finland (eriksson@finujo.bitnet).

Denis Feyel, Département de Mathématiques, Université Evry-Essonne, Boulevard des Coquibus, 91025 EVRY, France.

Bent Fuglede, Mat. Inst. Universitetsparken 5, DK-2100 Kobenhavn, Denmark (fuglede@math.ku.dk).

Stephen Gardiner, University College Dublin, Dublin 4, Ireland, UK (gardiner@ irlearn.ucd.ie).

Ronald Getoor, Department of Mathematics, University of California at San Diego, La Jolla, CA 92093-0112, USA (rgetoor@ucsd.edu).

Myron Goldstein, Department of Mathematics, Arizona State University, Tempe, Arizona 85287, USA.

Kohur GowriSankaran, Department of Mathematics and Statistics, McGill University, 805 Sherbrooke Street West, Montreal, Quebec, Canada, H3A 2K6 (gowri@ math.mcgill.ca).

Alexander Grigor'yan, Fakultät für Mathematik, Universität Bielefeld, Universitätsstrasse 1, 33501 Bielefeld, Germany (grigor@mathematik.uni-bielefed.de).

Wolfard Hansen, Fakultät für Mathematik, Universität Bielefeld, Universitätsstrasse 1, 33501 Bielefeld, Germany (hansen@mathematik.uni-bielefed.de).

Victor P. Havin, Department of Mathematics, Leningrad State University, Bibliotechnaia ploshcad 2, Stary Peterhof, St. Petersburg 198904, Russia.

Walter K. Hayman F.R.S., University of York, Heslington, York Y01 5DD, UK.

Walter Hengartner, Département de Mathématiques et Statistiques, Université Laval, Québec, Québec, Canada G1K 7P4 (walheng@mat.ulaval.ca).

Francis Hirsch, Département de Mathématiques, Université Evry-Essone, Boulevard des Coquibus, 91025 Evry, France.

Klaus Janssen, Mathematisches Institut, Heinrich Heine-Universität, Universitätsstrasse 1, 40225 Düsseldorf, Germany.

Heinz Leutwiler, Mathematisches Institut, Universität Erlangen-Nürnberg, Bismarckstrasse $1\frac{1}{2}$, 91054 Erlangen, Germany (leutwil@mi.uni-erlangen.de).

Peter A. Loeb, Department of Mathematics, University of Illinois at Urbana-Champaign, 1409 West Green Street, Urbana, IL 61801, USA (loeb@math.uiuc.edu).

Oleg Loukianov, Université Evry-Essone, Boulevard des Coquibus, 91025 Evry CEDEX, France.

Jaroslav Lukeš, Faculty of Mathematics and Physics, Charles University, Sokolovska 83, 186000 Praha 8, Republic of Czech (jlukes@cspguk11.bitnet).

Gabriel Mokobodzki, Equipe d'analyse, Université Paris VI, 4 Place Jussieu, 75252 Paris 5, France.

Takafumi Murai, College of Science, Nagoya University, Nagoya, Japan.

Tadashi Murazawa, Department of Mathematics, Kyoto Perfectural University, Shimogamo-Sakyo, Kyoto 606, Japan (a50436@jpnkudpc.bitnet).

Ivan Netuka, Faculty of Mathematics and Physics, Charles University, Sokolovska 83, 186000 Praha 8, Republic of Czech (netuka@cspguk11.bitnet).

Fernando Pérez-Gonzàlez, Departamiento de Anàlisis Matemàtico, Universitad de La Laguna, 38271 La Laguna, Tenerife, Spain (fpg@ull.es).

Arnaud de La Pradelle, Equipe d'analyse, Université Paris VI, 4 Place Jussieu, 75252 Paris 5, France.

Murali Rao, Department of Mathematics and Statistics, University of Florida, Gainesville, FL 32611-2082, USA (rao@math.ufl.edu).

Hermann Render, Fachbereich 11, Mathematik, Universität-GH Duisburg, Lotharstr. 65, D-47048 Duisburg, Germany (render@math.uni.duisburg.de).

Lee A. Rubel, Department of Mathematics, University of Illinois at Urbana-Champaign, 1409 West Green Street, Urbana, IL 61801, USA (rubel@symcom.math.uiuc.edu).

Jozef Siciak, Inst. Math. UJ, Reymonta 4, 30-059 Krakow, Poland (siciak@im.uj.edu.pl).

Emmanuel Smyrnelis, Equipe d'analyse, Université Paris VI, 4 Place Jussieu, 75252 Paris 5, France.

Shiqi Song, Université Evry-Essone, Boulevard des Coquibus, 91025 Evry CEDEX, France.

Jiri Vesely, Faculty of Mathematics and Physics, Charles University, Sokolovska 83, 186000 Praha 8, Republic of Czech (jvesely@cspguk11.bitnet).

Jang-Mei Wu, Department of Mathematics, University of Illinois at Urbana-Champaign, 1409 West Green Street, Urbana, IL 61801, USA (offic@symcom.math.uiuc.edu).

Xun Zhang, Department of Mathematics and Statistics, McGill University, 805 Sherbrooke Street West, Montreal, Quebec, Canada, H3A 2K6 (xun@zhapod.math.mcgill.ca).

Nonlinear PDE and the Wiener Test

D.R. ADAMS
Department of Mathematics
University of Kentucky
Lexington
Kentucky, 40506-0027

1. Introduction

In classical potential theory, the discovery of the Wiener Test for boundary regularity for harmonic functions with prescribed boundary values, was a singular achievement. It solved a long standing open problem of the time: find geometric conditions on the boundary of the domain that guarantee that the solution (continuous inside) takes on its boundary values continuously. This result is due to N. Wiener (1924) and first published in [W]. It can be expressed as the divergence of the Dini type integral

$$WT_2(x;\Omega) \equiv \int_0 r^{2-n} C_2(\Omega^c \cap B(x,r) \frac{dr}{r}$$

where Ω is the (bounded) domain (contained in $\boldsymbol{R}^n$, $n \geq 3$) in question, $B(x,r)$ is a (closed) ball of radius $r > 0$ centered at $x \in \partial\Omega =$ boundary of Ω, and C_2 is the usual variational capacity,

$$C_2(K) = \inf \left\{ \int |\nabla \phi|^2 \, dx : \ \phi \in C_0^\infty(\boldsymbol{R}^n), \ \phi > 1 \ \text{ on } K \right\}. \tag{1}$$

Here K is a compact subset of $\boldsymbol{R}^n$ and $\Omega^c = \boldsymbol{R}^n \setminus \Omega$. That is, if f is a given continuous function defined on $\partial\Omega$, and u is the solution to

$$\Delta u = 0, \quad \Omega$$

$$u = f, \quad \partial\Omega$$

is a generalized sense (e.g. the Perron solution or a weak $W^{1,2}(\Omega)$ solution), then

$$\lim_{\substack{y \to x \\ y \in \Omega}} u(y) = f(x) \tag{2}$$

for all such f if and only if $WT_2(x;\Omega) = +\infty$. Such a point $x \in \partial\Omega$ satisfying (2) for all such f is termed a *regular point* for the Laplace equation in Ω.
Since Wiener's original paper, many authors have extended, and expanded upon this result, first to more general linear elliptic equations (and parabolic equations), and

K. GowriSankaran et al. (eds.), Classical and Modern Potential Theory and Applications, 1–9.

more recently to certain nonlinear elliptic equations. The purpose of this article is to briefly review some of these highlights especially the most recent ones, and then to describe how some of these more recent ideas can be used to extend the Wiener Test to a very large family of nonlinear equations of the form

$$\operatorname{div} A(x,u,\nabla u) = B(x,u,\nabla u), \tag{3}$$

where A and B satisfy specific L^p-growth conditions:

$$(a) \quad A(x,u,\nabla u) = \langle a_{\alpha\beta}(x,u,\nabla u)|\nabla u|^{p-2}\partial_\alpha u\rangle$$

(with the usual summation convention). The $\langle \quad \rangle$ denotes a vector with n components, $\beta = 1,...,n$.

$$(a_1) \quad a_{\alpha\beta}(x,t,\zeta)\xi_\alpha\xi_\beta \geq \lambda|\xi|^2, \quad \lambda > 0$$

$$(a_2) \quad |a_{\alpha\beta}(x,t,\zeta)| \leq M < \infty,$$

for all $(x,t,\zeta) \in \Omega \times \boldsymbol{R}^{n+1}$, $\xi \in \boldsymbol{R}^n$, $\xi \neq 0$.

$$(b) \quad B(x,t,\zeta) \leq b_1|\zeta|^p + b_2,$$

b_1 and b_2 positive constants $1 < p < \infty$. (And as usual, $a_{\alpha\beta}$ and B should be measurable in x and continuous in t and ζ in the usual Caratheodory sense.) The prototype is the p-Laplace equation

$$\Delta_p u \equiv \operatorname{div}(|\nabla u|^{p-2}\nabla u) = 0.$$

Notice that $\Delta_2 = \Delta \equiv \partial_\alpha\partial_\alpha$, $\partial_\alpha = \partial/\partial x_\alpha$, $\alpha = 1,...,n$. Of course $\nabla u = \langle\partial_\alpha u\rangle$.
In the linear case, the definitive result of Littman-Stampacchia-Wienberger [LSW] is, in itself, remarkable. There the authors show that a point $x \in \partial\Omega$ is a regular point for the equation

$$\partial_\beta(A_{\alpha\beta}(y)\partial_\alpha u) = 0 \tag{4}$$

in Ω with $a_{\alpha\beta}(y)\xi_\alpha\xi_\beta \geq \lambda|\xi|^2$, $\lambda > 0$, and $|a_{\alpha\beta}(y)| \leq M \leq \infty$, for all $y \in \Omega$, $\xi \in \boldsymbol{R}^n \setminus \{0\}$ if and only if x is a regular point for the Laplace equation in Ω. Linear equations with lower order terms were subsequently considered in [S].
The theory for nonlinear second order elliptic equations began with V.G. Maz'ya [M1] when he found that a sufficient condition for a point $x \in \partial\Omega$ to be a regular point for the p-Laplace equation, it is: $WT_p(x;\Omega) = +\infty$, where

$$WT_p(x;\Omega) \equiv \int_0 \{r^{p-n}C_p(\Omega^c \cap B(x,t)\}^{p'-1}\frac{dt}{t}$$

with $p' = p/(p-1)$, $1 \leq p \leq n$. $C_p(\cdot)$ is the variational capacity obtained by replacing $|\nabla\phi|^2$ by $|\nabla\phi|^p$ in (1). This result has been extended to a class of equations that includes those of (3) $(a), (a_1), (a_2), (b)$ by R. Gariepy-W.P. Ziemer [GZ], but since there is in general no maximum principle for solutions to (3), the authors restrict attention to the class bounded (weak) solutions to (3) with specified boundary values.

However, it took several years before effective tools were developed to deal with the question of the necessity of $WT_p(x;\Omega) = +\infty$ for x a regular point even for the p-Laplace equation $p \neq 2$. The first step in this direction was taken by P. Lindqvist - O. Martio [LM]. They established this necessity for $p = n$ (and then subsequently it was realized that their argument worked for $n-1 < p \leq n$. That is, they showed equivalently that if $WT_p(x;\Omega) < \infty$, then there is a continuous function f on $\partial\Omega$ such that the corresponding (generalized) solution to

$$\Delta_p u = 0, \quad \text{in } \Omega$$

$$u = f, \quad \text{on } \partial\Omega$$

(which is necessary continuous in Ω), has a discontinuity at $x \in \partial\Omega$.

And then finally in 1993, T. Kilpelanien - J. Maly [KM1] found the long sought after solution to the question that intrigued us all: is $WT_p(x;\Omega) = +\infty$ necessary for $x \in \partial\Omega$ to be a regular point for the p-Laplace equation in Ω, for all $1 < p \leq n$? The answer is yes, furthermore and the methods they develop are quite technical and frankly rather astonishing - perhaps explaining why it took so long! Actually, [KM1] treats a slightly more general equation than $\Delta_p u = 0$, namely

$$\operatorname{div} A(x, \nabla u) = 0$$

where $A(x,\xi)\cdot\xi \geq \lambda|\xi|^p$, $|A(x,\xi)| \leq M|\xi|^{p-1}$, and $A(x,t\xi) = t|t|^{p-1}A(x,\xi)$ for $t \neq 0$.

For equations like (3), esp. with lower order terms that exhibit the critical growth in the B term, i.e. condition (b), little is known with regard to the necessity of $WT_p(x;\Omega) = +\infty$, $x \in \partial\Omega$, when x is a regular point for Ω. In the semilinear case, there are several relevant papers: [N], [Sk], [AH]. Here, the main thing we do is to investigate the necessity of $WT_p(x;\Omega) = +\infty$ when x is a regular point for (3) $(a), (a_1), (a_2), (b)$ on Ω. But first, we need to introduce the concept of *local regular point.* We say $x \in \partial\Omega$ is a local regular point for Ω if it is a regular point for any subdomain $\Omega' \subset \Omega$ such that $x \in \partial\Omega \cap \partial\Omega$. Notice that if x is irregular (i.e. not regular), then there exists a subdomain Ω' and a continuous function f on $\partial\Omega'$ for which the solution on Ω' has a boundary discontinuity. Also notice that if $WT_p(x;\Omega) = +\infty$, then $WT_p(x,\Omega') = +\infty$, for all $\Omega' \subset \Omega$, hence the sufficiency of $WT_p(x;\Omega) = +\infty$ actually implies that x is a local regular point - cf. the results of [M1] or [GZ]. Our result is

Theorem 1:
Assume that the Dirichlet problem for equation (3) $(a)(a_1)(a_2)(b)$ has a unique $W^{1,p}(\Omega')\cap L^\infty(\Omega')$ solution $u = u_f$ for each $f \in C^\infty(\bar{\Omega}')$ where $\Omega' \subset \Omega\,(x \in \partial\Omega' \cap \partial\Omega$, diam Ω' sufficiently small), then if x is a local regular point for Ω, it follows that $WT_p(x;\Omega) = +\infty$.

Remarks.
(1) The precise meaning of solvability above is: there is a unique $u \in W^{1,p}(\Omega'0 \cap L^\infty(\Omega')$ such that

$$\int_{\Omega'} A(x,u,\nabla u)\nabla\eta\,dx + \int_{\Omega'} B(x,u,\nabla u)\eta\,dx = 0$$

for all $\eta \in W_0^{1,p}(\Omega')$ and $u - f \in W_0^{1,p}(\Omega')$.
(2) The ideas for the proof of our theorem are mainly those of [KM1] with a little [AH] thrown in.

Proof.
We suppose that $WT_p(x;\Omega) < \infty$ and then produce a subdomain $\Omega' \subset \Omega$ of the required type for which x is irregular. In fact we take $\Omega' = \Omega \cap B(x,R)$ for some sufficient small R.

STEP 1.
With f yet to be determined, we let $u = u_f$ and consider $v = e^{cu} - 1$, $c > 0$. Clearly $\nabla v = ce^{cu}\nabla u$, so

$$\partial_\beta(a_{\alpha,\beta}(x,u,\nabla u)|\nabla v|^{p-2}\partial_\alpha v) = c^{p-1}\partial_\beta(a_{\alpha,\beta}e^{c(p-1)u}|\nabla u|^{p-2}\partial_\alpha u)$$

$$= c^{p-1}\{\partial_\beta(a_{\alpha,\beta}|\nabla u|^{p-2}\partial_\alpha u)e^{c(p-1)u} + a_{\alpha,\beta}|\nabla u|^{p-2}\partial_\alpha u\cdot c(p-1)\partial_\beta u\cdot e^{c(p-1)u}\}$$

$$\geq c^{p-1}e^{c(p-1)u}\{-b_1|\nabla u|^p - b_2 + c(p-1)\lambda|\nabla u|^p\}$$

$$= c^{p-1}e^{c(p-1)u}\{-b_2\} = -c^{p-1}b_2(v+1)^{p-1},$$

upon choosing $c = b_1/\lambda(p-1)$.
Next, we set $\mathcal{L}_p(u)z \equiv -\partial_\beta(A_{\alpha\beta}(u)|\nabla z|^{p-2}\partial_\alpha z)$, where

$$A_{\alpha\beta}(y) = \begin{cases} a_{\alpha,\beta}(y,u(y),\nabla u(y)), & y \in \Omega' \\ \delta_{\alpha\beta}, \quad y \notin \Omega'. \end{cases}$$

Then notice that we have $\mathcal{L}_p(u)v \leq \sigma(v+1)^{p-1}$ in Ω', and $v = e^{cf} - 1 \equiv \Phi$ on $\partial\Omega'$, $\sigma = c^{p-1}b_2$.

STEP 2.
We now specify Φ (and hence f) by taking $\Phi \in C_0^\infty(\boldsymbol{R}^n)$ such that $\Phi \equiv 1$ on $B(x,\rho/2)$, $\Phi = 0$ on $\partial B(x,\rho)$, and $0 \le \Phi \le 1$; $0 < \rho < R/4, R$ yet to be determined. And we solve the auxilliary problem

$$\min \int_{B(x,R)} [A_{\alpha\beta}(y)|\nabla w|^{p-2}\partial_\alpha w \partial_\beta w - \sigma(w+1)^p]\, dy \tag{5}$$

where the minimum is taken over all $w \in W_0^{1,p}(B(x,R))$ such that $w = 1$ on $\Omega^c \cap B(x,\rho)$. By standard variational arguments, esp. when $\Omega' = \Omega \cap B(x,R)$ is sufficiently small, problem (5) has a unique solution w which is a weak solution to the differential equation

$$-\partial_\beta(A_{\alpha\beta}(y)|\nabla w|^{p-2}\partial_\alpha w) = \sigma(w+1)^{p-1} + \mu$$

where μ is a positive Borel measure with support in $\Omega^c \cap B(x,\rho)$. (The restriction on R above is necessary in order to insure that the eigenvalue problem $-\mathrm{div}(A(x)|\nabla u|^{p-2}\nabla u) = \sigma u^{p-1}$ in B, $u = 0$ on ∂B, $\sigma > 0$, has a unique solution, and consequently the integral in (5) is bounded below by a negative constant depending only on λ and R.)
Next, we notice that although $w = 1$ quasi-everywhere (i.e. almost everywhere with respect to $C_p(\cdot)$) on $\Omega^c \cap B(x,\rho)$, we have $w(x) < 1/2$ for R sufficiently small. To see this, we need two facts, expressed below as lemmas.

Lemma 1:
If μ is the above Borel measure, then $\mu(e) \le$ const. $C_p(e)$, for any Borel set e.

Proof.
If we denote by μ^e the restriction of μ to e, then

$$\begin{aligned}\mu(e) &\le \int I_1 * g \, d\mu^e = \int g I_1 * \mu^e \le ||g||_p \, ||I_1 * \mu^e||_{p'} \\ &\le R_{1,p}(e)^{1/p} \cdot \text{const.} \left[\int W_{1,p}^{\mu^e} \, d\mu^e\right]^{1/p'}\end{aligned} \tag{6}$$

where $I_1(y) = |y|^{1-n}$, the Riesz kernel of order 1, and $R_{1,p}$ is the corresponding Riesz capacity on $\boldsymbol{R}^n$. $W_{1,p}^\nu(y)$ is the Wolf potential of the measure ν, i.e.

$$W_{1,p}^\nu(y) = \int_0^\infty [r^{p-n}\nu(B(y,r))]^{p'-1} \frac{dr}{r}$$

and the last inequality in (6) is Wolff's inequality; see [HW]. Now, since supp μ is compact, it follows that $W_{1,p}^\mu(y) \le W_{1,p}^\mu(y, R/4) +$ const. $\le$ const.$w+$const., because the solution to (5) is bounded and from [KM1] (Theorem 1.6), we know that

$$W_{1,p}^\mu(y, R/4) \le \text{const. } w(y)$$

for all $y \in \text{supp}\,\mu$. Thus

$$\mu(e) \leq \text{const. } R_{1,p}(e) \leq \text{const. } C_p(e).$$

Lemma 2:
If z is a $W_0^{1,p}(B(x_0,2r))$ solution to $\partial_\beta(A_{\alpha\beta}(x)\partial_\alpha z) = \nu$, a Borel measure supported in $B(x_0,r)$, then

$$z(x_0) \leq \text{const. } W_{1,p}^{\nu}(x_0,2r).$$

Proof.
See [KM], Theorem 1.6.

From these two facts we can conclude:

$$\begin{aligned} w(x) &\leq \text{const. } W_{1,p}^{[(w+1)^{p-1}+\mu]}(x;2R) \\ &\leq \text{const. } W_{1,p}^{\mu}(x;2R) + \text{const. } R \end{aligned}$$

Consequently, $w(x) < 1/2$ for R sufficiently small since $WT_p(x;\Omega) < \infty$.

STEP 3.
Our final step is to compare the two functions v and w. But this is easy since $v \leq w$ q.e. on $\partial(\Omega \cap B(x,r))$, because $v = \Phi$ there and $\Phi \leq w$ q.e. there. The desired conclusion now follows from the basic comparisons principles for sub and super solutions to

$$\mathcal{L}_p(u)z = \sigma(z+1)^{p-1} \text{ in } \Omega \cap B(x,r).$$

In fact one can easily modify the proof of Theorem 9.5 of [GT] to cover this situation, provided, again, that the parameter R is chosen sufficiently small. For similar such comparison results, see also [KM2]. Thus we conclude by arguing as in [KM], to get

$$\liminf_{\substack{y \to x \\ y \in \Omega \cap B(x,r)}} v(y) \leq \liminf_{\substack{y \to x \\ y \in \Omega \cap B(x,r)}} w(y) = w(x) < 1/2 < \Phi(x) = 1 \qquad (7)$$

Notice that we have used the lower semi-continuity of the solution w in (7). Hence v has a boundary discontinuity at x which implies that u must also. This concludes our proof.
As a corollary to our argument above, we notice that we can also obtain a partial converse of a boundary regularity result of Paulik [P] for certain nonlinear elliptic systems satisfying the "smallness condition". Here we are referring to the systems

$$-\partial_\beta(a_{\alpha,\beta}(y,u,\nabla u)\partial_\alpha u^i) = B^i(y,u,\nabla u) \text{ in } \Omega \qquad (8)$$

where $\alpha, \beta = 1, ..., n$, $y \in \Omega \subset \boldsymbol{R}^n$, Ω bounded, $i = 1, ..., N$, and $u^i = f^i$ on $\partial\Omega$.

Our assumptions are: $a_{\alpha,\beta}(y,t,\zeta)$ satisfies (a_1) and (a_2), and

$$(b^*) \qquad t\cdot B(y,t,\zeta)\le b_1^*|\zeta|^2+b_2^*.$$

The smallness assumption is:

$$(c) \qquad b_1^*<\lambda.$$

¿From $(a_1),(a_2),(b^*),(c)$, we have a uniform estimate on the L^2-norm of the gradient, locally, in terms of $||u||_\infty$, but more to our point, we have (as before)

$$\mathcal{L}_2(u)v-2cb_2^*v\le 2cb_2^* \text{ in } \Omega \tag{9}$$

for any $c>0$, when u solves (8) and

$$v=e^{c|u|^2}-1.$$

Again, we set $\Phi=e^{c|f|^2}-1$. $u=\langle u^1,...,u^N\rangle$, $f=\langle f^1,...,f^N\rangle$. In (9), we have used (c) in a crucial way.
Thus, as we have done already, we next solve the auxilliary problem

$$\begin{aligned}\mathcal{L}_2(u)w-2cb_2^*w&=2cb_2^* \text{ in } \Omega\\ w&=\Phi \text{ on } \partial\Omega\end{aligned} \tag{10}$$

and now choose c small enough so that it has a unique solution, i.e. less that the first eigenvalue of $\mathcal{L}_2(u)$ on Ω divided by λ. The maximum principle - in this case - then provides the needed comparison principle for $v\le w$ as before yielding the discontinuous nature of v and hence of u at the boundary point x. Thus we have

Corollary 1:
Assuming that equation (8), satisfying $(a_1),(a_2),(b^*),(c)$ has a unique $W^{1,2}(\Omega)\cap L^\infty(\Omega)$ vector solution for each $f^i\in C^\infty(\bar\Omega)$, $i=1,...,N$, then if $x\in\partial\Omega$ is a regular point for Ω, it follows that $WT_2(x;\Omega)=+\infty$.
Next, we note that linear elliptic systems (8) always satisfies the above conditions of Corollary 1, i.e. when $a_{\alpha,\beta}=a_{\alpha,\beta}(x)$ and $B^i=b_{ij}u^j+f^i(x)$. In particular the system

$$\Delta u=Bu+f(x) \text{ in } \Omega,\ u=0 \text{ on } \partial\Omega,\ \ f^2\in L^\infty(\Omega),\ f^1=0$$

$u=(u^1,u^2)$, $B=\begin{pmatrix}0&1\\0&0\end{pmatrix}$, is always solvable in $W_0^{1,2}(\Omega)$. Thus by Corollary 1, $x\in\partial\Omega$ is a regular point for the biharmonic problem

$$\Delta^2 z=f^2 \text{ in } \Omega$$

$$z=0 \text{ on } \partial\Omega$$

$$\Delta z=0 \text{ on } \partial\Omega$$

if and only if $WT_2(x;\Omega) = +\infty$. However, in contrast to this, V.G. Maz'ya [M2] has shown that a sufficient condition for $x \in \partial\Omega$ to be a regular point for the usual Dirichlet problem for the biharmonic operator

$$\Delta^2 z = f \quad \text{in } \Omega$$

$$z = 0 \quad \text{on } \partial\Omega$$

$$\frac{\partial}{\partial n} z = 0 \quad \text{on } \partial\Omega$$

$z \in W_0^{2,2}(\Omega)$, is that

$$\int_0 r^{4-n} C_{2,2}(\Omega^c \cap B(x,r)) \frac{dr}{r} a = +\infty$$

when $n = 4,5,6,7$. Here $C_{2,2}(K)$ is defined by replacing the $|\nabla\phi|^2$ in (1) by $|\Delta\phi|^2$. Thus in a sense, this last result shows that the Wiener Test for higher order elliptic equations at a boundary point depends very strongly on the nature of the boundary value problem being considered.

Bibliography

[AH] D.R. Adams, A. Heard, *The necessity of the Wiener Test for some semi-linear elliptic equations*, Ind. U. Math. J. **41** (1992), 109-124.

[GZ] R. Gariepy, W. Ziemer, *A regularity condition at the boundary for solutions of quasilinear elliptic equations*, Arch. Rat. Mech. Anal. **67** (1977), 25-39.

[GT] D. Gilbarg, N. Trudinger, *Elliptic Partial Differential Equations of Second Order*, Second Ed., Springer-Verlag, NY 1983.

[HW] L.I. Hedberg, T. Wolff, *Thin sets in nonlinear potential theory*, Ann. Inst. Fourier **33** (1983), 161-187.

[KM1] T. Kilpeläinen, J. Malý, *The Wiener Test and potential estimates for quasilinear elliptic equations*, preprint Univ. Jyväskylä, 1993, to appear.

[KM2] T. Kilpeläinen, J. Malý, *Generalized Dirichlet problem in nonlinear potential theory*, Manscripta Math. **66** (1989), 25-44.

[LM] P. Lindqvist, O. Martio, *Two theorems of N. Wiener for solutions of quasilinear elliptic equations*, Acta Math. **155** (1985), 153-171.

[LSW] W. Littman, G. Stampacchia, H. Wienberger, *Regular points for elliptic equations with discontinuous coefficients*, Ann. Scuola Norm. Sup. Pisa, **17** (1963), 43-77.

[M1] V.G. Maz'ya, *On the continuity at a boundary point of solutions of quasilinear elliptic equations*, Vestnik Leningrad Univ. Math. **3** (1976), 225-242. (Russian orginal - Vestnik Leningrad Univ. **25** (1970), 42-55.)

[M2] V.G. Maz'ya, *Behavior of solutions to the Dirichlet problem for the biharmonic operator at the boundary point*, Wissenschaftliche Inform. **10**, Tech. Hoch. Karl-Marx-Stadt, 1979.

[P] G. Paulik, *A regularity condition at the boundary for weak solutions of some nonlinear elliptic systems*, Proc. Symp. Pure Math., Amer. Math. Soc., **44** (1986), 353-357.

[S] G. Stampacchia, *Le probleme de Dirichlet pour les equations elliptique du second order a coefficients discontinuous*, Ann. Inst. Fourier **15** (1965), 189-258.

[Sk] I. Skrypnik, *Nonlinear elliptic boundary value problems*, Teubner-Texte zur Math., Liepzig, 1986.

[W] N. Wiener, *The Dirichlet problem*, J. Math. Physics, **3** (1924), 127-146.

k-SUPERHARMONIC FUNCTIONS AND L. KELVIN'S THEOREM

ÖMER AKIN
Ankara University Faculty of Science
Department of Mathematics
06100 Tandoğan-Ankara
Turkey

ABSTRACT. In this work we have given the definition of "k-superharmonic" functions on the hyper octant related to the operator L_k. After that we examined some properties of them such as the concavity property of the weighted mean value and the boundary behaviour of k-superharmonic functions.

AMS (1993) subject classification: 31, 31B, 31D

1. Introduction

The operator

$$L_k := \sum_{i=1}^{n} \frac{\partial^2}{\partial x_i^2} + \sum_{i=1}^{t} \frac{k_i}{x_{\ell+i}} \frac{\partial}{\partial x_{\ell+i}} \quad ; \ell + t = n, \ k := (k_1, k_2, \dots, k_t) \tag{1.1}$$

has a fundamental role in the researches of the generalized poly-axially symmetric potential theory (GPASPT). The equation of $L_k f=0$ is an equation of the GPASPT and it has been researched by some authors such as N.S.Hall-D.W.Quinn-R.J.Weinacht [8], A.Huber [9], A.O.Çelebi [7], Ö.Akın [1,4]. The solutions of the inequality $L_k f \geq 0$ were examined in [2,5,11] in different points of views. In our work we examined the solutions of $L_k f \leq 0$ and we gave the correspondence principle, the concavity property of these solutions and the boundary behaviour of them. Moreover, by using the radial form of these solutions we established the new form of Lord Kelvin's theorem for them.

2. The Definition Of k-Superharmonic Functions And The Correspondence Principle

In this part firstly we present some notations which we will use in the sequel.

$$K(x,z) = \frac{2(r^2 - |x|^2)\Gamma(p)}{r\pi^{n/2} \cdot \prod_{i=1}^{t} \Gamma\left(\frac{k_i}{2}\right)} \int_0^{\pi} \cdots \int_0^{\pi} \sigma^{-2p} \prod_{i=1}^{t} \sin^{k_i - 1}\theta_i \, d\theta_i$$

$$\sigma := \left[|x - z|^2 + 2 \cdot \sum_{i=1}^{t} x_{\ell+i} \, z_{\ell+i} (1 - \cos\theta_i) \right]^{1/2}$$

K. GowriSankaran et al. (eds.), Classical and Modern Potential Theory and Applications, 11–17.

$$2p := n + \sum_{i=1}^{t} k_i := n + |k_i| \cdot t \ ; \ n,\ k_i \in Z^+, z \in P_r, x \in B_r$$

$t+\ell=n\geq 2,\ 1\leq t\leq n,\ 0\leq \ell<n\ \ k_i\geq 1,\ B_r:=\{x:|x|<r;\ x\in R^n,\ r>0\}$

$P_r:=\{x:|x|=r,\ x_{\ell+i}>0,\ i=1,2,\ldots,t,\ x\in R^n\}\ k:=(k_1,k_2,\ldots,k_t).$

$P:=\{x:\ x\in R^n,\ x_{\ell+i}>0,\ i=1,2,\ldots,t\}$

$S:=\{x:\ x=(x_1,x_2,\ldots,x_\ell,\ x_{\ell+1},\ldots,x_n)\in R^n$, at least for one i $x_{\ell+i}=0$, $i=1,2,\ldots,t,\ \ell+t=n\}$.

S is called singular hyperplanes for the operator L_k,

Definition 2.1. Let f be a lower-semicontinuous function in B_r and satisfies the inequality

$$f(x) \geq \int_{P_r} \prod_{i=1}^{t} z_{\ell+i}^{k_i}\ K(x,z) f(z) dP_r : \ = I_k(f,x,r)$$

Then we call the function f as "k-superharmonic" in B_r.

Remark: We can obtain the operator L_k from the Laplace operator Δ by using the map which will be given here by the formula (5.4). (see also [5]).

2.2. The Correspondence Principle. If the function f is k-superharmonic of class C^2 then the following function g is 2-k-superharmonic; i.e. $L_{2-k}g\leq 0$;

$$f(x) = g(x) \cdot \prod_{i=1}^{t} x_{\ell+i}^{1-k_i}$$

where $2\text{-}k:=(2-k_1,\ 2-k_2,\ldots,2-k_t)$, $x_{\ell+i}\geq 0$, $k_i<1$.

This property can be shown as

$$f\{k_1,\ k_2,\ \ldots,\ k_t\}=g(2-k_1,\ 2-k_2,\ldots,2-k_t\}\cdot\prod_{i=1}^{t} x_{\ell+i}^{1-k_i}$$

and it is called the correspondence principle related to the L_k.

We can prove this property in the similar way of [7]. It can be seen from [7] that we can obtain different results regarding to the different values of k_i; $i=1,2,\ldots,t$.

3. The Lord Kelvin's Theorem For k-Superharmonic Functions

It is known from the Lord Kelvin's theorem [10] that if f is harmonic then

$$g(x_1,\ x_2,\ldots,x_n) := r^{2-n}\ f(x_1/r^2,\ x_2/r^2,\ldots,x_n/r^2) \tag{3.1}$$

is also harmonic. It is also known that the generalized forms of this theorem for k-harmonic functions and k-subharmonic functions were given in [7] and in [3] orderly. We have given a new form of these theorem for k-superharmonic functions of class C^2 by getting help from [7], [3,10]:

Theorem 3.1. Let f be a positive valued and homogen of p-degree k-superharmonic of class C^2. Moreover,

$$g(x_1,\ x_2,\ldots,x_n); = r^q\ f(z_1,\ z_2,\ldots,z_n) \tag{3.2}$$

is given, where

$r=(x_1^2+x_2^2+\ldots+x_n^2)^{1/2}$, $q=\alpha+2-n-\prod_{i=1}^{t} k_i$, $0\le\alpha\le n+\prod_{i=1}^{t} k_i -2$

$z_i := x_1/r^2$, $i=1,2,\ldots,n$, $p\ge q/2$.

Under these conditions the function g is also k-superharmonic.

Proof. If we take the necessary derivatives of g and put them in (1.1) then we obtain $L_k g$ as follows

$$L_k g = r^{q-4}\cdot L_k f + r^{q-2}\cdot(q-\alpha)\cdot\sum_{i=1}^{t} z_i f_z + q(\alpha f - 2\sum_{i=1}^{t} z_i f_z) \tag{3.3}$$

By using the hypotheses of the theorem we can see that $L_k g\le 0$ holds good. This completes the proof.

Some Remarks:

a. It is easy to see that $L_k r^q \le 0$ holds; i.e. a k-superharmonic function in radial form

b. If all $k_i=0$ then r^q is superharmonic; i.e. $\Delta r^q \le 0$

c. If $\alpha=0$ then r^q is k-harmonic; i.e. $L_k r^q = 0$

4. The Concavity Property Of The Weighted Mean Values Of k-Superharmonic Functions

Theorem 4.1. Let f be a k-superharmonic function in the annulus P_{r_1,r_2} and

$$M_k := \frac{1}{W_n\cdot r^{2p-1}}\int_{P_r}\prod_{i=1}^{t}\gamma_{\ell+i}\cdot f(y)\cdot dP_r;\quad t+\ell=n,\quad \gamma\in P_r$$

Then if $0<r_1<r_2$, M_k is a concave function of r^{2-2p} i.e.

$$M_k = I_k(f,x,r) \ge \frac{(r_2^*)^{2-2p}-r^{2-2p}}{(r_2^*)^{2-2p}-(r_1^*)^{2-2p}} I_k(f,x,r_1^*) + \frac{r^{2-2p}-(r_1^*)^{2-2p}}{(r_2^*)^{2-2p}-(r_1^*)^{2-2p}} I_k(f,x,r_2^*)$$

where $0<r_1^*<r<r_2^*\le r_2$, $P_{r_1,r_2}=\{x: r_1<|x|<r_2,\ x_{\ell+i}>0,\ i=1,2,\ldots,t,\ 0\le r_1<r_2<\infty\}$

$$W_n = \frac{\pi^{(n-t)/2}\cdot\prod_{i=1}^{t}\Gamma[(k_i+1)/2]}{2\Gamma(p)},\quad k_i\ge 1$$

M_k is also a decreasing function of r when $r_1=0$.

Proof. For the proof use a similar method as in [3]

Theorem 4.2. If a function G is k-harmonic, $L_k G=0$, and the function $g=f(G)$ is k-superharmonic in class C^2 then f is a concave function of G.

Proof. If we calculate the necessary derivates of g in (1.1) and put them in $L_k g\le 0$ then by using the hypotheses we can see that $f''(G)\le 0$ is satisfied, which is required to prove.

Remark: The converse of this theorem is also true.

5. The Behaviour Of k-Superharmonic Functions Near The Singular Hyperplanes

The main result in this paper is summarized in the following:

Theorem 5.1. If the function f given in the above definition is in class $C^2[B_r] \cap C^1[P_r]$ and has the behaviour given below near the singular hyperplanes related to L_k

$$\lim_{\xi\to\xi_o} \rho(\xi_{\ell+1};k_1)\cdot\rho(\xi_{\ell+2};k_2)\ldots\rho(\xi_{\ell+t};k_t)\cdot f(\xi)=0;\quad \xi\in P_r,\ \ \xi_o\in\bar{P}_r\cap\partial P \tag{5.1}$$

then it is a k-superharmonic function in class $C^2[B_r\text{-}S]$ and for all $x\in B_r^+$, $x_o\in\bar{B}_r\cap\partial P$

$$\liminf_{x\to x_o} \rho(x_{\ell+1};k_1)\cdot\rho(x_{\ell+2};k_2)\ldots\rho(x_{\ell+t};k_t)\cdot f(x)=0 \tag{5.2}$$

is also satisfied; where S is given in Section 2 and ∂P is the boundary of P.

$$\rho(m;\lambda):=\begin{cases} m^{\lambda-1} & ;\ \lambda>1\\ (\log m)^{-1} & ;\ \lambda=1\end{cases} \tag{5.3}$$

Proof. To prove the first part of the theorem we take the map given in [5] as follows

$$\begin{aligned} &x_i=y_i,\ \ i=1,2,\ldots,\ell\\ &x_{\ell+1}=[y_{\ell+1}^2+y_{\ell+2}^2+\ldots+y_{\ell+k_1}^2]^{1/2}\\ &x_{\ell+2}=[y_{\ell+k_1+1}^2+1+y_{\ell+k+2}^2+\ldots+y_{\ell+|k_i|_2}^2]^{1/2}\\ &\ldots\ldots\ldots\ldots\ldots\ldots\ldots\ldots\ldots\ldots\\ &x_{\ell+t}=[y_{\ell+|k_i|_{t-1}+1}^2+y_{\ell+|k_i|_{t-1}+2}^2+\ldots+y_{\ell+|k_i|_t}^2]^{1/2}\end{aligned} \tag{5.4}$$

where $x\in R^n$ and $y\in R^{2p}$. By using the map (5.4) and the hypothesis $f\in C^2[B_r]\cap C^1[P_r]$ we can see that f is k-superharmonic, i.e. it satisfies $L_k f\le 0$ in class C^2. So the first part of the theorem is obvious. Hence, we have to prove the second part of it. We treat the case where for all i $k_i>1$. Since (5.1) and $f\in C[P_r]$ the function $\prod_{i=1}^{t}\xi_{\ell+i}^{k_i-1}\cdot f(\xi)$ can be extended continuously to P_r. So there exists a $\delta_1=\delta_1(\in)$ for every $\in>0$ such that $\delta_1<\frac{r}{2}$ and δ_1, is independent of ξ. While $|\xi-\xi_o|<\delta_1$ $f(\xi)\le\in\cdot\sum_{i=1}^{t}\xi_{\ell+i}^{1-k_i}$ is satisfied.

Moreover we can find a positive number $\delta_2=\delta_2(\in)<\delta_1$ such that for all $\xi_{\ell+i}\ge\delta_1>\delta_2$

$$f(\xi)\le\in\prod_{i=1}^{t}\delta_j^{1-k_i};\quad j=1,2 \tag{5.5}$$

is satisfied. Now choose $x\in B_r^+=B_r\cap P$ such that for all i=1,2,...,t $x_{\ell+i}<2\delta_2$. Then $f(\xi)\le\varphi(\xi)$ is satisfied on P_r; Where

$$\varphi(\xi) = \begin{cases} \in \prod_{i=1}^{t} \xi_{\ell+i}^{1-k_i} & \text{on } P_1 = P_r \cap \{\xi : \xi_{\ell+i} \le \frac{x_{\ell+i}}{2}, \ i = 1,2,\ldots,t\} \\ \in \prod_{i=1}^{t-1} \xi_{\ell+i}^{1-k_i} \left(\frac{x_n}{2}\right)^{1-k_t} & \text{on } P_2 = P_r \cap \{\xi : \xi_{\ell+i} \le \frac{x_{\ell+i}}{2}, \ i = 1,2,\ldots,t-1, \xi_n > \frac{x_n}{2}\} \\ \cdots & \cdots \\ \in \prod_{i=1}^{t-1} \left(\frac{x_{\ell+i}}{2}\right)^{1-k_i} \xi_{1+t}^{1-k_t} & \text{on } P_{2^{n-\ell}-1} = P_r \cap \{\xi : \xi_{\ell+i} > \frac{x_{\ell+i}}{2}, \ i = 1,2,\ldots,t_{-1}, \xi_n \le \frac{x_n}{2}\} \\ \in \prod_{i=1}^{t} \left(\frac{x_{\ell+i}}{2}\right)^{1-k_i} & \text{on } P_{2^{n-\ell}} = P_r \cap \{\xi : \xi_{\ell+i} > \frac{x_{\ell+i}}{2}, \ i = 1,2,\ldots,t\} \end{cases} \tag{5.6}$$

Then it follows that

$$f(x) \le \int_{P_j} \varphi(\xi) := \sum_{j=1}^{2^{n-\ell}} I_j \tag{5.7}$$

where

$$I_j = \int_{P_j} \prod_{i=1}^{t} \xi_{\ell+i}^{k_i} \ K(x,\xi) \ \varphi(\xi) \ dP_j \ , \ j = 1,2,\ldots,2^{n-\ell} \tag{5.8}$$

and I_j=0 when P_j is an empty set. Now, we would like to estimate I_j integrals. Let us take

$$k(x,\xi) = \int_0^{\pi} \ldots \int_0^{\pi} \sigma^{-2p} \prod_{i=1}^{t} \sin^{k_i-1}\theta_i d\theta_i \ , \ \gamma = \frac{2}{r \cdot \pi^{n/2} \prod_{i=1}^{t} \Gamma(k_i/2)} \tag{5.9}$$

Then $K(x,\xi)$ can be written as $K(x,\xi) = (r^2 - |x|^2) \ \gamma \ k(x,\xi)$ (See part 1).
We see that on P_1

$$I_1 = \gamma \cdot \in (r^2 - |x|^2) \int_{P_1} \prod_{i=1}^{t} \xi_{\ell+i} k(x,\xi) dP_1 \quad \text{and} \quad k(x,\xi) \le |x-\xi|^{-2p} \int_0^{\pi} \ldots \int_0^{\pi} \prod_{i=1}^{t} \sin^{k_i-1}\theta_i d\theta_i$$

because of (4.1). So we have

$$I_1 \le \gamma_1 \cdot \in \int_{P_1} \prod_{i=1}^{t} x_{\ell+i}^{1-k_i} K_n(x,\xi) dP_1 \le \gamma_1 \cdot \in \cdot \prod_{i=1}^{t} x_{\ell+i}^{1-k_i} \tag{5.10}$$

Where K_n is the Poisson kernel for the sphere and for the n-dimensional Laplacian. So its integral value over B_r is 1(see [8]). γ_1, is a constant depending only on r, n, k_i.

We now consider I_2 on P_2(See (5.6)).

$$k(x,\xi) \le |x-\xi|^{-\sum_{i=1}^{t-1} k_i} \int_0^\pi \ldots \int_0^\pi \prod_{i=1}^{t-1} \sin^{k_i-1}\theta_i d\theta_i \quad \left[|x-\xi|^2 + 2x_n\xi_n(1-\cos\theta_t)\right]^{\frac{-n-k_t}{2}} \sin^{k_t-1}\theta_t d\theta_t$$

and then

$$I_2 = \gamma \in \left(\frac{x_n}{2}\right)^{1-k_t} (r^2 - |x|^2) \int_{P_2} \prod_{i=1}^{t-1} \xi_{\ell+1}\, \xi_n^{k_t}\, k(x,\xi)dP_2$$

$$\le \gamma_2 \cdot \in \cdot \prod_{i=1}^{t} x_{\ell+i}^{1-k_i} \int_{P_2} \xi_n^{k_t} k_{k_t}(x,\xi)dP_2 \le \gamma_2 \in \prod_{i=1}^{t} x_{\ell+i}^{1-k_i} \tag{5.11}$$

is satisfied, where $\xi_n^{k_t} k_{k_t}(x,\xi)$ is the Poisson kernel for the sphere and for the n-dimensional GASPT (Generalized axially symmetric potential theory) operator with parameter k_t and the integral of these kernels less than 1 on P_2 (see [8] and [9]).

The other integrals can be estimated in the same way. Hence we hold

$$I_j \le \gamma_j \in \prod_{i=1}^{t} x_{\ell+i}^{1-k_i}; \; j = 1,2,\ldots,2^{n-\ell} - 1 \tag{5.12}$$

Since

$$\int_{P_r} \prod_{i=1}^{t} \xi_{\ell+i}^{k_i} K(x,\xi)dP_r \equiv 1 \text{ and } (5.3)$$

$$I_{2^{n-m}} \le \gamma_{n-m} \in \prod_{i=1}^{t} \xi_{\ell+i}^{1-k_i} \tag{5.13}$$

can be held, where γ_j, $j=1,2,\ldots,2^{n-\ell}$ are only depending on r, n, k_i. Finally we can hold from the above inequlities, (5.7-5.13), that

$$f(x) \le \sum_{i=1}^{2^{n-\ell}} \gamma_i \cdot \in \cdot \prod_{i=1}^{t} \xi_{\ell+i}^{1-k_i}$$

So that we obtain the result required to prove

$$\liminf_{x \to x_0} \prod_{i=1}^{t} x_{\ell+i}^{k_i-1} f(x) \le 0$$

The remaining cases where all $k_1=1$ or some $k_j=1$, and the other $j<1$, $i=1,2,\ldots,t$, can be treated in a similar way with the help of the correspondence principle (see 2.2.)

Final Remark: It can be seen easily that the space of $SL_k:=\{S: S \text{ is } k\text{-superharmonic}\}$ is a harmonic space, see [6].

References.

[1] Ö.Akın, *The integral representation of the positive solutions of the generalized Weinstein equation on a quarter space*, SIAM Journal of Mathematical Analysis, Vol.19, No.6, November (1988).

[2] Ö.Akın, A.O.Çelebi, *Generalized Cauchy-Riemann equations and the positive k-subharmonic functions*, the proceedings of the conference on potential theory, 7-11, (1988), held in Prague, 1987. Plenum Press New York.

[3] Ö.Akın, *On the convexity of the weighted mean values of k-subharmonic functions and Lord Kelvin's theorem*, Journal Mathematics and Statistics of the Faculty of Arts an Science, Gazi University, Vol.2, 27-34, (1989).

[4] Ö.Akın, *Boundary Behaviour of solutions to the generalized Weinstein equation*, Seminaire de Theorie du Potentiel, Paris, Vol.9, 24-38, (1989).

[5] Ö.Akın, *Completely k-subharmonic functions*, The Journal K.T.U., Faculty of Arts an Sciences Series of Mathematics-Physics, Vol.XI, 219-224, (1989).

[6] C.Constantinescu and A.Cornea, *Potential theory on harmonic spaces* (Springer-Verlag, Berlin, 1972).

[7] A.O.Çelebi, *On the generalized Tricomi's equation*, Comm. Fac. Sci. Univ. Ankara Serie A, 1-31, tome 17 (1968).

[8] N.S.Hall, D.W.Quinn and R.J.Weinacht, *Poisson integral formulas in generalized bi-axially symmetric potential theory*, SIAM J. Math. Anal. Vol.5, No.1, 1974.

[9] A.Huber, *On the uniqueness of generalize axially symmetric potentials*, Ann. of Math. 60, 1954, pp. 20-38.

[10] L.Kelvin, J.de Math. Pure et Appl., 12, 256 (1847).

[11] S.V.Parter, *On the existence and uniqueness of symmetric axially potentials*, Arc. Rat. Mech. Anal., Vol.20, 279-286, (1965).

ON THE INVARIANCE OF THE SOLUTIONS OF THE WEINSTEIN EQUATION UNDER MÖBIUS TRANSFORMATIONS

Ömer Akin
Ankara University
Faculty of Science
Department of Mathematics
06100 Tandoĝan–Ankara
Turkey

Heinz Leutwiler
Mathematisches Institut
Universität Erlangen–Nürnberg
Bismarckstr. 1 1/2
D-91054 Erlangen
Germany

ABSTRACT. The invariance of the Weinstein equation under Möbius transformations and a mean–value type property are used to derive the corresponding Poisson formula in an elementary, function–theoretic way.

AMS (1991) subject classification: primary 31B10, 31C12;
secondary 53A30

1. Introduction

In this paper we are concerned with the invariance – under Möbius transformations – of the solutions of the Weinstein type equation

$$\sum_{i=1}^{n} \frac{\partial^2 u}{\partial x_i^2} + \frac{k}{x_n}\frac{\partial u}{\partial x_n} + \frac{\ell}{x_n^2} u = 0, \tag{1.1}$$

$(-\infty < \ell \leq (k-1)^2/4; k, \ell \in \mathbb{R})$, considered on the upper half space $\mathbb{R}^n_+ = \{x : x = (x_1, \ldots, x_n) \in \mathbb{R}^n, x_n > 0\} \quad (n \geq 3)$.

In case $\ell = 0$ the equation (1.1) can be viewed as the Laplace–Beltrami equation associated with the Riemannian metric

$$ds^2 = x_n^{2k/(n-2)} \sum_{1}^{n} dx_i^2. \tag{1.2}$$

Besides the Euclidian case ($k = 0$), where the solutions of (1.1) (for $\ell = 0$) are the classical harmonic functions, (1.1) also incorporates the non–Euclidian, so–called hyperbolic harmonic ones ($k = 2 - n, \ell = 0$). In fact (1.1) even includes the

K. GowriSankaran et al. (eds.), Classical and Modern Potential Theory and Applications, 19–29.

eigenfunctions u, corresponding to eigenvalues $\lambda \geq -(n-1)^2/4$, of the Laplace–Beltrami operator associated with the hyperbolic metric, i.e. the metric (1.2) with $k = 2-n$.

Transplanted to the unit ball $B_n = \{x \in \mathbb{R}^n, ||x|| < 1\}(||x||^2 = \sum_1^n x_i^2)$ by means of the special Möbius transformation $T_B : B_n \to \mathbb{R}^n_+ : x \to x'$, where

$$\begin{cases} x_i' = \dfrac{2x_i}{||x-e_n||^2} & (i = 1, \dots, n-1) \\ x_n' = \dfrac{1-||x||^2}{||x-e_n||^2}, \end{cases} \tag{1.3}$$

and $e_n = (0, \dots, 0, 1)$, the equation (1.1) reads as follows:

$$\sum_{i=1}^n \frac{\partial^2 v}{\partial x_i^2} - \frac{2k}{1-||x||^2} \sum_{i=1}^n x_i \frac{\partial v}{\partial x_i} + \left[\frac{k(2-n-k)}{1-||x||^2} + \frac{4\ell}{(1-||x||^2)^2} \right] v = 0, \tag{1.4}$$

for $x \in B_n$. A twice continuously differentiable function $u : \mathbb{R}^n_+ \to \mathbb{R}$ is a solution of (1.1) if and only if the function $v : B_n \to \mathbb{R}$, with

$$v(x) = ||x-e_n||^{2-n-k} u(T_B(x)), \tag{1.5}$$

solves (1.4) (see [8]).

In section 2 we generalize the classical Kelvin transformation (see, e.g., [5], p. 36), defined for solutions u of (1.1) by B. Brelot–Collin and M. Brelot in [2], to arbitrary Möbius transformations in $\mathbb{R}^n_+$ (see section 2 for their definition). In section 3 we shall deal with a generalization of the mean–value property, adequate to solutions of the equation (1.4). In section 4, finally, we combine the results of sections 2 and 3 in order to give an elementary, function–theoretic derivation of the Poisson formula, for solutions of (1.4), in case $\ell = 0$. Here one has to distinguish between the cases $k < 1$ and $k \geq 1$.

2. The invariance theorem.

Recall that by definition the group of Möbius transformations, denoted by $M(\mathbb{R}^n)$, is the group of sense–preserving transformations (under composition) on $\mathbb{R}^n \cup \{\infty\}$, generated by the similarities and the reflection in the unit sphere. In this paper we shall primarily be interested in the subgroups $M(\mathbb{R}^n_+)$, resp. $M(B_n)$, of Möbius transformations mapping $\mathbb{R}^n_+$, resp. B_n, bijectively onto itself.

According to Ahlfors [1] it is useful to associate to every Möbius transformation $T \in M(\mathbb{R}^n)$ and every point $x \in \mathbb{R}^n$ with $T(x) \neq \infty$ the positive real number $|T'(x)|$, uniquely determined by the requirement that the normalized Jacobian matrix $T'(x)/|T'(x)|$ lies in the orthogonal group $O(n)$. Consequently

$$|T'(x)| = |\det T'(x)|^{1/n}, \tag{2.1}$$

for all $x \in \mathbb{R}^n$ with $T(x) \neq \infty$. Note that in case of the reflection J in the unit sphere, i.e. the mapping $J : x \to x/||x||^2$, we have

$$|J'(x)| = ||x||^{-2}, \tag{2.2}$$

for all $x \in \mathbb{R}^n \setminus \{0\}$, whereas for the transformation T_B, given by (1.3),

$$|T_B'(x)| = 2||x - e_n||^{-2}, \tag{2.3}$$

for all $x \in B_n$.

With these notions at hand, we are now in position to formulate the invariance theorem.

Theorem 2.1. *Let $T \in M(\mathbb{R}^n_+)$ be an arbitrary Möbius transformation, mapping $\mathbb{R}^n_+$ onto itself. Then, for every solution $u : \mathbb{R}^n_+ \to \mathbb{R}$ of* (1.1) *the function*

$$\tilde{u}(x) = |T'(x)|^{(n+k-2)/2} u(T(x)), \tag{2.4}$$

$x \in \mathbb{R}^n_+$, defines a solution of (1.1).

Let us mention that in case of the reflection J, due to the property (2.2), the transformation (2.4) is exactly the generalization of the classical Kelvin transformation, given by B. Brelot–Collin and M. Brelot in [2]. In case $k = \ell = 0$, but $T \in M(\mathbb{R}^n_+)$ arbitrary, Theorem 2.1 is already contained in ([7], Lemma 2.1). Finally, let us also observe that in the case $k = 2 - n$ of the hyperbolic metric the factor $|T'(x)|^{(n+k-2)/2}$ disappears.

Proof of Theorem 2.1. First observe that if T_1 and T_2 are Möbius transformations with the property that (2.4) maps solutions u of (1.1) into solutions, the same is true for the composition $T_1 \circ T_2$. Indeed, all there remains to be seen is that

$$|(T_1 \circ T_2)'(x)| = |T_1'(T_2(x))| \cdot |T_2'(x)| \tag{2.5}$$

holds for all $x \in \mathbb{R}^n_+$. This, however, is a simple consequence of the chain rule.

The second observation is that any $T \in M(\mathbb{R}^n_+)$ is representable in the form

$$T = T_1 \circ T_2 \circ T_3, \tag{2.6}$$

with T_1 denoting a translation of the form $x \to x - a, a = (a_1, \ldots, a_{n-1}, 0)$, T_2 a dilation, and T_3 a Möbius transformation keeping fixed the point $e_n = (0, \ldots, 0, 1)$. Since Theorem 2.1 trivially holds for T_1 and T_2, it suffices to look at T_3.

Transplanting T_3 – with the help of T_B – to the unit ball B_n, i.e. setting

$$R = T_B^{-1} \circ T_3 \circ T_B, \tag{2.7}$$

we obtain a Möbius transformation $R \in M(B_n)$ with the property that $R(0) = 0$. Hence (see, e.g., [1], p. 21) R is a rotation and thus

$$|R'(x)| = 1 \tag{2.8}$$

for all $x \in B_n$. Assuming now that u is a solution of (1.1) the statement

$$\tilde{u}(x) = |T_3'(x)|^{(n+k-2)/2} u(T_3(x)) \tag{2.9}$$

being a solution of (1.1) is equivalent to the claim that

$$\tilde{v}(x) = ||x - e_n||^{2-n-k} \tilde{u}(T_B(x)) \tag{2.10}$$

is a solution of (1.4). By hypothesis

$$v(x) = ||x - e_n||^{2-n-k} u(T_B(x)) \tag{2.11}$$

is a solution of (1.4). Since R is a rotation, the composition $v \circ R$ also defines a solution of (1.4). From $T_B \circ R = T_3 \circ T_B$ and (2.8) we conclude, applying the chain rule, that

$$|T_B'(R(x))| = |T_3'(T_B(x))| \cdot |T_B'(x)|.$$

Inserting (2.3) we obtain

$$||R(x) - e_n||^{-2} = |T_3'(T_B(x))| \cdot ||x - e_n||^{-2}, \tag{2.12}$$

for all $x \in B_n$. Drawing again on (2.7), we now conclude from (2.9) - (2.12) that

$$\begin{aligned} v(R(x)) &= |T_3'(T_B(x))|^{(n+k-2)/2} ||x - e_n||^{2-n-k} u(T_3(T_B(x))) \\ &= ||x - e_n||^{2-n-k} \tilde{u}(T_B(x)) = \tilde{v}(x), \end{aligned}$$

showing that $\tilde{v}$ is indeed a solution of (1.4). Thus the proof of Theorem 2.1 is complete.

We would expect Theorem 2.1 to have an analogue in B_n. In order to derive it, observe first that on account of (2.3) the transformation (1.5) can be written in the form

$$v(x) = |T_B'(x)|^{(n+k-2)/2} u(T_B(x)), \tag{2.13}$$

provided we drop the constant factor $2^{(2-n-k)/2}$. Solving this equation for u, we see that, conversely, given any solution v of (1.4), the function

$$u(x) = |(T_B^{-1})'(x)|^{(n+k-2)/2} v(T_B^{-1}(x)) \tag{2.14}$$

represents a solution of (1.1).

Suppose now that $T \in M(B_n)$ and set $S = T_B \circ T \circ T_B^{-1}$. On account of Theorem 2.1 the function

$$\tilde{u}(x) = |S'(x)|^{(n+k-2)/2} u(S(x))$$

is a solution of (1.1). Consequently

$$\begin{aligned} \tilde{v}(x) &= |T_B'(x)|^{(n+k-2)/2} \tilde{u}(T_B(x)) \\ &= |T'(x)|^{(n+k-2)/2} |T_B'(T(x))|^{(n+k-2)/2} u(T_B(T(x))) \\ &= |T'(x)|^{(n+k-2)/2} v(T(x)) \end{aligned}$$

is a solution of (1.4) We have thus proved

Theorem 2.2. *Let $T \in M(B_n)$ be an arbitrary Möbius transformation, mapping B_n onto itself. Then, for every solution $v : B_n \to \mathbb{R}$ of (1.4), the function*

$$\tilde{v}(x) = |T'(x)|^{(n+k-2)/2} v(T(x)), \tag{2.15}$$

$x \in B_n$, defines a solution of (1.4).

Of particular interest, in what follows, is the Möbius transformation $T_a : B_n \to B_n$ ($a \in B_n$) defined by

$$T_a(x) = \frac{(1 - ||a||^2)(x - a) - ||x - a||^2 a}{[x, a]^2}, \quad x \in B_n, \tag{2.16}$$

where $[x, a]^2 = (1 - ||x||^2)(1 - ||a||^2) + ||x - a||^2$. It maps a into 0 and satisfies

$$|T_a'(x)| = \frac{1 - ||a||^2}{[x, a]^2} \tag{2.17}$$

([1], p. 26). From Theorem 2.2 we therefore get the following result, already contained in [8] (but unfortunately with a misprint):

Corollary 2.3. *If $v : B_n \to \mathbb{R}$ is a solution of (1.4), so is the function $\tilde{v} : B_n \to \mathbb{R}$, defined by*

$$\tilde{v}(x) = \frac{1}{[x, a]^{n+k-2}} v(T_a(x)), \tag{2.18}$$

for $x \in B_n$.

3. The mean–value property

The particular solution $x \to x_n^{(1-k+s)/2}$, $s = \sqrt{(k-1)^2 - 4\ell}$ of the equation (1.1), transplanted to B_n by means of (1.5), yields the following solution of (1.4.):

$$x \longrightarrow \frac{(1 - ||x||^2)^\alpha}{||x - e_n||^{2\beta}} \tag{3.1}$$

($x \in B_n$), where $\alpha = \frac{1}{2}(1 - k + s)$ and $\beta = \frac{1}{2}(n - 1 + s)$. Applying a rotation R to (3.1) we deduce that for arbitrary y of the unit sphere $S_{n-1} = \partial B_n$ the function

$$P_y(x) = \frac{(1 - ||x||^2)^\alpha}{||x - y||^{2\beta}} \tag{3.2}$$

($x \in B_n$) represents a solution of (1.4). Integrating P_y over the unit sphere S_{n-1} we obtain the solution

$$v(x) = \frac{1}{\sigma_n} \int_{S_{n-1}} \frac{(1 - ||x||^2)^\alpha}{||x - y||^{2\beta}} d\sigma(y), \tag{3.3}$$

with σ denoting the Euclidian surface measure on S_{n-1} and $\sigma_n = \sigma(S_{n-1})$.

Obviously v only depends on $||x||$. Consequently, setting

$$\varphi(r) = (1-r^2)^{\alpha} \frac{1}{\sigma_n} \int_{S_{n-1}} \frac{d\sigma(y)}{||re_n - y||^{2\beta}} \tag{3.4}$$

for $0 < r < 1$, the function φ is a solution of the ordinary second order differential equation

$$(1-r^2) r \frac{d^2 f}{dr^2} + \left[n - 1 - (n-1+2k) r^2\right] \frac{df}{dr} + \left[k(2-n-k) + \frac{4\ell}{1-r^2}\right] rf = 0. \tag{3.5}$$

The basic ingredient for the derivation of the mean–value property of the solutions (1.4) will be - as in the classical situation - Green's 1st identity:

$$\int_\Omega u \Delta v \, dx + \int_\Omega \nabla u \cdot \nabla v \, dx = \int_{\partial\Omega} u \frac{\partial v}{\partial n} d\sigma \tag{3.6}$$

where $\nabla = \left(\frac{\partial}{\partial x_1}, \ldots, \frac{\partial}{\partial x_n}\right)$, $\Delta = \sum_1^n \frac{\partial^2}{\partial x_i^2}$ and $\frac{\partial v}{\partial n} = \nabla v \cdot \underline{n}$ ($\underline{n}$ = outer normal).
It will however be necessary to adapt it to the situation of a metric of type $ds^2 = \rho^2(x)(dx_1^2 + \ldots + dx_n^2)$. Replacing u in (3.6) by $\rho^{n-2} u$ and setting

$$dx^\rho = \rho^n dx, d\sigma^\rho = \rho^{n-1} d\sigma, \nabla^\rho = \frac{1}{\rho}\nabla \quad \text{and} \quad \frac{\partial v}{\partial n^\rho} = \frac{1}{\rho}\frac{\partial v}{\partial n},$$

the identity (3.6) transforms into the following one, involving the Laplace–Beltrami operator

$$\Delta^\rho v = \rho^{-2}\left[\Delta v + (n-2)\rho^{-1} \nabla\rho \cdot \nabla v\right] \tag{3.7}$$

associated with the metric $ds^2 = \rho^2(dx_1^2 + \ldots + dx_n^2)$:

$$\int_\Omega u \Delta^\rho v \, dx^\rho + \int_\Omega \nabla^\rho u \cdot \nabla^\rho v \, dx^\rho = \int_{\partial\Omega} u \frac{\partial v}{\partial n^\rho} d\sigma^\rho. \tag{3.8}$$

Specializing $\rho(x) = 2(1 - ||x||^2)^{k/(n-2)}$, choosing $u \equiv 1$ and, for Ω, the ball $B_n(r)$ with center 0 and radius $r, 0 < r < 1$, we obtain (3.8) in the following Euclidian form:

$$\int_{B_n(r)} \left(\Delta v - \frac{2k}{1-||x||^2} \sum_1^n x_i \frac{\partial v}{\partial x_i}\right)(1-||x||^2)^k dx = \int_{S_{n-1}(r)} \frac{\partial v}{\partial n}(1-||x||^2)^k d\sigma_r$$

where σ_r denotes the Euclidian surface measure on the sphere $S_{n-1}(r) = \partial B_n(r)$ of center 0 and radius r.

Assuming further that v is a solution of (1.4) we obtain

$$\int_{B_n(r)} \left[k(n+k-2) - \frac{4\ell}{1-||x||^2}\right] v(x)(1-||x||^2)^{k-1} dx = (1-r^2)^k \int_{S_{n-1}(r)} \frac{\partial v}{\partial n} d\sigma_r.$$

Introducing polar coordinates and setting

$$\Phi_v(r) = \frac{1}{\sigma_n r^{n-1}} \int_{S_{n-1}(r)} v(x) d\sigma_r(x) = \frac{1}{\sigma_n} \int_{S_{n-1}} v(ry) d\sigma(y) \tag{3.9}$$

$(S_{n-1} = S_{n-1}(1))$ we find that

$$\int_0^r \left[k(n+k-2) - \frac{4\ell}{1-s^2} \right] \Phi_v(s)(1-s^2)^{k-1} s^{n-1} ds$$

$$= (1-r^2)^k r^{n-1} \frac{\partial}{\partial s} \Phi_v(s)|_{s=r},$$

for all $0 < r < 1$. Differentiating this equation on both sides with respect to r implies that the function $r \to \Phi_v(r)$ is a solution of (3.5).

The Wronskian determinant W associated with equation (3.5) being equal to

$$W(r) = c \exp \left[-\int \left(\frac{n-1}{r} - \frac{2kr}{1-r^2} \right) dr \right] = \frac{\text{const}}{r^{n-1}(1-r^2)^k}$$

it follows that one of the two solutions of any fundamental system of (3.5) will be unbounded at the (regular singular) point 0. Consequently, since both solutions $\Phi_v(r)$ and $\varphi(r)$ from (3.4) are bounded near 0 they must be proportional: $\Phi_v(r) = c_v \varphi(r)$, for some constant c_v and all $0 < r < 1$. Letting $r \to 0$ entails that $c_v = v(0)$. We thus proved

Theorem 3.1. *Every solution v of (1.4) has the following mean-value property:*

$$v(0)\varphi(r) = \frac{1}{\sigma_n r^{n-1}} \int_{S_{n-1}(r)} v d\sigma_r, \tag{3.10}$$

for all $0 < r < 1$, where φ is defined by (3.4).

4. The generalized Poisson formula

Throughout this section we assume that $\ell = 0$. Consequently $u \equiv 1$ is a particular solution of (1.1) and thus, by (1.5),

$$v(x) = ||x - e_n||^{2-n-k}$$

$(x \in B_n)$ is a solution of (1.4). Inserting v into (3.10) yields the following, alternative definition of φ, valid in case $\ell = 0$:

$$\varphi(r) = \frac{1}{\sigma_n} \int_{S_{n-1}} \frac{d\sigma(y)}{||ry - e_n||^{n+k-2}}. \tag{4.1}$$

It has the advantage that it allows one to pass to the limit easily, as $r \to 1$. Indeed, <u>in case $k < 1$:</u>

$$d_n^k = \lim_{r \to 1} \varphi(r) = \frac{1}{\sigma_n} \int_{S_{n-1}} \frac{d\sigma(y)}{||y - e_n||^{n+k-2}}, \tag{4.2}$$

in case $k \geq 1$:

$$\lim_{r \to 1} \varphi(r) = \infty.$$

Assuming now that v is an arbitrary solution of(1.4) and a an arbitrary point of B_n, we make use of the Möbius transformation $T_{-a} \in M(B_n)$, with $T_{-a}(0) = a$, defined by (2.16). According to Corollary 2.3 the function

$$\tilde{v}(x) = \frac{1}{[x, -a]^{n+k-2}} v(T_{-a}(x)) \tag{4.3}$$

is again a solution of (1.4). Applying Theorem 3.1 to $\tilde{v}$, we obtain

$$\frac{1}{\sigma_n} \int_{S_{n-1}} \tilde{v}(ry) d\sigma(y) = \tilde{v}(0) \varphi(r) \tag{4.4}$$

for all $0 < r < 1$. Passing to the limit $(r \to 1)$ we infer from (4.2) that in case $k < 1$

$$\frac{1}{\sigma_n} \int_{S_{n-1}} \tilde{v}(y) d\sigma(y) = \tilde{v}(0) d_n^k, \tag{4.5}$$

provided we assume that v - and hence also $\tilde{v}$ - extends continuously to the closure $\overline{B}_n$ of B_n. Observing that $[y, -a] = ||y + a||$, for $y \in S_{n-1}$, we thus have

$$\frac{1}{\sigma_n} \int_{S_{n-1}} \frac{1}{||y + a||^{n+k-2}} v(T_{-a}(y)) d\sigma(y) = v(a) d_n^k. \tag{4.6}$$

Applying next the substitution $z = T_{-a}(y)$ with $y = T_a(z)$ and

$$d\sigma(y) = \frac{(1 - ||a||^2)^{n-1}}{||z - a||^{2(n-1)}} d\sigma(z),$$

(4.6) will be transformed into

$$\frac{1}{\sigma_n} \int_{S_{n-1}} \frac{(1 - ||a||^2)^{n-1}}{||T_a(z) + a||^{n+k-2} ||z - a||^{2(n-1)}} v(z) d\sigma(z) = v(a) d_n^k.$$

From $||T_a(z) + a|| = (1 - ||a||^2)/||z - a||$ ([1], p.25) we thus conclude that

$$v(a) = \frac{1}{d_n^k \sigma_n} \int_{S_{n-1}} \frac{(1 - ||a||^2)^{1-k}}{||z - a||^{n-k}} v(z) d\sigma(z).$$

Setting $c_n^k = \sigma_n d_n^k$ and replacing a by x we obtain

Theorem 4.1. *Let $v : B_n \to \mathbb{R}$ be a solution of the equation (1.4) with $\ell = 0$ and $k < 1$. Assuming that v extends continuously to $\overline{B}_n$, we have the representation*

$$v(x) = \frac{1}{c_n^k} \int_{S_{n-1}} \frac{(1 - ||x||^2)^{1-k}}{||z - x||^{n-k}} v(z) d\sigma(z) \tag{4.7}$$

for all $x \in B_n$.

The method employed to derive this Poisson type formula resembles the derivation usually applied in complex analysis. It seems to be new even in case of the classical harmonic functions of B_n $(n \geq 3)$, where $k = 0$ and $c_n^0 = \sigma_n$.

In order to obtain a representation for v in case $k > 1$, we apply the so-called "correspondence principle" of A. Weinstein [9], which in our situation says (see [8]):

Given any solution v of (1.4) the function $\bar{v} : B_n \to \mathbb{R}$, defined by

$$\bar{v}(x) = \frac{v(x)}{(1 - ||x||^2)^\gamma}, \quad x \in B_n, \tag{4.8}$$

satisfies the equation (1.4), with k replaced by $\bar{k} = k + 2\gamma$ and ℓ replaced by $\bar{\ell} = \gamma^2 + (k-1)\gamma + \ell$.
In particular, if v is a solution of (1.4) with $\ell = 0$ and $k > 1$, the function

$$\bar{v}(x) = (1 - ||x||^2)^{k-1} v(x) \tag{4.9}$$

solves (1.4) with $\ell = 0$ and k replaced by $2 - k$ (< 1). Hence, assuming that $\bar{v}$ extends continuously to $\bar{B}_n$, we may apply Theorem 4.1 and conclude that

$$\bar{v}(x) = \frac{1}{c_n^k} \int_{S_{n-1}} \frac{(1 - ||x||^2)^{k-1}}{||z - x||^{n+k-2}} \bar{v}(z) d\sigma(z).$$

Substituting the right hand side of (4.9) for $\bar{v}$, we obtain

Theorem 4.2. *Let* $v : B_n \to \mathbb{R}$ *be a solution of* (1.4) *with* $\ell = 0$ *and* $k > 1$. *Assuming that the function* $\bar{v}$, *given by* (4.9) *extends continuously to* $\bar{B}_n$, *we have*

$$v(x) = \frac{1}{c_n^k} \int_{S_{n-1}} \frac{1}{||z - x||^{n+k-2}} \bar{v}(z) d\sigma(z), \tag{4.10}$$

for all $x \in B_n$.

Concluding remarks. Let us first recall ([8], Theorem 3.2) that any positive solution v of (1.4) can be represented in terms of the kernels P_z from (3.2), i.e., there is a uniquely determined, positive measure μ on the unit sphere S_{n-1} with the property that

$$v(x) = \int_{S_{n-1}} \frac{(1 - ||x||^2)^{(1-k+s)/2}}{||x - z||^{n-1+s}} d\mu(z), \tag{4.11}$$

where $s = \sqrt{(k-1)^2 - 4\ell}$. The proof is based on the corresponding theorem for the positive solutions of (1.1), given by A. Huber [6] for integer k and B. Brelot-Collin and M. Brelot [2] for arbitrary real k. Since the kernels in (4.7), resp. (4.10), are exactly of the form (3.2), our representations are in accordance with (4.11).

Secondly, let us mention that in [3] – in case of the upper half space – B. Brelot–Collin and M. Brelot studied the nontangential approach of the solutions u of (1.1), using the limit theorem of K. Gowrisankaran [4]. Analogous results hold for the equation (1.4)

Acknowledgement

This research was started while the second author was visiting Ankara University. He would like to take the opportunity to express his gratitude to Professor Hilmi Hacisalihoĝlu and the many Turkish mathematicians who were so graciously hospitable.

Both authors would like to thank the Scientific Affairs Division of the NATO.

References.

[1] Ahlfors, L.V., *Möbius transformations in several dimensions*, Ordway Lectures in Mathematics, University of Minnesota, 1981.

[2] Brelot–Collin, B. and M. Brelot: *Représentation intégrale des solutions positives de l'equation*

$$L_k(u) = \sum_1^n \frac{\partial^2 u}{\partial x_i^2} + \frac{k}{x_n}\frac{\partial u}{\partial x_n} = 0 \quad (k \quad \text{constante réelle})$$

dans le demi–espace $E(x_n > 0)$, de $\mathbb{R}^n$. Bull. Acad. Royale de Belg. 58 (1972), 317 - 326.

[3] Brelot–Collin, B. and M. Brelot: *Allure à la frontière des solutions positives de l'equation de Weinstein*

$$L_k(u) = \Delta u + \frac{k}{x_n}\frac{\partial u}{\partial x_n} = 0$$

dans le demi–espace $E(x_n > 0)$ de $\mathbb{R}^n$ $(n \geq 2)$. Bull. Acad. Royale de Belg. 59 (1973), 1100 - 1117.

[4] Gowrisankaran, K., *On minimal positive harmonic functions*, Séminaire Brelot–Choquet–Deny (Théorie du potential) 1966/67, n° 18.

[5] Helms, L.L. *Introduction to potential theory*, Wiley–Interscience, New York, 1969.

[6] Huber, A., *Some results on generalized axially symmetric potentials*, Proceed. Conf. on Diff. Equ., Univ. of Maryland, Bookstore, College Park, MD, 1955, 147–155.

[7] Leutwiler, H., *On a distance invariant under Möbius transformations in* $\mathbb{R}^n$, Ann. Acad. Sci. Fenn. Ser. A.I. Math. 12 (1987), 3 - 17.

[8] Leutwiler, H., *Best constants in the Harnack inequality for the Weinstein equation*, Aequationes Math. 34 (1987), 304 - 315.

[9] Weinstein, A., *Discontinuous integrals and generalized potential theory*, Trans. Amer. Math. Soc., 63 (1948), 342 - 354.

RADIAL LIMITING BEHAVIOUR OF HARMONIC AND SUPER-HARMONIC FUNCTIONS

D.H. Armitage
Queen's University
Belfast BT7 1NN
Northern Ireland.

Abstract. A theorem of W.J. Schneider [17] states that there is no harmonic function h in the plane such that $r^{-\mu}|h(re^{i\theta})| \to +\infty$ for all $\mu > 0$ and for a second category set of θ. Neither the growth rate nor the category condition can be relaxed. This article surveys various analogues and extensions of Schneider's result. The role of holomorphic and harmonic approximation theorems in the production of counterexamples is indicated. Sections 1 and 2 are largely a summary of joint work with M. Goldstein [6]; Section 4 mainly concerns joint work with C.S. Nelson [7]; Sections 3 and 5 contain brief remarks on hitherto unpublished results of the author.

1. Decay of Harmonic Functions on Rays

If f is an entire function and

$$f(re^{i\theta}) \to 0 \qquad (r \to +\infty) \tag{1.1}$$

uniformly for all $\theta \in [0, 2\pi)$, then, by Liouville's theorem, $f \equiv 0$. Uniformity is crucial here: there exist non-constant entire functions f satisfying (1.1) for all θ non-uniformly (see, for example, Lindelöf [14, p.122]). The real part of such a function provides an example of a plane harmonic function exhibiting the same limiting behaviour. It is natural to ask how rapidly a non-constant entire function or harmonic function on $\mathbf{R}^N$ can decay on all rays (or on a large class of rays) emanating from the origin 0. Writing S_N for the unit sphere $\{\xi \in \mathbf{R}^N : ||\xi|| = 1\}$, where $N \geq 2$, we have the following result [6, Theorem 1]; notions of category refer to the relative topology on S_N.

THEOREM 1.1 *Let h be harmonic on $\mathbf{R}^N$. If there is a second category subset E of S_N such that*

$$h(r\xi) = o(\exp(-r^{\mu})) \qquad (r \to +\infty) \tag{1.2}$$

for all ξ in E and all positive numbers μ, then $h \equiv 0$.

K. GowriSankaran et al. (eds.), Classical and Modern Potential Theory and Applications, 31–40.

The proof may be outlined as follows. A Baire category argument shows that h is bounded on some open cone K of vertex 0 such that $K \cap E$ is second category in S_N. Let $e_1 = (1, 0, \ldots, 0)$. We suppose, without loss of generality, that $e_1 \in K \cap E$. It is not difficult to show that there is a bounded holomophic function g on some sector $\{re^{i\theta} : r > 0, |\theta| < \alpha\}$ such that $g(t) = h(te_1)$ for all $t > 0$. Hence $t^{-\mu} \log |g(t)| \to -\infty$ as $t \to +\infty$ for every $\mu > 0$. It follows easily that $g \equiv 0$, and hence $h(te_1) = 0$ for all $t > 0$. The same argument shows that $h(t\xi) = 0$ for all $\xi \in K \cap E$ and all $t > 0$, and it then follows that $h \equiv 0$.

In relation to Theorem 1, the main interest is perhaps in the construction of counterexamples to show its sharpness with respect to both the category condition and the decay rate (1.2); harmonic approximation is important in these constructions. Our first example [6, Example 1] concerns the decay rate.

EXAMPLE 1.2 *If μ is a fixed positive number, then there exists a non-constant harmonic function h on $\mathbf{R}^N$ such that* (1.2) *holds for each $\xi \in S_N$.*

This depends upon another example which we shall require again later.

EXAMPLE 1.3 *If μ is a fixed positive number, then there exists a harmonic function v on $\mathbf{R}^2$ such that*

$$r^{-\mu} v(a + r\xi) \to +\infty \qquad (r \to +\infty) \tag{1.3}$$

for all $\xi \in S_2$ and all $a \in \mathbf{R}^2$.

We illustrate the construction of the function v in the case where $0 < \mu < 2$. Define

$$F_1 = \{(x_1, x_2) : |x_1 x_2| \geq 1\}, \quad F_2 = \{(x_1, x_2) : x_1 x_2 = 0\}, \quad F = F_1 \cup F_2.$$

It is easy to find explicity a continuous function $v_0 : F \to \mathbf{R}$ which is harmonic on the interior of F and satisfies $v_0(x) \geq 1 + \|x\|^{(\mu+2)/2}$ for all $x \in F$. A harmonic approximation theorem [5, Theorem 1.3] yields the existence of a harmonic function v on $\mathbf{R}^2$ such that $|v - v_0| < 1$ on F, and hence $v(x) \geq \|x\|^{(\mu+2)/2}$ for all $x \in F$. If $a \in \mathbf{R}^2$ and $\xi \in S_2$, then $a + r\xi \in F$ for all sufficiently large r and therefore (1.3) holds.

The same argument works for larger values of μ but the set F is altered; the larger μ is, the more connected components F has.

To produce Example 1.2, replace μ by $\mu + 1$ in Example 1.3, take the corresponding function v, and define h on $\mathbf{R}^N$ by

$$h(x_1, \ldots, x_N) = \mathrm{Re}\{(x_1 + ix_2) e^{-(v + i\tilde{v})(x_1, x_2)}\},$$

where $\tilde{v}$ is a conjugate function to v. It is routine to verify that h has the required properties.

At least in the plane case, Theorem 1 fails badly if E is any first category set: Schneider [18, Example 10] indicated how Mergelyan's theorem can be used to prove the following result.

EXAMPLE 1.4 *Let E be a first category subset of S_2. If $\epsilon : [0, +\infty) \to (0, +\infty)$ is decreasing, then there exists a non-constant harmonic function h on $\mathbf{R}^2$ such that*

$$h(r\xi) = o(\epsilon(r)) \qquad (r \to +\infty) \tag{1.4}$$

for all $\xi \in E$.

In higher dimensions the position is less satisfactory. Example 1.4 is easily modified to show that there *exists* a first category subset E of S_N ($N \geq 3$), having full measure, such that for any ϵ (as in Example 1.4) there exists a harmonic function h on $\mathbf{R}^N$ satisfying (1.4) for all $\xi \in E$. However, it is not clear whether the same is true for *all* first category subsets of S_N when $N \geq 3$.

In Theorem 1.1 it is not essential that h be harmonic on the whole of $\mathbf{R}^N$; it suffices that h is harmonic on some open cone K of vertex 0 (and then $E \subseteq K \cap S_N$). In particular, if h is harmonic on the half-space $D_N = \{(x_1, \ldots, x_N) : x_N > 0\}$ and (1.2) holds for all $\xi \in D_N \cap S_N$ and all $\mu > 0$, then $h \equiv 0$. This may be compared with a similar result in which a uniform decay hypothesis is assumed [8], [2]: if h is harmonic on D_N and $h(r\xi) = O(e^{-\alpha r})$ uniformly for all $\xi \in D_N \cap S_N$ and for some $\alpha > 0$, then $h \equiv 0$.

An analogue of Theorem 1.1 in which rays are replaced by parallel lines is as follows (Nelson [15, Theorem 4.8]).

THEOREM 1.5 *Let h be harmonic on D_N. If there is a second category subset E of $\mathbf{R}^{N-1}$ such that*

$$h(x_1, \ldots, x_N) = o(\exp(-e^{\mu x_N})) \qquad (x_N \to +\infty) \tag{1.5}$$

for all $(x_1, \ldots, x_{N-1}) \in E$ and all $\mu > 0$, then $h \equiv 0$.

This is sharp in the same senses as Theorem 1.1. There is also a uniform version [3] of Theorem 1.5 in which (1.5) is replaced by the hypothesis that, for all $\mu > 0$,

$$h(x_1, \ldots, x_N) = o(e^{-\mu x_N}) \qquad (x_N \to +\infty)$$

uniformly for $(x_1, \ldots, x_{N-1}) \in \mathbf{R}^{N-1}$.

We can also consider normal approach to the finite boundary of D_N [15, Theorem 4.17]: if h and E are as in Theorem 1.5 and

$$h(x_1, \ldots, x_N) = o(x_N) \qquad (x_N \to 0+)$$

for all $(x_1, \ldots, x_{N-1}) \in E$, then $h \equiv 0$.

2. Growth of Harmonic Functions on Rays

In Example 1.3 we saw that a plane harmonic function can grow at a polynomial rate on every semi-infinite line in $\mathbf{R}^2$. Although Example 1.3 seems to be recent (see [6, Example 7]), Schneider [17, Theorem 2] had much earlier proved the following result in the case $N = 2$; the general result is [6, Theorem 2].

THEOREM 2.1 *Let E be a second category subset of S_N. There is no harmonic function h on $\mathbf{R}^N$ such that*

$$r^{-\mu}|h(r\xi)| \to +\infty \qquad (r \to +\infty)$$

for all $\xi \in E$ and all $\mu > 0$.

For the moment we defer consideration of the proof, which is very simple. First we indicate how Example 1.3 may be generalized to higher dimensions by constructing a corresponding example in $\mathbf{R}^3$. The trivial device of taking the plane harmonic function v of Example 1.3 and defining h_1 by

$$h_1(x_1, x_2, x_3) = v(x_1, x_2)$$

almost succeeds: h_1 is harmonic on $\mathbf{R}^3$ and $h_1(x)$ grows more rapidly than $\|x\|^\mu$ on every semi-infinite line, with the exception of those lines parallel to the x_3-axis. To deal with the exceptional set of lines, we modify h_1 as follows. Let

$$P = \{(x_1, x_2, x_3) : |x_3| \geq 1 + x_1^2 + x_2^2\},$$
$$Q = \{x \in \mathbf{R}^3 : \ \operatorname{dist}(x, P) \geq 1\},$$

and let ω_1 and ω_2 be disjoint open neighbourhoods of P and Q respectively. Let h_2 be a harmonic function on $\omega_1 \cup \omega_2$ such that $h_2 = h_1$ on ω_2 and $h_2(x) \geq \|x\|^{\mu+1}$ when $x \in P$. (The latter property is easily arranged, since P is "thinner" than any cone in $\mathbf{R}^3$.) By a theorem on harmonic approximation [13, Theorem 1], there exists a harmonic function h on $\mathbf{R}^3$ such that $|h - h_2| < 1$ on $P \cup Q$. Clearly $h(x)$ grows more rapidly than $\|x\|^\mu$ on any semi-infinite line in $\mathbf{R}^3$. This argument forms the basis for an induction on the dimension N which yields corresponding examples in $\mathbf{R}^N$ [6, Example 4].

The position regarding the sharpness of the category condition is the same for Theorem 2.1 as for Theorem 1.1; for future reference, we state this formally.

EXAMPLE 2.2 (i) *If E is a first category subset of S_2 and $\epsilon : [0, +\infty) \to (0, +\infty)$ is decreasing, then there exists a harmonic function h on $\mathbf{R}^2$ such that*

$$\epsilon(r)h(r\xi) \to +\infty \qquad (r \to +\infty) \tag{2.1}$$

for all $\xi \in E$.

(ii) *There exists a first category set E of full measure in $S_N (N \geq 3)$ such that* *for each function* ϵ *(as in* (i)*) there is a harmonic function* h *on* $\mathbf{R}^N$ *satisfying* (2.1) *for all* $\xi \in E$.

Part (i) is given in work of Schneider [1, Second Example] who quotes Bagemihl and Seidel [9, pp. 187-190]. Part (ii) is derived from part (i) without difficulty [6, Example 6], but we do not know whether (ii) is valid for *all* first category subsets of S_N.

We now sketch the proof of Theorem 2.1. We may assume that $r^{-\mu}h(r\xi) \to +\infty$ as $r \to +\infty$ for all $\xi \in E$ and all $\mu > 0$. This hypothesis and a Baire category argument yield that h is bounded below on some cone K of vertex 0 such that $K \cap E \neq \emptyset$. By Harnack's inequality, if $\xi \in K \cap S_N$, then $h(r\xi) = O(r^\mu)$ as $r \to +\infty$ for some $\mu > 0$, and this contradicts our hypothesis when $\xi \in K \cap E$.

In this sketch-proof the hypothesis was invoked twice, but in neither instance was its full strength required. Examination of a detailed proof reveals the following strengthening of Theorem 2.1.

THEOREM 2.3 *Let* E *be a second category subset of* S_N. *There is no harmonic function* h *on* $\mathbf{R}^N$ *such that*

$$\liminf_{r\to+\infty} h(r\xi) > -\infty$$

for all $\xi \in E$ *and such that*

$$E \subseteq \bigcap_{\mu>0} \overline{U(h,\mu)}, \tag{2.2}$$

where

$$U(h,\mu) = \{\xi \in S_N : \limsup_{r\to+\infty} r^{-\mu}h(r\xi) = +\infty\}. \tag{2.3}$$

A curious feature of this result is that there is no *a priori* assumption that there exists even one $\xi \in S_N$ such that $\limsup_{r\to+\infty} r^{-\mu}h(r\xi) = +\infty$ for all $\mu > 0$.

Again, the results of this section have analogues in which rays are replaced by parallel lines. For example, the first part of the following theorem [15, Theorems 4.11, 4.18] is analogous to Theorem 2.1.

THEOREM 2.4 *Let* E *be a second category subset of* $\mathbf{R}^{N-1}$.

(i) *There is no harmonic function* h *on* D_N *such that*

$$e^{-\mu x_N}|h(x_1,\ldots,x_N)| \to +\infty \qquad (x_N \to +\infty)$$

for all $(x_1,\ldots,x_{N-1}) \in E$ *and all* $\mu > 0$.

(ii) *If* $\epsilon : (0,1] \to (0,1]$ *is an increasing function such that* $\epsilon(t) \to 0+$ *as* $t \to 0+$, *then there is no harmonic function* h *on* D_N *such that*

$$\epsilon(x_N)h(x_1,\ldots,x_N) \to +\infty \qquad (x_N \to 0+)$$

for all $(x_1, \ldots, x_{N-1}) \in E$.

3. Growth of Superharmonic Functions on Rays: the Plane Case

We first point out that Theorem 1.1 on the decay of harmonic functions seems to have no straightforward superharmonic analogue, for it is not difficult to show that there exists a superharmonic function u on $\mathbf{R}^N$, unbounded above and below, with the property that u is eventually 0 on every ray. (We can arrange, for example, for u to vanish everywhere on the x_1-axis and outside an "asymptotic" neighbourhood of the x_1-axis.)

Next we consider the possibility of extending the growth theorems of Section 2 to superharmonic functions. It is convenient to work first in the plane. Theorem 2.3 fails for superharmonic functions, as the following example shows; this failure is related to the fact that a polar set can be second category (in $\mathbf{R}^N$) or even residual (that is, the complement is first category). Recall the notation (2.3).

EXAMPLE 3.1 *There exists a superharmonic function* u *on* $\mathbf{R}^2$ *such that* $u(r\xi) \to +\infty$ *as* $r \to +\infty$ *for all* $\xi \in S_2$ *and such that* $\bigcap_{\mu>0} U(u,\mu)$ *is a residual subset of* S_2 *(and hence* $\bigcap_{\mu>0} \overline{U(u,\mu)} = S_2$*).*

To verify this, we start by working in the half-plane D_2. Let F be a countable dense subset of $D_2 \cap S_2$, and let u_0 be a positive superharmonic function on D_2 such that $u_0 = +\infty$ on the polar set $T = \{n\xi : n \in \mathbf{N}, \xi \in F\}$. Define $G_n = \{\xi \in S_2 \cap D_2 : u_0(n\xi) = +\infty\}$. Since $u_0 = +\infty$ on T and u is lower semi-continuous, it is easy to verify that each set G_n is residual in $S_2 \cap D_2$, and hence so also is $\bigcap_{n=1}^{\infty} G_n = G$, say. Note that $u_0(n\xi) = +\infty$, when $n \in \mathbf{N}$ and $\xi \in G$.

It is easy to construct a superharmonic function u on $\mathbf{R}^2$ such that $u(x_1, x_2) = u_0(x_1, |x_2|) + |x_2|$ if $|x_2| > 2$ and $u(x_1, x_2) = \cosh x_1 \cos x_2$ if $|x_2| < 1$. This function u has the required properties; in particular, $\{(\xi_1, \pm\xi_2) : (\xi_1, \xi_2) \in G\} \subset \bigcap_{\mu>0} U(u,\mu)$.

To obtain a superharmonic version of Theorem 2.3, the hypotheses must be strengthened. Theorem 2.3 becomes true for plane superharmonic functions if, in the definition (2.3) of $U(\cdot,\mu)$, the limit superior is replaced by the limit as $r \to +\infty$. This is proved by a Baire category argument and an application of the classical Ahlfors-Heins theorem [1, p.341]. In this instance the category argument is somewhat complicated by the fact that superharmonic functions are not generally continuous.

Similar techniques lead to an alternative modification of Theorem 2.3 for plane superharmonic functions. Note that condition (2.2) in Theorem 2.3 may be paraphrased as follows: every open subset of S_N which meets E has non-empty intersection with $U(h,\mu)$ for each μ. A plane superharmonic version of Theorem 2.3 is obtained by requiring the corresponding intersections to be not merely non-empty but non-polar:

THEOREM 3.2 *Let E be a second category subset of S_2. There is no superharmonic function u on $\mathbf{R}^2$ such that*

$$\limsup_{r\to+\infty} u(r\xi) > -\infty$$

for all $\xi \in E$ and such that $\omega \cap U(u,\mu)$ is a non-polar subset of $\mathbf{R}^2$ for each $\mu > 0$ and for each open subset ω of S_2 which meets E.

4. Growth of Superharmonic Functions in Higher Dimensions

None of the theorems already mentioned concerning growth along rays generalizes in a straightforward manner to the case of superharmonic functions on $\mathbf{R}^N$ when $N \geq 3$. The underlying reason is that, when $N \geq 3$, there exists a set of rays whose union is both polar and residual in $\mathbf{R}^N$ (and intersects S_N in a residual subset of S_N). The following example merely expresses this fact.

EXAMPLE 4.1 *There exist a residual subset E of S_N, where $N \geq 3$, and a positive superharmonic function u on $\mathbf{R}^N$ such that $u(r\xi) = +\infty$ for all $\xi \in E$ and all $r > 0$.*

The construction is most conveniently illustrated with $N = 4$. Let $\{\xi_1, \xi_2, \ldots\}$ be a countable dense subset of S_4. For each j let w_j be a positive superharmonic function on $\mathbf{R}^4$, homogeneous of degree -1, such that $w_j(r\xi_j) = +\infty$ for all $r > 0$. (If, for example, $\xi_1 = (1,0,0,0)$, take $w_1(x_1,\ldots,x_4) = 1/\sqrt{(x_2^2+x_3^2+x_4^2)}$.) Let $u = \sum_{j=1}^{\infty} \alpha_j w_j$, where the positive coefficients α_j are chosen to make the series converge at some point. Then u is positive, superharmonic and homogeneous on $\mathbf{R}^4$. Let $E = \{\xi \in S_4 : u(\xi) = +\infty\}$. It is easy to check that E is residual, and by the homogeneity of u, we have $u(r\xi) = +\infty$ for each $r > 0$ and each $\xi \in E$.

By adding together the harmonic function of Example 2.2(ii) and the positive superharmonic function of Example 4.1, we obtain the following result.

EXAMPLE 4.2 *There exists a residual set E of full measure in S_N, where $N \geq 3$, such that for each decreasing function $\epsilon : [0,+\infty) \to (0,+\infty)$ there is a superharmonic function u on $\mathbf{R}^N$ satisfying*

$$\epsilon(r)u(r\xi) \to +\infty \qquad (r \to +\infty) \tag{4.1}$$

for all $\xi \in E$.

This example suggests the intruiging question whether it is possible for a superharmonic function to satisfy (4.1) for *all* $\xi \in S_N$. A similar question was raised by Rippon in 1978 in relation to his result [16, Theorem 6] that there exists a superharmonic function on the half-space D_N $(N \geq 3)$ which has normal limit $+\infty$ at a set of points which is both residual and of full measure in the boundary of D_N.

To obtain growth theorems, like those in Sections 2 and 3, for superharmonic functions on $\mathbf{R}^N$ ($N \geq 3$), we must replace rays by rather "thicker" sets. The main idea that allows us to do this is due to Rippon [16]. It is convenient to work here in the context of parallel lines in a half-space (as in Theorems 1.5 and 2.4). Modifying a definition of Rippon's, we say that a subset A of D_N ($N \geq 2$) belongs to the class $\mathcal{T}$ if

(i) for each $\eta > 0$, there exists $X > 0$ such that $x_1^2 + \ldots + x_{N-1}^2 < \eta^2$ whenever $(x_1, \ldots, x_N) \in A$ and $x_N > X$,

(ii) there exists $X_0 \geq 0$ such that

$$\{(x_1, \ldots, x_N) \in A : X_1 < x_N < X_2\}$$

is non-polar whenever $X_0 < X_1 < X_2$.
For each $z = (z_1, \ldots, z_{N-1}) \in \mathbf{R}^{N-1}$, we write

$$A_z = \{(z_1 + x_1, \ldots, z_{N-1} + x_{N-1}, x_N) : (x_1, \ldots, x_N) \in A\}.$$

It is significant that the x_N-axis is of class $\mathcal{T}$ when $N = 2$ but not when $N \geq 3$. In [7, Theorem 3] we give the following superharmonic analogue of Theorem 2.4(i).

THEOREM 4.3 *If $A \in \mathcal{T}$ and E is a second category subset of $\mathbf{R}^{N-1}$, then there is no superharmonic function u on D_N such that*

$$e^{-\mu x_N} u(x_1, \ldots, x_N) \to +\infty \qquad (x_N \to +\infty, (x_1, \ldots, x_N) \in A_z)$$

for all $z \in E$ and all $\mu > 0$.

The main step in the proof is to show that $u > 0$ on some half-cylinder parallel to the x_N-axis. This is achieved by modifying a fine topology version, due to Rippon [16, Theorem 1], of a classical result of Collingwood [11, p.76] on cluster sets.

A construction [7, Theorem 5], based on work of Gardiner [12, Example 2] and again using a harmonic approximation theorem [5, Theorem 1.3], shows that, at least when A is closed, Theorem 4.3 fails if A does not satisfy condition (ii) in the definition of the class $\mathcal{T}$.

5. Decay of Harmonic Functions in an Arbitrary Unbounded Domain

Throughout this section Ω denotes an unbounded domain in $\mathbf{R}^N$, where $N \geq 2$. Answering a question of Gauthier and Hengartner [10, Problem 3.27], Armitage, Bagby and Gauthier [4] showed that a harmonic function on Ω which decays uniformly and sufficiently rapidly must be identically 0:

THEOREM 5.1 *There exists a continuous function $\epsilon : [0, +\infty) \to (0, 1]$ with the following property: if h is harmonic on Ω and $|h(x)| \leq \epsilon(||x||)$ for all $x \in \Omega$, then $h \equiv 0$.*

Theorems 1.1 and 1.5 show that, at least for certain special domains, the uniform decay hypothesis of Theorem 5.1 can be replaced by a requirement that h should decay rapidly, but not *a priori* uniformly, on a suitable set of paths. Here we indicate how a non-uniform version of Theorem 5.1 may be obtained. Let M be an $(N-1)$-dimensional manifold, and let $\Gamma : M \times [0, +\infty) \to \Omega$ be a continuous function such that (i) $\Gamma(\omega \times (0, +\infty))$ is open for each open subset ω of M, (ii) for each $\xi \in M$ the set $\gamma_\xi = \{\Gamma(\xi, t) : t \geq 0\}$is closed and unbounded. Such a situation can always be realised for any Ω; indeed, it can be arranged that M is an open $(N-1)$-dimensional ball lying in Ω and that each γ_ξ is an unbounded polygonal path starting at ξ. There is no requirement that Γ be surjective, nor that $||\Gamma(\xi, t)|| \to +\infty$ as $t \to +\infty$ (which would be impossible for some domains Ω).

THEOREM 5.2 *There exists a continuous function* $\eta : [0, +\infty) \to (0, 1]$ *with the following property: if h is harmonic on Ω and*

$$h(x) = O(\eta(\|x\|)) \qquad (\|x\| \to +\infty, x \in \gamma_\xi)$$

for a second category set of ξ in M, then $h \equiv 0$.

The proof proceeds as follows. First a sequence (Ω_k) of unbounded domains, each contained in Ω, is chosen in an appropriate way, and the function η is chosen so that for each k we have $\eta(t) = o(\epsilon_k(t))$ as $t \to +\infty$, where ϵ_k is the function obtained by replacing Ω by Ω_k in Theorem 5.1. A Baire category argument is then used to show that $h(x) = O(\eta(||x||))$ uniformly as $||x|| \to +\infty$ with $x \in \Omega_k$ for some k, and the choice of η then allows us to invoke Theorem 5.1 and conclude that $h = 0$ on Ω_k and hence on Ω.

This sketch-proof suggests that generally the function η will be much smaller than the function ϵ. This is confirmed in concrete examples: if Ω is a half-space, then it suffices to take $\epsilon(r) = e^{-\alpha r}$ for some $\alpha > 0$ (see, for example, [2]), while Example 1.2 shows that if $\{\gamma_\xi\}$ is a system of rays in a half-space (or a cone), then even taking $\eta(r) = \exp(-r^\mu)$, with any fixed μ, is not enough.

In [4] we proved Theorem 5.1 not just for harmonic functions but for solutions of a large class of elliptic partial differential equations; Theorem 5.2 is also valid in the same generality.

References

1. L.V. Ahlfors and M. Heins, "Questions of regularity connected with the Phragmén-Lindelöf principle", Ann. Math. **50** (1949), 341-346.
2. D.H. Armitage, "A new proof of a uniqueness theorem for harmonic functions in half-spaces", Bull. London Math. Soc. **9** (1977), 317-320.
3. D.H. Armitage, "A uniqueness theorem for harmonic functions on half-spaces", Glasgow Math. J. **31** (1989), 189-191.

4. D.H. Armitage, T. Bagby and P.M. Gauthier, "Note on the decay of solutions of elliptic equations", Bull. London Math. Soc. 17 (1985), 554-556.
5. D.H. Armitage and M. Goldstein, "Better than uniform approximation on closed sets by harmonic functions with singularities", Proc. London Math. Soc. (3), 60 (1990), 319-343.
6. D.H. Armitage and M. Goldstein, "Radial limiting behaviour of harmonic functions in cones", Complex Variables 22 (1993) (to appear).
7. D.H. Armitage and C.S. Nelson, "The growth of superharmonic functions on sets of parallel lines", (manuscript).
8. I.S. Aršon and M.A. Pak, "A uniqueness theorem for harmonic functions in a half-space", (Russian) Mat. Sborn. 68 (110) (1965), 148-151.
9. F. Bagemihl and W. Seidel, "Some boundary properties of analytic functions", Math. Z. 61(1954), 186-199.
10. K.F. Barth, D.A. Brannan and W.K. Hayman, "Research problems in complex analysis", Bull. London Math. Soc. 16 (1984), 490-517.
11. E.F. Collingwood and A.J. Lohwater, The Theory of Cluster Sets. Cambridge University Press, 1966.
12. S.J. Gardiner, "Uniqueness and extension theorems for subharmonic functions", J. London Math. Soc. (to appear).
13. P.M. Gauthier, M. Goldstein and W.H. Ow, "Uniform approximation on closed sets by harmonic functions with Newtonian singularities", J. London Math. Soc. (2) 28 (1983), 71-82.
14. E. Lindelöf, Le Calcul des Residus, Gauthier-Villars, Paris, 1905.
15. C.S. Nelson, Some Special Sets Occurring in Potential Theory, Ph.D. thesis, Queen's University, Belfast, 1992.
16. P.J. Rippon, "The boundary cluster sets of subharmonic functions", J. London Math. Soc. (2), 17 (1978), 469-479.
17. W.J. Schneider, "On the growth of entire functions along half-rays", Entire Functions and Related Parts of Analysis (Proceedings of a symposium on pure mathematics, La Jolla, 1966) Amer. Math. Soc., Providence, R.I., 1968, 377-385.
18. W.J. Schneider, "Approximation and harmonic measure", Aspects of Contemporary Complex Analysis, D.A. Brannan and J.G. Clunie, eds. Academic Press, London, 1980, 333-349.

MULTIPARAMETER PROCESSES ASSOCIATED WITH ORNSTEIN–UHLENBECK SEMIGROUPS

JÜRGEN BAUER
Fachbereich Mathematik, Universität Frankfurt
Robert Mayer–Strasse 6–10
D–60054 Frankfurt am Main
Germany

ABSTRACT. We consider Sobolev spaces and capacities associated with a generalized Ornstein–Uhlenbeck semigroup. In order to obtain a probabilistic counterpart, we construct and investigate certain multiparameter processes. These are used to characterize polar sets and quasi–continuous functions.

0. Introduction

Let E be a separable Fréchet space and let m be a centered Gaussian measure on E. The Ornstein–Uhlenbeck semigroup on E can be defined by the so–called Mehler formula

$$\Gamma(e^{-t})f(x) = \int f(e^{-t}x + \sqrt{1-e^{-t}}y)\, m(dy),$$

where f is a bounded Borel function on E.

In [5], D. Feyel and A. de La Pradelle were able to give a sense to this formula when e^{-t} is replaced by a contraction semigroup $(e^{-tA})_{t\geq 0}$ on the Cameron–Martin space $\mathcal{H}$ of (E,m). Assuming that A satisfies a suitable condition, they determined in their paper [6] the associated Dirichlet form and proved the existence of a corresponding continuous Markov process.

These "generalized Ornstein–Uhlenbeck semigroups" give rise to Sobolev spaces $W_A^{r,p}$ $(r \geq 0, p \geq 1)$ and related potential theoretic concepts (see e.g. [9], where a general theory is developped). For the standard case $A = I$, S. Song established in [10],[11],[12] a probabilistic counterpart to the potential theory of the spaces $W^{n,2}$ $(n \in \mathbb{N})$ by means of n–parameter processes.

The aim of this paper is to construct continuous multiparameter processes associated with generalized Ornstein–Uhlenbeck semigroups. These are used to estimate functional $(n,2)$–capacities in terms of a probabilistic quantity (theorem 12). Applying this main result, we easily obtain characterizations of sets of $(n,2)$–capacity zero (corollary 13) and of $(n,2)$–quasi–continuous functions (theorem 14).

In the first two sections, we describe the analytical framework. It will be shown that the investigated Sobolev spaces are "regular" (Proposition 3). This and the known result

K. GowriSankaran et al. (eds.), Classical and Modern Potential Theory and Applications, 41–55.

that the associated capacities are tight assures that we can make full use of the potential theory developped in [9].

In section 3, we introduce the multiparameter processes that will be used for probabilistic interpretations later on. The basic idea is that of Song, but since a possibly unbounded operator A occurs, the considerations that lead to the existence of continuous processes are much more involved. Section 4 is devoted to some essential properties of the processes previously constructed. Finally, we relate probabilistic and analytic notions in section 5.

This article details and improves our paper [1].

1. Generalized Ornstein–Uhlenbeck semigroups

Let E be a separable Fréchet space equipped with a centered Gaussian (Radon) measure m. We recall first some basic notions and results of [3], [5]. Denote by $\mathcal{H}'$ the closure of E' in $L^2(E,m)$ endowed with its inherited Hilbertian structure, by $\mathcal{H}$ the dual space of $\mathcal{H}'$ and by $\kappa : \mathcal{H}' \to \mathcal{H}$ the Riesz isomorphism. The elements of $\mathcal{H}'$ can be identified with (classes of) m–measurable linear forms on E. $\mathcal{H}$ is contained in E, and an element h of E lies in $\mathcal{H}$ if and only if translation of m by h yields a measure which has a density w.r.t. m, this density being given by

$$k_h(x) = e^{\langle h,x\rangle - \frac{1}{2}\|h\|^2_{\mathcal{H}}} \quad , x \in E,$$

where $\langle h, x\rangle = \kappa^{-1}(h)(x)$ denotes the pseudo scalar product on $\mathcal{H} \times E$.

We consider a symmetric contraction semigroup $C_t = e^{-tA}$ of class (C_0) on the Hilbert space $\mathcal{H}$, where A is the (self–adjoint and nonnegative) generator. Throughout this paper, we will suppose that A is subject to the following condition:

(1) For all $n \in \mathbb{N}$, $\kappa(E')$ is contained in the domain of $\sqrt{A}^{\,n}$, and $(\kappa(E'), \sqrt{A}^{\,n})$ can be extended to an m–measurable linear map $E \to E$.

The extension theorem of Feyel–de La Pradelle [5] assures that (1) is satisfied for bounded operators A. Note that (1) entails in particular that the measure $\tilde{m} = \sqrt{2A}(m)$ is again a centered Gaussian Radon measure on E and that $(C_t)_{t\geq 0}$ operates on $\tilde{\mathcal{H}}$, the associated Cameron–Martin space (cf. [6]).

The Ornstein–Uhlenbeck semigroup on $L^2(E,m)$ with drift A is defined by the "generalized Mehler formula"

$$Q_t f(x) = \int f(C_t x + S_t y)\, m(dy), \tag{2}$$

where $S_t = \sqrt{I - C_t^2}$ and where C_t and S_t are to be understood as m–measurable extensions.

For $f \in L^2(E,m)$ let

$$Uf = \frac{1}{\sqrt{\pi}} \int_0^\infty t^{-1/2} e^{-t} Q_t f\, dt, \tag{3}$$

where the integral is taken in the Bochner sense. The operator U is an injective, symmetric contraction of $L^2(E,m)$ and coincides with $(I+A)^{-1/2}$, as is seen by spectral decomposition. Moreover, $U^n = \int_0^\infty t^{n/2-1} e^{-t} Q_t\, dt / \Gamma(\frac{n}{2})$ for $n \in \mathbb{N}$.

2. Sobolev spaces and capacities

We introduce Sobolev spaces associated with $(Q_t)_{t\geq 0}$ by a inductive procedure close to the one indicated in [4]. For the purpose of this paper it is sufficient to treat the Hilbertian case ($p=2$).

Define the set of elementary functions on E by

$$\mathcal{F}C_0^\infty(E) = \{\varphi(f_1,\ldots,f_n):\ n\in\mathbb{N}, f_i\in E', \varphi\in C_0^\infty(E)\},$$

and for $f\in\mathcal{F}C_0^\infty(E)$, let $f'(x,y)=\lim_{t\to 0}\frac{1}{t}(f(x+ty)-f(x))$, which defines a function on $E\times E$, linear in the second variable.

$(Q_t)_{t\geq 0}$ acts as a symmetric and Markovian contraction semigroup of class (C_0) on $L^2(E,m)$, hence it is associated with a Dirichlet form, which is known to be the closure of $\mathcal{F}C_0^\infty(E)\ni f\mapsto \|f'\|^2_{L^2(E\times E,m\otimes\tilde{m})}$ (cf. [6]). Let $(W_A^1(E,m),\|\cdot\|_1)$ be the resulting Dirichlet space. Note that $\mathcal{F}C_0^\infty(E)\ni f\mapsto f'$ obviously extends by continuity to a mapping $W_A^1(E,m)\to L^2(E\times E,m\otimes\tilde{m})$.

$(C_t)_{t\geq 0}$ gives rise to a semigroup $(C_t\otimes C_t)_{t\geq 0}$ on $\mathcal{H}\times\tilde{\mathcal{H}}$ with generator $A_2=A\otimes I+I\otimes A$. By Mehlers formula one obtains a semigroup $(\check{Q}_t)_{t\geq 0}$ on $L^2(E\times E,m\otimes\tilde{m})$ and by a formula analogous to (3) an associated operator $\check{U}$. Obviously $\sqrt{A_2}=\sqrt{I+A}\otimes\sqrt{I+A}$, hence [5] Rem. 23b implies that A_2 satisfies (1) w.r.t. the space $(E\times E,m\otimes\tilde{m})$. This allows to define the Sobolev spaces $(W_A^n(E,m),\|\cdot\|_n)$ for integers $n\geq 2$ by induction, namely

$$\begin{aligned} W_A^n(E,m) &= \{f\in W_A^{n-1}(E,m):\ f'\in W_{A_2}^{n-1}(E\times E,m\otimes\tilde{m})\},\\ \|f\|_n^2 &= \|f\|_{n-1}^2+\|f'\|_{n-1,A_2}^2. \end{aligned}$$

It is convenient to set $W_A^0(E,m)=L^2(E,m)$. The spaces $W_A^n(E,m)$ are clearly Hilbert spaces. We denote by $W_A^{-n}(E,m),\|\cdot\|)$ the dual space of $W_A^n(E,m)$.

It is a standard result of Dirichlet form theory that U is an isometry of $L^2(E,m)$ onto $W_A^1(E,m)$. Moreover, for every $t\geq 0$, Q_t maps $L^2(E,m)$ continuously to $W_A^1(E,m)$. Consequently, $\lim_{t\to 0}\|Q_tu-u\|_1=0$ for $u\in W_A^1(E,m)$ (namely, if $u=Uf$ with $f\in L^2(E,m)$, then as $t\to 0$, $\|Q_tu-u\|_1=\|UQ_tf-Uf\|_1=\|Q_tf-f\|_{L^2(E,m)}\longrightarrow 0$). Our first aim is to generalize these facts to orders $n>1$.

1. Lemma. *Let $f\in W_A^1(E,m)$ and $t>0$. Then $(Q_tf)'=\check{Q}_t(f')$ and $(Uf)'=\check{U}(f')$ $m\otimes\tilde{m}$–a.e.*

Proof. For $f\in\mathcal{F}C_0^\infty(E)$, an easy computation yields

$$(Q_tf)'(x,y)=\int f'(C_tx+S_tz,C_ty)\,m(dz).$$

Since f' is linear in the second variable, integration over $\tilde{m}$ yields $(Q_tf)'=\check{Q}_t(f')$, hence $(Uf)'=\check{U}(f')$ by formula (3). $\mathcal{F}C_0^\infty(E)$ being dense in $W_A^1(E,m)$, the two equalities subsist for general $f\in W_A^1(E,m)$. □

2. Proposition. *For every $n \in \mathbb{N}$, U is an isometry of $W_A^{n-1}(E,m)$ onto $W_A^n(E,m)$. Moreover, for every $t > 0$, Q_t is a continuous map from $L^2(E,m)$ to $W_A^n(E,m)$. In particular, $\lim_{t\to 0} \|Q_t u - u\|_n = 0$ for $u \in W_A^n(E,m)$.*

Proof. Let $n \geq 2$ and suppose the proposition to be true up to order $n-1$ for every triple (E,m,A) satisfying (1). If $f \in W_A^{n-1}(E,m)$, then $f' \in W_{A_2}^{n-2}(E \times E, m \otimes \tilde{m})$ by definition, and the preceding lemma implies that $(Uf)' = \check{U}f' \in W_{A_2}^{n-1}(E \times E, m \otimes \tilde{m})$, hence $Uf \in W_A^n(E,m)$. Moreover,

$$\|Uf\|_n^2 = \|Uf\|_{n-1}^2 + \|\check{U}f'\|_{A_2,n-1}^2 = \|f\|_{n-2}^2 + \|f'\|_{A_2,n-2}^2 = \|f\|_{n-1}^2.$$

Similarly, one obtains that for $t > 0$, Q_t is a continuous map from $L^2(E,m)$ to $W_A^n(E,m)$.

It remains to show that $U : W_A^{n-1}(E,m) \to W_A^n(E,m)$ is onto. For $g \in W_A^n(E,m)$ the induction hypothesis entails the existence of functions $f \in W_A^{n-2}(E,m)$ and $h \in W_{A_2}^{n-2}(E \times E, m \otimes \tilde{m})$ such that $g = Uf$ and $g' = \check{U}h$. By lemma 1, $\check{U}(Q_t f)' = \check{U}(\check{Q}_t h)$, hence $(Q_t f)' = \check{Q}_t h$ for every $t > 0$, since $\check{U}$ is injective on $L^2(E \times E, m \otimes \tilde{m})$. However, $\lim_{t\to 0} \|Q_t f - f\|_{n-2} = \lim_{t\to 0} \|\check{Q}_t h - h\|_{A_2,n-2} = 0$ by the induction hypothesis, thus $(Q_{1/k} f)_{k\in\mathbb{N}}$ is a Cauchy sequence in $W_A^{n-1}(E,m)$. Necessarily, its limit has derivative h and equals f, which shows that $f \in W_A^{n-1}(E,m)$ and therefore $g \in W_A^n(E,m)$. □

3. Proposition. *For every $n \in \mathbb{N}$, $\mathcal{F}C_0^\infty(E)$ is dense in $W_A^n(E,m)$.*

Proof. Let $k \in \mathbb{N}$ and define $\delta_k \in C_0^\infty(\mathbb{R})$ by $\delta_k(t) = \exp(-\frac{kt^2}{k^2 - t^2})$ if $|t| < k$ and $\delta_k(t) = 0$ elsewhere. If $h \in \mathcal{H}$ and $(f_k)_{k\in\mathbb{N}} \subset E'$ such that $\lim_{k\to\infty} f_k = \kappa^{-1}(h)$ in $L^2(E,m)$, then it is seen by induction that for every $n \in \mathbb{N}$,

$$\lim_{k\to\infty} e^{-\frac{1}{2}\|h\|^2} \delta_k(f_k) e^{f_k} = k_h \quad \text{in} \quad W_A^n(E,m).$$

Therefore, we only have to prove that the set $\{k_h : h \in \mathcal{H}\}$ is total in $W_A^n(E,m)$. Note that this is well known if $n = 0$. Let $u \in W_A^n(E,m)$ such that for every $h \in \mathcal{H}$, $\langle u, k_h\rangle_n = 0$. Then, in particular $\langle u, Q_t k_h\rangle_n = \langle u, k_{C_t h}\rangle_n = 0 \ \forall h \in \mathcal{H}, t \geq 0$, and this implies for $h \in \mathcal{H}$,

$$0 = \langle u, \frac{1}{\Gamma(\frac{n}{2})} \int_0^\infty t^{n-1} e^{-t} Q_t k_h \, dt\rangle_n = \langle u, U^{2n} k_h\rangle_n = \langle u, k_h\rangle_{L^2(E,m)}.$$

Hence $u = 0$. □

Fix $n \in \mathbb{N}$. We introduce a functional capacity c_n in the following manner:

$$\begin{array}{lll} \text{if } f \geq 0 \text{ is l.s.c. on } E, & c_n(f) = \inf\{\|g\|_n^2 : g \in W_A^n(E,m),\ g \geq f \ m-\text{a.e.}\}, \\ \text{if } f : E \to \bar{\mathbb{R}}, & c_n(f) = \inf\{c_n(g) : g \text{ l.s.c. and } g \geq |f|\}. \end{array}$$

The n–capacity of sets $A \subset E$ is defined by $c_n(A) = c_n(1_A)$. A property is said to hold n–quasi everywhere (n–q.e.) if it holds outside of a set of n–capacity zero.

It is known (see [5], Corollaire 22) that the capacity c_n is tight. This and the preceding proposition show that the Sobolev space $W_A^n(E,m)$ satisfies the conditions (A.1)–(A.3) of [9]. Consequently, we can apply the theory developped in [9]. Let's recall some notions and results used in the sequel.

The positive elements of $W_A^{-n}(E,m)$ are finite measures on E, called measures of finite n–energy. An element $u \in W_A^n(E,m)$ which can be written as $u = U^{2n}\nu$ with $\nu \in (W_A^{-n}(E,m))^+$ is called an n–potential.

A function u on E is called n–quasi–continuous if it has the Lusin property w.r.t. the capacity c_n, i.e. if for every $\varepsilon > 0$, there exists a closed set $F \subset E$ such that $c_n(E \setminus F) < \varepsilon$ and such that $u|_F$ is continuous. It is known that every function $u \in W_A^n(E,m)$ possesses an n–quasi continuous modification, i.e. there exists an n–quasi–continuous function $\tilde{u}$ which equals u m–almost everywhere.

If f is a function on E of finite n–capacity, then there exists a unique element $u \in W_A^n(E,m)$ satisfying

$$\tilde{u} \geq |f| \quad n\text{–q.e. and} \quad c_n(f) = \|u\|_n^2.$$

Moreover, u is an n–potential. We call u the n–equilibrium function of f and the measure $\nu = U^{-2n}u$ its n–equilibrium measure. We have $\int u\, d\nu = \int |f|\, d\nu$, as is easily seen by, for example, [9], lemma 5.9.

Finally remark that for every positive $\nu \in W_A^{-n}(E,m)$, there exists a sequence $(p_k)_{k\in\mathbb{N}}$ in $L^2(E,m)^+$ converging to ν in $W_A^{-n}(E,m)$. In fact, this is an immediate consequence of the bipolar theorem.

3. Construction of multiparameter processes

Let $(\mathcal{C}(\mathbb{R}_+,E), W_t, \mu)$ be an E–valued Brownian motion as constructed in [6], i.e. let $(W_t)_{t\geq 0}$ be an E–valued process on $\mathcal{C}(\mathbb{R}_+,E)$ with a.s. continuous paths, and μ a centered Gaussian measure on the latter space such that, for every $f \in E'$, $(f(W_t))_{t\geq 0}$ is a real Brownian motion with variance $\mathbb{E}[f(W_t)^2] = t\|f\|^2_{\mathcal{H}'}$. We will suppose that $(W_t)_{t\geq 0}$ is the coordinate process. The Cameron–Martin space of $(\mathcal{C}(\mathbb{R}_+,E),\mu)$ is given by

$$\mathcal{F} = \{u :\ u(t) = \int_0^t \dot{u}(s)\,ds,\ \dot{u} \in L^2(\mathbb{R}_+,\mathcal{H})\}; \quad \|u\|_{\mathcal{F}} = \|\dot{u}\|_{L^2(\mathbb{R}_+,\mathcal{H})}.$$

Note that the process $\tilde{W}_t = \sqrt{2A} \circ W_t$ is a Brownian motion on $(\mathcal{C}(\mathbb{R}_+,E),\mu)$ with variance $\mathrm{E}[f(\tilde{W}_t)^2] = t\|f\|^2_{\tilde{\mathcal{H}}'}$ (cf. [6], Cor. 10). If we denote its law by $\tilde{\mu}$, then the Cameron–Martin space of $(\mathcal{C}(\mathbb{R}_+,E),\tilde{\mu})$ is the space $\tilde{\mathcal{F}}$ which is defined in the same way as $\mathcal{F}$, but using $(E,\tilde{m})$ as state space.

In [6] it is shown that the process

$$X_t(x,\omega) = C_t x + \sqrt{2A}\int_0^t C_{t-s}\,dW_s(\omega) \tag{4}$$

on the probability space $(E \times \mathcal{C}(\mathbb{R}_+,E), m \otimes \mu)$ has a.s. continuous paths and that it is Markovian with transition semigroup $(Q_t)_{t\geq 0}$. Since $(Q_t)_{t\geq 0}$ is symmetric, the law of each random variable X_t is m.

In order to construct n–parameter processes, the preceding construction will be iterated. The basic idea (for n=2) is as follows: The law $m^{(2)}$ of the process X is a centered Gaussian measure on $\mathcal{C}(\mathbb{R}_+,E)$. We will define a semigroup on the corresponding Cameron–Martin space which in turn gives rise to a "Mehler semigroup" on $L^2(\mathcal{C}(\mathbb{R}_+,E), m^{(2)})$. The initial structure will therefore be recovered (where E is replaced by $\mathcal{C}(\mathbb{R}_+, E)$ and m by $m^{(2)}$), and consequently there exists a continuous process $X^{(2)}$ with state space $\mathcal{C}(\mathbb{R}_+, E)$ which can be canonically identified with an E–valued two–parameter process whose paths (as functions on $\mathbb{R}^2_+$) are a.s. continuous.

If necessary, we will distinguish the objects constructed so far by a superscript $^{(1)}$ ($E^{(1)} = E$, $\mathcal{H}^{(1)} = \mathcal{H}$, $m^{(1)} = m$ etc.). For $n \geq 2$, we set $E^{(n)} = \mathcal{C}(\mathbb{R}_+, E^{(n-1)})$. By induction, we will construct the following objects:

1. A centered Gaussian measure $m^{(n)}$ on $E^{(n)}$, whose Cameron–Martin space will be denoted by $\mathcal{H}^{(n)}$.
2. A canonical $E^{(n)}$–valued Brownian motion $(E^{(n+1)}, W_t^{(n)}, \mu^{(n)})$ associated with $m^{(n)}$.
3. A symmetric contraction semigroup $(C_t^{(n)})_{t\geq 0}$ of class (C_0) on $\mathcal{H}^{(n)}$, such that the square root of its infinitesimal generator, $\sqrt{A}^{(n)}$, satisfies the condition (1) w.r.t. the space $(E^{(n)}, m^{(n)})$.

Let $n \in \mathbb{N}$ such that $(E^{(n)}, \mathcal{H}^{(n)}, m^{(n)})$, $(E^{(n+1)}, W_t^{(n)}, \mu^{(n)})$ and $(C_t^{(n)}, \sqrt{A}^{(n)})$ are already constructed. By Mehler's formula, one obtains a semigroup $(Q_t^{(n)})_{t\geq 0}$ on $L^2(E^{(n)}, m^{(n)})$ which is the transition semigroup of a continuous process $(X_t^{(n)})_{t\geq 0}$ (given by a formula of type (4)) on $(E^{(n)} \times E^{(n+1)}, m^{(n)} \otimes \mu^{(n)})$ with state space $E^{(n)}$. Let $\tilde{m}^{(n)}$ be the centered Gaussian measure on $E^{(n)}$ which is the image of $m^{(n)}$ by $\sqrt{2A}^{(n)}$, and let $\tilde{\mathcal{H}}^{(n)}$ be the Cameron–Martin space of $(E^{(n)}, \tilde{m}^{(n)})$.

4. PROPOSITION. – *The law $m^{(n+1)}$ of the process $X^{(n)}$ is a centered Gaussian measure on $E^{(n+1)}$. The Cameron–Martin space $\mathcal{H}^{(n+1)}$ of $(E^{(n+1)}, m^{(n+1)})$ consists of the functions $u_{h,\tilde{\varphi}} \in E^{(n+1)}$ which can be written as*

$$u_{h,\tilde{\varphi}}(t) = C_t^{(n)} h + \int_0^t C_{t-s}^{(n)} \tilde{\varphi}(s)\, ds \ ,$$

where $h \in \mathcal{H}^{(n)}$ and $\tilde{\varphi} \in L^2(\mathbb{R}_+, \tilde{\mathcal{H}}^{(n)})$. The norm on $\mathcal{H}^{(n+1)}$ is given by

$$\|u_{h,\tilde{\varphi}}\|^2_{\mathcal{H}^{(n+1)}} = \|h\|^2_{\mathcal{H}^{(n)}} + \int_0^\infty \|\tilde{\varphi}\|^2_{\tilde{\mathcal{H}}^{(n)}}\, ds.$$

Proof. According to the induction hypothesis, we are in the initial situation if we choose $(E^{(n)}, m^{(n)})$ as underlying state space. We can thus suppose that $n = 1$ and consequently the superscript $^{(n)}$ will be dropped. We will use the construction of the process X as specified in the proofs of theorems 12 and 13 of [6].

1. Let X^0 be the $\tilde{\mu}$–measurable linear extension of the injective bounded operator

$$\tilde{\mathcal{F}} \longrightarrow \tilde{\mathcal{F}} \ , \ u \mapsto (t \mapsto \int_0^t C_{t-s} \dot{u}\, ds).$$

The measure $m_0 = X^0(\tilde{\mu})$ on $\mathcal{C}(\mathbb{R}_+, E)$ is Gaussian and centered and coincides with the law of the process

$$X_t^0 = \sqrt{2A} \int_0^t C_{t-s}\, dW_s = \int_0^t C_{t-s}\, d\tilde{W}_s.$$

Let $\mathcal{G}'$ be the closure of $\mathcal{C}(\mathbb{R}_+, E)'$ in $L^2(m_0)$, which is nothing else but the space of m_0-measurable linear formes on $\mathcal{C}(\mathbb{R}_+, E)$. Since $X^0|_{\tilde{\mathcal{F}}}$ is injective, the natural isometry

$$(X^0)^* : \mathcal{G}' \longrightarrow \tilde{\mathcal{F}}' \ , \ G \mapsto G \circ X^0$$

is surjective. This shows that the Cameron–Martin space of $(\mathcal{C}(\mathbb{R}_+, E), m_0)$ equals $X^0(\tilde{\mathcal{F}})$.

2. The mapping

$$Y : E \ni x \mapsto (t \mapsto C_t x)$$

is m-measurable and a.s. $\mathcal{C}(\mathbb{R}_+, E)$-valued, hence $\nu = Y(\mu)$ is a centered Gaussian measure on $\mathcal{C}(\mathbb{R}_+, E)$. Y is clearly injective on $\mathcal{H}$, and as above one can conclude that the Cameron–Martin space of (E, ν) is $Y(\mathcal{H})$.

3. According to our definitions, we have

$$X_t(x, \omega) = C_t x + X_t^0(\omega) \ , \ (x, \omega) \in E \times \mathcal{C}(\mathbb{R}_+, E).$$

Consequently, the law of X is the convolution of ν and the law of X^0, i.e. $m^{(2)} = \nu * m_0$. In particular, $m^{(2)}$ is Gaussian and centered. Since ν is carried by $\Omega_1 = Y(E)$ and m_0 is carried by Ω_0, the space of continuous paths in E starting from zero, one concludes that $m^{(2)}$ is carried by $\Omega_1 \oplus \Omega_0$. It is easily verified that

$$\overline{(\Omega_1 \oplus \Omega_0)'}^{L^2(m^{(2)})} = \overline{(\Omega_1)'}^{L^2(\nu)} \oplus \overline{(\Omega_0)'}^{L^2(m_0)},$$

and by duality, it follows that the Cameron–Martin space of $(\mathcal{C}(\mathbb{R}_+, E), m^{(2)})$ is the direct sum of $Y(\mathcal{H})$ and $X^0(\tilde{\mathcal{F}})$. □

For $t \geq 0$, we define a linear operator $C_t^{(n+1)}$ on $\mathcal{H}^{(n+1)}$ by the formula

$$C_t^{(n+1)} u_{h,\tilde{\varphi}} = C_t^{(n)} \circ u_{h,\tilde{\varphi}} = u_{C_t^{(n)} h, C_t^{(n)} \circ \tilde{\varphi}},$$

where $h \in \mathcal{H}^{(n)}$, $\tilde{\varphi} \in L^2(\mathbb{R}_+, \tilde{\mathcal{H}}^{(n)})$. $(C_t^{(n+1)})_{t \geq 0}$ is easily seen to be a symmetric contraction semigroup on $\mathcal{H}^{(n+1)}$ of class (C_0). The domain of its infinitesimal generator $A^{(n+1)}$ consists of all functions $u_{h,\tilde{\varphi}} \in \mathcal{H}^{(n+1)}$ such that $h \in D(A^{(n)})$, $\tilde{\varphi}(s) \in D(A^{(n)})$ ds–a.e. and $A^{(n)} \circ \tilde{\varphi} \in L^2(\mathbb{R}_+, \mathcal{H}^{(n)})$. For these functions,

$$(5) \qquad (A^{(n+1)} u_{h,\tilde{\varphi}})(t) = A^{(n)} C_t^{(n)} h + \sqrt{2A^{(n)}} \int_0^t C_{t-s}^{(n)} (A^{(n)} \tilde{\varphi}(s))\, ds.$$

5. Proposition. – *For $t \geq 0$, the $m^{(n+1)}$-measurable linear extension of $C_t^{(n+1)}$ is given by*

$$C_t^{(n+1)} u = C_t^{(n)} \circ u \ , \ u \in E^{(n+1)}.$$

Moreover, the kth power ($k \in \mathbb{N}$) of the square root of the infinitesimal generator of $(C_t^{(n+1)})_{t \geq 0}$ extends to the $m^{(n+1)}$-measurable linear operator

$$(6) \qquad (\sqrt{A}^{(n+1)})^k : E^{(n+1)} \to E^{(n+1)} \ , \ u \mapsto (\sqrt{A}^{(n)})^k \circ u.$$

Proof. Once more we can assume $n = 1$ and drop the superscript $^{(1)}$.

1. Fix $t \geq 0$. According to [6], théorème 13, the process

$$C_t X_s(x,\omega) = C_{t+s}x + \sqrt{2AC_t^2} \int_0^s C_{t-r}\, dW_r$$

has $m \otimes \mu$–a.s. continuous paths. This amounts to say that the mapping $C_t^{(2)}$ on $\mathcal{C}(\mathbb{R}_+, E)$ is $m^{(2)}$–measurable with values lying $m^{(2)}$–a.s. in $\mathcal{C}(\mathbb{R}_+, E)$.

2. In order to prove the second assertion, we have to show that for fixed $k \in \mathbb{N}$, the process $(\sqrt{A}^k \circ X_t)_{t \geq 0}$ has a modification with $m \otimes \mu$–a.s. continuous paths. Using hypothesis (1), we conclude by [6], théorème 13, that the process $Z_t = \sqrt{A}^k \circ X_t^0$ is a solution of the stochastic Cauchy problem

$$dZ_t = \sqrt{A}^k dW_t - AZ_t dt \quad ; \quad Z_0 = 0$$

with μ–a.s. continuous paths.

Denote for the moment by ν the gaussian measure $\sqrt{A}^k(m)$ and by $\mathcal{H}^\nu$ the associated Cameron–Martin space. Condition (1) implies that C_t defines a bounded linear operator on $\mathcal{H}^\nu$; let C_t^ν be its ν–measurable linear extension. By [6], théorème 12, $t \to C_t^\nu x$ is continuous on $[0, +\infty[$ for ν–almost every $x \in E$, so $t \to C_t^\nu \circ \sqrt{A}^k x$ is continuous for m–almost every x. Now $\sqrt{A}^k \circ C_t = C_t \circ \sqrt{A}^k : \mathcal{H} \longrightarrow \mathcal{H}^\nu$ is a bounded operator, and therefore we obtain $\sqrt{A}^k \circ C_t = C_t^\nu \circ \sqrt{A}^k$ m–a.e. by the uniqueness of its m–measurable linear extension. Thus the desired continuity is proved.

Finally, the relation (5) implies that the operators defined by (6) actually extend the powers of the square root of the generator of $C_t^{(2)}$. □

For $t \geq 0$, let $l_t^{(n-1)}$ be the evaluation map at t on $E^{(n)} = \mathcal{C}(\mathbb{R}_+, E^{(n-1)})$. For $\mathbf{t} = (t_1, \ldots, t_{n-1}) \in \mathbb{R}_+^{n-1}$, we set $l_{\mathbf{t}}^{(n-1)} = l_{t_1}^{(1)} \circ \ldots \circ l_{t_{n-1}}^{(n-1)}$. If we identify $u \in E^{(n)}$ with the function

$$\mathbb{R}_+^{(n-1)} \ni \mathbf{t} \mapsto l_{\mathbf{t}}^{(n-1)}(u) = u(\mathbf{t}),$$

then we can identify $E^{(n)}$ and $\mathcal{C}(\mathbb{R}_+^{(n-1)}, E)$, the space of continuous paths in E with $n-1$ parameters. We define

$$X_{\mathbf{t}}^{(n)} = l_{(t_1,\ldots,t_{n-1})}^{(n-1)} \circ X_{t_n}^{(n)}\ , \ \mathbf{t} = (t_1, \ldots, t_n) \in \mathbb{R}_+^n,$$

which yields an E–valued n–parameter process whose paths are a.s. contained in $\mathcal{C}(\mathbb{R}_+^n, E) = E^{(n+1)}$. An induction argument shows that each random variable $X_{\mathbf{t}}^{(n)}$ follows the law m.

4. Properties of the n–parameter process $(X_{\mathbf{t}}^{(n)})_{\mathbf{t}\in\mathbb{R}_+^n}$

In this section, we fix $n \geq 2$. The starting point for the further investigation of the process $(X_{\mathbf{t}}^{(n)})_{\mathbf{t}\in\mathbb{R}_+^n}$ is the following crucial lemma, which relates the semigroups $(Q_t^{(n)})_{t\geq 0}$ and $(Q_t)_{t\geq 0}$.

6. LEMMA. *Let $t \in \mathbb{R}_+$ and $\mathbf{s} \in \mathbb{R}_+^{(n-1)}$. Then for every bounded Borel function f on E, the following relation holds*

$$Q_t^{(n)}(f \circ l_{\mathbf{s}}^{(n-1)}) = (Q_t f) \circ l_{\mathbf{s}}^{(n-1)}.$$

Proof. First remark that the semigroup $(C_t^{(n)})_{t\geq 0}$ is characterized by the property

$$\forall t \in \mathbb{R}_+, \forall \mathbf{s} \in \mathbb{R}_+^{(n-1)}, \quad l_{\mathbf{s}}^{(n-1)} \circ C_t^{(n)} = C_t \circ l_{\mathbf{s}}^{(n-1)}. \tag{7}$$

The same relation holds for $S_t^{(n)} = \sqrt{I - (C_t^{(n)})^2}$. Using the fact that the law of $X_{\mathbf{s}}^{(n-1)}$ is m, we obtain for $u \in E^{(n)}$

$$\begin{aligned} Q_t^{(n)}(f \circ l_{\mathbf{s}}^{(n-1)})(u) &= \int f(C_t \circ l_{\mathbf{s}}^{(n-1)}(u) + S_t \circ l_{\mathbf{s}}^{(n-1)}(v))\, m^{(n)}(dv) \\ &= \mathbb{E}[f(C_t \circ l_{\mathbf{s}}^{(n-1)}(u) + S_t \circ X_{\mathbf{s}}^{(n-1)})] \\ &= (Q_t f)(l_{\mathbf{s}}^{(n-1)}(u)). \end{aligned}$$

□

Let $\mathbf{t} = (t_1, \ldots, t_n) \in \mathbb{R}_+^n$. We denote $|\mathbf{t}| = t_1 + \ldots + t_n$ and $\pi(\mathbf{t}) = (t_{\pi(1)}, \ldots, t_{\pi(n)})$ if π is a permutation of $\{1, \ldots, n\}$. For $B \subset \{1, \ldots, n\}$ we define $\overline{B} = \{1, \ldots, n\} \setminus B$ and $\mathbf{t}_B = (t_i)_{i\in B}$. We use the natural order on $\mathbb{R}_+^n$, i.e. $\mathbf{s} \leq \mathbf{t} \iff s_i \leq t_i\ \forall i$. We introduce an n–parameter filtration by

$$\mathcal{F}_{\mathbf{t}}^{(n)} = \sigma\{X_{\mathbf{s}}^{(n)} : \mathbf{s} \leq \mathbf{t}\}\ ,\ \mathbf{t} \in \mathbb{R}_+^n.$$

More generally, we set for $B \subset \{1, \ldots, n\}$

$$\mathcal{F}_{\mathbf{t},B}^{(n)} = \sigma\{X_{\mathbf{s}}^{(n)} : \mathbf{s}_B \leq \mathbf{t}_B\}\ ,\ \mathbf{t} \in \mathbb{R}_+^n.$$

As usual, $\mathcal{F}_\infty^{(n)} = \sigma\{X_{\mathbf{s}}^{(n)} : \mathbf{s} \in \mathbb{R}_+^n\}$.

7. LEMMA. *Let π be a permutation of $\{1, \ldots, n\}$, $k \in \mathbb{N}$ and let $h_1, \ldots, h_k$ be bounded Borel functions on E. Then for $\mathbf{t}^1, \ldots, \mathbf{t}^k \in \mathbb{R}_+^n$,*

$$\mathbb{E}\,[h_1(X_{\mathbf{t}^1}^{(n)}) \cdot \ldots \cdot h_k(X_{\mathbf{t}^k}^{(n)})] = \mathbb{E}\,[h_1(X_{\pi(\mathbf{t}^1)}^{(n)}) \cdot \ldots \cdot h_k(X_{\pi(\mathbf{t}^k)}^{(n)})].$$

Proof. Since the law of the Gaussian process $X^{(n)}$ is determined by its covariance function, it is enough to prove the assumption for $k = 2$. Using the symmetry of $(Q_t^{(n)})_{t\geq 0}$ and the preceding lemma, we obtain

$$\begin{aligned} \mathbb{E}\,[h_1(X_{\mathbf{t}^1}^{(n)})h_2(X_{\mathbf{t}^2}^{(n)})] &= \mathbb{E}\,[h_1 \circ l_{(t_1^1,\ldots,t_{n-1}^1)}^{(n-1)}(X_0^{(n)}) Q_{|t_n^2 - t_n^1|}^{(n)}(h_2 \circ l_{(t_1^2,\ldots,t_{n-1}^2)}^{(n-1)})(X_0^{(n)})] \\ &= \mathbb{E}\,[h_1(X_{(t_1^1,\ldots,t_{n-1}^1)}^{(n-1)}) Q_{|t_n^2 - t_n^1|} h_2(X_{(t_1^2,\ldots,t_{n-1}^2)}^{(n-1)})] \end{aligned}$$

Repeating this calculation leads to

$$\mathbb{E}\,[h_1(X^{(n)}_{\mathbf{t}^1})h_2(X^{(n)}_{\mathbf{t}^2})] = \int h_1 Q_r h_2 \, dm,$$

where $r = \sum_{i=1}^n |t_i^2 - t_i^1|$, which is clearly invariant under permutations. □

8. PROPOSITION. *Let f be a bounded Borel function on E. If $B \subset \{1,\ldots,n\}$ and $\mathbf{s},\mathbf{t} \in \mathbb{R}^n_+$, set $C = \{i \in B :\ s_i \geq t_i\}$ and define $\mathbf{u} \in \mathbb{R}^n_+$ by $u_i = t_i$ if $i \in C$, $u_i = s_i$ otherwise. Then the following equality holds:*

$$\mathbb{E}\,[f(X^{(n)}_{\mathbf{s}})|\mathcal{F}^{(n)}_{\mathbf{t},B}] = Q_{|\mathbf{s}_C - \mathbf{t}_C|} f(X^{(n)}_{\mathbf{u}}).$$

Proof. The function on the right is clearly $\mathcal{F}^{(n)}_{\mathbf{t},B}$–measurable, and so we have to prove that for any function $H \in \mathcal{F}^{(n)}_{\mathbf{t},B}$,

$$\mathbb{E}\,[f(X^{(n)}_{\mathbf{s}})H] = \mathbb{E}\,[Q_{|\mathbf{s}_C - \mathbf{t}_C|} f(X^{(n)}_{\mathbf{u}})H].$$

We can assume that $H = h_1(X^{(n)}_{\mathbf{t}^1}) \cdot \ldots \cdot h_k(X^{(n)}_{\mathbf{t}^k})$, where $k \in \mathbb{N}$, h_i are bounded Borel functions and $\mathbf{t}^i \in \mathbb{R}^n_+$ are such that $\mathbf{t}^i_B \leq \mathbf{t}_B$ for $i = 1,\ldots,k$.

If $n \in C$, conditioning w.r.t. $X^{(n)}_{t_n}$ and lemma 6 yield

$$\mathbb{E}\,[f(X^{(n)}_{\mathbf{s}})H] = \mathbb{E}\,[Q_{s_n - t_n} f \circ l^{(n-1)}_{(s_1,\ldots,s_{n-1})}(X^{(n)}_{t_n})H].$$

If $n \notin C$, we leave the expression $\mathbb{E}\,[f(X^{(n)}_{\mathbf{s}})H]$ unchanged. In both cases we apply the cyclic permutation $\pi(1,\ldots,n) = (n,1,\ldots,n-1)$ and start anew. After n steps, the desired result is proved. □

9. PROPOSITION. *If Z is $\mathcal{F}^{(n)}_\infty$–measurable and bounded and if $B \subset C \subset \{1,\ldots,n\}$, $\mathbf{t} \in \mathbb{R}^n_+$, then*

$$\mathbb{E}\,[\,\mathbb{E}\,[Z\,|\,\mathcal{F}^{(n)}_{\mathbf{t},B}]\,|\,\mathcal{F}^{(n)}_{\mathbf{t},C\setminus B}] = \mathbb{E}\,[Z\,|\,\mathcal{F}^{(n)}_{\mathbf{t},C}].$$

Proof. We have to prove that for an arbitrary finite number of bounded Borel functions $h_1,\ldots,h_k$ on E and for $\mathbf{t}^1,\ldots,\mathbf{t}^k \in \mathbb{R}^n_+$ such that $\mathbf{t}^i_B \leq \mathbf{t}_B$ for $i \in \{1,\ldots,k\}$, the conditional expectation of $H = h_1(X^{(n)}_{\mathbf{t}^1}) \cdot \ldots \cdot h_k(X^{(n)}_{\mathbf{t}^k})$ w.r.t. $\mathcal{F}^{(n)}_{\mathbf{t},C\setminus B}$ is $\mathcal{F}^{(n)}_{\mathbf{t},C}$–measurable.

Fix $l \in C \setminus B$. Using relation (7), proposition 8 and the method of its proof, it is seen by successive conditioning on the σ–fields $\mathcal{F}_{\mathbf{t}^i,\{l\}}$ $(1 \leq i \leq k)$ that $\mathbb{E}\,[H\,|\,\mathcal{F}^{(n)}_{\mathbf{t},C\setminus B}] = \mathbb{E}[G\,|\,\mathcal{F}^{(n)}_{\mathbf{t},C\setminus B}]$, where G only depends on a finite number of coordinates $\mathbf{s}^j$, and where each of the $\mathbf{s}^j$ satisfies $s^j_B \leq \mathbf{t}_B$ and $s^j_l \leq t_l$. Proceeding this way for every $l \in C \setminus B$, the proof is finished after a finite number of steps. □

Note that the last proposition assures that the martingale inequality can be applied to n–parameter martingales in the filtration $(\mathcal{F}^{(n)}_{\mathbf{t}})_{\mathbf{t}\in\mathbb{R}^n_+}$ (cf. e.g. [2]).

5. Relations between analytic and probabilistic concepts

The two inequalities stated in the next theorem, as well as the subsequent representation of finite energy measures, are (in the case $A = I$) key results of S. Song [11]. The properties proved in the previous section allow to deduce them in the same manner in our more general setting. For that reason, we only outline the proofs and refer to Song's article for details.

10. THEOREM. *For any Borel function f on E*

(8) $$\mathbb{E}[(\sup_{\mathbf{t}\in\mathbf{R}_+^n} e^{-|\mathbf{t}|}|f|(X_{\mathbf{t}}^{(n)}))^2] \leq 4^n c_n(f),$$

(9) $$\mathbb{E}[(\sup_{\mathbf{t}\in\mathbf{R}_+^n} |\int_{\mathbf{0}}^{\mathbf{t}} e^{-|\mathbf{u}|} f(X_{\mathbf{u}}^{(n)})\, d\mathbf{u}|)^2] \leq \alpha_n \|f\|_{-n}^2,$$

where in (9) f is supposed to be square integrable, and where α_n is a constant only depending on n.

Proof. If $f \in L^1(E, m)$, define an n–parameter martingale by

(10) $$M_{\mathbf{t}}^f = \mathbb{E}[\int_{\mathbf{R}_+^n} e^{-|\mathbf{u}|} f(X_{\mathbf{u}}^{(n)})\, d\mathbf{u} \mid \mathcal{F}_{\mathbf{t}}^{(n)}].$$

Using proposition 8, one obtains the identity

(11) $$M_{\mathbf{t}}^f = \sum_{B\subset\{1,\ldots,n\}} e^{-|\mathbf{t}_{\overline{B}}|} \int_{\mathbf{0}}^{\mathbf{t}_B} e^{-|\mathbf{u}_B|} U^{2(n-|B|)} f(X_{(\mathbf{u}_B,\mathbf{t}_{\overline{B}})}^{(n)})\, d\mathbf{u}_B.$$

A similar computation yields for $f \in L^2(E, m)$

(12) $$\mathbb{E}[|M_\infty^f|^2] = \|f\|_{-n}^2.$$

In order to establish (8), suppose first that $I \subset \mathbb{R}_+^n$ is finite and that $f = U^{2n}g$ with positive $g \in L^2(E, m)$. Then by (10), (11) and the martingale inequality,

(13) $$\mathbb{E}[(\sup_{\mathbf{t}\in I} e^{-|\mathbf{t}|} f(X_{\mathbf{t}}^{(n)}))^2] \leq 4^n \|f\|_n^2$$

If f is an n–potential, there exists a sequence $(p_k)_{k\in\mathbb{N}} \subset L^2(E, m)^+$ with $U^{2n}p_k \longrightarrow f$ in $W_A^n(E, m)$. A passage to the limit shows that (13) subsists.

If f is a general function of finite n–capacity, let u be an n–equilibrium function for f. We apply (12) to obtain

$$\mathbb{E}[(\sup_{\mathbf{t}\in I} e^{-|\mathbf{t}|}|f|(X_{\mathbf{t}}^{(n)}))^2] \leq 4^n c_n(f).$$

This implies that (8) holds for l.s.c. functions f and therefore for general f by the definition of the functional capacity c_n.

In order to show the estimate (9), one can adopt the proof of [11], §5, théorème 2. □

11. THEOREM. *Let ν be a positive measure in $W_A^{-n}(E,m)$. There exists a sequence $(p_k)_{k\in\mathbb{N}} \subset L^2(E,m)^+$, converging to ν in $W_A^{-n}(E,m)$, such that the sequence $(e^{-|\mathbf{u}|}p_k(X_{\mathbf{u}}^{(n)})\,d\mathbf{u})_{k\in\mathbb{N}}$ of random measures on $\mathbb{R}_+^n$ converges for almost every ω narrowly. The limit random measure $N(d\mathbf{u})$ satisfies*

$$\int fU^{2n}g\,d\nu = \mathbb{E}[g(X_{\mathbf{0}}^{(n)})\int_{\mathbb{R}_+^n} f(X_{\mathbf{u}}^{(n)})N(d\mathbf{u})]$$

for all bounded Borel functions f, g on E.

Proof (cf. [8], theorem 3.3.1). Let $(p_k)_{k\in\mathbb{N}}$ be a sequence in $L^2(E,m)^+$ such that $p_k \to \nu$ in $W_A^{-n}(E,m)$ and let

$$F_k(\omega,\mathbf{t}) = \int_{\mathbf{0}}^{\mathbf{t}} e^{-|\mathbf{u}|}p_k(X_{\mathbf{u}}^{(n)}(\omega))\,d\mathbf{u}.$$

Inequality (9) implies

$$\mathbb{E}[(\sup_{\mathbf{t}\in\mathbb{R}_+^n} |F_k(\mathbf{t}) - F_l(\mathbf{t})|)^2] \le \alpha_n\|p_k - p_l\|_{-n} \longrightarrow 0$$

if $k,l \longrightarrow \infty$. Hence $(F_k)_{k\in\mathbb{N}}$ converges in the space $L^2(\Omega,\mathbb{P};\mathcal{C}_b(\mathbb{R}_+^n))$ and consequently a subsequence, still denoted by $(F_k)_{k\in\mathbb{N}}$, converges a.s. uniformly on $\mathbb{R}_+^n$. This entails the existence of a random measure N on $\mathbb{R}_+^n$ such that $e^{-|\mathbf{u}|}p_k(X_{\mathbf{u}}^{(n)}(\omega)) \to N(\omega)$ narrowly for almost every ω.

Now let $f,g \in \mathcal{FC}_0^\infty(E)^+$ and set $Y_k = g(X_{\mathbf{0}}^{(n)})\int_{\mathbb{R}_+^n} f(X_{\mathbf{u}}^{(n)})e^{-|\mathbf{u}|}p_k(X_{\mathbf{u}}^{(n)})\,d\mathbf{u}$. (9) implies $\sup_{k\in\mathbb{N}}\mathbb{E}[Y_k^2] < \infty$, hence

$$\lim_{k\to\infty}\mathbb{E}[Y_k] = \mathbb{E}[g(X_{\mathbf{0}}^{(n)})\int_{\mathbb{R}_+^n} f(X_{\mathbf{u}}^{(n)})N(d\mathbf{u})].$$

On the other hand, by proposition 8,

$$\mathbb{E}[Y_k] = \int_{\mathbb{R}_+^n} e^{-|\mathbf{u}|}\,\mathbb{E}\left[g(X_{\mathbf{0}}^{(n)})Q_{|\mathbf{u}|}(fp_k)(X_{\mathbf{0}}^{(n)})\right]d\mathbf{u} = \langle fU^{2n}g, p_k\rangle_{L^2(E,m)}.$$

Since $fU^{2n}g$ is an n–quasi–continuous element of $W_A^n(E,m)$ (cf. [5], corollaire 28), we obtain

$$\lim_{k\to\infty}\mathbb{E}[Y_k] = {}_{W_A^n(E,m)}\langle fU^{2n}g, p_k\rangle_{W_A^{-n}(E,m)} = \int fU^{2n}g\,d\nu.$$

Therefore, the desired formula holds for $f,g \in \mathcal{FC}_0^\infty(E)$, and by the monotone class theorem, it subsists for bounded Borel functions f,g. □

The following theorem is our key result.

12. THEOREM. *Let f be a Borel function on E and let $T > 0$. Define*

$$P_T(f) = \mathbb{E}[(\sup_{|\mathbf{t}|\leq T} |f|(X_{\mathbf{t}}^{(n)}))^2].$$

Then

$$(1 - e^{-T/n})^{2n}\, c_n(f) \leq P_T(f) \leq 4^n e^{2T}\, c_n(f).$$

Proof. The right–hand estimate is an immediate consequence of inequality (8). In order to prove the second estimate, suppose first that $f \in W_A^n(E, m) \cap \mathcal{C}_b(E)$. Let ν be the n–equilibrium measure of f and let $(p_k)_{k\in\mathbb{N}}$ and N be associated with ν in the sense of theorem 11. For $\varphi \in \mathcal{C}_b(\mathbb{R}_+^n)$ we obtain

$$\mathbb{E}[\int \varphi(\mathbf{u})|f|(X_{\mathbf{u}}^{(n)})\, N(d\mathbf{u})] = (\int \varphi(\mathbf{u})e^{-|\mathbf{u}|}d\mathbf{u}) \lim_{k\to\infty} \langle |f|, p_k\rangle_{L^2(E,m)}.$$

Since $f \in W_A^n(E, m) \cap \mathcal{C}_b(E)$, we have $\lim_{k\to\infty}\langle|f|, p_k\rangle_{L^2(E,m)} = \int |f|\, d\nu = c_n(f)$ (cf. [9], lemma 5.10), and therefore

$$\mathbb{E}[\int \varphi(\mathbf{u})|f|(X_{\mathbf{u}}^{(n)})\, N(d\mathbf{u})] = c_n(f) \int \varphi(\mathbf{u})e^{-|\mathbf{u}|}d\mathbf{u}.$$

Let $R_T = \{\mathbf{u} \in \mathbb{R}_+^n : |\mathbf{u}| \leq T\}$. Passage to the limit $\varphi \downarrow 1_{R_T}$ yields

$$\mathbb{E}[\int_{R_T} |f|(X_{\mathbf{u}}^{(n)})\, N(d\mathbf{u})] = \lambda_T\, c_n(f), \text{ where } \lambda_T = \int_{R_T} e^{-|\mathbf{u}|}\, d\mathbf{u}.$$

By the Schwarz inequality,

$$\lambda_T^2\, c_n(f)^2 \leq (\mathbb{E}[\sup_{|\mathbf{u}|\leq T} |f|(X_{\mathbf{u}}^{(n)}) \int_{R_T} N(d\mathbf{u})])^2 \leq P_T(f)\, \mathbb{E}[(\int_{R_T} N(d\mathbf{u}))^2]$$

On the other hand, equation (12) implies

$$\mathbb{E}\,[(\int_{\mathbb{R}_+^n} N(d\mathbf{u}))^2] \leq \liminf_{k\to\infty} \|p_k\|_{-n}^2 = c_n(f).$$

We have thus shown that

$$(1 - e^{-T/n})^{2n}\, c_n(f) \leq P_T(f) \leq 4^n e^{2T}\, c_n(f)$$

for $f \in W_A^n(E, m) \cap \mathcal{C}_b(E)$. If, more generally, $f \in \mathcal{C}_b(E)$, there exists by [9], proposition 5.6 a sequence $(f_k)_{k\in\mathbb{N}} \subset W_A^n(E, m) \cap \mathcal{C}_b(E)$ such that $c_n(f - f_k) \longrightarrow 0$ and $f_k \longrightarrow f$ n–q.e. This implies that the desired inequalities hold for $f \in \mathcal{C}_b(E)$ and therefore, by Choquet's theorem, they subsist for Borel functions. □

13. COROLLARY. *If $B \subset E$ is a Borel set, then $c_n(B)$ vanishes if and only if the process $(X_{\mathbf{t}}^{(n)})_{\mathbf{t}\in\mathbb{R}_+^n}$ does not hit B almost surely.* □

Let's remark that in one direction, theorem 12 even implies a stronger result: If the process $X^{(n)}$ does a.s. not hit the Borel set B in some finite time interval $[0, T]$, then $c_n(B) = 0$.

The subsequent probabilistic characterization of quasi–continuous functions was proved by S. Song [12] in the case $A = I$. For one implication we give, however, a much shorter prove by a completely different method which uses an idea of A. de La Pradelle.

14. THEOREM. *A Borel function u on E is n–quasi–continuous if and only if the random function*

$$\mathbb{R}_+^n \ni \mathbf{t} \longmapsto u(X_{\mathbf{t}}^{(n)})$$

is almost surely continuous.

Proof. In order to verify that the condition is necessary, it is enough to adapt the proof of [11], §4, Lemme 5. For the proof that it is sufficient, we can suppose that $(X_{\mathbf{t}}^{(n)})_{\mathbf{t}\in\mathbb{R}_+^n}$ is the canonical process on $\mathcal{C}(\mathbb{R}_+^n, E)$. Let $\Omega \subset \mathcal{C}(\mathbb{R}_+^n, E)$ be a set of probability one such that $\mathbf{t} \mapsto u(X_{\mathbf{t}}^{(n)}(\omega)) = u(\omega(\mathbf{t}))$ is continuous for every $\omega \in \Omega$. Then the mapping

$$\hat{u} : \Omega \longrightarrow \mathcal{C}(\mathbb{R}_+^n, \mathbb{R}), \omega \mapsto u \circ \omega$$

is well–defined, and standard arguments show that it is measurable.

Fix $\varepsilon > 0$ and $T > 0$. By Lusin's theorem, there exists a compact set $K \subset \Omega$ such that $\mathbb{P}(\Omega \setminus K) < \lambda_T\, \varepsilon$ and such that $\hat{u}|_k$ is continuous. Let $\Phi : \mathbb{R}_+^n \times \Omega \to E$ be the continuous map $(\mathbf{t}, \omega) \mapsto \omega(\mathbf{t})$ and set $R_T = \{\mathbf{t} \in \mathbb{R}_+^n : |\mathbf{t}| \leq T\}$. Then $F = \Phi(R_T \times K)$ is a compact set, and by theorem 12,

$$\lambda_T\, c_n(E \setminus F) \leq P_T(E \setminus F) \leq \mathbb{P}(\Omega \setminus K) \leq \lambda_T \varepsilon.$$

It remains to show that $u|_F$ is continuous. Let $\mathcal{U}$ be an ultrafilter on F with limit point $x \in F$. There exists an ultrafilter $\mathcal{V}$ on $R_t \times K$ such that $\Phi(\mathcal{V})$ generates $\mathcal{U}$. $R_T \times K$ being compact, $\mathcal{V}$ converges to an element $(\mathbf{t}, \omega) \in R_T \times K$, and necessarily, $\omega(\mathbf{t}) = x$. Finally, the continuity of $\hat{u}|_K$ implies that $u(\mathcal{U})$ converges to $u(x)$. □

References

[1] J. BAUER: *Sur les draps d'Ornstein–Uhlenbeck et les capacités associées.* C. R. Acad. Sci. Paris **317** série 1 (1993), 879–882.

[2] M. DOZZI: *Stochastic processes with a multidimensional parameter.* Pitman Research Notes in Mathematics Series **194**, Longman 1989.

[3] D. FEYEL and A. DE LA PRADELLE: *Capacités gaussiennes.* Ann. Inst. Fourier, Grenoble **41**, 1 (1991), 49–76.

[4] D. FEYEL and A. DE LA PRADELLE: *Opérateurs linéaires et espaces de Sobolev sur l'espace de Wiener.* C. R. Acad. Sci. Paris **313** série 1 (1991), 727–729.

[5] D. FEYEL and A. DE LA PRADELLE: *Opérateurs linéaires gaussiens.* To appear in Potential Analysis.

[6] D. FEYEL and A. DE LA PRADELLE: *Processus browniens de dimension infinie.* To appear.

[7] F. HIRSCH: *Capacité associée à une somme de processus indépendants à accroissements indépendants.* C. R. Acad. Sci. Paris **316** série 1 (1993), 925–928.

[8] F. HIRSCH: *Potential theory related to some multiparameter processes.* To appear.

[9] T. KAZUMI and I. SHIGEKAWA: *Measures of finite (r,p)-energy and potentials on a separable metric space.* Sém. Prob. XXVI, Springer LNM **1526** (1992), 415–444.

[10] S. SONG: *Processus d'Ornstein–Uhlenbeck et ensembles* $W^{2,2}$*-polaires.* Potential Analysis **2** (1993), 171–186.

[11] S. SONG: *Inégalités relatives aux processus d'Ornstein–Uhlenbeck à n paramètres et capacité gaussienne* $c_{n,2}$. To appear.

[12] S. SONG: *Some results on potential theory for n-parameter Ornstein–Uhlenbeck processes.* To appear.

ON THE PROBLEM OF HYPOELLIPTICITY ON THE INFINITE DIMENSIONAL TORUS

A. D. Bendikov
Mathematical Institute
Universität Erlangen–Nürnberg
Bismarckstr. 1 1/2
D-91054 Erlangen, Germany

We consider the differential operator $L = \sum_1^\infty a_k \partial_k^2, a_k > 0, \partial_k = \partial/\partial x_k$, which is well defined on the set of cylindric smooth functions D on $T^{\mathbb{N}}$. For any open set $\Omega \subset T^{\mathbb{N}}$ denote by $M(\Omega)$ a set of signed Borel measures on Ω and define

$$\mathcal{H}_L(\Omega) := \{u \in M(\Omega) : Lu = 0 \quad \text{in} \quad \Omega \quad \text{in a weak sense}\}$$

One of the main problem in the theory of harmonic functions of the operator L ([1], [2]) is the following

Problem. $^\forall\Omega \subset T^{\mathbb{N}}$ $\mathcal{H}_L(\Omega) \subset C(\Omega)$, that means, that any weak solution u of equation $Lu = 0$ is absolutely continuous with respect to Haar measure dx and has continuous density $u(x), x \in \Omega$.

The exhausting answer on this problem can be given in terms of the distribution function $N(\lambda) := \sharp\{k : a_k \leq \lambda\}$ of the coefficients of the operator L.

Theorem 1. The following properties are equivalent

1. $$\mathcal{H}_L(\Omega) \subset C(\Omega), \ ^\forall\Omega \subset T^{\mathbb{N}}.$$

2. $$N(\lambda) = o(\lambda), \ \lambda \uparrow \infty.$$

Let $\dot{T}^{\mathbb{N}} = R^1 \times T^{\mathbb{N}}$ and $\dot{L} = \partial_t + L$ is the parabolic operator corresponding to the operator L. We will denote by $\mathcal{H}_L^c$ and $\mathcal{H}_{\dot{L}}^c$ the sheaves of continuous weak solutions of the equations $Lu = 0$ and $\dot{L}u = 0$ correspondently.

We complete the list of the equivalent properties 1. and 2. in Theorem 1. by

3. $(\mathcal{H}_L^c, T^{\mathbb{N}})$ is a Brelot harmonic space [3]
4. $(\mathcal{H}_{\dot{L}}^c, \dot{T}^{\mathbb{N}})$ is a Bauer harmonic space with Doob convergence property [3], [4]

Proof of Theorem 1. We define for every $t > 0$ Gaussian measure μ_t on $(T^{\mathbb{N}}, \mathcal{B})$ by its Fourier transformation $\hat{\mu}_t(\theta), \theta \in \mathbb{Z}^{(\infty)}$

$$\hat{\mu}_t(\theta) = \exp\{-t\psi(\theta)\}, \psi(\theta) = \sum_1^\infty a_k \theta_k^2, \quad \theta = (\theta_k)_1^\infty \in \mathbb{Z}^{(\infty)} \tag{1}$$

K. GowriSankaran et al. (eds.), Classical and Modern Potential Theory and Applications, 57–59.

It is easy to see that (μ_t) is a weakly continuous convolution semigroup and the operator L coincides with its generator on D. According to [1, Theorem 3.3.1, Lemma 2.2.2] property 1. is equivalent to the following properties of the semigroup (μ_t)

$$(i) \qquad \forall t > 0 \ d\mu_t(x) = \mu_t(x)dx \quad .$$

$$(ii) \qquad \forall \epsilon > 0 \ \exp(-\frac{\epsilon}{t})\mu_t(0) = o(1), t \downarrow 0.$$

Now we are left to prove the equivalence of (i) and (ii) to the property 2.
According to [1, Theorem 2.2.3] the property (i) holds iff the series $\sum_1^\infty \exp(-\epsilon a_k)$ converges for any $\epsilon > 0$. Standard arguments show that this convergence holds iff

$$\forall \epsilon > 0 \int_0^\infty \exp(-\epsilon\lambda)N(\lambda)d\lambda < \infty. \tag{2}$$

It follows from (1) that μ_t is a product measure and

$$\mu_t(x) = \prod_1^\infty n_{a_k t}(x_k), \quad x = (x_k) \in T^n \tag{3}$$

where

$$n_t(x) = \sqrt{\frac{\pi}{t}} \sum_{-\infty}^{\infty} \exp\left\{-\frac{(x-2\pi k)^2}{4t}\right\} \tag{4}$$

is a Gaussian density on $T \simeq [= \pi, \pi]$.
Using Fourier transformation we write down

$$n_t(0) = \sum_{-\infty}^{\infty} \exp(-k^2 t) = 1 + 2\varphi(t). \tag{5}$$

According to (3) and (5) we will have

$$\log \mu_t(0) = \sum_1^\infty \log(1 + \varphi(a_k t)) = \int_0^\infty \log(1 + 2\varphi(\lambda t))dN(\lambda). \tag{6}$$

Using (2) and (6) and the following asymptotics of the function $\varphi(t)$ [2,2.6]:

$$\varphi(t) \sim \frac{1}{2}\sqrt{\pi} t^{-1/2}, t \downarrow 0, \ \varphi(t) \sim e^{-t}, t \uparrow \infty \tag{7}$$

we will get the following equivalence

$$\log \mu_t(0) \approx \int_0^{\frac{1}{t}} \frac{N(x)}{x} dx + \int_{\frac{1}{t}}^\infty N(x)e^{-xt} t \, dx, \ t \downarrow 0. \tag{8}$$

Now using (8) it is not hard to proof that (ii) is equivalent to property 2. ∎

Examples. Let us consider group $R^3 \times T^{\mathbb{N}}$ and let μ_t be the Gaussian measure with Fourier transformation $\hat{\mu}_t(\theta), \theta \in \mathbb{R}^3 \times \mathbb{Z}^{(\infty)}$ given by (1). Consider the corresponding potential kernel $k := \int_0^\infty \mu_t dt$. Then $Lk = -\epsilon_0$, where ϵ_0 is a Dirac measure supported on the set $\{0\}$. In particular, $Lk = 0$ for any domain $\Omega \subset R^3 \times T^{\mathbb{N}} \setminus \{0\}$, hence $k \in \mathcal{H}_L(\Omega)$ for any such Ω. Let $a_n = n^\alpha, \alpha > 0$, then we will have [1], [2].

1) $k \perp dx$ if $\alpha = 0$,
2) $k \sim dx$, but dk/dx is a discontinuous function if $0 < \alpha < 1$,
3) $k \sim dx$ and dk/dx is a continuous function if $\alpha > 1$.

For the sake of completeness we give the following important result [1, Th. 2.1.5].

Theorem 2. The following properties are equivalent

3. $(\mathcal{H}_L^c, \dot{T}^{\mathbb{N}})$ is an elliptic Bauer harmonic space.
4. $(\mathcal{H}_L^c, \dot{T}^{\mathbb{N}})$ is a Bauer harmonic space.
5. $\log N(\lambda) = o(\lambda), \ \lambda \uparrow \infty$.

Acknowledgements

This research has been done while the author was a fellow of the Alexander von Humboldt Foundation. He would like to express his sincere gratitude to Professor H. Bauer and the many German mathematicians who were so graciously hospitable. The author would like to thank the Scientific Affairs Division of the NATO.

References.

[1] A.D. Bendikov: Markov processes and differential equations on a group: the space homogenuous case, Russian Math. Surveys 42:5 (1987) 49-94.

[2] Chr. Berg: Potential theory on the infinite dimensional torus, Inventiones Math. 32, (1976) 49-100.

[3] C. Constantinescu and A. Cornea, Potential theory on harmonic spaces, Grundlehren der mathematischen Wissenschaften in Einzeldarstellungen, Band 158, Springer-Verlag, Berlin-Heidelberg-New York (1972).

[4] H. Bauer, Harmonische Räume und ihre Potentialtheorie, Lecture Notes in Math., 22, Springer-Verlag, Berlin-Heidelberg-New York-Tokyo, 1966.

[5] A.D. Bendikov, Functions that are harmonic for a class of diffusion processes on a group. Theory Probab. Appl. 20 (1975) 759-769.

[6] U. Schirmeier, Konvergenzeigenschaften in Harmonischen Räumen, Invent. math. 55 (1979), 91 - 95.

L'équation de Monge-Ampère dans un espace de Banach

E.M.J. BERTIN
Département de Mathématiques
Université d'Utrecht
Budapestlaan 8
3508 TA Utrecht
Pays-Bas.

RÉSUMÉ. On introduit un problème de Monge-Ampère homogène dans un espace de Banach et on examine sa théorie du potentiel.

AMS SUBJECT CLASSIFICATIONS (1991). 31B35, 31C99, 31D05.

MOTS CLÉS. Monge-Ampère, théorie du potentiel convexe.

1 Introduction

En principe, la théorie moderne des espaces harmoniques est une théorie linéaire dans une espace localement compact. Dans [BER79, BER93] nous avons développé une théorie axiomatique non-linéaire du potentiel, dite convexe, motivée par l'équation homogène de Monge-Ampère et un problème de Dirichlet apparenté en termes de fonctions convexes 'minimales'. Une généralisation de cette axiomatique est donnée par van Gool [GOO92], valable pour les fonctions A-harmoniques de l'école scandinave (Grandlund, Lindqvist et Martio [GRA86], Martio [MAR92]), et la classe de perturbations semi-linéaires d'espaces harmoniques (Maeda [MAE81]).

Le présent travail reprend l'étude de la théorie convexe du potentiel associée à l'équation de Monge-Ampére homogène, cette fois formulée pour un espace de Banach réflexif et donc sacrifiant la compacité locale. On donne les débuts modestes d'une telle théorie, fondée sur un opérateur $\mathcal{I}$ du type Dirichlet (Paragraphe 3), représentant la solution du problème de Dirichlet pour l'équation de Monge-Ampère homogène. L'opérateur $\mathcal{I}$ produit, de façon classique, un faisceau de fonctions dites MAD-harmoniques. Ce faisceau admet les modifications de Poisson (Proposition 4.5), grâce à une propriété de Baire déjà implicitement présente dans la méthode d'Alexandrow [ALE55] pour l'étude des solutions faibles dans le cas Euclidien. La même propriété entraîne l'existence d'une base d'ouverts réguliers (Théorème 5.2), ainsi que des propriétés de convergence monotone.

K. GowriSankaran et al. (eds.), Classical and Modern Potential Theory and Applications, 61–75.

2 Préliminaires

2.1 NOTATIONS

Dans ce travail, E désigne un espace de Banach réel réflexif, de dimension ≥ 1, muni de la topologie définie par la norme $\|\cdot\|$ de E. $\mathcal{B}_0$ dénote l'ensemble des boules fermés centrés à l'origine. La topologie affaiblie de E est notée σ, la topologie faible du dual topologique E' de E par σ'.

Si A est un espace topologique, $\mathcal{K}(A)$ est l'ensemble des parties compactes de A. On écrit $A \sqsubset U$ s'il existe un voisinage W de l'origine tel que $A + W \subset U$.

On note NUM(A) l'ensemble des fonctions numériques définies dans l'ensemble A, et on note $\mathrm{sci}_A(f)$ la régularisée semi-continue inférieurement (s.c.i.) dans $\overline{A}$ de la fonction $f \in \mathrm{NUM}(A)$. Si H est une partie de NUM(A), H_b est l'ensemble des fonctions bornées de A, et (pour $A \subset E$) H_a celui des fonctions de H, minorées par une fonction continue affine. Si $x \in A$, $H(x)$ est l'ensemble $\{h(x) : h \in H\}$.

Désignons par Aff(C) l'ensemble des fonctions affines et continues dans la partie convexe C de E, par Conv(C) l'ensemble des fonctions convexes finies continues dans C, par SCI(A) l'ensemble des fonctions semi-continues inférieurement et minorées par une fonction de $\mathrm{C_a}(A)$, et par SCS(A) l'ensemble des fonctions semi-continues supérieurement et majorées par une fonction de $\mathrm{C_a}(A)$. Évidemment, $\mathrm{scs}_A(f) = -\mathrm{sci}_A(-f)$ et $\mathrm{C_a}(A) = \mathrm{SCI}(A) \cap \mathrm{SCS}(A)$.

Pour toute fonction $f \in \mathrm{NUM}(A)$, une inégalité du type $f < c$ est un abus d'abréviation pour la relation $\forall x \in A \; f(x) < c$.

A quelques exceptions auto-explicatives près, les autres conventions notationelles seront celles de Constantinescu et Cornea [CON72, p. 4–8].

2.2 FONCTIONS CONVEXES

Rappelons les faits suivants bien connus:

REMARQUE 2.1. *Soit U une partie ouverte convexe de E et soit $f \in \mathrm{NUM}(U)$.*

1) *Soit f convexe dans l'intervalle $[a, b]$ de U et supposons que $f(a) = -\infty$. Alors $f = -\infty$ dans $[a, b]$ ou il existe $c \in [a, b]$ tel que $f = -\infty$ dans $[a, c)$ et $f = \infty$ dans $(c, b]$.*

2) *Si $f < \infty$ est convexe dans U, alors $f = -\infty$ dans U ou f est finie dans U.*

3) *Soient $x, h \in E$. Si f est convexe dans $[x - h, x + h] \subset U$ et $f(x) < \infty$, l'application $\lambda \mapsto (f(x + \lambda h) - f(x))/\lambda$ est croissante dans $[-1, 0) \cup (0, 1]$.*

REMARQUE 2.2. *Soit C une partie convexe et fermée de E. Toute fonction $f > -\infty$, convexe et semi-continue inférieurement dans C, est l'enveloppe supérieure de ses minorants dans* Aff(E).

Soient U une partie ouverte et convexe de E, x un point de U et H un ensemble de fonctions dans U.

PROPOSITION 2.3. *Supposons que les fonctions de H soient convexes et finies et que $H(x)$ soit borné. Alors:*

1) *H est équicontinu en x si et seulement si H est borné supérieurement dans un voisinage de x.*

2) *Si H est borné supérieurement pour la topologie de la convergence simple, H est borné pour la topologie de la convergence simple.*

3) *Soit H borné pour la topologie de la convergence simple. Alors H est équicontinu dans U si et seulement si H est borné supérieurement dans une partie ouverte non vide de U.*

Démonstration. La nécessité des conditions dans 1) et 3) est évidente.

1): Soient $\epsilon > 0$, V un voisinage absolument convexe de l'origine, $M > 0$, et $\alpha > 1$ de telle sorte que, quel que soit $u \in H$, $x + V \subset U$, $|u(x)| < M$, $u(y) \leq M$ pour tout $y \in x + V$, et $2M/\alpha \leq \epsilon$. Pour tout $z \in (1/\alpha)V$ on a $[x, x + \alpha z] \subset x + V$ et donc (Remarque 2.1,3):

$$\frac{-2M}{\alpha} \leq \frac{u(x - \alpha z) - u(x)}{-\alpha} \leq u(x + z) - u(x) \leq \frac{u(x + \alpha z) - u(x)}{\alpha} \leq \frac{2M}{\alpha}. \tag{1}$$

2): Soit $y \in U$. Il existe $z \in U$ tel que x soit une combinaison strictement convexe de z et de y. Comme H est borné supérieurement en z et inférieurement en x, H est nécessairement borné inférieurement en y.

3): Soit H dominé par M dans un voisinage $V \subset U$ de x et soit $y \in U$. Si $r > 1$ est tel que $z = x + r(y - x) \in U$, l'image de V par la fonction $v \mapsto (1 - 1/r)v + (1/r)z$ est un voisinage $W \subset U$ de y. Comme $u \leq (1 - 1/r)M + (1/r)u(z)$ dans W pour chaque $u \in H$, l'énoncé resulte de 1). ∎

En particulier, toute fonction convexe finie dans une partie convexe et ouverte de $\mathbf{R}^n$ est continue.

COROLLAIRE 2.4. *Soit V ouvert convexe, $V \sqsubset U$, et soit $H \subset$ Conv(U) uniformément borné dans U. Alors H est équicontinu uniformément dans V.*

Démonstration. D'après (1), étant donné $\epsilon > 0$, il existe $B \in \mathcal{B}_0$, tel que $V + B \subset U$ et $|u(x) - u(y)| \leq \epsilon$ quels que soient $u \in H$, $x \in V$ et $y \in x + B$. ∎

REMARQUE 2.5. Revenons à la Proposition 2.3. Comme E est un espace de Baire, la condition que H soit localement borné supérieurement est satisfaite si les fonctions de H sont semi-continues inférieurement et $\sup H < \infty$ (Bourbaki [BOU74, p. IX.56]). En particulier, toute fonction convexe finie et semi-continue inférieurement est continue.

Soit U maintenant une partie ouverte quelconque de E. Une fonction numérique u dans U est dite *localement convexe* dans U s'il existe un recouvrement $\mathcal{R}$ de U, de parties ouvertes et convexes de U, tel que la restriction de u à tout $V \in \mathcal{B}$ soit convexe.

LEMME 2.6. *Soit $u < \infty$ finie supérieurement et localement convexe dans U. Alors:*
1) *Dans toute composante connexe de U, u est finie ou $u = -\infty$.*
2) *u est convexe dans toute partie convexe de U.*

Démonstration. 1): L'ensemble $\{x \in U : u(x) = -\infty\}$ est ouvert et fermé dans U (Remarque 2.1,1).
2): Soit $[x, y]$ un intervalle de U. En vertu de 1), u est finie ou $u = -\infty$ dans $[x, y]$ et d'après la Proposition 2.3, la restriction de u à $[x, y]$ est continue. Soit g une fonction affine dans $[x, y]$ telle que $(u - g)(p) = (u - g)(q)$, $p, q \in [x, y]$. Soit λ le plus petit nombre dans $[0, 1]$ tel que la fonction $u - g$ soit maximale à $\lambda p + (1 - \lambda)q$. Comme la fonction $u - g$ est convexe dans un voisinage de $r = \lambda p + (1 - \lambda)q$, on a $r = q$ et $u - g \leq 0$ dans $[p, q]$. ∎

Les résultats de ce paragraphe sont valables dans tout espace localement convexe séparé, remplaçant $\mathcal{B}_0$ par un système fondamental de voisinages absolument convexes fermés. Dans la Remarque 2.5 il faut supposer en outre que E soit un espace de Baire.

2.3 LE SOUSDIFFÉRENTIEL D'UNE FONCTION CONVEXE

Dans ce paragraphe U est une partie ouverte et convexe de E et u est une fonction convexe continue dans U. Rappelons que $x' \in E'$ est un *sous-gradient* de u en $x \in U$, notation $x' \in \partial_u(x)$, si

$$\forall y \in U \quad u(y) \geq u(x) + x'(y) - x'(x). \tag{2}$$

Notons que x' est un sous-gradient de u en x dès que (2) est satisfait pour tous les points y d'un voisinage de x (Remarque 2.1,3). L'ensemble $\partial_u(x)$ des sous-gradients de u en x est appelé le *sous-différentiel* de u en x. La fonction multivoque $\partial_u : x \mapsto \partial_u(x)$ de U dans E' est appelée le *sous-différentiel* de u par rapport à U.

PROPOSITION 2.7.
1) $\lambda \geq 0 \Longrightarrow \partial_{\lambda u} = \lambda \partial_u$.
2) $h \in \mathrm{Aff}(E) \Longrightarrow \partial_{u+h} = \partial_u + h$.
3) $A \sqsubset U$ *σ-compact* $\Longrightarrow \partial_u\langle A\rangle$ *fortement fermé dans E'.*

Démonstration. 3): Soient $(x_n) \subset A$ et $x'_n \in \partial_u(x_n)$ tels que $x'_n \to x'$ fortement. On suppose que la suite (x_n) converge faiblement vers $x \in A$ (Šmulian). Démontrons que $x' \in \partial_u(x)$. En effet, soit $\epsilon > 0$; comme u est faiblement semi-continu inférieurement, on a $u(x_n) \geq u(x) - \epsilon$, $\|x'_n - x'\| \leq \epsilon$, et $|x'(x_n - x)| \leq \epsilon$ pour tout n assez grand.

Pour tout point y d'un voisinage ouvert de A, contenu dans U et de diamètre d, on a donc

$$\begin{aligned} u(y) - u(x) &\geq u(y) - u(x_n) - \epsilon \geq x'_n(y - x_n) - \epsilon \\ &\geq x'(y - x_n) - \epsilon d - \epsilon \geq x'(y - x) - \epsilon d - 2\epsilon. \blacksquare \end{aligned}$$

PROPOSITION 2.8. *Soit* $A \sqsubset U$ *et soit* $H \subset \text{Conv}(U)$ *un ensemble uniformément équicontinu dans* A. *Alors l'ensemble*

$$\partial_H\langle A\rangle = \bigcup\{\partial_u(x) : u \in H,\ x \in A\}$$

est uniformément équicontinu.

Démonstration. Étant donné $\epsilon > 0$, soit $V \in \mathcal{B}_0$ tel que $A + V \subset U$ et tel que $|u(z) - u(z + y)| \leq \epsilon/2$ quels que soient $u \in H$, $y \in V$ et $z \in A$. Pour tout $x \in A$, $x' \in \partial_u(x)$ et $y \in V$ on a $\epsilon \geq u(x + y) - u(x) =\geq x'(x + y) - x'(x)$, d'où $x'(y) \leq \epsilon$ et par suite $|x'(y)| \leq \epsilon$. $\blacksquare$

Une fonction multivoque ϕ de X dans Y est dite *cocontinue* (ou *s.c.s.*) si l'image réciproque $\{x \in X : \phi(x) \cap B \neq \emptyset\}$ d'une partie fermée B de Y est fermée dans X. Une correspondance est une fonction multivoque à valeurs non vides. Rappelons ([BER79, Th. 3.5, Cor. 3.7]):

THÉORÈME 2.9. *Soit* $H \subset \text{Conv}(U)$ *muni de la topologie de la convergence simple. Supposons que, quelque soit* $x \in U$, *il existe* $V \in \mathcal{B}$ *tel que* $\partial_H\langle(x + V) \cap U\rangle$ *soit équicontinu. Alors,* $\partial : (x, v) \mapsto \partial_v(x)$ *est une correspondance cocontinue de* $U \times H$ *dans* (E', σ'), *à valeurs convexes et compactes.*

COROLLAIRE 2.10. *Pour tout* $K \in \mathcal{K}(U)$ *et tout* $F \in \mathcal{K}(H)$, *l'ensemble* $\partial_F\langle K\rangle$ *est* σ'*-compact.*

COROLLAIRE 2.11. *Soit* u *la limite d'une base de filtre* $\mathcal{F}$ *sur* H *et soit* $K \in \mathcal{K}(U)$. *Pour chaque voisinage faible* O *de* $\partial_u\langle K\rangle$ *il existe* $F \in \mathcal{F}$ *et un voisinage* $W \subset U$ *de* K *tels que* $\partial_F\langle W\rangle \subset O$.

L'assertion duale du Corollaire 2.11 est plus délicate:

PROPOSITION 2.12. *Soit* $\mathcal{F}$ *une base de filtre sur* $\text{Conv}(U)$ *et soit* V *un voisinage convexe borné de* $x \in U$, *tel que* $\overline{V} \subset U$. *De plus, soit* $x' \in \partial_u(x)$ *tel que* $u(y) - u(x) - x'(y - x) > \epsilon > 0$ *quel que soit* $y \in \partial V$. *Alors il existe* $F \in \mathcal{F}$, *tel que* $x' \in \partial_v\langle V\rangle$ *quel que soit* $v \in F$, *dans les deux cas suivants:*

i) $\mathcal{F}$ *converge uniformément vers* u *dans* $\overline{V}$.

ii) $\mathcal{F}$ *est une base du filtre des sections d'un ensemble filtrant à gauche dans* $\text{Conv}(U)$, *convergeant simplement vers* u *dans* $\overline{V}$.

Démonstration. Soit $F \in \mathcal{F}$ tel que, pour tout $v \in \mathcal{F}$, $v(y)-v(x) \geq u(y)-u(x)-\epsilon/2$ quel que soit $y \in \partial V$ (cas i)), ou $v(x) \leq u(x)+\epsilon/2$ (cas ii)). Dans les deux cas on a $v(y)-v(x) > x'(y-x)$, donc la fonction faiblement s.c.i. $v - x'$ atteint sa valeur minimale dans $\overline{V}$ dans un point intérieur, d'où $x' \in \partial_v\langle V\rangle$. ∎

Les deux résultats suivants sont souvent utiles:

LEMME 2.13. *Soient $V \subset U$ un ouvert borné et $u, v \in$ Conv$(U) \cap$ C$(\overline{U})$ tels que $u \leq v$ dans V et $v \leq u$ dans $\overline{U} \setminus V$. Alors $\partial_v\langle V\rangle \subset \partial_u\langle V\rangle$.*

Démonstration. Soient $x \in V$ et $x' \in \partial_v(x)$. Pour tout $y \in \overline{U} \setminus V$ on a $u(y) - u(x) \geq v(y) - v(x) \geq x'(y-x)$. Par suite, la fonction convexe et faiblement s.c.i. $y \mapsto u(y) - u(x) - x'(y-x)$ est positive dans $\overline{U} \setminus V$. Elle atteint donc un minimum ≤ 0 dans un point z de V, d'où $x' \in \partial_u(z)$. ∎

LEMME 2.14. *Soit V un voisinage convexe de $x \in U$, de diamètre d et tel que $\overline{V} \subset U$. De plus, soit $x' \in E'$ un sous-gradient de u en x tel que* $\inf\{u(y)-u(x)-x'(y-x) : y \in \partial V\} = \epsilon > 0$ *Alors x' est un point intérieur fort de $\partial_u\langle V\rangle$.*

Démonstration. Soit $y' \in E'$, $\|y'\| \leq \epsilon/d$. La fonction affine $y \mapsto u(y) - u(x) - (x' + y')(y-x)$ est positive dans ∂V, s'annule en x et atteint donc un minimum ≤ 0 dans un point intérieur de V. ∎

3 Les opérateurs MAD

La *courbure* d'une fonction convexe continue u, définie dans une partie ouverte convexe de $\mathbf{R}^n$ est la fonction

$$\nu_u : K \mapsto \int_{\partial_u\langle K\rangle} d\gamma,\ K \in \mathcal{K}(U),$$

où γ est la mesure de Lebesgue. Comme l'ensemble des $x' \in E'$ qui sont sous-gradient de u en plusieurs points de U est γ-négligeable, on démontre aisément que ν_u est une mesure de Radon sur U. Une équation du type $\nu_u = \mu$, μ une mesure de Radon donnée sur U, est appelée une équation *faible* de Monge-Ampère, Bakelman [BAK86B].

Dans un espace de Banach on perd la mesure de Lebesgue. Cependant, quelle que soit la généralisation choisie, laissant de côté son existence, une solution convexe de l'équation homogène de Monge-Ampère généralisée doit avoir très peu de courbure. Soit donc u la solution unique du problème de Dirichlet classique pour un ouvert convexe borné U et soit v une deuxième fonction convexe continue dans $\overline{U}$, égale à u dans ∂U. D'après le Lemme 2.13, appliqué à v et $w = \sup(u, v)$, on a $\partial_w\langle U\rangle \subset \partial_u\langle U\rangle$, donc $w = u$ et u majore toute autre fonction convexe avec les mêmes valeurs dans ∂U. On arrive donc nécessairement à l'étude des opérateurs suivants:

DÉFINITION 3.1. Soit V ouvert convexe. Pour toute fonction $f \in \mathrm{NUM}(\partial V)$ on note $\mathcal{I}(V,f)$, ou parfois $\mathcal{I}(f)$, la fonction numérique convexe définie dans $\overline{V}$ par

$$\mathcal{I}(V,f) = \sup\{h \in \mathrm{Aff}(E) : h \leq f \text{ dans } \partial V\}. \tag{3}$$

Un ensemble ouvert convexe U est dit *épaulé* si $U = \emptyset$ ou si un point de U (donc tout point de U) est une combinaison convexe de deux points de ∂U.

Rappelons quelques propriétés de l'opérateur $\mathcal{I}$. ([BER79, Pages 397–399]):

REMARQUE 3.2.

1) $\mathcal{I}(V,f)$ est la plus grande fonction convexe et s.c.i. dans $\overline{V}$, $-\infty$ ou finie inférieurement, et minorant f dans ∂V.

2) On a $\mathcal{I}(V,f) = -\infty$ ou $\mathcal{I}(V,f) > -\infty$ dans $\overline{V}$.

3) Si la fonction f est dominée par une fonction dans $\mathrm{Aff}(E)$, $\mathcal{I}(V,f)$ est une fonction numérique continue dans V.

4) $f \leq g \Longrightarrow \mathcal{I}(V,f) \leq \mathcal{I}(V,g)$.

5) $\lambda > 0 \Longrightarrow \mathcal{I}(V,\lambda f) = \lambda\mathcal{I}(V,f)$ et $\mathcal{I}(V,0) = 0$ si et seulement si V est épaulé.

6) $\mathcal{I}(V,f+g) \geq \mathcal{I}(V,f) + \mathcal{I}(V,g)$ $(\infty + (-\infty) = -\infty)$.

7) $h \in \mathrm{Aff}(E) \Longrightarrow \mathcal{I}(V,f+h) = \mathcal{I}(V,f) + h$.

8) $u = \mathcal{I}(V,f) < \infty$, $B \subset V$ ouvert, convexe et épaulé $\Longrightarrow u = \mathcal{I}(B,u)$ dans $\overline{B}$.

9) Soit $u < \infty$ convexe s.c.i. dans $\overline{V}$ et soit $x \in V$. Alors $u(x) = \mathcal{I}(V,u)(x)$ si
$\forall \epsilon > 0,\ \exists x' \in E'\ x \in \overline{\mathrm{co}}(\{y \in \partial V : (u - x')(y) \leq (u - x')(x) + \epsilon\})$.

REMARQUE 3.3. *Soit V épaulé. Alors:*

1) L'opérateur $\mathcal{I}(V,\cdot)$ est continu pour les normes uniformes.

2) Soit $\mathcal{I}(V,f)$ borné dans un voisinage de $x \in \partial V$ et soit V épaulé. Alors
$\mathrm{scs}_{\overline{V}}\,\mathcal{I}(V,f) \leq \mathrm{scs}_{\partial V}\,\mathcal{I}(V,f)$.

La Remarque 2.5 entraîne:

PROPOSITION 3.4. *Soit V épaulé et soit $\infty > f \in \mathrm{NUM}(\partial V)$. Alors $\mathcal{I}(V,f) \in \mathrm{Conv}(V)$ ou $\mathcal{I}(V,f) = -\infty$ dans V.*

Voici un contre-exemple pour une variation sur la Remarque 3.2,1:

EXEMPLE 3.5. Soit U le premier quadrant de $\mathbf{R}^2$ et soit $f : \partial U \to \overline{\mathbf{R}}$ définie par $f(0,y) = -\infty$ pour tout $y \geq 0$ et $f(x,0) = \infty$ pour tout $x > 0$. On a $\mathcal{I}(U,f) = -\infty$, tandis que la fonction u, définie par $u(0,y) = -\infty$, $u(x,y) = \infty$ si $x > 0$, est la plus grande fonction convexe s.c.i. dans $\overline{U}$ majorée par f dans ∂U.

PROPOSITION 3.6. *Soit $\mathcal{G} \subset \mathrm{NUM}(\partial U)$ une famille non vide filtrante à gauche. Posons $f_0 = \inf\mathcal{G}$ et $u = \inf\{\mathcal{I}(U,f) : f \in \mathcal{G}\}$. Supposons que la fonction u soit bornée localement supérieurement dans U et soit finie supérieurement dans ∂U. Alors*

$$\mathcal{I}(U,f_0) = u \quad \text{dans } U \quad \text{et } \mathcal{I}(U,f_0) = \mathrm{sci}_U(u) \quad \text{dans } \partial U.$$

Démonstration. On a $u = -\infty$ ou u est finie continu dans U (Remarque 2.1,2 et Proposition 2.3). La fonction $\mathrm{sci}_U(u)$ est convexe et s.c.i. dans $\overline{U}$ et minore f_0 dans ∂U. L'assertion résulte alors de la Remarque 3.2,1. ∎

La version duale de la Proposition 3.6 ne tient pas:

EXEMPLE 3.7. En effet, soit U la boule unité de l'espace de Hilbert $l^2(\mathbf{N})$ et soit $u_n(x) = (\sum_0^n x_i^2)^{1/2}$. On a $u_n \uparrow 1$ dans ∂U, $\mathcal{I}(1) = 1$, et $\mathcal{I}(u_n) = u_n$. Par suite, $\mathcal{I}(u_n) \not\to \mathcal{I}(1)$.

Récapitulant:

DÉFINITION 3.8. Soit U un ouvert convexe épaulé et soit $f \in \mathrm{C_a}(\partial U)$. La restriction de $\mathcal{I}(U, f)$ à U est notée $I(U, f)$ (ou $I(f)$). L'application $I(U, \cdot)$ de $\mathrm{C_a}(\partial U)$ dans $\mathrm{NUM}(U)$ est appelé l'*opérateur* MAD de U.

PROPOSITION 3.9. *Soit U convexe épaulé. Alors:*

1) *L'opérateur* MAD *de U est une application positive surlinéaire de* $\mathrm{C_a}(\partial U)$ *dans* $\mathrm{Conv}(U)$.

2) *Soit $f \in \mathrm{C_a}(\partial U)$ la borne inférieure d'une famille non vide filtrante à gauche $\mathcal{G} \subset \mathrm{C}(\partial U)$. Alors $I(f) = \inf\{I(g) : g \in \mathcal{G}\}$.*

Démonstration. 1): Proposition 3.4 et Remarque 3.2,4-6.
2): C'est un cas particulier de la Proposition 3.6 ∎

DÉFINITION 3.10. Soient U un ouvert convexe épaulé, I l'opérateur MAD dans U et $f \in \mathrm{NUM}(\partial U)$. On définit:

1) $I_\star(U, f) = \inf\{I(U, g) : f \leq g \in \mathrm{C_a}(\partial U)\}$ si $f \in \mathrm{SCS}(\partial U)$.

2) $I_\star(U, f) = \sup\{I_\star(U, g) : f \geq g \in \mathrm{SCS}(\partial U)\}$.

3) $I^\star(U, f) = \sup\{I(U, g) : f \geq g \in \mathrm{C_a}(\partial U)\}$ si $f \in \mathrm{SCI}(\partial U)$.

4) $I^\star(U, f) = \inf\{I^\star(U, g) : f \leq g \in \mathrm{SCI}(\partial U)\}$.

5) $I_\star(U, A) = I_\star(U, 1_A)$ et $I^\star(U, A) = I^\star(U, 1_A)$ quel que soit $A \subset \partial U$.

6) La fonction f est dite *I-intégrable* si $I_\star(f) = I^\star(f)$. Cette valeur commune est alors notée $I(f)$.

PROPOSITION 3.11. *Soit I l'opérateur* MAD *du convexe épaulé U. Alors:*

1) *$I_\star$ est un prolongement positif et surlinéaire de I.*

2) $\inf\{I_\star(f) : f \in \mathcal{G}\} = I_\star(\inf \mathcal{G})$ *pour toute partie non vide et filtrante à gauche $\mathcal{G}$ de* $\mathrm{SCS}(\partial U)$.

Démonstration. Il est immédiat que $I_\star$ est un prolongement monotone et positivement homogène de I.

2): Posons $g = \inf \mathcal{G}$. On a (Proposition 3.9):

$$\begin{aligned}
I_\star(g) &= \inf\{I(h) : g \le h \in \mathrm{C_a}(\partial U)\} \\
&= \inf\{\inf\{I(\sup(h,f)) : \exists k \in \mathcal{G}\ k \le f \in \mathrm{C_a}(\partial U)\} : g \le h \in \mathrm{C_a}(\partial U)\} \\
&= \inf\{\inf\{I(\sup(h,f)) : g \le h \in \mathrm{C_a}(\partial U)\} : k \le f \in \mathrm{C_a}(\partial U), k \in \mathcal{G}\} \\
&= \inf\{I(f) : k \le f \in \mathrm{C_a}(\partial U),\ k \in \mathcal{G}\} \\
&\ge \inf\{I_\star(k) : k \in \mathcal{G}\} \ge I_\star(g).
\end{aligned}$$

1): Soient $f, g \in \mathrm{NUM}(\partial U)$. Il suffit à considérer le cas $I_\star(f)$, $I_\star(g) > -\infty$. Supposons d'abord que f et g appartiennent à $\mathrm{SCS}(\partial U)$. Alors $I_\star(f) + I_\star(g) = \inf\{I(f') : f \le f' \in \mathrm{C_a}(\partial U)\} + \inf\{I(g') : g \le g' \in \mathrm{C_a}(\partial U)\} \le \inf\{I(f'+g') : f', g' \in \mathrm{C_a}(\partial U),\ f \le f',\ g \le g'\} \le I_\star(f+g)$. Pour f et g arbitraires, l'assertion résulte des inégalités $I_\star(f) + I_\star(g) = \sup\{I_\star(f') + I_\star(g') : f \ge f' \in \mathrm{SCS}(\partial U), g \ge g' \in \mathrm{SCS}(\partial U)\} \le \sup\{I_\star(h') : f + g \ge h' \in \mathrm{SCS}(\partial U)\} = I_\star(f+g)$. ∎

PROPOSITION 3.12. *Soit I l'opérateur* MAD *du convexe épaulé U. Alors:*

1) *$I^\star$ est un prolongement isotone et positivement homogène de I.*

2) *$I_\star(f) \le I^\star(f)$ pour tout $f \in \mathrm{NUM}(\partial U)$.*

3) *$I_\star(f) = I^\star(f)$ pour tout $f \in \mathrm{SCI}(\partial U) \cup \mathrm{SCS}(\partial U)$.*

4) *$\mathcal{I}(f) \le I_\star(f)$ dans U pour toute fonction $f \in \mathrm{NUM}(\partial U)$.*

5) *$I^\star(f) = \mathcal{I}(f)$ dans U pour toute fonction $f \in \mathrm{SCI}(\partial U)$ et pour toute fonction $f \in \mathrm{NUM}(\partial U)$, majorée par une fonction $g \in \mathrm{C}(\partial U)$.*

Démonstration. 2): Soit d'abord $h \in \mathrm{SCI}(\partial U)$. Comme ∂U est normal, pour toute fonction $f \in \mathrm{SCS}(\partial U)$, $f \le h$, il existe une fonction $g \in \mathrm{C_a}(\partial U)$ telle que $f \le g \le h$ ([BOU74, p. 102]). On a donc $I_\star(h) = \sup\{I_\star(f) : h \ge f \in \mathrm{SCS}(\partial U)\} = \sup\{I(g) : h \ge g \in \mathrm{C_a}(\partial U)\} = I^\star(h)$. Pour $f \in \mathrm{NUM}(\partial U)$ on a maintenant $I^\star(f) = \inf\{I^\star(h) : f \le h \in \mathrm{SCI}(\partial U)\} = \inf\{I_\star(h) : f \le h \in \mathrm{SCI}(\partial U)\} \ge I_\star(f)$.
3): C'est déjà démontré pour $f \in \mathrm{SCI}(\partial U)$. Pour $f \in \mathrm{SCS}(\partial U)$ on a $I_\star(f) = \inf\{I(g) : f \le g \in \mathrm{C_a}(\partial U)\} \ge \inf\{I^\star(g) : f \le g \in \mathrm{SCI}(\partial U)\} \ge I^\star(f)$.
4): Soit $h \in \mathrm{Aff}(E)$ dominée par f dans ∂U. On a $h = \mathcal{I}(h) = I(h) \le I_\star(f)$.
5): En effet, pour $f \in \mathrm{SCI}(\partial U)$, l'assertion résulte de l'inégalité $I^\star(f) = \sup\{\mathcal{I}(g) : f \ge g \in \mathrm{C_a}(\partial U)\} \le \mathcal{I}(f)$, tandis que, pour f majorée dans $\mathrm{C}(\partial U)$, $I^\star(f) = \inf\{I^\star(g) : f \le g \in \mathrm{SCI}(\partial U)\} = \inf\{\mathcal{I}(g) : f \le g \in \mathrm{SCI}(\partial U)\} = \mathcal{I}(f)$ (Proposition 3.4, Proposition 3.6). ∎

En particulier, toute fonction numérique majorée par une fonction continue est I-intégrable.

La proposition suivante donne un premier résultat de comportement à la frontière:

PROPOSITION 3.13. *Soit I l'opérateur* MAD *du convexe épaulé U. Alors:*

1) *Si* $\mathrm{sci}_U(I(g)) \ge g$ *pour toute fonction* $g \in \mathrm{C_a}(\partial U)$, $\mathrm{sci}_U(I_\star(f)) \ge \mathrm{sci}_{\partial U}(f)$ *pour toute fonction* $f \in \mathrm{NUM_a}(\partial U)$.

2) *Si* $\text{scs}_U(I(g)) \leq g$ *pour toute fonction* $g \in \text{C}_\mathbf{a}(\partial U)$, $\text{scs}_U(I^\star(f)) \leq \text{scs}_{\partial U}(f)$ *pour toute fonction* $f \in \text{NUM}(\partial U)$, *majorée par une fonction* $g \in \text{C}(\partial U)$.

Démonstration. 1): Les restrictions à ∂U satisfont à

$$\begin{aligned} \text{sci}_U(I_\star(f)) &= \text{sci}_U(\sup\{I(h) : f \geq h \in \text{SCS}(\partial U\}) \\ &\geq \text{sci}_U(\sup\{I(g) : f \geq g \in \text{C}_\mathbf{a}(\partial U)\}) \\ &\geq \sup\{g : f \geq g \in \text{C}_\mathbf{a}(\partial U)\} = \text{sci}_{\partial U}(f). \blacksquare \end{aligned}$$

EXEMPLE 3.14. Quelque condition de finitude est nécessaire dans la Remarque 3.3,2 et les propositions 3.12 et 3.13. En effet, soit I l'opérateur MAD de l'intervalle réel $U = (0,1)$. Posons $f(0) = -\infty$ et $f(1) = 0$. On a $\text{sci}_U(I_\star(f))(1) = -\infty < \text{sci}_{\partial U}(f)(1) = 0$. D'autre part, avec $g(0) = -\infty$, $g_a(0) = a$, et $g_a(1) = g(1) = \infty$ pour $a \in \mathbf{R}$, on a $I(g_a) = \infty$, $\mathcal{I}(g) = I_\star(g) = -\infty$, et $I^\star(g) = \inf\{I(g_a) : a \in \mathbf{R}\} = \infty$. Par suite, la fonction g n'est pas I-intégrable et $\text{scs}_{\overline{U}}(\mathcal{I}(g_a))(0) = \text{scs}_U(I^\star(g_a))(0) = \infty > \text{scs}_{\partial U}(\mathcal{I}(g_a))(0) = \text{scs}_{\partial U}(g_a)(0)$=a.

4 Fonctions MAD-harmoniques

Désormais, on suppose que E soit de type dénombrable.

Dans le cas Euclidien, le faisceau $\mathcal{H}$ des fonctions MA-harmoniques, c'est-à-dire les solutions de l'équation de Monge-Ampère généralisée homogène, est constitué par les fonctions localement convexes u 'sans courbure': $\nu_u = 0$ dans U. Le faisceau $\mathcal{H}$ est le même que le faisceau $\mathcal{H}'$ des fonctions localement convexes continues u qui satisfont localement à la relation $u = \mathcal{I}(V,u)$, V convexe, $\overline{V} \subset U$ ([BER93, p. 179]). En effet, soit u localement convexe. Si $\mathcal{I}(V,u) \neq u$, l'intèrieur de $\partial_u\langle V\rangle$ n'est pas vide (Proposition 4.5 ci-dessous), donc $\nu_u(V) \neq 0$. D'autre part, si $u = \mathcal{I}(V,u)$, tout $x' \in \partial_u\langle V\rangle$ est sous-gradient en plusieurs points de V. L'ensemble de ces formes linéaires à contacts multiples est un ensemble à la fois maigre, donc d'intérieur vide, et γ-négligeable.

Dans un espace de Banach non-trivial, outre la mesure de Lebesgue, on perd aussi la propriété que tout sous-gradient de $\mathcal{I}(V,f)$ possède plusieurs points de contact. Ceci complique la démonstration de la *propriété de Poisson* pour le faisceau $\mathcal{H}'$:

$$u \in \mathcal{H}'(U),\ V \ \text{ ouvert convexe borné, } \overline{V} \subset U \Longrightarrow I(V,u) = u \ \text{ dans } V.$$

EXEMPLE 4.1. Soit $E = l^2$ l'espace des suites $x = (x_n)_{1 \leq n \in \mathbf{N}}$ de carré sommable, identifié à son dual E' et soit U la boule unité fermée de E. Les fonctions u_i, définies ci-dessous, appartiennent à $\mathcal{H}'(U)$ et possèdent la propriété de Poisson. De plus, les ensembles $\partial_{u_i}\langle U\rangle$ sont sans points intérieurs.

1) Soit $u_1 \in \text{Conv}(E)$ la fonction définie par

$$u_1(x) = \sum_i \frac{|x_i|}{i}$$

([BER79, Exemple 3.10]). On a $\mathcal{I}(V, u_1) = 1 \neq u_1$ si $V = \{u_1 < 1\}$. L'image $\partial_{u_1}\langle U \rangle$ est l'ensemble de toutes les formes linéaires $x' = (x'_i)$, où $|x'_i| \leq 1/i$. Notons que que (x'_i) est sous-gradient de u_1 en 0, et seulement en 0, si et seulement si chaque $|x'_i| < 1/i$. La fonction u_1 est faiblement continue dans toute partie bornée de E.

2) Soit $u_2 \in \mathrm{Conv}(E)$ la fonction définie par

$$u_2(x) = \sum_{i \geq 1} (x_{2i}^2/(2i) + x_{2i-1}^2).$$

On a $\mathcal{I}(V, u_2) = 1 \neq u_2$ pour $V = \{u_2 < 1\}$. Le sous-gradient de u_2 en x est la seule forme linéaire (x'_i), où $x'_i = 2x_i/i$ pour i pair, $x'_i = 2x_i$ pour i impair. Tout sous-gradient a un seul point de contact et la fonction u_2 n'est pas faiblement continue.

3) **(Cator)** Soit u_3 une fonction convexe faiblement continue dans U.

Donnons maintenant une définition alternative du faisceau $\mathcal{H}$:

DÉFINITION 4.2. On dit qu'une fonction localement convexe continue u, définie dans l'ouvert U est MAD-*harmonique* (ou simplement *harmonique*) dans U, notation $u \in \mathcal{H}(U)$, s'il existe un recouvrement ouvert $\mathcal{R}$ de U, tel que tout $V \in \mathcal{R}$ soit un ouvert convexe borné, $\overline{V} \subset U$, et $\partial_u\langle\overline{V}\rangle$ soit sans points intérieurs.

THÉORÈME 4.3. *La transformation $\mathcal{H} : U \mapsto \mathcal{H}(U)$ est un faisceau de fonctions continues reélles. Chaque ensemble $\mathcal{H}(U)$ est un cone et $u + h \in \mathcal{H}(U)$ pour tout $u \in \mathcal{H}(U)$ et $h \in \mathrm{Aff}(E)$.*

PROPOSITION 4.4. *Soit $u \in \mathcal{H}(U)$ et soit V ouvert, convexe et borné, $\overline{V} \subset U$. Alors $\partial_u\langle\overline{V}\rangle$ est un ensemble rare.*

Démonstration. On se ramène à un recouvrement $\mathcal{R}$ dénombrable ([BOU74, p. IX.19]). L'assertion découle alors de la Proposition 2.7 et du Théorème de Baire. ∎

PROPOSITION 4.5. *Soit V un ouvert convexe et borné et soit $u \in \mathcal{H}(V) \cap \mathrm{C}(\overline{V})$. Alors $\mathcal{I}(V, u) = u$ dans $\overline{V}$.*

Démonstration. Soit $v = \mathcal{I}(V, u)$. On a $v = u$ dans ∂V et $u \leq v$ dans V (Remarque 3.2,1). Supposons que z est un point de V tel que $u(z) < v(z) - \epsilon < v(z)$. Posons $h = v + \epsilon\left(\frac{\|\cdot - z\|}{d} - 1\right)$, où d est le diamètre de V, et $W = \{h > u\} \cap \overline{V}$. On a $h < u = v$ dans ∂V, donc W est ouvert, $\overline{W} \subset V$ et $z \in W$. De plus, pour tout $x' \in \partial_v(z)$ et $y' \in E'$, $\|y'\| \leq \epsilon/d$, comme $(x' + \epsilon y')(y - z) \leq v(y) - v(z) + \epsilon \|y - z\|/d \leq h(y) - h(z)$, on a $x' + \epsilon y' \in \partial_h(z)$. Par suite, $\partial_h(z)$ contient un point intérieur. D'autre part, $\partial_h(z) \subset \partial_h\langle W \rangle \subset \partial_u\langle W \rangle$ (Lemma 2.13), donc l'intérieur de l'ensemble maigre $\partial_u\langle V \rangle$ ne serait pas vide. ∎

D'autre part:

PROPOSITION 4.6. **(Cator)** *Soient U convexe ouvert épaulé, $f \in \mathrm{NUM}_a(\partial U)$ finie, et $u = \mathcal{I}(U, f)$. Alors, pour tout ouvert convexe borné V, $V \sqsubset U$, l'ensemble $\partial_u\langle\overline{V}\rangle$ n'a aucun point intérieur.*

Démonstration. Si l'on suppose le contraire, il existe un point intérieur x' de $\partial_u\langle\overline{V}\rangle$, une boule fermée $B' \subset \partial_u\langle\overline{V}\rangle$ de rayon $\epsilon > 0$ et de centre x', et un point $x \in \overline{V}$ tel que $x' \in \partial_u(x)$. Elargissant au besoin V, on se ramène à $x \in V$, et ensuite à $x = 0$, $u(0) = 0$ et $x' = 0$ (Proposition 2.7).

Soit y_0 un point quelconque de ∂U et soit y le point unique dans $[0, y_0] \cap \partial V$. Notons que $\sup\{\|y\|/\|y_0\| : y_0 \in \partial U\} = \alpha < 1$. Soit enfin $y' \in E'$, tel que $\|y'\| = \epsilon$ et $y'(y) = \sup\{y'(z) : z \in \overline{V}\} = \beta$ ([BOU81, p. II.40]). Evidemment, β/ϵ majore la distance de 0 à ∂V. D'après l'hypothèse, on a $y' \in \partial_u(x)$ pour un $x \in \overline{V}$. Pour tout $z \in \overline{U}$ on a donc

$$\begin{aligned} u(z) &\geq u(x) + y'(z - x) \geq y'(z - y), \\ f(y_0) &\geq u(y_0) \geq y'(y_0 - y) \geq (\frac{1}{\alpha} - 1)y'(y) \geq \beta(\frac{1}{\alpha} - 1). \end{aligned}$$

Comme y_0 est arbitraire, on aurait donc $u(0) \geq \beta/\alpha - \beta$, contrairement à l'hypothèse. ∎

Bref, les opérateurs MAD sont réhabilités, on a $\mathcal{H} = \mathcal{H}'$, et $\mathcal{H}$ possède la propriété de Poisson:

THÉORÈME 4.7. *Soit u une fonction localement convexe continue dans l'ensemble ouvert U. Les propriétés suivantes sont équivalentes:*

(i) *$u \in \mathcal{H}(U)$.*

(ii) *Il existe un recouvrement ouvert $\mathcal{R}$ de U, tel que tout $V \in \mathcal{R}$ soit un ouvert convexe borné, $\overline{V} \subset U$, et $I(V, u) = u$ dans V.*

COROLLAIRE 4.8. *Soit $\mathcal{F}$ une famille non vide filtrante à gauche de fonctions harmoniques dans un ouvert connexe U. Alors $\inf \mathcal{F} \in \mathcal{H}(U)$ ou $\inf \mathcal{F} = -\infty$.*

Démonstration. Proposition 3.6. ∎

COROLLAIRE 4.9. *Soit $\mathcal{F}$ une famille non vide filtrante à droite de fonctions harmoniques dans un ouvert U, convergeant localement uniformément vers une fonction u. Alors $u \in \mathcal{H}(U)$.*

Démonstration. Remarque 3.3,1. ∎

Donnons encore une condition commode pour la MAD-harmonicité d'une fonction. Soient U un ouvert convexe borné, $x \in U$ et $e \in E$, de norme 1. On désigne par y_e^+ et par e_e^- les deux points de ∂U tels que les $y_e^\pm - x$ soient un multiple de e. Alors:

PROPOSITION 4.10. *Soit $u \in \mathrm{Conv}(\overline{U})$. Supposons que, pour tout $x \in U$ et pour tout $\epsilon > 0$, il existe un vecteur e, de norme 1, tel que $u(y_e^{\pm}) - u(x) \leq \epsilon$. Alors $u \in \mathcal{H}(U)$.*

Démonstration. Il suffit à démontrer $\mathcal{I}(U, u)(x) \leq u(x)$ pour tout $x \in U$. Choisisons x, ϵ et e. Alors $\epsilon \geq u(y_e^{\pm}) - u(x) \geq x'(y_e^{\pm} - x)$ pour $x' \in \partial_u(x)$. Comme le rapport des deux vecteurs $y_e^{\pm} - x$ est borné par un nombre K, indépendamment de e, on a $u(y_e^{\pm}) - x'(y_e^{\pm}) \leq u(x) - x'(x) + (K+1)\epsilon$. L'assertion résulte donc de la Remarque 3.2,9. ∎

Ce critère s'applique en particulier aux fonctions u_i de l'Exemple 4.1.

5 Ensembles réguliers

DÉFINITION 5.1. On dit qu'une partie ouverte bornée non vide U de E est régulière si, quelle que soit la fonction $f \in \mathrm{C_a}(\partial U)$, il existe une fonction unique $u \in \mathcal{H}(U) \cap \mathrm{C_a}(\overline{U})$, dont la restriction à U est notée $I(U, f)$ (ou $I(f)$), telle que:

1) $u = f$ dans ∂U.
2) $f \geq 0 \Longrightarrow I(U, f) \geq 0$.
3) $I(U, f + g) \geq I(U, f) + I(U, g)$.

L'application $f \mapsto I(U, f)$ de $\mathrm{C_a}(\partial U)$ dans $\mathcal{H}_\mathrm{a}(U)$ est donc isotone et surlinéaire et $I(U, f + h) = I(U, f) + h$ pour tout $f \in \mathrm{C_a}(\partial U)$ et tout $h \in \mathrm{Aff}(E)$. Notons que tout ensemble régulier convexe est nécessairement strictement convexe.

La notation $I(U, f)$ est permise. En effet, si l'ensemble régulier U est convexe, on a $I(U, f) \leq \mathcal{I}(U, f)$ dans U (Remarque 3.2,1), donc $\mathcal{I}(U, f) = f$ dans ∂U et, par suite, $\mathcal{I}(U, f) = I(U, f)$ dans U (Thóréme 4.7).

THÉORÈME 5.2. *Il existe une base d'ouverts réguliers convexes.*

Démonstration. En effet, comme E est réflexif, il existe une base d'ouverts bornés et localement uniformément convexes (Day, [DAY73, p. 160]). Prouvons qu'un tel ensemble B est régulier.

Démontrons d'abord l'existence de $I(B, f)$. D'après la Remarque 3.2,8 et le Theórème 4.7, il suffit à vérifier que la fonction $u = \mathcal{I}(B, f)$ est un prolongement continu de f à $\overline{B}$. Soient $x \in \partial B$, $\epsilon > 0$, et V un voisinage de x, tel que $|f - f(x)| < \epsilon$ dans $V \cap \partial B$. Comme B est localement uniformément convexe, il existe une fonction $q \in \mathrm{Aff}(E)$ telle que $q(x) = 1$, $q \leq 1$ dans $\overline{B}$ et $q \leq 0$ dans $\overline{B} \setminus V$ ([BER79, Lemme 6.5]). Si a est un mombre tel que $f + a \geq 0$ dans ∂B, on voit sans peine que la fonction affine $h = (f(x) + a)q - \epsilon - a$ est dominée par f dans ∂B et que $\lim_{y \to x} h(y) = f(x) - \epsilon$. D'autre part, pour tout $n > 0$, tout point z de $(x + (1/n)(V - x)) \cap B$ est une combinaison convexe de x et d'un point soit de $\partial B \cap V$, soit de $B \cap \partial V$. D'où

$u(z) \leq \max(f(x) + \epsilon, (1 - (1/n))$. Compte tenu de la Remarque 3.3,2, ceci prouve la continuité de u dans $\overline{B}$.

L'unicité de la solution $\mathcal{I}(f)$ résulte immédiatement de la Proposition 4.5. ∎

COROLLAIRE 5.3. *Soit l'ouvert U convexe et régulier et soit $f \in \mathrm{NUM_b}(U)$. Alors $I(f) \in \mathcal{H}(U)$, $I(f) = \mathcal{I}(f)$ dans U et*

$$\mathrm{scs}_{\partial U}(f) \geq \mathrm{scs}_U\, I(f) \geq \mathrm{sci}_U\, I(f) \geq \mathrm{sci}_{\partial U}(f).$$

Démonstration. C'est un cas particulier des Propositions 3.12,5 et 3.13, compte tenu de l'harmonicité de $\mathcal{I}(f)$. ∎

Références

[ALE55] A.D. Alexandrow. *Die Innere Geometrie der Konvexen Flächen.* Akademie-Verlag, Berlin, 1955.

[BAK86B] I.J. Bakelman. Generalized elliptic solutions of the Dirichlet problem for n-dimensional Monge-Ampère equations. In f.e. Browder, editor, *Nonlinear Functional Analysis and its Applications, Berkeley, 1983*, Proc. Sympos. Pure Math. 45-1, pages 73–102. American Mathematical Society, Providence, 1986.

[BER79] E.M.J. Bertin. Fonctions convexes et théorie du potentiel. *Indag. Math.*, 41:385–409, 1979.

[BER93] E.M.J. Bertin. Axiomatic convex potential theory. In I.J. Bakelman, editor, *Geometric Inequalities and Convex Bodies, Denton, Texas, 1990*, pages 149–180. Marcel Dekker, New York, 1993.

[BOU74] N. Bourbaki. *Topologie générale,* chapitres V–X, *Éléments de mathématique.* Hermann, Paris, 1974.

[BOU81] N. Bourbaki. *Espaces vectoriels topologiques, Éléments de mathématique.* Masson, Paris, 1981.

[CON72] C. Constantinescu and A. Cornea. *Potential Theory on Harmonic Spaces.* Springer-Verlag, Berlin, 1972.

[DAY73] M.M. Day. *Normed Linear Spaces.* Springer-Verlag, Berlin, third edition, 1973.

[GOO92] F. van Gool. *Topics in Non-Linear Potential Theory.* Phd thesis, Universiteit Utrecht, Mathematisch Instituut, 1992.

[GRA86] S. Grandlund, P. Lindqvist, and O. Martio. Note on the PWB-method in the non-linear case. *Pacific J. Math.*, 125:381–395, 1986.

[MAE81] F.-Y Maeda. Semi-linear perturbation of harmonic spaces. *Hokkaido Math. J.*, 10:464–493, 1980.

[MAR92] O. Martio. Potential theory and quasiconformal mappings. In Masanori Kishi, editor, *Potential Theory. Proc. of the International Conference on Potential Theory, Nagoya, 1990*, pages 55–64. Walter de Gruyter, Berlin, 1992.

EXCESSIVE FUNCTIONS AND EXCESSIVE MEASURES: HUNT'S THEOREM ON BALAYAGES, QUASI–CONTINUITY

LUCIAN BEZNEA
Katholische Universität Eichstätt
Lehrstuhl für Analysis, Ostenstr. 26-28
85072 Eichstätt, Germany; Permanent address: *Institute of Math. of the Romanian Academy, P.O. Box 1-764*
RO-70700 Bucharest, Romania

NICU BOBOC
Faculty of Mathematics
University of Bucharest
str. Academiei 14
RO-70109 Bucharest
Romania

ABSTRACT. We show that the balayage operation on measurable sets exists in the class of all universally measurable excessive functions, giving an analytic version of the well-known fundamantal result of G. A. Hunt and C. T. Shih. The quasi–continuous elements in H-cones are introduced and characterized. (In the classical situation these are the countable sums of bounded continuous potentials with compact carrier.) In the particular case of excessive measures we prove that the quasi-continuous elements are exactly the potentials of measures which does not charge the semi-polar sets.

Introduction

The aim of this paper is to study the quasi–continuity for excessive measures associated to a proper submarkovian resolvent $\mathcal{U} = (U_\alpha)_{\alpha>0}$ on a Lusin measurable space. The starting point is the construction (in the first section) of the balayage on measurable sets of excessive functions, including on analytic version (see Theorem 1.3) of the well known fundamental result of G. A. Hunt and C. T. Shih (see e.g. [5] and [7]). An $\mathcal{U}$–excessive measure ξ is called *quasi–continuous* if $\bigwedge_{i\in I} R(\xi - \xi_i) = 0$ for any family $(\xi_i)_{i\in I}$ of excessive measures increasing to ξ , where R denotes the usual reduit operation in the cone of potential Exc of all $\mathcal{U}$– excessive measures. Having in mind that Exc is an H–cone, we developp in Section 2 the quasi–continuity in the frame of general H–cone. We prove that if S is an H–cone such that its dual S^* separates S then an element $s \in S$ will be quasi–continuous iff $\left(\bigwedge_{i\in I} \nu_i\right)(s) = \inf_{i\in I} \nu_i(s)$ for every decreasing family $(\nu_i)_{i\in I} \subset S^*$. Applying this characterization we show in the last section that an excessive measure of potential type $\mu \circ U$ is quasi–continuous iff μ does not charge the semi–polar sets. This result was known for the case when $\mathcal{U}$ has a reference measure (see [3], Theorem

K. GowriSankaran et al. (eds.), Classical and Modern Potential Theory and Applications, 77–92.

5.5.10 and Proposition 5.6.1). We discuss also the problem of approximation with quasi-continuous elements in H-cones.

1. Balayages on excessive functions

Let $\mathcal{U} = (U_\alpha)_{\alpha>0}$ be a submarkovian resolvent of kernels on the Lusin measurable space $(X, \mathcal{X})$ such that the initial kernel U is proper. We denote by $\mathcal{E}_\mathcal{U}$ the convex cone of all $\mathcal{X}$-measurable $\mathcal{U}$-excessive functions on X which are finite $\mathcal{U}$-a.s. and we suppose that $\mathcal{E}_\mathcal{U}$ is min-stable, $1 \in \mathcal{E}_\mathcal{U}$ and $\sigma(\mathcal{E}_\mathcal{U})$, the σ-algebra generated by $\mathcal{E}_\mathcal{U}$, coincides with $\mathcal{X}$.

Let us denote by $\mathcal{X}^*$ the σ-algebra of all universally measurable subsets of X. The resolvent $\mathcal{U} = (U_\alpha)_{\alpha>0}$ is canonically extended to $(X, \mathcal{X}^*)$ and we denote by $\mathcal{E}^*_\mathcal{U}$ the convex cone of all $\mathcal{X}^*$-measurable $\mathcal{U}$-excessive functions on X which are finite $\mathcal{U}$-a.s.

Remark. 1. Let f be a positive universally measurable function and μ a measure on $(X, \mathcal{X})$ with $\mu(Uf) < \infty$. Then there exists a pair of two $\mathcal{X}$-measurable functions f' and f'' such that $f' \leq f \leq f''$ and $\mu(Uf'' - Uf') = 0$.
2. If $s \in \mathcal{E}^*_\mathcal{U}$ and μ is a measure on $(X, \mathcal{X})$ with $\mu(s) < \infty$ then there exist $s', s'' \in \mathcal{E}_\mathcal{U}$ such that $s' \leq s \leq s''$ and $\mu(s'' - s') = 0$. Indeed, let $(f_k)_{k\in\mathbb{N}}$ be a sequence of $\mathcal{X}^*$-measurable functions with $Uf_k \nearrow s$. From the preceding remark there exist two sequences $(f'_k)_{k\in\mathbb{N}}$ and $(f''_k)_{k\in\mathbb{N}}$ of $\mathcal{X}$-measurable functions such that $f'_k \leq f_k \leq f''_k$ and $\mu(Uf''_k - Uf'_k) = 0$, for all $k \in \mathbb{N}$. If we put $s' := \bigvee_{i\in\mathbb{N}} \bigwedge_{k\geq i} Uf'_k$, $s'' := \bigvee_{i\in\mathbb{N}} \bigwedge_{k\geq i} Uf''_k$ then $s' \leq s \leq s''$ and $\mu(s'' - s') = 0$.
3. The convex cone $\mathcal{E}^*_\mathcal{U}$ is min-stable and if $s \in \mathcal{E}^*_\mathcal{U}$, $A \subset X$ and μ is a measure on $(X, \mathcal{X})$ then

$$\inf\{\mu(t)/t \in \mathcal{E}^*_\mathcal{U}, s \leq t \text{ on } A\} = \inf\{\mu(t)/t \in \mathcal{E}_\mathcal{U}, s \leq t \text{ on } A\}.$$

Definition. For any subset A of $X, s \in \mathcal{E}^*_\mathcal{U}$ and any positive σ-finite measure μ on $(X, \mathcal{X})$ for which there exists $v \in \mathcal{E}_\mathcal{U}, v > 0$ with $\mu(v) < \infty$ we put

$$R^A_s(\mu) := \inf\{\mu(t)/t \in \mathcal{E}_\mathcal{U}, s \leq t \text{ on } A\}.$$

We denote by $R^A s : X \to \overline{\mathbb{R}_+}$ the function on X defined by

$$R^A s(x) := R^A_s(\varepsilon_x), x \in X$$

The following propositon collects some properties of the reduit operation R^A. The proof beeing as in the case of H-cones of functions (cf. [3], §3.2) we omit it.

Note also that the hypothesis that $(X, \mathcal{X})$ is a Lusin measurable space is not used.

Proposition 1.1. *Let $A, A_k \subset X$ and $s, s_k \in \mathcal{E}_{\mathcal{U}}$ for all $k \in \mathbb{N}$. If μ is a measure on $(X, \mathcal{X})$ such that there exists $v \in \mathcal{E}_{\mathcal{U}}, v > 0$ with $\mu(v) < \infty$, then the following assertions hold:*

1. *If $s_1 \leq s$ then $R^A_{s_1}(\mu) \leq R^A_s(\mu) \leq \mu(s)$.*
2. *If A is fine open then $B^A s := R^A s$ is an $\mathcal{U}$-excessive function and $B^A s = s$ on A.*
3. *For all $\alpha > 0$ we have $R^A_{\alpha s_1 + s_2}(\mu) = \alpha R^A_{s_1}(\mu) + R^A_{s_2}(\mu)$.*
4. *$R^{A_1 \cup A_2}_s(\mu) + R^{A_1 \cap A_2}_s(\mu) \leq R^{A_1}_s(\mu) + R^{A_2}_s(\mu)$.*
5. *If $s_k \nearrow s$ and $A_k \nearrow A$ then $R^{A_k}_{s_k}(\mu) \nearrow R^A_s(\mu)$.*

Remark. It was shown in [1], Theorem 1.2 that if $s \in \mathcal{E}_{\mathcal{U}}, A \in \mathcal{X}$ and $A^* := \left\{x \in A / \liminf_{n\to\infty} nU_n(\chi_A) = 1\right\}$ then the function $B^{A^*}s := R^{A^*}s$ is $\mathcal{U}$-excessive and $B^A s = s$ on A^*.

We remember now some considerations concerning the Ray compactification:

We suppose that the initial kernel U of $\mathcal{U}$ is bounded. A *Ray cone* associated with $\mathcal{U}$ will be a subcone $\mathcal{R}$ of $(\mathcal{E}_U)_b$ (:= the bounded excessive functions) which is separable in the uniform norm and min-stable, separates the points of X and moreover $1 \in \mathcal{R}, U((\mathcal{R} - \mathcal{R})_+) \subset \mathcal{R}, \alpha \geq 0$ and $\sigma(\mathcal{R}) = \mathcal{X}$. We consider the Ray compactification Y of X with respect to $\mathcal{R}$. For any $p \in \mathcal{R}$ we denote by $\tilde{p}$ the real continuous extension of p to Y. The set $\widetilde{\mathcal{R}} := \{\tilde{p}/p \in \mathcal{R}\}$ separates the points of Y, $1 \in \widetilde{\mathcal{R}}$ and it is min-stable. Since $(X, \mathcal{X})$ is a Lusin space, it follows that X is a Borel subset of Y and $\mathcal{B}(Y)|_X = \mathcal{X}$, where $\mathcal{B}(Y)$ denotes the σ-algebra of all Borel subsets of Y.

Proposition 1.2. *Suppose that U is bounded and let $\mathcal{R}$ be a Ray cone associated with $\mathcal{U}$. Then for any $s \in \mathcal{R}$ and any Ray compact subset A of X the function $R^A s$ is $\mathcal{X}$-measurable and*

$$R^A s(x) = \widehat{R^A s}(x), \quad (\forall) x \in X \setminus A$$

Moreover the function $R^A s$ is upper semi-continuous and we have

$$R^A_s(\mu) = \inf\left\{R^G_s(\mu)/G \text{ open}, \ A \subset G\right\} = \mu\left(R^A s\right),$$

for any measure μ on $(X, \mathcal{X})$ such that there exists $t_\circ \in \mathcal{R}$, $t_\circ > 0$ on X, $s \leq t_\circ$ with $\mu(t_\circ) < \infty$. (Recall that if f is $\mathcal{U}$-supermedian function, $\widehat{f}$ denotes its excessive regularization.)

Proof. Let us put

$$s_0 := \inf\left\{\tilde{t}/\tilde{t} \in \widetilde{\mathcal{R}}, \tilde{s} \leq \tilde{t} \text{ on } A\right\}.$$

From [2], Lemma 1.3 it follows that for any $x \in X$ there exists a measure μ_x on A such that $\mu_x(\tilde{t}) \leq \tilde{t}(x)$ for any $t \in \mathcal{R}$ and

$$s_0(x) = \mu_x(s \cdot \chi_A).$$

Since $U((\mathcal{R}-\mathcal{R})_+) \subset \mathcal{R}$ and the monoton class generated by $(\mathcal{R}-\mathcal{R})_+$ coincides with $\mathcal{F} := \{f : X \to \overline{I\!R}_+ / f$ is $\mathcal{X}$–measurable$\}$, it follows that $\mu_x(t) \leq t(x)$, for any $t \in \mathcal{E}_{\mathcal{U}}$. Hence if $t \in \mathcal{E}_{\mathcal{U}}$ and $s \leq t$ on A then

$$t(x) \geq \mu_x(t) \geq \mu_x(s \cdot \chi_A) = s_0(x)$$

and therefore

$$R^A s(x) = s_0(x)\,, \quad (\forall)\, x \in X\,.$$

Consequently $R^A s$ is upper semi–continuous.

We prove now the following lemma:

Lemma. *Let D be an open subset of X, $G := X \setminus \overline{D}^f$ ($\overline{D}^f$ denotes the fine closure of D). If we define the kernel W on $D' := X \setminus \overline{G}^f$ by*

$$Wf := \left(Uf - R^G Uf\right)|_{D'}\ ,\ f \geq 0\,,$$

then the following assertions hold:

1. *If $M \subset D, M \in \mathcal{X}$ is such that $W(\chi_M) = 0$ then $U(\chi_M) = 0$.*
2. *If $s \in \mathcal{E}_{\mathcal{U}}, s < \infty$ then $\left(s - R^G s\right)\big|_{D'} \in \mathcal{E}_{\mathcal{W}}$, where $\mathcal{W} = (W_\alpha)_{\alpha > 0}$ is the resolvent on D' having the initial kernel W.*

Proof of Lemma. Let $M \in \mathcal{X}, M \subset D$ be such that $W(\chi_M) = 0$. We may suppose that M is compact. Let us put

$$\mathcal{T} := \left\{\tilde{p} - \alpha \tilde{U}(\chi_M) / p \in \mathcal{R}, \alpha \in I\!R_+\right\},$$

where $\tilde{U}$ denotes the initial kernel of the Ray resolvent on Y induced by $\mathcal{U}$. Therefore $\mathcal{T}$ is a convex cone of lower semi–continuous functions on Y separating the points. We consider the Šilov boundary K with respect to $\mathcal{T}$. Since $U(\chi_M) = R^G U(\chi_M)$ it follows that if $t \in \mathcal{T}$ and $t \geq 0$ on G then $t \geq 0$ on Y. Therefore K is a subset of $\overline{G}$. On the other hand K is a subset of M. We conclude that $K = \emptyset$ and hence $U(\chi_M) = 0$. Let now $s \in \mathcal{E}_{\mathcal{U}}, s < \infty$. If we put $t := \left(s - R^G s\right)\big|_{D'}$ then t is a $\mathcal{W}$–supermedian function on D' and we denote by $\hat{t}^{\mathcal{W}}$ its $\mathcal{W}$–excessive regularization. Let $(f_k)_{k \in I\!N}$ be a sequence of positive $\mathcal{X}$–measurable functions such that $\left(Uf_k - R^G Uf_k\right)\big|_{D'} \nearrow \hat{t}^{\mathcal{W}}$. We may suppose that $Uf_k \curlywedge R^G Uf_k = 0$ and therefore $(Uf_k)_{k \in I\!N}$ is an increasing sequence on X, $Uf_k \leq s$, for all $k \in I\!N$. If we put $s' := \sup\limits_{k \in I\!N} Uf_k$ then $s' \in \mathcal{E}_{\mathcal{U}}, s' \leq s$ and

$$\hat{t}^{\mathcal{W}} = \left(s' - R^G s'\right)\big|_{D'}\,.$$

Since $t = \hat{t}^{\mathcal{W}}$ $\mathcal{W}$–a.s. from the first assertion of Lemma we deduce that

$$s' - R^G s' = s - R^G s \quad \mathcal{U}\text{–a.s. on } D$$

and therefore, D beeing finely dense in D',

$$\hat{t}^{\mathcal{W}} = \left(s - R^G s\right)\Big|_{D'}$$

or equivalently $\left(s - R^G s\right)\Big|_{D'} \in \mathcal{E}_{\mathcal{W}}$, which completes the proof of Lemma.

From the preceding considerations there exists a decreasing sequence $(G_k)_{k\in\mathbb{N}}$ of open subsets of X such that

$$R^A s = \inf_{k\in\mathbb{N}} R^{G_k} s$$

and such that $\bigcap_{k\in\mathbb{N}} \overline{G_k} = A$. For any $x \notin A$ there exists $k_x \in \mathbb{N}$ such that $x \notin \overline{G_{k_x}}$. Let us denote by G_x the fine interior of $\overline{G_{k_x}}$. Then for any $k \geq k_x$ and $t \in (\mathcal{E}_{\mathcal{U}})_b$ we have: $\left(s - R^{G_k} s\right) \wedge \left(t - R^{G_x} t\right) = \left(s + R^{G_x} t\right) \wedge \left(t + R^{G_k} s\right) - \left(R^{G_k} s + R^{G_x} t\right) = u - R^{G_x} u$, where $u := \left(s + R^{G_x} t\right) \wedge \left(t + R^{G_k} s\right)$. Hence if we take $f \in \mathcal{F}_b, f > 0$ then

$$Uf - R^{G_x} Uf > 0 \qquad \text{on } D'_x := X \setminus \overline{G_x}^f$$

and therefore

$$\sup_{\substack{n,k\in\mathbb{N}\\ k\geq k_x}} \left[n(Uf - R^{G_x} Uf) \wedge \left(s - R^{G_k} s\right)\right] = s - R^A s \text{ on } D'_x .$$

On the other hand for any $k, n \in \mathbb{N}$ there exists $u_{n,k} \in (\mathcal{E}_{\mathcal{U}})_b$ such that

$$n\left(Uf - R^{G_x} Uf\right) \wedge \left(s - R^{G_k} s\right) = u_{n,k} - R^{G_x} u_{n,k} .$$

By the above lemma it follows that the function $\left(u_{n,k} - R^{G_x} u_{n,k}\right)\Big|_{D'_x}$ is $\mathcal{W}$– excessive on D'_x, where $\mathcal{W} = (W_\alpha)_{\alpha>0}$ is the resolvent on D'_x having the initial kernel $W := \left(U - R^{G_x} U\right)\Big|_{D'_x}$. As a consequence $\left(s - R^A s\right)\Big|_{D'_x}$, is also $\mathcal{W}$–excessive. Since the resolvent $\mathcal{W}$ is subordinate to $\mathcal{U}$ $(i.e. W_\alpha \leq U_\alpha, (\forall)\alpha > 0)$ and $s\Big|_{D'_x} \in \mathcal{E}_{\mathcal{W}}$ we have $s(x) - R^A s(x) = \widehat{(s - R^A s)}^{\mathcal{W}}(x) = s(x) - \widehat{R^A s}^{\mathcal{W}}(x) \geq s(x) - \widehat{R^A} s(x)$ and we conclude that $R^A s(x) = \widehat{R^A} s(x)$ for all $x \notin A$.

The last assertion follows from the fact that there exists a decreasing sequence $(t_n)_{n\in\mathbb{N}}$ in $\mathcal{R}, t_n \geq s$ on $A, t_n \leq s$ for all $n \in \mathbb{N}$ such that

$$R^A s = \inf_{n\in\mathbb{N}} t_n$$

and from the fact that $\mu(t_n) \leq \mu(s) < \infty$. Therefore the proof of Proposition 1.2 is complete.

Theorem 1.3. *For any* $s \in \mathcal{E}_{\mathcal{U}}^*$ *and any* $A \in \mathcal{X}$ *the function* $R^A s$ *is universally measurable and:*

1. $R_s^A(\mu) = \mu\left(R^A s\right)$, *for any measure* μ *on* $(X, \mathcal{X})$ *for which there exists* $v \in \mathcal{E}_{\mathcal{U}}, v > 0$ *on* X *with* $\mu(v) < \infty$ *;*
2. $R^A s(x) = \widehat{R^A} s(x)$, *for any* $x \in X \setminus A$.

Proof. Let μ be a measure on $(X, \mathcal{X})$ such that there exists $v \in (\mathcal{E}_{\mathcal{U}})_b, v > 0$ with $\mu(v) < \infty$. We may suppose that $s \leq v$ and U is bounded. Consider first the case when $s \in \mathcal{E}_{\mathcal{U}}$. Let $\mathcal{R}$ be a Ray cone associated to $\mathcal{U}$ such that $s, v \in \mathcal{R}$. From Proposition 1.1 and Propositon 1.2 it follows that the map

$$A \longmapsto \inf\left\{R_s^{G\cap X}(\mu)/A \subset G,\ G \subset Y \text{ open }\right\}$$

is a capacity on Y and

$$R_s^A(\mu) = \inf\left\{R_s^G(\mu)/G \text{ open },\ A \subset \dot{G}\right\}.$$

for any compact subset A of X. Hence for any $A \in \mathcal{X}$ we have

$$R_s^A(\mu) = \inf\left\{R_s^G(\mu)/G \text{ open },\ A \subset G\right\} = \sup\left\{R_s^K(\mu)/K \text{compact},\ K \subset A\right\}.$$

Therefore there exists an increasing sequence $(K_n)_{n\in\mathbb{N}}$ of compact subsets of X, $K_n \subset A$ and a decreasing sequence $(G_n)_{n\in\mathbb{N}}$ of open subsets of $X, G_n \supset A$ such that

$$\sup_{n\in\mathbb{N}} R_s^{K_n}(\mu) = \inf_{n\in\mathbb{N}} R_s^{G_n}(\mu) = R_s^A(\mu).$$

From

$$R_s^{K_n}(\mu) = \mu\left(R^{K_n} s\right),\ R_s^{G_n}(\mu) = \mu\left(R^{G_n} s\right),\ R^{K_n} s \leq R^A s \leq R^{G_n} s$$

it follows that the function $R^A s$ is μ–measurable and

$$\mu\left(R^A s\right) = R_s^A(\mu).$$

If we take for any $x \in X \setminus A$ the measure ε_x instead of μ, we have by Proposition 1.2

$$R^{K_n} s(x) = \widehat{R^{K_n} s}\,(x) \leq \widehat{R^A} s(x) \leq R^A s(x) \leq R^{G_n} s(x)$$

and since

$$\sup_{n\in\mathbb{N}} R^{K_n} s(x) = \inf_{n\in\mathbb{N}} R^{G_n} s(x) = R^A s(x)$$

we conclude that

$$\widehat{R^A} s(x) = R^A s(x).$$

Let now $s \in \mathcal{E}_{\mathcal{U}}^*$. We may suppose that $s = Uf$ with f universally measurable, positive and bounded. Then there exist the $\mathcal{X}$–measurable functions f' and f'' such that $f' \leq f \leq f''$ and $\mu(Uf'' - Uf') = 0$. From the first part of the proof the functions $R^A Uf'$ and $R^A Uf''$ are $\mathcal{X}^*$–measurable and $\mu\left(R^A Uf'\right) = R^A_{Uf'}(\mu)$, $\mu\left(R^A Uf''\right) = R^A_{Uf''}(\mu)$. Since

$$R^A_{Uf'}(\mu) = R^A_{Uf}(\mu) = R^A_{Uf''}(\mu) \ , \ \ R^A Uf' \leq R^A Uf \leq R^A Uf''$$

we conclude that the function $R^A Uf$ is also μ–measurable and moreover

$$\mu\left(R^A Uf\right) = R^A_{Uf}(\mu) \, ,$$

which completes the proof.

Definition. For any $A \in \mathcal{X}$ and $s \in \mathcal{E}_{\mathcal{U}}^*$ we set

$$B^A s := \widehat{R^A s} \, ,$$

(i.e. the excessive regularization of the $\mathcal{X}^*$–measurable $\mathcal{U}$–supermedian function $R^A s$). Hence $B^A s \in \mathcal{E}_{\mathcal{U}}^*$ and the map $B^A : \mathcal{E}_{\mathcal{U}}^* \to \mathcal{E}_{\mathcal{U}}^*$
is additive, increasing, contractive and σ–continuous in order from below (i.e. $s_n \nearrow s \Rightarrow B^A s_n \nearrow B^A s$).

As usually we are now able to define the semi–polar sets. The set $A \in \mathcal{X}$ is called *thin at the point* $x \in X$ if there exists $s \in \mathcal{E}_{\mathcal{U}}$ such that $B^A s(x) < s(x)$. The set A is called *totally thin* if it is thin at any point of X. Countable unions of totally thin sets are called *semi–polar*. Note that if $(s_n)_{n \in \mathbb{N}}$ is a sequence in $\mathcal{E}_{\mathcal{U}}$ then the set

$$[\bigwedge_{n \in \mathbb{N}} s_n < \inf_{n \in \mathbb{N}} s_n]$$

is semi–polar.

2. Quasi–continuous elements in *H*–cones

The results of this section are established in the frame of H–cones (see [3] for the theory of H–cones). *However, the reader interested only in the particular case of excessive measures, may always think to the H–cone of all excessive measures with respect to a proper submarkovian resolvent of kernels on a measurable space.*

Let S be a fixed H–cone. As usually, $\leq$ (resp.$\vee, \wedge$) denotes the order relation (resp. the lattice operations) on S and for $f = s - t \in (S - S)_+$, the *reduit of* f is the element Rf of S defined by

$$Rf := \bigwedge \{v \in S / f \leq v\} \, .$$

Note that $Rf \preccurlyeq s$, where $\preccurlyeq$ is the specific order on S.

Definition. An element $s \in S$ is called *quasi–continuous* if for any increasing family $(s_i)_{i\in I}$ in S such that $\bigvee_{i\in I} s_i = s$ we have

$$\bigwedge_{i\in I} R(s - s_i) = 0 .$$

We denote by $Q_c(S)$ the set of all quasi–continuous elements of S.

It is easy to see that $Q_c(S)$ is a convex subcone of S which is a specific band in S (i.e. a band with respect the the specific order on S).

Remark. If S is a standard H–cone then an element $s \in S$ is quasi–continuous iff s is nearly–continuous (i.e. $s = \sum_{n\in I\!N} s_n \, , s_n$ universally continuous for all $n \in I\!N$); see [3], Proposition 5.6.1 .

Recall that a map $\mu : S \to \overline{I\!R_+}$ which is additive, increasing, continuous in order from below (i.e. $s_i \nearrow s \Rightarrow \mu(s_i) \nearrow \mu(s)$) and increasingly dense finite is called *H–integral* on S. The ordered convex cone S^* of all H–integrals (endowed with the usual algebraic operations and the pointwise order relation) is also an H–cone (see [3]), called the *dual of* S.

A *balayage on* S is a map $B : S \to S$ which is additive, increasing, contractive (i.e. $Bs \le s$, for all $s \in S$), idempotent and continuous in order from below.

Lemma 2.1 . *Let L be a specifically solid subcone of $Q_c(S)$ and let $\theta : L \to I\!R_+$ be a map which is additive, increasing and such that there exists an H–integral $\mu \in S^*$ with*

$$\theta(s) \le \mu(s) < \infty \, , \, (\forall) s \in L .$$

Then there exists $\widetilde{\theta} \in S^*$ such that

$$\widetilde{\theta}(s) = \theta(s) \, , \, (\forall) s \in L$$

and for all $\nu \in S^*$ with $\theta \le \nu\big|_L$ we have

$$\widetilde{\theta} \le \nu .$$

Proof. For any $s \in S$ we put

$$\widetilde{\theta}(s) := \sup \{\theta(t) / t \in L \, , \, t \le s\} .$$

Since L is a specifically solid subcone of S it follows that $\widetilde{\theta}$ is additive. Obviously $\widetilde{\theta}$ is increasing. Let now $(s_i)_{i\in I}$ be an increasing family in S with $\bigvee_{i\in I} s_i = s$. If $t \in L$ is such that $t \le s$ then $t \wedge s_i \nearrow t$ and therefore

$$R(t - t \wedge s_i) \searrow 0 .$$

Seting $r_i := R(t - t \wedge s_i)$ we get $t - r_i \in L$, $t - r_i \nearrow t$ and therefore

$$\infty > \mu(t) = \mu(r_i) + \mu(t - r_i) = \inf_{i\in I} \mu(r_i) + \mu(t),$$

$$\inf_{i\in I} \mu(r_i) = 0, \ \inf_{i\in I} \theta(r_i) = 0.$$

From $\theta(t) = \theta(r_i) + \theta(t - r_i)$ and $t - r_i \leq s_i \wedge t \leq s_i$ we deduce now

$$\theta(t) = \sup_{i\in I} \theta(t - r_i) \leq \sup_{i\in I} \widetilde{\theta}(s_i),$$

$$\widetilde{\theta}(s) = \sup\{\theta(t)/t \in L, t \leq s\} \leq \sup_{i\in I} \widetilde{\theta}(s_i).$$

By the above considerations it follows that $\widetilde{\theta} \in S^*$. Obviously $\widetilde{\theta} \leq \nu$, for any $\nu \in S^*$ with $\theta \leq \nu\big|_L$.

Proposition 2.2. *Let $s \in S$ be a quasi-continuous element of S. Then for any decreasing family $(\mu_i)_{i\in I}$ in S^* such that $\inf_{i\in I} \mu(s_i) < \infty$ we have*

$$\left(\bigwedge_{i\in I} \mu_i\right)(s) = \inf_{i\in I} \mu_i(s).$$

Proof. We consider the specifically solid subcone L of S generated by s. Obviously we have

$$L = \{t \in S/(\exists)\alpha > 0 \text{ with } t \preccurlyeq \alpha s\}.$$

If we define the map $\theta : L \to \mathbb{R}_+$ by

$$\theta(t) := \inf_{i\in I} \mu_i(t), \ t \in L,$$

it follows that θ is additive, increasing with respect to the natural order and

$$\theta(t) \leq \mu_{i_0}(t) < \infty, \ (\forall)t \in L$$

for a suitable $i_0 \in I$. By Lemma 2.1 there exists $\widetilde{\theta} \in S^*$ such that

$$\widetilde{\theta} \leq \mu_i, \ (\forall)i \in I$$

and such that

$$\widetilde{\theta}(t) = \theta(t), \ (\forall)t \in L.$$

Consequently $\widetilde{\theta} \leq \bigwedge_{i\in I} \mu_i$ and

$$\left(\bigwedge_{i\in I} \mu_i\right)(t) = \inf_{i\in I} \mu_i(t), \quad (\forall)t \in L.$$

Particularly we get $\left(\bigwedge_{i\in I} \mu_i\right)(s) = \inf_{i\in I} \mu_i(s)$.

Theorem 2.3. *Suppose that* S^* *separates* S *and let* $s \in S$. *Then* s *is quasi–continuous iff for any decreasing family* $(\mu_i)_{i\in I}$ *in* S^* *with* $\inf_{i\in I} \mu(s_i) < \infty$ *we have*

$$\left(\bigwedge_{i\in I} \mu_i\right)(s) = \inf_{i\in I} \mu_i(s) .$$

Proof. By Proposition 2.2 is suffices to justify only the "if" part of the assertion. Let $(s_i)_{i\in I}$ be an increasing family in S with $\bigvee_{i\in I} s_i = s$ and let

$$r_i := R(s - s_i) \;\; , \;\; r := \bigwedge_{i\in I} r_i .$$

To show that $r = 0$, it will be sufficient to prove that $\mu(r) = 0$ for any $\mu \in S^*$ with $\mu(s) < \infty$. Let now fix $\mu \in S^*$ such that $\mu(s) < \infty$. For every $f \in (S - S)_+$ the balayage B_g on S is defined by

$$B_g(s) := \bigvee_{n\in\mathbb{N}} R(s \wedge nf) \, , \, (\forall) s \in S$$

and we note that (cf. [4]) we have

$$B_g\left(R_g\right) = R_g \, ,$$

where $g \in (S - S)_+$ and $f := (\alpha g - Rg)_+$ with $\alpha > 1$.
Therefore if $\alpha > 1$ is fixed we get

$$R(s - s_i) = B_{f_i}(R(s - s_i)) = B_{g_i}(R(s - s_i))$$

where

$$f_i := \left(\alpha(s - s_i) - R(s - s_i)\right)_+ \, , \, g_i := \left(\alpha(s - s_i) - r\right)_+$$

If $j \le i$ then $r_i = B_{g_i}(r_i) \le B_{g_j}(r_i) \le r_i$ and hence $B_{g_j}(r_i) = r_i$. Let us prove that

$$B_{g_j}(r) = r \, , \, (\forall) j \in I .$$

Indeed, since for all $i \in I$ we have $s = r_i + u_i$ with $u_i \in S$ and $r_i \searrow r$, we deduce that $s = r + u$, where $u_i \nearrow u \in S$. Consequently if $j \le i$ then

$$B_{g_j}(s) = B_{g_j}(r_i) + B_{g_j}(u_i) = r_i + B_{g_j}(u_j) = r + B_{g_j}(u) \, ,$$

$$B_{g_j}(s) = B_{g_j}(r) + B_{g_j}(u)$$

and therefore $B_{g_j}(r) = r$.

If for any $j \in I$ we put

$$\mu_j := \mu \circ B_{g_j} ,$$

then $(\mu_j)_{j\in I}$ is a decreasing family in S^* and let us define

$$\nu := \bigwedge_{j\in I} \mu_j .$$

From $r = B_{g_j}(r) \leq B_{g_j}\, (\alpha(s - s_j))$ it follows that for any $j \in I$ we have

$$\mu(r) = \mu\left(B_{g_j}(r)\right) \leq \mu\left(B_{g_j}(\alpha(s - s_j))\right) = \alpha\left(\mu_j(s) - \mu_j(s_j)\right) \leq$$

$$\leq \alpha\left(\mu_j(s) - \nu(s_j)\right) .$$

Since $\mu_j \searrow \nu$, by hypothesis we have $\inf_{j\in I} \mu_j(s) = \nu(s)$ and since $\sup_{j\in I} \nu\,(s_j) = \nu(s)$ we conclude that $\mu(r) = 0$.

Theorem 2.4. *Let B be a balayage on S and let $s \in S$ be such that $s \curlywedge Bs = 0$. Then s is quasi-continuous in S iff $s - Bs$ is quasi-continuous in $S_B := \{s - Bs / s \in S\}$ the localized of S with respect to B.*

Proof. Suppose that $s \in Q_c(S)$ and let $(s_i)_{i\in I}$ be a family in S such that

$$s_i - Bs_i \nearrow s - Bs .$$

Obviously we may suppose that $s_i \curlywedge Bs_i = 0$, for all $i \in I$. In this case the family $(s_i)_{i\in I}$ is also increasing and we have $\bigvee_{i\in I} s_i = s$. Hence $\bigwedge_{i\in I} R\,(s - s_i) = 0$ and therefore from

$$^{S_B}R\Big((s - Bs) - (s_i - Bs_i)\Big) \leq R\,(s - s_i) \wedge (s - Bs)$$

we deduce

$$^{S_B}R\Big(s - Bs) - (s_i - Bs_i)\Big) \searrow 0 ,$$

where ^{S_B}R denotes the reduit in the H-cone S_B.

Conversely, suppose that $s - Bs \in Q_c(S_B)$ and let $(s_i)_{i\in I}$ be an increasing family in S with $\bigvee_{i\in I} s_i = s$. We set

$$t_i - Bt_i := \bigwedge_{j\geq i} (s_j - Bs_j) ,$$

where $t_i \in S$ is such that $t_1 \curlywedge Bt_i = 0$. Obviously the family $(t_i - Bt_i)_{i\in I}$ is increasing and

$$s_i - Bs \leq t_i - Bt_i \leq s - Bs .$$

As a consequence

$$\bigvee_{i\in I} (t_i - Bt_i) = s - Bs .$$

and therefore the family $(t_i)_{i\in I}$ increases to s and

$$t_i \le s_i\ ,\ (\forall) i \in I\ .$$

Let us put

$$u_i - Bu_i :=^{S_B} R\Big((s-Bs)-(t_i-Bt_i)\Big)\ ,$$

with $u_i \in S$ such that $u_i \curlywedge Bu_i = 0$. By hypothesis $u_i - Bu_i \searrow 0$ and by definition for every $i \in I$ there exists $v_i \in S$ with

$$v_i \curlywedge Bv_i = 0\ ,\ v_i - Bv_i + u_i - Bu_i = s - Bs\ ,\ v_i - Bv_i \le t_i - Bt_i\ .$$

We deduce

$$s = u_i + v_i\ ,\quad v_i \le t_i$$

and therefore

$$u_i \ge s - t_i\ ,\ u_i \ge R(s-t_i)\ .$$

Since $(u_i)_{i\in I}$ is decreasing it follows that $(v_i)_{i\in I}$ is increasing, $Bv_i \nearrow Bv$ and

$$s = u + v\ ,$$

where $u := \bigwedge_{i\in I} u_i\ ,\ v := \bigvee_{i\in I} v_i$. As a consequence $Bu_i \searrow Bu$ and since $u_i - Bu_i \searrow 0$ we obtain $u = Bu$ and so

$$s - Bs = v - Bv\ ,\ s = v\ \ ,\ u = 0\ .$$

Hence $u_i \searrow 0$ and we conclude that $R(s-s_i) \searrow 0$ and the proof is complete.

Recall now the following result from [3]:

Proposition ([3], Proposition 2.2.10). *Let L be a specifically solid subcone of $Q_c(S)$. Then the map $B_L : S \to S$ defined by*

$$B_L s := \bigvee\{t \in L / t \le s\}\ , s \in S$$

is a balayage on S.

Corollary 2.5. *If $B_c : S \to S$ is the map defined by*

$$B_c s := \bigvee\{t \in Q_c(S)/t \le s\}\ \ , s \in S$$

then B_c is the smallest balayage B on S such that $Bs = s$ for all $s \in Q_c(S)$.

Corollary 2.6. *Let B_c be the balayage on S definied by (see Corollary 2.5)*

$$B_c s := \bigvee\{t \in Q_c(S)/t \le s\}\ .$$

Then the H-cone S_{B_c}, the localized of S with respect to B_c, has no nonzero quasi-continuous elements and $Q_c(S) = Q_c(B_c(S))$ is increasingly dense in $B_c(S) := \{B_c(s)/s \in S\}$.

Remark. 1. It is not difficult to see that in the H–cone of al excessive measures, the quasi–continuous elements are increasingly dense or equivalent $B_c = I$.
2. It is still open the question if $B_c = I$ for every H–cone.

3. Quasi–continuous excessive measures

Let $\mathcal{U} = (U_\alpha)_{\alpha>0}$ be a submarkovian resolvent of kernels on a Lusin measurable space $(X, \mathcal{X})$ such that the initial kernel U is proper. We suppose, as in Section 1, that the convex cone $\mathcal{E}_\mathcal{U}$ of all $\mathcal{X}$–measurable $\mathcal{U}$– a.s. finite $\mathcal{U}$–excessive function is min–stable, $1 \in \mathcal{E}_\mathcal{U}$ and the σ–algebra generated by $\mathcal{E}_\mathcal{U}$ coincides with $\mathcal{X}$. We denote by Exc the convex cone of all $\mathcal{U}$–excessive measures on $(X, \mathcal{X})$ i.e. the set of all σ–finite measures m on X for which $m(\alpha U_\alpha) \leq m$, for all $\alpha > 0$. It was shown in [6] (see also [8]) that Exc is an H–cone with respect to the usual order relation in the set of all positive measures on $(X, \mathcal{X})$.

Our aim is to describe the quasi–continuous elements of the H–cone Exc, using the results established in the preceding section. We recall therefore from [1] some results concerning the dual of Exc.

We denote by $\overline{\overline{\mathcal{E}_\mathcal{U}}}$ the set of all families

$$\mathbf{s} = (\overline{\overline{s}}_\xi)_{\xi \in Exc},$$

where
a) for any $\xi \in Exc$ s_ξ is an $\mathcal{U}$–excessive function on X finite ξ–a.s. and $\overline{\overline{s}}_\xi$ denotes the class of all positive $\mathcal{X}$–measurable functions which coincide ξ–a.s. with s_ξ;
b) for any $\xi_1, \xi_2 \in Exc$, $\xi_1 \leq \xi_2$ we have

$$s_{\xi_1} = s_{\xi_2}, \ \xi_1 - a.s..$$

Obviously $\overline{\overline{\mathcal{E}_\mathcal{U}}}$ has a natural structure of ordered convex cone and in fact it is an H–cone.

Remark 3.1. For any $s \in \mathcal{E}_\mathcal{U}$ the element $(\overline{\overline{s}}_\xi)_{\xi \in Exc}$, where $s_\xi := s$ for any $\xi \in Exc$ belongs to $\overline{\overline{\mathcal{E}_\mathcal{U}}}$. In this way $\mathcal{E}_\mathcal{U}$ is identified with an ordered convex subcone of $\overline{\overline{\mathcal{E}_\mathcal{U}}}$. More generally, any universally measurable $\mathcal{U}$–excessive function s, finite $\mathcal{U}$–a.s., may be identified with the element $(\overline{\overline{s}}_\xi)_{\xi \in Exc}$ of $\overline{\overline{\mathcal{E}_\mathcal{U}}}$, where for any $\xi \in Exc$ s_ξ is an $\mathcal{U}$–excessive function equal s ξ–a.s.

As usually the *energy functional* associated with $\mathcal{U}$ is the map $L :\ Exc \times \mathcal{E}_\mathcal{U} \to \overline{I\!R_+}$ defined by

$$L(\xi, s) := \sup\{\mu(s)/\mu \circ U \leq \xi\}, \ \xi \in Exc, \ s \in \mathcal{E}_\mathcal{U}.$$

Theorem 3.2. ([1], Theorem 3.4 and Theorem 3.2). *For any* $\mathbf{s} = (\overline{\overline{s}}_\xi)_{\xi \in Exc} \in \overline{\overline{\mathcal{E}_\mathcal{U}}}$ *the map* $\tilde{\mathbf{s}} : Exc \to \overline{I\!R_+}$ *defined by*

$$\tilde{\mathbf{s}}(\xi) := L(\xi, s_\xi), \ \xi \in Exc$$

belongs to Exc^* *and the correspondence*

$$\mathbf{s} \mapsto \tilde{\mathbf{s}}$$

is an isomorphism between the H*-cones* $\overline{\overline{\mathcal{E}_{\mathcal{U}}}}$ *and* Exc^*. *Moreover if* $(\mathbf{s}^i)_{i\in I}$ *is a family in* $\overline{\overline{\mathcal{E}_{\mathcal{U}}}}$, $\mathbf{s}^i = (\overline{\overline{s^i_\xi}})_{\xi\in Exc}$ *and* $\bigwedge_{i\in I} \mathbf{s}^i = \mathbf{s} := (\overline{\overline{s}}_\xi)_{\xi\in Exc}$ *then*

$$\overline{\overline{s}}_\xi = \bigwedge_{i\in I} \overline{\overline{s^i_\xi}}\,, \quad (\forall)\, \xi \in Exc\,.$$

(i.e. $s_\xi \leq s^i_\xi$ ξ*-a.s. for any* $i \in I$ *and there exists a sequence* $(i_n)_{n\in\mathbb{N}}$ *such that* $s_\xi = \inf_{n\in\mathbb{N}} s^{i_n}_\xi$ ξ*-a.s.)*

Theorem 3.3. *Let* $\xi = \mu \circ U \in Exc$. *Then the following assertions are equivalent:*
1) ξ *is quasi-continuous.*
2) μ *does not charge the semi-polar subsets of* X.

Proof. "1) $\Rightarrow$ 2)". Let A be a totally thin subset of X and let f be a bounded positive $\mathcal{X}$-measurable function, $f > 0$ such that Uf is bounded. Hence $B^AUf < Uf$ on X. Since B^AUf is an universally measurable excessive function and therefore, by Remark 3.1, an element of $\overline{\overline{\mathcal{E}_{\mathcal{U}}}}$, it follows that

$$B^AUf = \bigwedge_{\overline{\overline{\mathcal{E}_{\mathcal{U}}}}} \{t \in \mathcal{E}_{\mathcal{U}} / Uf \leq t \text{ on } A\}\,.$$

From Theorem 3.2 we have $\overline{\overline{\mathcal{E}_{\mathcal{U}}}} = Exc^*$. As a consequence, by Theorem 2.3 we get

$$\mu\left(B^AUf\right) = L\left(\xi, B^AUf\right) = \inf\{L(\xi,t)/t \in \mathcal{E}_{\mathcal{U}}\,,\, Uf \leq t \text{ on } A\} =$$
$$= \inf\{\mu(t)/t \in \mathcal{E}_{\mathcal{U}}\,,\, Uf \leq t \text{ on } A\} = \mu\left(R^AUf\right)\,.$$

We conclude that $\mu\left(B^AUf\right) = \mu\left(R^AUf\right)$ which implies $\mu(A) = 0$

"2) $\Rightarrow$ 1)". Again by Theorem 2.3 it is sufficient to show that for any decreasing family $(\mathbf{s}^i)_{i\in I}$ in $\overline{\overline{\mathcal{E}_{\mathcal{U}}}}$ with $\inf_{i\in I} \widetilde{\mathbf{s}^i}(\xi) < \infty$ we have

$$\left(\widetilde{\bigwedge_{i\in I} \mathbf{s}^i}\right)(\xi) = \inf_{i\in I} \widetilde{\mathbf{s}^i}(\xi)\,.$$

From Theorem 3.2 there exists a decreasing family $(s_i)_{i\in I}$ of excessive functions finite ξ-a.s. and an excessive function s finite ξ-a.s. such that $\widetilde{\mathbf{s}^i}(\xi) = L(\xi, s_i)$, $\left(\widetilde{\bigwedge_{i\in I} \mathbf{s}^i}\right)(\xi) = L(\xi, s)$ and $\overline{\overline{s}} = \bigwedge_{i\in I} \overline{\overline{s_i}}$. Since $L(\xi, s_i) = \mu(s_i)$ and $\inf_{i\in I} \widetilde{\mathbf{s}^i}(\xi) < \infty$ we may suppose that $\mu(s_i) < \infty$, for all $i \in I$. Moreover there exists a sequence $(i_n)_{n\in\mathbb{N}}$ such that

$$s = \inf_{n\in\mathbb{N}} s_{i_n} \quad \xi\text{-a.s.}$$

Therefore $s = \bigwedge_{n\in\mathbb{N}} s_{i_n}$ ξ–a.s. and consequently

$$\mu(s) = L(\xi, s) = L\left(\xi, \bigwedge_{n\in\mathbb{N}} s_{i_n}\right) = \mu\left(\bigwedge_{n\in\mathbb{N}} s_{i_n}\right) .$$

Since the set $\left\{x \in X / \inf_{n\in\mathbb{N}} s_{i_n}(x) \neq \left(\bigwedge_{n\in\mathbb{N}} s_{i_n}\right)(x)\right\}$ is semi–polar, by hypothesis we get

$$\mu\left(\bigwedge_{n\in\mathbb{N}} s_{i_n}\right) = \mu\left(\inf_{n\in\mathbb{N}} s_{i_n}\right) = \inf_{n\in\mathbb{N}} \mu\left(s_{i_n}\right) = \inf_{n\in\mathbb{N}} \widetilde{\mathbf{s}^{i_n}}(\xi) \geq \inf_{i\in I} \widetilde{\mathbf{s}^i}(\xi) .$$

We conclude that $\left(\widetilde{\bigwedge_{i\in I} \mathbf{s}^i}\right)(\xi) = \mu(s) = \inf_{i\in I} \widetilde{\mathbf{s}^i}(\xi)$.

Definition. Let $\xi \in Exc$. A set $A \in \mathcal{X}$ is called *ξ–polar* if $B^A 1 = 0$ ξ–a.s. A *ξ–semi–polar* set is a union of a semi–polar set with a ξ–polar set (cf.[8]).

Corollary 3.4. *Let $m \in Exc$ and $\xi = \mu \circ U \in Exc$, ξ absolutely continuous with respect to m. Then the following assertions are equivalent:*

1) *ξ is quasi–continuous.*

2) *μ does not charge the m–semi–polar subsets of X .*

Proof. By Theorem 3.3 it remains to show that if ξ is quasi–continuous then μ does not charge any m–polar subset of X . Indeed, if M is an m–polar subset of X then m and ξ are carried by the absorbent set

$$A := \{x \in X / t_0(x) = 0\} ,$$

where $t_0 \in \mathcal{E}_{\mathcal{U}}$ is such that $t_0 \leq B^M 1$ and $t_0 = B^M 1$ m–a.s. From $\mu\left(B^M 1\right) = L\left(\xi, B^M 1\right) = 0$ we deduce that μ is also carried by A . On the other hand the set $M \cap A$ it totally thin and therefore it is semi–polar. By the above theorem we get $\mu(M \cap A) = 0$ and consequently $\mu(M) = \mu(M \cap A) = 0$.

References

[1] L. Beznea, N. Boboc, *Duality and biduality for excessive measures.* (submited)

[2] N. Boboc, Gh. Bucur, *Conuri convexe de funcţii continue pe spaţii compacte.* Ed. Academiei, Bucureşti, 1976.

[3] N. Boboc, Gh. Bucur, A. Cornea, *Order and Convexity in Potential Theory: H–Cones.* Lecture Notes in Math. **853**, Springer, 1981.

[4] N. Boboc, A. Cornea, *Cônes convexes ordonnés. H–cônes et biadjoints des H–cônes.* C. R. Acad. Sc. Paris, t. **270** (1970), Série A, 1679–1682.

[5] R. M Blumenthal, R. K. Getoor, *Markov Processes and Potential Theory.* Academic Press, New York, 1968.

[6] A. Cornea, G. Licea, *Order and Potential Resolvent Families of Kernels.* Lecture Notes in Math. **494**, Springer, 1975.

[7] R. K. Getoor, *Markov Processes: Ray Processes and Right Processes.* Lecture Notes in Math. **440**, Springer, 1975.

[8] R. K. Getoor, *Excessive measures.* Birkhäuser, Boston, 1990.

THE WIENER TEST FOR POINCARÉ-DIRICHLET FORMS

M. BIROLI
Department of Mathematics
Politecnico di Milano
Piazza Leonardo da Vinci, 32
20133 Milano
Italy

ABSTRACT. We give a necessary and sufficient condition of Wiener type for the regularity of boundary points for Poincaré-Dirichlet forms; moreover estimates on the modulus of continuity and on the energy decay of a solution at a regular boundary point are studied.

1. Poincaré-Dirichlet forms.

We will precise at first that all the results given in this paper has been obtained jointly with U. Mosco.

Let X be a connected locally compact topological Hausdorff space and m a Radon measure on X; we denote by H the space $L^2(X,m)$ and by a(u,v) a regular *Dirichlet form* with domain D[a] contained in H for which the *strong locality property* holds (see [2][3] for definitions).

Under the above assumptions the form a(u,v) has the following integral representation

(1.1) $$a(u,v) = \int_X \alpha(u,v)(dx)$$

where $\alpha(u,v)$ is a Radon-measure-valued bilinear form on D[a], called the *energy measure* of a. We refer to [3][12] for the properties of $\alpha(u,v)$. This representation allow us to define the domain D[a,O] and the local domain $D_{loc}[a,O]$ of the form a with respect to the open set O in X.

We assume that $D[a] \cap C_0(X)$ is separating, that is for every $x,y \in X$, $x \neq y$, there exists a function ϕ in $D[a] \cap C_0(X)$ such that $\alpha(\phi,\phi) \leq m$ and $\phi(x) \neq \phi(y)$. Under this assumption a *distance* function can be defined as

K. GowriSankaran et al. (eds.), Classical and Modern Potential Theory and Applications, 93–104.

(1.2) $d(x,y) = \sup \{ \phi(x)-\phi(y) ; \ \phi \in D[a] \cap C_o(X) \text{ with } \alpha(\phi,\phi) \le m \}$

and we assume that the topology defined on X by d is equivalent to the initial one. We denote by B(x,r) the ball for d with center x and radius r.

Remark 1.1. We will prove here that d(x,.) is in $D_{loc}[a] = D_{loc}[a,X]$. To prove the result it is enough to prove that for every $y \in X$ there exists a ball B(y,r) such that d(x,.) is in $D_{loc}[a,B(y,r)]$. Suppose $x \notin$ cl (B(y,r)) (cl(E) = closure of the set E) and that B(y,r) is relatively compact . We can cover B(y,r) by a finite by a finite number of balls B(y \S(,i),n\S(-1)) , i = 1,2,...,m \S(,n) , y \S(,i)∈ B(y,r), and n is an integer such that $n > d(x,B(y,r))^{-1}$. By the same methods used in the Lemma 3.2 of [3] we can construct a function ϕ_n such that $\alpha(\phi_n,\phi_n) \le m$, $\phi_n(x)=0$, $\phi_n(z) \ge d(x,z) - n^{-1}$ for z in B(y,r). Let now ϕ be the supremum of ϕ_n with respect to n , we have $\phi = d(x,.)$ on B(y,r) and $\alpha(\phi,\phi) \le m$, then d(x,.) is in $D_{loc}[a,B(y,r)]$. We observe that an independent proof of this result has been given by K. Th. Sturm [20].

We say that a(u,v) is a *Dirichlet-Poincaré form* on an open set X_o in X if the following properties hold :

(1) a *duplication property* , that is for every ball with center y in X_o and radius $r \le R_o$ we have

(1.3) $$m(B(y,2r)) \le c_o \, m(B(y,r))$$

with a constant c_o possibly depending on X_o and R_o.

We observe that from the duplication property it follows that for y in X_o and 0< s < 2-1)r ≤ ≤ $2^{-1}R_o$ we have

(1.4) $$m(B(y,r)) \le 2\, m(B(y,s)) \left(\frac{r}{s}\right)^{\nu}$$

where $\nu = \log_2 c_o$.

(2) a *Poincaré inequality* of the type

(1.5) $$\int_{B(y,r)} |f - f_{y;r}|^2 m(dx) \le c_1 r^2 \int_{B(y,\kappa r)} \alpha(f,f)(dx)$$

for all f in $D_{loc}[a]$, $r \le R_o$ and y in X_o , where $f_{y;r}$ is the average of f on B(y,r).

We recall that in [4] has been proved that from the Poincaré inequality also a *Sobolev inequality* follows for f in D[a] with support in B(y,r)

(1.6_1) $$⨍_{B(y,r)} |f|^{\frac{2\nu}{\nu-2}} m(dx) \le \left(⨍_{B(y,r)} \alpha(f,f)(dx)\right)^{\frac{\nu}{\nu-2}}, \text{ if } \nu > 2$$

$$(1.6_2)\qquad ⨍_{B(y,r)} \exp(\theta\,|f| - 1)\, m(dx) \le \frac{\theta}{\xi-\theta}\ ,\ \text{if } \nu = 2$$

where $\theta < \xi = c\,[\,r\,(⨍_{B(y,r)} \alpha(f,f)(dx)\,)]^{-\frac{1}{2}}$.

$$(1.6_3)\qquad \sup_{B(y,r)} |f| \le c\, r\, (⨍_{B(y,r)} \alpha(f,f)(dx)\,)^{\frac{1}{2}}\ ,\ \text{if } \nu < 2$$

$(⨍_{B(y,r)} = \frac{1}{m(B(y,r))} \int_{B(y,r)}\ )$.

Finally we will observe that if the assumptions (1) and (2) hold for a regular, strong local Dirichlet form a(u,v) with domain D[a], then (1) and (2) hold again (with possibly different constants c_o, c_1, κ) for every regular, strong local Dirichlet form b(u,v) with domain D[b] = D[a] and for which the following inequality holds

$$(1.7)\qquad \lambda\, a(u,u) \le b(u,u) \le \Lambda\, a(u,u)\ ,\ \forall u \in D[a].$$

Now we will describe some examples of degenerate elliptic operators, which satisfy as Dirichlet form our assumptions.

(a) *Elliptic degenerate operators with a weight* - Consider the form

$$(1.8)\qquad a(u,v) = \int \sum_{ij=1}^{n} D_{x_i} u(x)\, D_{x_j} v(x)\, a^{ij}(x)\, dx$$

on $C_o^1(R^n)$ where $a^{ij}(x) = a^{ji}(x)$, i,j =1,...,n , are Lebesgue measurable functions satisfying the condition

$$\lambda\, |\xi|^2 w(x) \le \sum_{ij=1}^{n} \xi_i\, \xi_j\, a^{ij}(x) \le \Lambda\, |\xi|^2 w(x)$$

a. e. in R^n and for every $\xi \in R^n$. Here w is a weight in the Muckenhoupt class A_2, or a weight $w(x) = |\det F'|^{1-2/n}$ associated with a quasiconformal map F in R^n. The domain of the form is the completion of $C_o^1(R^n)$ with respect to the norm $(a(u,u)+(u,u))^{1/2}$ where (.,.) denotes the usual inner product in $L^2(R^n, w(x)dx)$. In this case we choose m(dx) = w(x)dx, so the distance d is equivalent to the usual euclidean distance . The assumption (1) derives from the analogous property of the above weights with respect to the euclidean distance and the scaled Poincaré inequalities (assumption (2)) have been proved in [8].
We observe that the usual unform elliptic forms with bounded measurable coefficients are obtained as a particular case with w=1.

(b) *Operators of uniform Hörmander type* - Consider m smooth vector fields X_h , h= 1,...,m, on R^n that satisfy the Hörmander condition ; in this case we have a distance d* associated to the vector fields, [9][17][19], which is known to satisfy the condition

$$\frac{1}{c}\, |x - y| \le d^*(x,y) \le c\, |x - y|^{\varepsilon}$$

where $\varepsilon = \frac{1}{q+1}$ if q is the number of commutators in the Hörmander condition.

Let $\alpha(x) = (\alpha^{hk}(x))$ be a symmetric $m \times m$ matrix of measurable functions on R^n such that

$$\lambda\, |\xi|^2\, w(x) \le \sum_{ij=1}^{n} \xi_h \xi_k\, \alpha^{hk}(x) \le \Lambda\, |\xi|^2\, w(x)$$

a. e. in R^n and for every $\xi \in R^m$, where w is in the Muckenhoupt class A_2 with respect to the balls relative to d*. We consider the form

$$a(u,v) = \int \sum_{hk=1}^{m} X_h u(x)\,)X_k v(x)\, \alpha^{hk}(x)\, dx \tag{1.9}$$

Again the domain of the form is defined as the completion of $C_o^1(R^n)$ with respect to the norm $(a(u,u)+(u,u))^{1/2}$ where (.,.) denotes the usual inner product in $L^2(R^n, w(x)dx)$. We choose m(dx) = w(x)dx, so the distance d turns out to be equivalent to the distance d* (precisely d = d* if $\alpha^{hk} = \delta^{hk}w$),[14]. The duplication property (assumption (1)) derives from the analogous property of the above weights with respect to the distance d* and the scaled Poincaré inequalities (assumption (2)) are consequence of the analogous inequalities in the case $\alpha^{hk} = \delta^{hk}w$ that have been proved in [16] (see also [14] for the case w = 1).

We end this section recalling that local regularity properties (Harnack inequality, Hölder continuity) of harmonics with respect to the Dirichlet form has been proved in [2][3] under the above assumptions as well as estimates for the Green function relative to an (intrinsic ball) with singularity at the center of the ball.

2. The Wiener test for regular points at the boundary.

Let us consider (under the assumption of section 1) the following problem

$$a(u,v) = 0 \text{ for every } v \text{ in } D[a, X_o] \tag{2.1}$$

$$u - g \in D[a, X_o]$$

where X_o is a reletively compact open set in X and $g \in D[a]$.

It is easy to see that problem (2.1) has a unique solution.

We will observe at first that there a natural notion of *capacity* associated to the Dirichlet form a. Denote by O an open set in X wih closure contained in the open set A, then we define

$$cap(O,A) = \inf\{a(u,u) \mid u \in D[a,A] \text{ with } u \ge 1 \text{ m-a.e. on } O\} \tag{2.2}$$

as the *capacity of O with respect to A*. Let now E be a set with closure in A, then we define the *capacity of E with respect to A* by

$$cap(E,A) = \inf\{cap(O,A) \mid O \text{ open set with closure contained in } A\}. \tag{2.3}$$

We observe that (2.2)(2.3) define a Choquet capacity and that the sets of zero capacity in A have also zero capacity with respect to every open set A' containing A.

We call a function u defined on A *quasicontinuous* if for any $\varepsilon > 0$ there exists a an open set G in A such that cap(G,A) $< \varepsilon$ and $u|_{A-G}$ is continuous.

We recall that given a function u in D[a] there is a a *quasicontinuous representative* $\tilde{u}$ of u, this means that $\tilde{u}$ is quasicontinuous on X and $\tilde{u} = u$ m-a.e.(see [12] for the proof of this property). As a consequence of this property we can say that every function in D[a] is defined up to sets of zero capacity (in X).

Now we will give the notion of regular point of the boundary :

Definition 2.1 - *A point x_o of the boundary of X_o is said a regular point (for (2.1)) if for every boundary data g in D[a] continuous at x_o , the corresponding solution u is continuous at x_o .*

The problem of a characterization of regular points of the boundary was studied starting from the twenties for uniform elliptic operators; in particular N. Wiener , [21], had proved a necessary and sufficient condition in terms of capacity (the so called Wiener criterion) . This characterization was extended by Littman,Stampacchia and Weinberger, [15], in 1963 to uniform elliptic operators (in divergence form) with bounded measurable coefficients, proving in particular that the regular points of the boundary for this class of operators are the same as in the case of the Laplace operator (referring to example (a) we are in the case w=1 and the case of the Laplace of operator is obtained by choosing $a^{hk} = \delta^{hk}$).

The case of degenerate elliptic operator with weight (see example (a) with w $\neq$ constant) was studied in 1982 by Fabes,Jerison and Kenig, [7]; they extended the Wiener criterion to this case proving in particular that the regular points of the boundary for this class of operators are the same as in the case of the operator obtained in example (a) by choosing $a^{hk} = \delta^{hk}$ w. Then the regularity of a point of the boundary depends essentially on the weights w; an interesting feature of the weighted case is that it is possible to have points of positive capacity and that those points are always regular.

The case of degenerate elliptic operators of Hörmander type (example(b)) was studied only, at my knowledge, in [13][18] where the Wiener criterion is proved only for the case $\alpha^{hk} = \delta^{hk}$ and w=1. At my knowledge no results concerning the general problem for regular, strong local Diriclet forms have been proved (also under the assumptions (1)(2)).

Moreover in all the preciding papers only a qualitative proof of the Wiener criterion is given, without a study of the modulus of continuity at a regular boundary points. This problem was studied for the case of degenerate elliptic operators with a weight in [1], where an estimate of the modulus of continuity at a regular boundary point is given in terms of the so called Wiener integral.

The aim of this paper is to give the extention of the Wiener criterion to our general framework (see [5] for the proofs).

Let $x_o \in \partial X_o$ we denote

$$\delta(r,x_o) = \frac{\mathrm{cap}(B(x_o,r) \cap (X_o)^c, B(x_o, k^* r))}{\mathrm{cap}(B(x_o,r), B(x_o,k^* r))},$$

where $(X_o)^c$ denotes the set $X\text{-}X_o$ and k^* is a constant depending only on κ in the assumption (2) ; we say that x_o is a *a Wiener point* for ∂X_o if

$$\lim_{r \to 0} \int_r^1 \delta(\rho,x_o)\frac{d\rho}{\rho} = +\infty \tag{2.4}$$

Theorem 2.2 - *Let the assumptions (1) and (2) hold . A point $x_o \in \partial X_o$ is a regular point for (2.1) if and only if x_o is a Wiener point for ∂X_o..*

We are also able to give an estimate of the modulus of continuity of the solution of (2.1) at a regular point x_o in terms of the rate of divergence of the integral (2.4); precisely :

Theorem 2.3 - *Let the assumptions (1) and (2) hold . Let $x_o \in \partial X_o$ be a regular point for (2.1.) , u be the solution of (2.1) and assume that g is continuous at the point x_o; then*

$$\mathrm{osc}_{B(x_o,r)}\,\tilde{u} \le c\,\exp\left(-\alpha \int_r^R \delta(\rho,x_o)\frac{d\rho}{\rho}\right)\mathrm{osc}_{B(x_o,R)}))\,\tilde{u} + \mathrm{osc}_{B(x_o,R)}\,g\,,$$

where the constants c and α depends only on the constants c_o and c_1 in the assumptions (1) and (2) and $\tilde{u}$ is the quasicontinuous representative of u .

To prove the results of Theorem 2.2 , 2.3 we imbed our problem in a more general problem, which has an interest in itself (for ex. for Schrödinger type equations or in the homogenization with holes).

We denote by M_o the set of all nonegative Borel measures on X that vanishe on all sets of null capacity (with respect to X) but may be $+\infty$ on some Borel subsets of R^n. We consider the following problem

$$a(u,v) + \int_\Omega (u-g)\, v\, \mu(dx) = 0 \tag{2.5}$$

$$u \in D_{loc}[a\,;\Omega] \cap L^2_{loc}(\Omega,\mu)\,;$$

for every $v \in D[a\,;\Omega] \cap L^2(\Omega,\mu)$ with compact support in Ω,

where Ω is a bounded open set and $\mu \in M_o$.

Our problem is to give necessary and sufficient condition for the continuity of u at a point x_o of Ω with $u(x_o) = g(x_o)$, where $g \in D[a]$ is assumed to be a continuos at x_o.

We recall that this problem has been solved in [6] for the case of uniformly elliptic forms proving a necessary and sufficient condition that we will extend to our general framework.

To solve our problem we define a new type of capacity called *μ-capacity* .

For every Borel set E contained in Ω we denote by μ_E the Borel measure on Ω defined by $\mu_E(B) = \mu(B \cap E)$ for every Borel set B contained in Ω, where $\mu \in M_o$. We observe that $\mu_E \in M_o$ for every Borel set E in Ω. We say that a Borel subset E of Ω is *μ-admissible* if there exists w in $D[a\,;\Omega]$ such that $(w-1)$ is in $L^2(\Omega,\mu)$. The *μ-capacity* of a μ-admissible set E in Ω is defined by

$$\text{(2.6)} \qquad \mathrm{cap}_\mu(E,\Omega) = \min\,\{a(v,v) + \int_\Omega (v-1)^2\,\mu_E(dx)\,;\, v \in D[a\,;\Omega]\,\}$$

Definition 2.4 - *We say that x_o in Ω is a regular Dirichlet point for (2.5) if for every g continuous at x_o and u a function satisfying (2.5) a neighbourhood of x_o , u is continuous at x_o and $u(x_o) = g(x_o)$.*

Let now

$$\delta(r,x_o) = \frac{\mathrm{cap}_\mu(B(x_o,r),B(x_o,k^*\,r))}{\mathrm{cap}(B(x_o,r),B(x_o,k^*\,r))}$$

where k^* is a constant depending only on κ in the assumption (2) ; we say that x_o ia a *Wiener point* for (2.5) if

$$\lim_{r\to 0} \int_r^1 \delta(\rho, x_o)\frac{d\rho}{\rho} = +\infty$$

We are able to prove the following Wiener type criterion :

Theorem 2.5 - *Let the assumptions (1) and (2) hold . A point $x_o \in \Omega$ ia regular point for (2.5) if and only if x_o is a Wiener point..*

Moreover we can estimate the modulus of continuity of u at a regular point :

Theorem 2.6 - *Let the assumptions (1) and (2) hold . Let $x_o \in \Omega$ be a regular point for (2.5) and u a function satisfying (2.5) then*

$$osc_{B(x_o,r)}\ \tilde{u}\ \le c\, exp\,(-\alpha \int_r^R \delta(\rho,x_o)\frac{d\rho}{\rho}\ )\, osc_{B(x_o,R)}\ ))\ \tilde{u} +\ osc_{B(x_o,R)}$$

g , where the constants c and α depends only on the constants c_o and c_1 in the assumptions (1) and (2) and $\tilde{u}$ is the quasicontinuous representative of u .

We don't give here the proofs of Theorems 2.5 , 2.6; we only remark that the proofs uses an adaptation to our framework of the methods in [1][6] using the techniques developped in [2][3].

Now we will explain as Theorems 2.2, 2.3 follow from Theorems 2.5, 2.6 .

Let Ω be an open set containing the closure of X_0 we define the measure μ by

$$\mu(E) = 0 \ \text{ if } \ cap(E \cap (X_0)^c) = 0$$

$$\mu(E) = +\infty \ \text{ if } \ cap(E \cap (X_0)^c) > 0$$

where E is a Borel set in X; if u is a solution of (2.1) (and we denote again by u the extension of u to Ω by 0) then u is a solution of (2.5) . Moreover in this case we have that

(2.7) $$cap_\mu\ (B(x_0,r),B(x_0,k^*\, r)) =$$

$$= \min\ \{a(v,v) + \int_\Omega (v-1)^2\, \mu_{B(x_0,r)}\ (dx)\ ;\ v\in\ D[a\ ;\ B(x_0\ ,k^*\, r)]\} =$$

$$= \min\ \{a(v,v)\ ;\ v\in\ D[a\ ;\ B(x_0,k^*\, r)]\ ,\ \ v \ge 1 \text{ q.e. on } \ B(x_0,r) \cap (X_0)^c\ \} =$$

$$= cap\ (\ B(x_0,r) \cap (X_0)^c,\ B(x_0,k^*\, r)).$$

Theorems 2.2, 2.3 then follow from Theorem 2.5, 2.6 with the above choice of the measure μ using the relation (2.7).

3. Consequences and examples .

In this section we will derive some intersting consequences from Theorem 2.2, 2.3 and deal with some examples, which point out the influence of the weight on the existence of points of positive capacity (which are always regular points) and the influence of the anysotrpy of degeneration on the regularity of points and in particulare on the validity of cone condition as sufficient condition for regularity.

(a) *Invariance of regular points in a class of Dirichlet forms* - As observed in section 1. if the assumptions (1) and (2) hold for a regular, strong local Dirichlet form a(u,v) with domain D[a], then (1) and (2) hold again (with possibly different constants c_0, c

$_1$,κ) for every regular, strong local Dirichlet form b(u,v) with domain D[b] = D[a] and for which the following inequality holds

(3.1) $$\lambda\, a(u,u) \le b(u,u) \le \Lambda\, a(u,u)\,,\ \forall u \in D[a].$$

Moreover it is easy to see that for every open set O and every set E with closure contained in O we have

(3.2) $$\lambda\, cap_a(E,O) \le cap_b(E,O) \le \Lambda\, cap_a(E,O).$$

From (3.2) and Theorem 2.2 it follows that the regular points for ∂X_o are the same for all forms b in the class described by (3.1).

In particular consider the class of the forms described by (1.8) then the regular points for ∂X_o in this class of forms are the same as for the form with $\alpha^{hk} = \delta^{hk}w$. The above property is an extension of the property, proved in [15], that the regular points for ∂X_o in the class of uniformly elliptic forms with bounded measurable coefficients are the same as for the form corrsponding to the Laplace operator.

In the following we bound ourself to consider the form associated with weighted Laplacian on Heisenberg group. More precisely we consider the space R^{2n+1}, whose coordinates we denote by (x,y,z) where $x,y \in R^n$, $z \in R$. We recall that the Heisenberg group of degree n denoted by H^n is the Lie group whose underlying manifold is R^{2n+1} endowed with the group law

$$(x,y,z) \circ (x',y',z') = (x+x',\ y+y',\ z+z'+ 2(x'y-xy'))\ .$$

We consider the following vector fields

$$X_j = D_{x_j} + 2y_j D_z\ ,\quad Y_j = D_{y_j} - 2x_j D_z\ ;$$

we observe that $[X_j, Y_k] = \delta^{jk} D_z$, then the vector fields X_j, Y_j satisfy an Hörmander condition of degree 1. Consider the form defined by

(3.3) $$a(u,v) = \int \sum_{j=1}^{n} (X_j u(x,y,z)\ X_j v(x,y,z)+Y_j u(x,y,z)\ Y_j v(x,y,z))\ w(x,y,z)\ dxdydz$$

with domain given by $D[a] = \{\ v \mid v, X_j v, Y_j v \in L^2(R^{2n+1},w)\ \}$.

Taking as measure on R^{2n+1} w(x,y,z)dxdydz , the form (3.3) defines a distance d (independent of w) ; moreover we have

(3.4) $$c\, d((0,0,0),(x,y,z)) \le d^*((x,y,z)) \le \frac{1}{c}\ d((0,0,0),(x,y,z))$$

where

$$d^*((x,y,z)) = (\ |(x,y)|^4 + z^2\ )\ .$$

We assume that w is in the A_2 Muckenhoupt's class for the metric d .

(b) *Points of positive capacity in the weighted case* - We are interested to weights of the type $w(x,y,z)= d^{\beta}((0,0,0),(x,y,z))$ using the equivalence between d and d* in (3.4) we easily obtain that w is in the A_2 Muckenhoupt's class for the metric d if $-\nu < \beta < \nu$ where $\nu = 2(n+1)$. Moreover the condition of positive capacity of (0,0,0) is (recalling the estimates of the Green function and its relation with the capacity of balls)

$$\int_0^1 \frac{r^2}{w(B((0,0,0),r))}\frac{dr}{r} < +\infty \tag{3.5}$$

where B((0,0,0),r) is the ball for the distance d with center (0,0,0) and radius r and

$$w(B((0,0,0),r)) = \int_{B((0,0,0),r)} d^{\beta}((0,0,0),(x,y,z))\, dxdydz\,.$$

Using again the equivalence between d((0,0,0),(x,y,z)) and d*(x,y,z) we obtain that

$$w(B((0,0,0),r)) \approx r^{2(n+1)+\beta}$$

Then (3.5) holds for $-2(n+1) < \beta < -2n$. Finally we observe that that if $\nu = 2(n+1)$ then we have $|B((0,0,0),r)| \approx r^{\nu}$ (|E| denotes the Lebesgue measure of the set E); so we can say that ν is the *intrinsic dimension at (0,0,0)*. The above condition becomes

$$-\nu < \beta < -(\nu - 2)$$

which is formally the same as in the euclidean case for degenerate elliptic operators with a weight on R^{ν}, [1][7] .

(c) *The effect of anisotropy* - We consider again the form defined by (3.3) on a bounded open set Ω with $n > 2$ and $w = d^{\beta}$ where $-(\nu - 4) < \beta \le 0$. Take as Ω the set $B((0,0,0),1)-(C_{\alpha} \cap B((0,0,0),1))$ where C_{α} is the closed cone with vertex (0,0,0), axis z and openess $0 < \alpha < \pi$. We observe that

$$B_E((0,0,0), (\tfrac{r}{c})^2) \supseteq B^*(\tfrac{r}{c}) \cap C_{\alpha} \supseteq B((0,0,0),r) \cap (\Omega)^c$$

where $B^*(r) = \{(x,y,z) \mid d^*(x,y,z) < r\}$ and $B_E((0,0,0), r)$ is the euclidean ball of center (0,0,0) and radius r , c is a suitable constant and $(\Omega)^c = X - \Omega$. Then

$$\begin{aligned} &\mathrm{cap}_a(B((0,0,0),r) \cap (\Omega)^c, B((0,0,0), k^*r)) \le \\ &\le \mathrm{cap}_a(B_E((0,0,0), (\tfrac{r}{c})^2), B((0,0,0), k^*r)) \le \\ &\le \mathrm{cap}\,(B_E((0,0,0), (\tfrac{r}{c})^2), B_E((0,0,0), (\tfrac{k^*r}{c})^2)) \end{aligned}$$

where cap_a denotes the capacity with respect to the form a and cap denotes the usual Newtonian capacity. Then

$$\mathrm{cap}_a(B((0,0,0),r) \cap (\Omega)^c, B((0,0,0), k^*r)) \le k_1\, r^{4n+2\beta}$$

with k_1 suitable constant . We have also

$$\mathrm{cap}_a(B((0,0,0),r)\,,\, B((0,0,0), k^*r)) \ge k_2 \frac{w(B((0,0,0),r))}{r^2} \ge k_3\; r^{2(n+1)+\beta}.$$

Then the Wiener integral in Theorem (2.2) converges, so (0,0,0) is an irregular point for $\partial\Omega$.

Take as $\Omega = B((0,0,0),1)-(C'_\alpha \cap B((0,0,0),1))$ where C'_α is the closed cone with vertex (0,0,0) , axis x_1 and openess $0 < \alpha < \pi$. It is easy to see that

$$B((0,0,0),r) \cap (\Omega')^c \supseteq B(P, (\frac{r}{c}))$$

where P is a suitable point with $x_1 = c'r$ $x_i = 0$, $i = 2,...,n$, $y_i = 0$, $i = 1,...,n$, z=0 with c, c' suitable constants. Then using the same methods as above we obtain that (0,0,0) is a regular point for $\partial\Omega'$.

For other connected examples of the effect of the anisotropy see [13].

References

[1] Biroli M., Marchi S. - *Wiener estimates for degenerate elliptic equations* - Diff. Int. Eq., 2 (1989), 511-523.

[2] Biroli M., Mosco U. - *Formes de Dirichlet et estimations structurelles dans les milieux discontinus* - Comptes Rendus Acad. Sc. Paris, 315, Sèr. I (1991), 193-198.

[3] Biroli M., Mosco U. - *A Saint-Venant principle for Dirichlet forms on discontinuous media* - Preprint Laboratoire d' Analyse Numerique, Université Paris VI (1993).

[4] Biroli M., Mosco U. - *Sobolev inequalities for Dirichlet forms on homogeneous spaces.* - "Boundary value problems for partial differential equations" essais for 70 birthday of E. Magenes, ed. C. Baiocchi and J.L. Lions, Masson, Paris, 1993.

[5] Biroli M., Mosco U. - Forthcoming paper

[6] Dal Maso G., Mosco U. - *Wiener criteria and energy decay for relaxed Dirichlet problems* - Arch. Rat. Mech. An., 95 (1986), 345-387.

[7] Fabes E., Jerison D., Kenig C. - *The Wiener test for degenerate elliptic equations* - Ann. Inst. Fourier, 3 (1982), 151-183.

[8] Fabes E., Kenig C., Serapioni R. - *The local regularity of solutions of degenerate elliptic equations* - Comm. in P.D.E., 7 (1982), 77-116.

[9] Fefferman C.L., Phong D. - *Subelliptic eigenvalue problems* - Conference on Harmonic Analysis, Chicago, W. Beckner etc., Wadsworth, 1981, 590-606 .

[10] Fefferman C.L., Sanchez Calle A. - *Fundamental solution for second order subelliptic operators* - Ann. of Math., 124 (1986), 247-272.

[11] Folland G.B., Stein E.M. - *Hardy spaces on homogeneous groups* - Princeton University press, Princeton, 1982.

[12] Fukushima M. - *Dirichlet forms and Markov processes* - North Holland Math. 23, North Holland and Kodansha, Amsterdam, 1980.

[13] Hansen W., Hueber H. - *The Dirichlet problem for sublaplacian on nilpotent Lie groups - Geometric criteria for regularity* - Mat. Zeit., 276 (1987), 537-547.

[14] Jerison D., Sanchez Calle A. - *Subelliptic second order differential operators* - Lec. Notes in Math. 1277, Springer Verlag, Berlin-Heidelberg-New York, 1987, 46-77.
[15] Littman W., Stampacchia G., Weinberger H. - *Regular points for elliptic equations with discontinuous coefficients* - Ann. Sc. Norm. Sup. Pisa, 17 (1963), 45-79.
[16] Lu G. - *Weighted Poincaré and Sobolev inequalities for vector fields satisfying Hörmander's condition* - Rev. Iberoamericana, 8 (1992),367-440.
[17] Nagel A., Stein E.M., Weinger S. - *Balls and metrics defined by vector fields I: Basic properties* - Acta Math., 155 (1985), 103-147.
[18] Negrini P., Scornazzani V. - *Wiener criterion for degenerate a class of degenerate elliptic operators* - J. Diff. Eq., 66 (1987), 151-164.
[19] Sanchez Calle A. - *Fundamental solution and geometry of square of vector fields* - Inv. Math., 78 (1984), 143-160.
[20] Sturm K. Th. - *Analysis on local Dirichlet spaces* - Preprint.
[21] Wiener N. *The Dirichlet problem* - J. Math. Phys., 3 (1924).

THE BEST APPROACH FOR BOUNDARY LIMITS

JÜRGEN BLIEDTNER
Fachbereich Mathematik
Universität Frankfurt
Postfach 11 19 32
60054 Frankfurt, Germany

PETER A. LOEB
Department of Mathematics
University of Illinois
1409 West Green Street
Urbana, Ill. 61801, USA

1. Introduction

In [5], J. L. Doob showed that for positive harmonic functions on the unit disk, there is no coarsest approach neighborhood system, i.e. filter, for which a Fatou boundary limit theorem holds when the filter is copied by rotation at all points of the unit circle. In [4]the authors used the principal result from [3] to show that there is a coarsest filter when the problem is suitably normalized. The normalization assigns to each positive harmonic function a zero set; this is a set of boundary points at which the function must vanish. Known limits such as those provided by the Lebesgue Differentiation Theorem or the Fine Limit Theorem force consistency in this assignment. The zero sets on the boundary are then used in constructing approach neighborhoods which are level sets in the disk. These neighborhoods form the coarsest filters for which a Fatou boundary limit theorem holds and the required zero limits achieved. The construction is new for the unit disk, but it is also valid for very general settings. It shows that any limit theorem for positive harmonic functions can be replaced with one which is at least as good (in terms of the coarseness of the filters) where the approach neighborhoods are generated by level sets of harmonic functions.

In this note, we prove in terms of our special case the general theorem needed from [3]; it is a necessary and sufficient condition for almost everywhere convergence. We then sketch and extend the result from [4] showing that when zeros are specified, the corresponding approach neighborhoods are the coarsest ones producing the appropriate limits. The interest of this result rests in part on the fact that sets from other filters, such as nontangential neighborhoods, work because they are subsets of elements of the filters constructed here. This containment, for example, can be viewed as the heart of our short proof of the Fine Limit Theorem given in [2]. The relationship between our filters and minimal fine filters is discussed in [4], as is the relationship of our limit theorem and ratio limit theorems for superharmonic functions. Along with an extension of the principal result from [4], this note contains additional results on the construction of zero sets.

K. GowriSankaran et al. (eds.), Classical and Modern Potential Theory and Applications, 105–112.

2. General setting

Let $(X, \mathcal{B})$ be a measurable space and W a domain of definition for "harmonic functions" in a linear potential theoretic setting. Examples are obtained by letting X be a compactifying boundary for a domain in a Euclidean space or a Riemann surface with ordinary harmonic functions. Of course, W can also be a domain in an axiomatic, potential theoretic setting (see [6], [1], and [8]). In these cases, $\mathcal{B}$ would usually be the σ-algebra of Borel subsets of X. Given any finite measure ν on $(X, \mathcal{B})$ and a set $E \in \mathcal{B}$, we write ν_E for the "restriction" of ν to E; i.e., $\nu_E(A) = \nu(A \cap E)$ for each $A \in \mathcal{B}$.

Generalizing the discussion in [4], we do not assume here that X is a boundary of W. We do assume that a connection exists between $(X, \mathcal{B})$ and W in terms of the following integral representation:

1) There is a convex cone M of finite measures on $(X, \mathcal{B})$, with $\nu_E \in M$ when $\nu \in M$ and $E \in \mathcal{B}$.

2) To each $\nu \in M$, there corresponds a harmonic function h_ν on W so that the map $\nu \mapsto h_\nu$ is a positive affine map on M.

3) There is a fixed "reference measure" $\sigma \in M$ with h_σ strictly positive on W.

4) If $\nu \in M$ and $h_\sigma \leq h_\nu$, then $\sigma \leq \nu$; i.e., for each $E \in \mathcal{B}$, $\sigma(E) \leq \nu(E)$.

2.1 Examples. For a simple example of such an integral representation, let X and W equal the unit circle C and the unit disk D, respectively, in the complex plane. Here, h_ν denotes the Poisson integral of ν for each Borel measure ν. Normalized Lebesgue measure λ is a reference measure with $h_\lambda \equiv 1$. Integral representations for more general settings can be found by letting X be the Martin boundary

[6], or the boundary constructed by the second author in [8]; the space of harmonic functions may be quite general. (See [1], [6], and [8]). One can also let X be the set of minimal harmonic functions without any attachment to W and use the integral representation given by Choquet theory. In this, as well as in other cases, the set M may consist of all finite Borel measures. One can also restrict M to measures absolutely continuous with respect to the reference measure σ or to measures generated by densities in $L^p(\sigma)$ for some p with $1 \leq p \leq \infty$.

Given $\nu \in M$, we will write $d\nu/d\sigma$ for the Radon-Nikodym derivative of the absolutely continuous part of ν with respect to σ.

3. Fatou-filter mappings

A boundary approach neighborhood system, i.e. filter, is like a neighborhood system in a topological space except the point in question is not in any of the sets. We work with a base for such a system. That is, the empty set is not in the base, and the intersection of any two elements of the base contains a third element of the base. The filter itself consists of all supersets of sets in the base. A function h has limit L along a filter if for any $\epsilon > 0$, there is a set in the filter (and therefore also a set in the base for the filter) such that

$|h(w) - L| < \epsilon$ for each point w in the set. The lim sup of h along a filter $\mathcal{F}$ on W is $\inf_{F \in \mathcal{F}}(\sup_{w \in F} h(w))$.

A filter is coarser than another if it is contained in the other; that is, it has fewer elements. In terms of Fatou limit theorems, the coarser the filter, the better the theorem. For the unit disk, for example, the radial limit theorem provides less information than the nontangential limit theorem.

When working with a compactifying boundary, it is usual to assume that a boundary point is in the closure of each set in the corresponding approach neighborhood system. In this note, we will not use any topological relationship between X and W even when one exists. We will use a correspondence between points of X and filter bases on W with the following additional property.

3.1 Definition. *By a* Fatou filter mapping *we mean a function $\mathcal{G} : x \mapsto \mathcal{G}_x$ associating with each point $x \in X$ a filter base $\mathcal{G}_x$ on W so that for each $\nu \in M$, the ratio h_ν / h_σ has limit $d\nu/d\sigma(x)$ along $\mathcal{G}_x$ (i.e., along the filter generated by $\mathcal{G}_x$) for σ-a.e. $x \in X$. That is,*

$$\lim_{\mathcal{G}_x} \frac{h_\nu}{h_\sigma} = \frac{d\nu}{d\sigma}(x) \quad \text{for } \sigma\text{-a.e. } x \in X.$$

Given a filter mapping initially defined only σ-a.e. on X, we will assume that the mapping has been extend to the rest of X with the filter containing only the set W. We say that a Fatou filter mapping $\mathcal{F}$ is coarser than a Fatou filter mapping $\mathcal{G}$ if at each $x \in X$ the filter generated by $\mathcal{F}_x$ is contained in the filter generated by $\mathcal{G}_x$; we do not require the containment to be strict. For harmonic functions on the unit disk, the radial approach, nontangential approach, and minimal fine approach are all examples of Fatou filter mappings.

What must be established to show that a filter mapping is a Fatou filter mapping? A simple answer is given by the general reduction technique in [3]. We now state and prove a special case of that result using the following notation: $\{h_\nu < h_\sigma\}$ denotes the set $\{w \in W : h_\nu(w) < h_\sigma(w)\}$; $f\sigma$ denotes the measure generated by σ and a density f; CE denotes the complement of a set E in X; $\mathbf{N}$ denotes the set of natural numbers. We will write $\sigma(A) = 0$ if this is true for A in the σ-completion of $\mathcal{B}$.

3.2 Theorem. *Let $\mathcal{G} : x \mapsto \mathcal{G}_x$ be a mapping associating with each point $x \in X$ a filter base $\mathcal{G}_x$ on W. Then the following are equivalent:*

(1) $\mathcal{G}$ is a Fatou filter mapping.

(2) For each $E \in \mathcal{B}$ with $\sigma(E) > 0$ and each $\nu \in M$ with $\nu(E) = 0$,

$$\lim_{\mathcal{G}_x} \frac{h_\nu}{h_\sigma} = 0 \quad \text{for } \sigma\text{-a.e. } x \in E.$$

(3) For each $E \in \mathcal{B}$ with $\sigma(E) > 0$ and each $\nu \in M$ with $\nu(E) = 0$,

$$\{h_\nu < h_\sigma\} \in \mathcal{G}_x \quad \text{for } \sigma\text{-a.e. } x \in E.$$

Proof. (1 $\Rightarrow$ 3) Assume 1 holds and fix $E \in \mathcal{B}$ with $\sigma(E) > 0$ and $\nu \in M$ with $\nu(E) = 0$. Then $\lim_{\mathcal{G}_x} \frac{h_\nu}{h_\sigma} = \frac{d\nu}{d\sigma}(x) = 0$ for σ-a.e. $x \in E$, so

$$\{h_\nu < h_\sigma\} \in \mathcal{G}_x \text{ for } \sigma\text{-a.e. } x \in E.$$

(3 $\Rightarrow$ 2) Given $E \in \mathcal{B}$ with $\sigma(E) > 0$ and $\nu \in M$ with $\nu(E) = 0$, we have for each $k \in \mathbf{N}$, $k\nu \in M$ and $k\nu(E) = 0$, whence

$$\{h_\nu/h_\sigma < 1/k\} = \{h_{k\nu}/h_\sigma < 1\} \in \mathcal{G}_x$$

for σ-a.e. $x \in E$. Therefore,

$$\lim_{\mathcal{G}_x} \frac{h_\nu}{h_\sigma} = 0 \text{ for } \sigma\text{-a.e. } x \in E.$$

(2 $\Rightarrow$ 1) Given $\mu \in M$, fix a nonnegative, integrable, real-valued function f on X and a measure $\nu \perp \sigma$ so that $\mu = f\sigma + \nu$. The Lebesgue decomposition of X provides disjoint measurable sets A and B with $X = A \cup B$ such that $\sigma(B) = \nu(A) = 0$. Since μ_A and μ_B are both in M, $f\sigma$ and ν are both in M. By assumption, $\lim_{\mathcal{G}_x} \dfrac{h_\nu}{h_\sigma} = 0$ for σ-a.e. $x \in A$, and thus for σ-a.e. $x \in X$. We need only show, therefore, that for some set $U \in \mathcal{B}$ with $\sigma(U) = 0$ and for all $x \in X \setminus U$, $\lim_{\mathcal{G}_x} \dfrac{h_{f\sigma}}{h_\sigma} = f(x)$.

Choose $n \in \mathbf{N}$, and partition the nonnegative real line into intervals of length $1/4n$. Let $E \subseteq X$ be the inverse image with respect to f of one of the intervals $[r, r + 1/4n]$. If $\sigma(E) > 0$, then for each $x \in E$ and each $w \in W$ we have

$$h_\sigma(w) \cdot |h_{f\sigma}(w)/h_\sigma(w) - f(x)| = |h_{f\sigma}(w) - f(x) \cdot h_\sigma(w)|$$

$$\leq |h_{f\sigma}(w) - r \cdot h_\sigma(w)| + |r \cdot h_\sigma(w) - f(x) \cdot h_\sigma(w)|$$

$$\leq |h_{f\sigma_{CE}}(w) - h_{r\sigma_{CE}}(w)| + |h_{f\sigma_E}(w) - h_{r\sigma_E}(w)| + (f(x) - r)h_\sigma(w)$$

$$\leq h_{f\sigma_{CE}}(w) + h_{r\sigma_{CE}}(w) + \tfrac{1}{4n} \cdot h_{\sigma_E}(w) + \tfrac{1}{4n} \cdot h_\sigma(w).$$

Since $h_\sigma(w) = h_{\sigma_E}(w) + h_{\sigma_{CE}}(w)$, we have $h_{\sigma_E}(w)/h_\sigma(w) \leq 1$, so

$$\left|\frac{h_{f\sigma}(w)}{h_\sigma(w)} - f(x)\right| \leq \frac{h_{f\sigma_{CE}}(w)}{h_\sigma(w)} + \frac{h_{r\sigma_{CE}}(w)}{h_\sigma(w)} + \frac{1}{2n}.$$

By assumption, for σ-almost all $x \in E$,

$$\lim_{\mathcal{G}_x} \frac{h_{f\sigma_{CE}}(w)}{h_\sigma(w)} = 0 \quad \text{and} \quad \lim_{\mathcal{G}_x} \frac{h_{r\sigma_{CE}}(w)}{h_\sigma(w)} = 0.$$

Therefore, for σ-a.e. $x \in E$, there is a set G_x^n in $\mathcal{G}_x$ such that $|h_{f\sigma}(w)/h_\sigma(w) - f(x)| < 1/n$ for all $w \in G_x^n$. Let U_n contain the exceptional points for each partition set E with $\sigma(E) > 0$ and every point of any partition set E with $\sigma(E) = 0$. The set $U = \bigcup_{n \in \mathbf{N}} U_n$ is the desired set of σ-measure 0 in X. □

4. Zero sets

To apply the previous result, we specify for each measure $\nu \in M$ the set of points of X where a zero limit is required. That is, we assume the existence of a mapping $Z : \nu \mapsto Z_\nu$ from M into the σ-completion of $\mathcal{B}$; we call each Z_ν the *zero set* for ν. These sets only contain points where zeros must occur.

There are many ways that a zero-set mapping Z can be obtained. We assume in general that the following properties are satisfied for all ν, $\mu \in M$, all $E \in \mathcal{B}$, and all $\alpha > 0$ in $\mathbf{R}$:

$$\text{i) } Z_\nu \cap Z_\mu \subseteq Z_{\nu+\mu}, \text{ ii) } Z_{\alpha\nu} = Z_\nu, \text{ iii) } Z_\nu = \emptyset \text{ if } \sigma \leq \nu, \text{ iv) } Z_0 = X,$$

$$\text{v) if } \nu(E) = 0, \text{ then } \sigma(E \setminus Z_\nu) = 0.$$

4.1 Example. One can produce zero sets by using a Fatou filter mapping such as the mapping for radial, nontangential, or minimal fine limits on the disk. Here for each $\nu \in M$ and $x \in X$, one takes the lim sup of the ratio h_ν / h_σ along the filter associated with x; the set Z_ν contains x if the result is 0. This example illustrate how, in practice, the selection of a zero set mapping Z takes into account the singular part of ν. We will consider other examples in Sections 6 and 7.

5. The best Fatou-filter mappings

Before enlarging on the construction of zero sets, we sketch the proof from [4] showing that when such sets are given, there is a best Fatou filter mapping which produces zeros at the points of Z_ν for each $\nu \in M$. These filters are defined in terms of level sets of harmonic functions.

5.1 Theorem. *Given a zero-set mapping Z, for each $x \in X$, set*

$$\mathcal{F}_x := \{\{h_\nu < h_\sigma\} : \nu \in M,\, x \in Z_\nu\}.$$

The mapping $\mathcal{F}$ is the coarsest Fatou filter mapping on X with the property that for each $\nu \in M$,

$$Z_\nu \subseteq \{x \in X : \lim_{\mathcal{F}_x} \frac{h_\nu}{h_\sigma} = 0\}.$$

If for $\nu \in M$, h_ν / h_σ has limit 0 along $\mathcal{F}_y$ for some $y \in X$, then $\{h_\nu < h_\sigma\}$ is in the filter generated by $\mathcal{F}_y$.

Proof. First, fix $\mathcal{F}_x$. By Condition iii, $\emptyset \notin \mathcal{F}_x$. It follows from Condition i that the intersection of elements $\{h_\nu < h_\sigma\}$ and $\{h_\mu < h_\sigma\}$ is again in $\mathcal{F}_x$ since $\{h_{\nu+\mu} < h_\sigma\} \subseteq \{h_\nu < h_\sigma\} \cap \{h_\mu < h_\sigma\}$. This means that $\mathcal{F}_x$ is a filter base. Now, fix $E \in \mathcal{B}$ and $\nu \in M$ with $\nu(E) = 0$. Since $\sigma(E \setminus Z_\nu) = 0$, $\{h_\nu < h_\sigma\} \in \mathcal{F}_x$ for σ-a.e. $x \in E$. By Theorem 3.2, $\mathcal{F}$ is a Fatou filter mapping. If $\nu \in M$ and $x \in Z_\nu$, then for each $n \in \mathbf{N}$, $\{nh_\nu < h_\sigma\} \in \mathcal{F}_x$, so h_ν / h_σ has limit 0 along $\mathcal{F}_x$. If $\mathcal{G}_x$ is also a filter associated with x along which h_ν / h_σ has limit 0, then $\{h_\nu < h_\sigma\} \in \mathcal{G}_x$. Therefore, $\mathcal{F}$ is the coarsest Fatou filter mapping on X yielding the desired zeros. If $y \in X$ and h_ν / h_σ has limit 0 along

$\mathcal{F}_y$, then there is a set $F \in \mathcal{F}_y$ on which $h_\nu/h_\sigma < 1$. In this case, the set $\{h_\nu < h_\sigma\}$ is at least in the filter generated by $\mathcal{F}_y$. □

5.2 Remark. Suppose the map $\mathcal{F}$ of Theorem 5.1 is used to produce new zero sets $\widetilde{Z}_\nu$ using the construction of Example 4.1. Then for each $x \in X$, the corresponding filter base $\widetilde{\mathcal{F}}_x$ may be larger than $\mathcal{F}_x$, but the filters these bases generate will be the same. Therefore, the zero sets and the filter bases will not change if the process is repeated.

6. Zero sets and measure differentiation

A way to construct zero sets when X has a differentiation basis is to use the Lebesgue Differentiation Theorem to find a representative of $d\nu/d\sigma$ for each $\nu \in M$ (see [3]). For this construction, $Z_0 = X$, and for each nonzero $\nu \in M$, the set Z_ν consists of those points in X where the derivative of ν is defined and equals 0.

For an appropriate metric space, the Lebesgue Differentiation Theorem can be applied using closed metric balls. To show this in general, we consider "τ-satellite configurations" of balls. All one needs to know here is that there is a version of the Besicovitch covering theorem when the number of balls which can form such a configuration is bounded. Also, all the balls in such a configuration intersect a "central ball". For the interested reader, here is a full definition with ρ the metric on X.

6.1 Definition. *Fix $\tau > 1$ and $I_n = \{1, 2, \cdot\cdot, n\} \subset \mathbf{N}$. Let $\{B(\mathbf{c}_i, r_i) : i \in I_n\}$ be an ordered set of closed balls with one of the balls $B(\mathbf{c}_{i_0}, r_{i_0})$ being distinguished and called the central ball. We say that this set of balls is in τ-satellite configuration if the following conditions hold for each $i \in I_n$:*

(1) $B(\mathbf{c}_{i_0}, r_{i_0}) \cap B(\mathbf{c}_i, r_i) \neq \emptyset$,

(2) $\tau \cdot r_i > r_{i_0}$,

(3) if $1 \leq i < j \leq n$, then $\rho(\mathbf{c}_i, \mathbf{c}_j) > r_i > r_j/\tau$.

6.2 Theorem. *Assume that X is a metric space such that for some $n \in \mathbf{N}$, each $x \in X$ is the center of a closed metric ball $B(x, p(x))$ which can be isometrically imbedded into an n-dimensional normed vector space (the norm may vary with the point.) Let $Y \subseteq X$ be the support of the reference measure σ. Then for each Borel measure μ on X and for σ-almost every $y \in Y$,*

$$\lim_{r \to 0} \frac{\mu(B(y,r))}{\sigma(B(y,r))} = \frac{d\mu}{d\sigma}(y).$$

Proof. If X is compact, or just separable, one can use the fact that the result is true in a ball about each point, and put sets of measure 0 together in a single null set. In general, one can show directly that balls form a differentiation basis on X. Choose a number $\tau > 1$ sufficiently close to 1 so that the bounds given in [7] for the Besicovitch Covering Theorem are applicable. This means that in an n -dimensional normed vector space there can be at most 5^n closed balls in τ-satellite configuration. Let $\mathcal{S} = \{B(x_i, r_i) \subseteq X : 1 \leq i \leq n\}$ be a τ-satellite configuration of closed balls in X. Assume that for each index i, $r_i < p(x_i)/5$.

Let i_m be the first index such that r_{i_m} is maximum for the collection $\mathcal{S}$. Since the balls in $\mathcal{S}$ all intersect the central ball of $\mathcal{S}$, they are all contained in $B(x_{i_m}, p(x_{i_m}))$. Therefore, there are at most 5^n balls in $\mathcal{S}$. The result now follows from Theorem 5.7 of [3]. □

6.3 Examples. For certain Martin boundaries, it is easier to find a local rather than a uniform imbedding into a finite dimensional normed vector space. For example, let W be the open unit disk $\{|z| < 1\}$ with the closed disk $\{|z| \leq \frac{1}{2}\}$ and the line segment joining $\frac{1}{2}$ and 1 removed. The Martin compactification $\overline{W}$ of W replaces the line segment from $\frac{1}{2}$ to 1 with two such segments. A length metric on $\overline{W}$ sets the distance between two points equal to the infimum of the lengths of paths joining the points; only paths which do not cross the positive x-axis are considered. Locally, this is the same as the metric in the plane.

A modification of this example employs a length metric on just the boundary; only paths lying in the boundary are considered. With this metric, the boundary is locally isometric to the real line. This construction is also suitable for the Martin boundary of the open unit disk with the closed line segment joining 0 and 1 removed.

7. Zero sets for densities

In this concluding section, we work with nonnegative, quasibounded harmonic functions; these are the monotone limits of nonnegative bounded harmonic functions. We assume that h_σ is bounded and perhaps even 1. We use f, and g to denote nonnegative, σ-integrable, real-valued functions on X. Often, we will write h_f for $h_{f\sigma}$ and Z_f for $Z_{f\sigma}$.

To start, we assume that zero sets have been selected for the harmonic functions with bounded densities so that Conditions i-v of Section 4 are satisfied. We also assume that the bigger the density, the smaller the zero set; i.e.,

$$h_f \leq h_g \Rightarrow Z_f \supseteq Z_g.$$

This assumption is satisfied if we use boundary limits to define the zero sets. Another possibility is to choose an appropriate lifting of $L^\infty(\sigma)$ and let the zero set for each bounded density be the set of points in the support of σ where the corresponding lifting is 0. Again, however, we set $Z_0 = X$.

7.1 Definition. *For each nonnegative, quasibounded harmonic function with density f, let $f_n = f \wedge n$, and set $Z_f = \bigcap_n Z_{f_n}$.*

It is clear that for an essentially bounded f, this definition of Z_f agrees with the initial one since as n increases, the zero sets Z_{f_n} decrease and eventually equal the original Z_f. It is also clear that if $f = g$ σ-a.e., then $Z_f = Z_g$. We now show that the mapping Z is a zero-set mapping.

7.2 Theorem. *Let $M = \{f\sigma : f \geq 0$ and σ-integrable$\}$. The mapping $f\sigma \mapsto Z_f$ on M satisfies Conditions i-v of Section 4.*

Proof. Let $f \geq 0$ be a σ-integrable function on X.

i) For any σ-integrable $g \geq 0$, we have

$$Z_f \cap Z_g = \left(\bigcap_n Z_{f_n}\right) \cap \left(\bigcap_n Z_{g_n}\right) = \bigcap_n (Z_{f_n} \cap Z_{g_n}) \subseteq \bigcap_n Z_{f_n+g_n} \subseteq \bigcap_n Z_{(f+g)_n} = Z_{f+g}.$$

ii) For any positive number α and $n \in \mathbf{N}$, $Z_{(\alpha f)\wedge n} = Z_{\alpha(f \wedge \frac{n}{\alpha})} = Z_{f\wedge\frac{n}{\alpha}}$. For each $n \in \mathbf{N}$, there is an $m \in \mathbf{N}$, so that $\frac{n}{\alpha} \leq m$. Also, for each $m \in \mathbf{N}$, there is an $n \in \mathbf{N}$ so that $m \leq \frac{n}{\alpha}$. It follows that

$$Z_{\alpha f} = \bigcap_{n \in \mathbf{N}} Z_{f\wedge\frac{n}{\alpha}} = \bigcap_{m\in\mathbf{N}} Z_{f\wedge m} = Z_f.$$

iii) If $\sigma \leq f\sigma$, then $f \geq 1$ σ-a.e., so $(f \wedge 1)\sigma = \sigma$. It follows that $Z_f = \emptyset$.
iv) This is clear.
v) If $f = 0$ σ-a.e. on E, then for each $n \in \mathbf{N}$, $\sigma(E \setminus Z_{f_n}) = 0$, so

$$\sigma(E \setminus Z_f) = \sigma(E \setminus \bigcap_n Z_{f_n}) = \sigma(\bigcup_n E \setminus Z_{f_n}) = 0. \quad \square$$

The second author's work was supported in part by a grant from the U.S. National Science Foundation. The authors thank Richard Bishop for a helpful conversation.

References

[1] J. Bliedtner and W. Hansen, *Potential Theory,* Springer-Verlag, Berlin, 1986.

[2] J. Bliedtner and P. A. Loeb, A measure-theoretic boundary limit theorem, *Archiv der Math.* 43(1984), 373-376.

[3] ———, A reduction technique for limit theorems in analysis and probability theory, *Arkiv för Mat.* 30(1992), No. 1, 25-43.

[4] ———, Best filters for the general Fatou boundary limit theorem, to appear in *Proc. Amer. Math. Soc.*

[5] J. L. Doob, Boundary Approach Filters for Analytic Functions, *Ann. Inst. Fourier, Grenoble*, 23(1973), No. 3, 187-213.

[6] ———, *Classical potential theory and its probabilistic counterpart,* Springer-Verlag, Berlin, 1984.

[7] Z. Füredi and P. A. Loeb, On the best constant for the Besicovitch covering theorem. To appear in *Proc. Amer. Math. Soc.*

[8] P. A. Loeb, A regular metrizable boundary for solutions of elliptic and parabolic differential equations, *Math. Annalen* 251(1990), 43-50.

Fine Behaviour of Balayages in Potential Theory

NICU BOBOC
Faculty of Mathematics
University of Bucharest, str. Academiei 14
RO-70109 Bucharest, Romania

ABSTRACT. The behaviour of the balayage on the complement of a fine open set is investigated near an irregular boundary point. We extend also the classical Harnack inequality. We generalise essentially, in the frame of excessive functions, results of M. Brelot, E. Smyrnelis, H. Bauer, W. Hansen and I. Netuka obtained in the context of harmonic spaces.

Introduction

We consider a standard H-cone of functions $\mathcal{S}$ on a set X (i.e. $\mathcal{S}$ is the set of all excessive functions with respect to a submarkovian resolvent absolutely continuous on the metrisable space X). For any fine open set U of X we denote by $\widetilde{U}$ the set of all points $x \in X$ for which $X \setminus U$ is thin at x and by $\mathcal{S}(U)$ the localization of $\mathcal{S}$ on U (i.e. the set of all positive functions t on U such that t is finite on a fine dense subset of U and such that there exists a sequence $(s_n)_n$ in $\mathcal{S}$, s_n finite for all $n \in \mathbb{N}$ for which the sequence $(s_n - B^{X \setminus U} s_n)_n$ increases to t on U). Any function $t \in \mathcal{S}(U)$ may be extended to a fine continuous function on $\widetilde{U}$ which is also denoted by t. We remark that if X a harmonic space and $\mathcal{S}$ is the convex cone of all positive superharmonic functions on X then $\mathcal{S}(U)$ means the set of all positive fine superharmonic functions on U.

It is proved that for any positive Borel function f on X the function $B^{X \setminus U} f$ is fine continuous at any point $x \in U$.

Let now $(x_n)_n$ be a sequence of U which converges in X to a point $z \in U$. The sequence $(x_n)_n$ is called maximal if we have

$$\lim_{n \to \infty} B^{X \setminus U} s(x_n) = B^{X \setminus U} s(z)$$

for any universally continuous element s in $\mathcal{S}$. It is proved that the sequence $(x_n)_n$ will be maximal iff $(x_n)_n$ converges to z with respect to the natural topology on U associated with the H-cone $\mathcal{S}(U)$. Therefore for any positive Borel function f on X

K. GowriSankaran et al. (eds.), *Classical and Modern Potential Theory and Applications*, 113–123.

dominated by an element $s \in \mathcal{S}$ which is finite continuous at z we have

$$\lim_{n \to \infty} B^{X \setminus U} f(x_n) = B^{X \setminus U} f(z).$$

Particularly we have this last relation for any positive bounded Borel function f on X.

Finally if $z \in \widetilde{U}$ then it is proved that if U is a Doob set then there exists a fine neighbourhood V of z and a positive real function c on V such that

$$\varepsilon_y^{X \setminus U} \leq c(y) \cdot \varepsilon_z^{X \setminus U}, \ (\forall) y \in V$$

where $\varepsilon_y^{X \setminus U}$ means the balayage of the Dirac measure ε_y on the set $X \setminus U$. Such an inequality extend in a more general frame the well known Harnack inequality.

These assertions extend some similar results obtained in the classical potential theory by M. Brelot ([4], [5]) and in the frame of harmonic spaces, under various generality, by E. Smyrnelis ([9]), H. Bauer ([1]), W. Hansen ([6]) and I.Netuka ([8]).

1. Preliminaries

In all this paper $\mathcal{S}$ will be a standard H-cone of functionon a set X. ([2]).
We recall the following:

a) On X are distinguished two topologies which are strongly related with $\mathcal{S}$. The first one, $\tau_o(X)$, called the natural topology, is the coarsest topology on X such that any universally continuous element of $\mathcal{S}$ is continuous. The second one, called the fine topology, is the coarsest topology on X such that any $s \in \mathcal{S}$ is continuous. The set X endowed with the natural topology is a metrisable space with countable basis. The set X is called saturated if any H-integral on $\mathcal{S}$ which is finite on the function 1 is represented as a finite measure on X; the set X is called semisaturated if any H-integral on $\mathcal{S}$ which is dominated by an universally continuous H-measure is represented as a measure on X. In fact X will be semisaturated iff the set of all H-integrals on $\mathcal{S}$ which are represented as measures on X (i.e. the set of all H-measures) is solid with respect to the natural order in the set $\mathcal{S}^*$ of all H-integrals on $\mathcal{S}$. The set X is called nerly saturated if any universally continuous H-integral on $\mathcal{S}$ is represented as a measure on X. It is known that always there exists a set $X_1 \supset X$ such that $\mathcal{S}$ is a standard H-cone of functions on X_1 and such that X_1 is saturated; generally the set X is fine dense in X_1.

b) If A is a subset of X we denote by B^A the map on $\mathcal{S}$ into $\mathcal{S}$ defined by

$$B^A s = \bigwedge \{s' \in \mathcal{S} / \ s \leq s' \text{ on } A\}.$$

It is known that for any $x \in X$ the map

$$x \mapsto B^A s(x)$$

is an H-integral on $\mathcal{S}$ dominated by ε_x. Hence if X is semisaturated then the above H-integral is represented as a measure on X denoted by ε_x^A. In the sequel if f is a positive Borel function on X we denote by $B^A f$ the function on X given by

$$B^A f(x) = \varepsilon_x^A(f).$$

c) If A is a subset of X then A is called polar if $B^A s = 0$ for any $s \in \mathcal{S}$. The set A is called thin at x if there exists $s \in \mathcal{S}$ such that $B^A s(x) < s(x)$. The set A is called semipolar if it is a countable union of totally thin sets.

Let now X be a nearly saturated set (with respect to $\mathcal{S}$) and let X_1 be the saturated set (with respect to $\mathcal{S}$) such that $X \subset X_1$. Then any Borel measurable subset of $X_1 \setminus X$ is semipolar and this property characterizes the fact that X is nearly saturated. Moreover in this case a subset A of X will be semipolar iff A is semipolar as subset of X_1. If X is semisaturated then any Borel measurable subset of $X_1 \setminus X$ is polar and this property characterizes the fact that X is semisaturated ([3]).

d) Suppose now that X is nearly saturated and let U be a fine open subset of X, $U \neq \emptyset$. We denote by $\mathcal{S}(U)$ the convex cone of all positive functions t on U such that t is finite on a fine dense subset of U and such that

$$t = \sup\{s - B^{X\setminus U}s / \ s \in \mathcal{S}_b, \ s - B^{X\setminus U}s \leq t\}$$

where $\mathcal{S}_b = \{s \in \mathcal{S} / \ s \text{ bounded }\}$. It is known ([3]) that $\mathcal{S}(U)$ is a standard H-cone of functions on U. If X is semisaturated (with respect to $\mathcal{S}$) then U is also semisaturated (with respect to $\mathcal{S}(U)$). The H-cone $\mathcal{S}(U)$ is called the localization of $\mathcal{S}$ on U. Moreover if $s \in \mathcal{S}$, $t \in \mathcal{S}$, $t < \infty$ and $t \leq s$ then

$$(s - B^{X\setminus U}t)_{|U} \in \mathcal{S}(U).$$

Particularly $s_{|U} \in \mathcal{S}(U)$ for any $s \in \mathcal{S}$. We remark([7]) that if $t \in \mathcal{S}(U)$ and $s \in \mathcal{S}_b$ then the element from $\mathcal{S}(U)$ given by $t \wedge (s - B^{X\setminus U}s)$ is of the form $s' - B^{X\setminus U}s'$ where $s' \in \mathcal{S}_b$, $s' \leq s$.

e) Suppose that $\mathcal{S}$ is a standard H-cone of functions on a set X. Then for any increasing family $(s_i)_i$ from $\mathcal{S}$ the function $\sup_i s_i$ is fine continuous. Indeed, for any $n \in \mathbb{N}$ we have

$$\inf_n(\sup_{i\in I} s_i, n) = \sup_{i\in I}(\inf_n(s_i, n))$$

and the assertion follows from the fact that

$$\sup_{i\in I}(\inf_n(s_i, n)) \in \mathcal{S}.$$

If the elements of $\mathcal{S}$ are called usually superharmonic on X the function of the form $\sup_{i\in I} s_i$, where $(s_i)_{i\in I}$ is an increasing family in $\mathcal{S}$, is called hyperharmonic on X.

2. The standard H-cone $\mathcal{S}(U)$ as a standard H-cone of functions on the extension set $\widetilde{U}$

In this paragraph if U is a fine open subset of X we denote by $\widetilde{U}$ the set of all $z \in X$ such that $X \setminus U$ is thin at z. Obviously $\widetilde{U}$ is fine open and U is a fine dense subset $\widetilde{U}$. We intend to represent $\mathcal{S}(U)$ as a standard H-cone of functions on $\widetilde{U}$. In fact any element $t \in \mathcal{S}(U)$ is represented as the fine continuous extension on $\widetilde{U}$ of the function t.

Theorem 1. *Any element $t \in \mathcal{S}(U)$ may be extended to a fine continuous positive function $\tilde{t}$ on $\widetilde{U}$. In this way $\mathcal{S}(U)$ is represented as a standard H-cone of functions on the set $\widetilde{U}$ such that the subset $\widetilde{U}\setminus U$ is polar with respect to this standard H-cone of functions.*

Proof. The first part of the theorem follows using the fact that the convex cone of the form $(s - B^{X\setminus U}s)_{|U}$ where $s \in \mathcal{S}_b$ is a solid subcone of $\mathcal{S}(U)$ which is increasingly dense in $\mathcal{S}(U)$ and contains a strictly positive element, namely $(s_o - B^{X\setminus U}s_o)_{|U}$, where s_o is a bounded continuous generator of $\mathcal{S}$. Indeed any element of the form $(s - B^{X\setminus U}s)_{|U}$ where $s \in \mathcal{S}_b$ has a fine continuous extension on $\widetilde{U}$ (ie. $(s - B^{X\setminus U}s)_{|\widetilde{U}}$ and the function $(s_o - B^{X\setminus U}s_o)_{|\widetilde{U}}$ is a strictly positive function on $\widetilde{U}$. Now if $t \in \mathcal{S}$ we have

$$t(x) = \sup_n(\inf(t, n(s_o - B^{X\setminus U}s_o)(x)))\ ,\ (\forall)x \in U.$$

Since there exists $s_n \in \mathcal{S}_b$, $s_n \leq ns_o$ such that

$$\inf(t, n(s_o - B^{X\setminus U}s_o))(x) = (s_n - B^{X\setminus U}s_n)(x),\ (\forall)x \in U$$

then the element

$$t_n = \inf(t, n(s_o - B^{X\setminus U}s_o))$$

has a fine continuous extension $\tilde{t}_n$ on $\widetilde{U}$. The function

$$\tilde{t} = \sup_n \tilde{t}_n$$

is a fine continuous extension of t because

$$\inf(\tilde{t}, n(s_o - B^{X\setminus U}s_o)) = \tilde{t}_n \text{ on } \widetilde{U}.$$

From the above definition of $\tilde{t}$ it follows immediately that we have

$$\widetilde{t_1 + t_2} = \tilde{t}_1 + \tilde{t}_2\ ,\ \widetilde{\inf(t_1, t_2)} = \inf(\tilde{t}_1, \tilde{t}_2)\ ,\ t_i \nearrow t \Rightarrow \tilde{t}_1 \nearrow \tilde{t}$$

and therefore replacing t by $\tilde{t}$ the H-cone $\mathcal{S}(U)$ is represented as a standard H-cone of functions on the set $\widetilde{U}$.

Let now s_o be a bounded continuous generator of $\mathcal{S}$. To prove that $\widetilde{U}\setminus U$ is polar in $\widetilde{U}$ with respect to the standard H-cone of functions $\mathcal{S}(U)$ on $\widetilde{U}$ it will be sufficient to show that

$$^{\widetilde{U}}B^{\widetilde{U}\setminus U}(s_o - B^{X\setminus U}s_o) = 0$$

where ${}^{\widetilde{U}}B^A$ means the balayage on the subset A of $\widetilde{U}$ with respect to the H-cone of functions $\mathcal{S}(U)$ on $\widetilde{U}$.

Let $(G_n)_n$ be a sequence of fine open subsets of X such that $G_n \supset X \setminus U$ and such that

$$\bigwedge B^{G_n} s_o = B^{X\setminus U} s_o.$$

We have

$$B^{G_n} s_o - B^{X\setminus U} s_o = B^{G_n} s_o - B^{X\setminus U}(B^{G_n} s_o) \in \mathcal{S}(U),$$
$$B^{G_n} s_o - B^{X\setminus U} s_o \geq s_o - B^{X\setminus U} s_o \text{ on } G_n \cap \widetilde{U}$$

and therefore

$${}^{\widetilde{U}}B^{\widetilde{U}\setminus U}(s_o - B^{X\setminus U} s_o) \leq B^{G_n} s_o - B^{X\setminus U} s_o, (\forall) n \in \mathbb{N}\ ,\ {}^{\widetilde{U}}B^{\widetilde{U}\setminus U}(s_o - B^{X\setminus U} s_o) = 0.$$

Theorem 2. *Suppose that X is semisaturated and let f be a positive Borel function on X. Then for any fine open subset U of X the function $B^{X\setminus U} f$ is fine continuous at any point of $\widetilde{U}$. If moreover f is dominated by a element $s \in \mathcal{S}$ then we have*

$$B^{X\setminus U} f|_{\widetilde{U}} \in \mathcal{S}(U)\ ,\ B^{X\setminus U} f|_{\widetilde{U}} \preccurlyeq_{\mathcal{S}(U)} s|_{\widetilde{U}}.$$

Proof. Suppose that f is of the form $f = t' - t''$ where $t', t'' \in \mathcal{S}$, $t'' \leq t'$, t'' finite. We want to show that $B^{X\setminus U} f|_{\widetilde{U}} \in \mathcal{S}(U)$. Indeed let $s \in \mathcal{S}$ be such that $f \leq s$. For any $s' \in \mathcal{S}$ such that $s' \geq t'$ on $X \setminus U$ we have

$$s' \geq s' \wedge t'\ ,\ B^{X\setminus U}(s' \wedge t'') = B^{X\setminus U} t''$$

and therefore $(s' - B^{X\setminus U} t'')|_U \in \mathcal{S}(U)$. Since s' is arbitrary we get

$$B^{X\setminus U} t''|_U \preccurlyeq_{\mathcal{S}(U)} \bigwedge\{s'|_U /\ s' \in \mathcal{S},\ s' \geq t' \text{ on } X \setminus U\} = B^{X\setminus U} t'|_U.$$

Hence $B^{X\setminus U} f|_U \in \mathcal{S}(U)$. From $B^{X\setminus U} f = B^{X\setminus U} t' - B^{X\setminus U} t''$ it follows that $B^{X\setminus U} f$ is fine continuous on X and therefore on $\widetilde{U}$.

Let now F_o be the set of all bounded functions f on X such that there exists an increasing sequence $(f_n)_n$ in $(\mathcal{S}_b - \mathcal{S}_b)_+$ with $f = \sup_n f_n$. If $f \in F_o$ and $(f_n)_n$ is as above then $B^{X\setminus U} f_n \nearrow B^{X\setminus U} f$. From the previous considerations we have $B^{X\setminus U} f_n|_{\widetilde{U}} \in \mathcal{S}(U)$ and since $B^{X\setminus U} f$ is bounded we deduce that $B^{X\setminus U} f|_{\widetilde{U}} \in \mathcal{S}(U)$. Particularly $B^{X\setminus U} f$ is fine continuous on $\widetilde{U}$. We remark also that if $f, g \in F_o$, $f \leq g$ then

$$B^{X\setminus U}(g - f)|_{\widetilde{U}} \in \mathcal{S}(U).$$

Indeed let $(f_n)_n$ be an increasing sequence in $(\mathcal{S}_b - \mathcal{S}_b)_+$ such that $\sup_n f_n = f$. Since $g - f_n \in F_o$ we have $B^{X\setminus U}(g - f_n)|_{\widetilde{U}} \in \mathcal{S}(U)$ and $(B^{X\setminus U}(g - f_n)|_{\widetilde{U}})_n$ is specifically decreasing in $\mathcal{S}$. Hence

$$B^{X\setminus U}(g - f)|_{\widetilde{U}} = \inf_n B^{X\setminus U}(g - f_n)|_{\widetilde{U}} \in \mathcal{S}(U).$$

Suppose now that f is a positive bounded Borel function on X. We have $B^{X\setminus U}f = \inf\{B^{X\setminus U}g/\ g \in F_o,\ g \geq f\}$ and since the family $(B^{X\setminus U}g|_{\tilde{U}})_{g\in F_o, g\geq f}$ is specifically decreasing in $\mathcal{S}(U)$ we get

$$B^{X\setminus U}f|_{\tilde{U}} = \inf\{B^{X\setminus U}g|_{\tilde{U}}/\ g \in F_o,\ g \geq f\} \in \mathcal{S}(U).$$

Particularly $B^{X\setminus U}f$ is fine continuous on $\tilde{U}$.

Suppose that f is a positive Borel function on X. We have

$$B^{X\setminus U}f = \sup_n B^{X\setminus U}(\inf(f,n)).$$

From the previous considerations it follows that $B^{X\setminus U}f|_{\tilde{U}}$ is hyperharmonic on U (with respect to the H-cone $\mathcal{S}(U)$ and therefore it is fine continuous on U. Moreover if f is dominated by an element s of $\mathcal{S}$ then $B^{X\setminus U}f|_{\tilde{U}}$ is is dominated by $s|\tilde{U}$ and therefore belongs to $\mathcal{S}(U)$. In this case we have also

$$B^{X\setminus U}(\inf(f,n))|_{\tilde{U}} \preccurlyeq_{\mathcal{S}(U)} B^{X\setminus U}(\inf(s,n))|_{\tilde{U}} \preccurlyeq_{\mathcal{S}(U)} \inf(s,n)|_{\tilde{U}}$$

and therefore $B^{X\setminus U}f|_{\tilde{U}} \preccurlyeq_{\mathcal{S}(U)} s|_{\tilde{U}}$.

3. The natural topology in the standard H-cone of functions $\mathcal{S}(U)$ on the set U and the maximal sequences in U

If U is an open subset of X then it is known that the natural topology on U induced by $\mathcal{S}(U)$ coincides with the restriction to U of the natural topology on X induced by $\mathcal{S}$. This assertion is not true if U is fine open. Also even if U is open the natural topology induced by $\mathcal{S}(U)$ on the fine open set $\tilde{U}$ does not coincides with the restriction to $\tilde{U}$ of the natural topology on X induced by the H-cone $\mathcal{S}$.

Proposition 3. *Suppose that U is a fine open subset of X. Then the natural topology on U induced by the standard H-cone $\mathcal{S}(U)$ on U is the coarsest topology on U for which any function of the form $s - B^{X\setminus U}s$ is continuous where s runs the set of all universally continuous elements from $\mathcal{S}$.*

Proof. Let $\mathcal{S}_o$ be the set of all universally continuous elements from $\mathcal{S}$. We show that for any $s \in \mathcal{S}_o$ the element $s - B^{X\setminus U}s$ is 1-continuous with respect to the standard H-cone $\mathcal{S}(U)$. Indeed let $(t_n)_n$ be an increasing sequence from $\mathcal{S}(U)$ such that $\sup_n t_n = s - B^{X\setminus U}s$ on U. We have

$$t_n + B^{X\setminus U}s \in \mathcal{S} \text{ and } t_n + B^{X\setminus U}s \nearrow s.$$

Hence the sequence $(t_n + B^{X\setminus U}s)_n$ converges uniformly to s on X and therefore the sequence $(t_n)_n$ converges uniformly to $s - B^{X\setminus}s$ on U.

Conversely, suppose that $t \in \mathcal{S}(U)$ is an universally continuous element from $\mathcal{S}(U)$ and let s_o be a bounded continuous generator from $\mathcal{S}$. We want to show that t is of the form $t = s - B^{X\setminus U}s$ where $s \in \mathcal{S}$ is s_o-continuous. Indeed since $s_o - B^{X\setminus U}s_o$ is strictly positive on U it follows that there exists $\alpha > 0$ such that $t \leq \alpha(s_o - B^{X\setminus U}s_o)$ and therefore there exists $s \in \mathcal{S}$ such that $t = s - B^{X\setminus U}s$. Moreover we may suppose that s is such that if $s' \in \mathcal{S}$ has the property

$$s - B^{\setminus U}s \leq s' - B^{X\setminus U}s'$$

then $s \leq s'$.Particularly we deduce that $s \leq \alpha s_o$. We prove now that s is s_o-continuous. Indeed let $(s_n)_n$ be an increasing sequence in $\mathcal{S}$ such that $s = \sup\limits_n s_n$ on X. We have

$$B^{X\setminus U}s_n \nearrow B^{X\setminus U}s$$

and therefore

$$\lim_{n\to\infty}(s_n - B^{X\setminus U}s_n) = s - B^{X\setminus U}s\ ,\ \sup_n(\bigwedge_{k\geq n}(s_k - B^{X\setminus U}s_k)) = s - B^{X\setminus U}s.$$

Since $t = s - B^{X\setminus U}s$ is universally continuous in $\mathcal{S}(U)$ then for any $\varepsilon > 0$ there exists $n_\varepsilon \in \mathbb{N}$ such that

$$n \geq n_\varepsilon \Longrightarrow\ s - B^{X\setminus U}s \leq \bigwedge_{k\geq n}(s_k - B^{X\setminus U}s_k) + \varepsilon(s_o - B^{X\setminus U}s_o),$$

$$n \geq n_\varepsilon \Longrightarrow\ s - B^{X\setminus U}s \leq (s_n - B^{X\setminus U}s_n) + \varepsilon(s_o - B^{X\setminus U}s_o)$$

and therefore

$$n \geq n_\varepsilon \Longrightarrow\ s \leq s_n + \varepsilon s_o$$

which means that s is s_o-continuous.

The assertion from Proposition follows now using the fact that any $s \in \mathcal{S}$ which is s_o-continuous is the uniform limit of an inccreasing sequence of universally continuous elements of $\mathcal{S}$.

Definition. The natural topology on $\widetilde{U}$ (induced by the standard H-cone of functions $\mathcal{S}(U)$ is denoted by $\tau_o(U,\widetilde{U})$. The topology $\tau_1 = \tau_1(U,\widetilde{U})$ which is the coarsest topology on $\widetilde{U}$ which is finer than $\tau_o(U,\widetilde{U})$ and $\tau_o(X)|_{\widetilde{U}}$ is called the *maximal topology on* $\widetilde{U}$ associated with the standard H-cone $\mathcal{S}(U)$.

Corollary 4. *The maximal topology on $\widetilde{U}$ is the coarsest topology on $\widetilde{U}$ for which any function on $\widetilde{U}$ of the form $s|_{\widetilde{U}}$ and $B^{X\setminus U}s|_{\widetilde{U}}$ is continuous, where s runs the set of all universally continuous elements of $\mathcal{S}$.*

Corollary 5. *Let $z \in U$ and let $(x_n)_n$ be a sequence in $\widetilde{U}$. Then $(x_n)_n \longrightarrow z$ in the maximal topology iff $(x_n)_n \longrightarrow z$ in $\tau_o(X)$ and for any universally continuous element s of $\mathcal{S}$ we have*

$$B^{X\setminus U}s(z) = \lim_{n\to\infty} B^{X\setminus U}s(x_n).$$

Remark. If $z \in \widetilde{U}\setminus U$ then a sequence $(x_n)_n$ in U which converges to z in the maximal topology on U is called *maximal with respect to z*. This terminology was introduced by Brelot ([5]) and was used by Smyrnelis ([9]), Bauer ([1]), Hansen ([6]) and Netuka ([8]) in the case of the theory of harmonic spaces.

Theorem 6. *Let U be a fine open subset of X and let $z \in \widetilde{U}$. Then for any $t \in \mathcal{S}(U)$ for which there exists $s \in \mathcal{S}$ finite continuous at z such that $t \preccurlyeq_{\mathcal{S}(U)} s|_U$ we have*

$$fine\text{–}\lim_{U \ni x \to z} t(x) = \tau_1\text{–}\lim_{\widetilde{U} \ni x \to z} t(x)$$

or equivalent

$$fine\text{–}\lim_{U \ni x \to z} t(x) = \lim_{n \to \infty} t(x_n)$$

for any sequence $(x_n)_n$ in $\widetilde{U}$ which is maximal with respect to z.

Proof. Any element $t' \in \mathcal{S}(U)$ is considered as a function on $\widetilde{U}$. By hypothesis there exists $t' \in \mathcal{S}(U)$ such that $t + t' = s$ on $\widetilde{U}$. By the definition of the topology $\tau_1(U, \widetilde{U})$ the functions t, t' and $s|_{\widetilde{U}}$ are lower semicontinuous with respect to $\tau_1(U, \widetilde{U})$. By hypothesis s is continuous at z with respect to $\tau_o(X)|_{\widetilde{U}}$ and therefore with respect to $\tau_1(U, \widetilde{U})$. Hence we have

$$\text{fine–}\lim_{U \ni x \to z} t(x) = \tau_1\text{–}\lim_{\widetilde{U} \ni x \to z} t(x).$$

Proposition 7. *Let U be a fine open subset of X and let $z \in \widetilde{U}$. Then for any positive Borel function f on X for which there exists $s \in \mathcal{S}$ finite continuous at z such that $f \leq s$, the function $B^{X \setminus U} f|_{\widetilde{U}}$ is τ_1-continuous at z.*

Proof. The assertion follows from Theorem 6 since by Theorem 2 we deduce $B^{X \setminus U} f|_{\widetilde{U}} \preccurlyeq_{\mathcal{S}(U)} s|_{\widetilde{U}}$.

Corollary 8. *Let U be afine open subset of X. Then for any positive bounded function f on X the function $B^{X \setminus U} f|_{\widetilde{U}}$ is τ_1-continuous.*

We recall that an element t of an H-cone $\mathcal{T}$ is called *subtractible* if we have

$$t' \in \mathcal{T},\ t \leq t' \Longrightarrow t \preccurlyeq t'.$$

For instance if $(X, {}^*\mathcal{H})$ is a harmonic space and U is an open subset of X then any positive harmonic function is subtractible in the H-cone $\mathcal{S}(U)$ of all positive superharmonic functions on U. Moreover any positive superharmonic function on U which is harmonic outside a polar set is also subtractible.

Proposition 9. *Let U be a fine open subset of X and let $z \in \widetilde{U}$. Then for any subtractible element $t \in \mathcal{S}(U)$ for which there exists $s \in \mathcal{S}$ finite continuous at z such that $t \leq s|_U$ we have*

$$fine\text{–}\lim_{U \ni x \to z} t(x) = \tau_1\text{–}\lim_{\widetilde{U} \ni x \to z} t(x).$$

Proof. The assertion follows from Theorem 6 since we have $t \preccurlyeq_{\mathcal{S}(U)} s|_U$.

Corollary 10. *Let u be a fine open subset of X. Then for any bounded subtractible element $t \in \mathcal{S}(U)$ we have that t is a τ_1-continuous function on $\widetilde{U}$.*

4. Doob sets and Harnack generalized inequality

In this section U will be a fine open subset of X and z a point in X such that $X \setminus U$ is thin at z. We denote as in the preceding sections by $\widetilde{U}$ the set of all points $y \in X$ such that $X \setminus U$ is thin at y. We suppose that X is semisaturated with respect to the H-cone $\mathcal{S}$.

Lemma 11. *Let F_z be the set of all positive Borel functions f on X such that $\varepsilon_z^{X\setminus U}(f) < \infty$. Then there exists $f_o \in F_z$ such that for any $f \in F_z$ we have*

$$\{y \in \widetilde{U} / \ \varepsilon_z^{X\setminus U}(f_o) < \infty\} \subset \overline{\{y \in \widetilde{U} / \ \varepsilon_z^{X\setminus U}(f) < \infty\}}^{fine}$$

where $\overline{A}^{fine}$ means the fine closure of A.

Proof. For any $f \in F_z$ we denote

$$A_f = \{y \in \widetilde{U} / \ \varepsilon_y^{X\setminus U}(f) < \infty\}.$$

Obviously A_f is fine open and $z \in A_f$. Moreover if $(f_n)_n$ is a sequence in F_z then then there exists $f \in F_z$ such that

$$A_f = \bigcap_n A_{f_n}.$$

Indeed let $(a_n)_n$ be a sequence of strictly positive real numbers such that

$$\sum_n a_n \varepsilon_z^{X\setminus U}(f_n) < \infty.$$

Then

$$f := \sum_n a_n f_n \in F_z$$

and we have

$$\varepsilon_z^{X\setminus U}(f) < \infty \Longrightarrow \varepsilon_z^{X\setminus U}(f_n) < \infty \quad , \quad (\forall) n \in \mathbb{N}$$

and therefore $A_f \subset \bigcup_n A_{f_n}$.

Now we remark that for any $f \in F_z$ the set $\overline{A_f}^{fine}$ is absorbent in $\widetilde{U}$ with respect to $\mathcal{S}(U)$. From this fact we deduce that there exists a sequence $(f_n)_n$ in F_z such that

$$\bigcup_{f \in F_z} \overline{A_f}^{fine} = \bigcup_n \overline{A_{f_n}}^{fine}.$$

Definition. A fine open subset U of X is called a *Doob set* (with respect to $\mathcal{S}$) if for any fine open subset V of U there exists a fine open subset V_o of V such that $V \setminus V_o$ is totally thin and such that for any positive Borel function f on X for which $B^{X\setminus U} f$ is finite dense on V then $B^{X\setminus U} f$ is finite on V_o.

Remark. 1. Suppose that $(\mathbb{R}^{n+1}, {}^*\mathcal{H})$, $n \geq 0$ is the harmonic space associated with the heat equation on $\mathbb{R}^{n+1}$. Then any open subset of $\mathbb{R}^{n+1}$ is a Doob set with respect to the H-cone $\mathcal{S}$ of all positive superharmonic functions s on this harmonic space. Indeed let U be an open subset of $\mathbb{R}^{n+1}$, V be a fine open subset of U and let f be a positive Borel function on X such that $B^{X\setminus U}f$ is finite dense on V. If $x_o = (x', t_o) \in V$ and $D \times I$ is a rectangle with the center in x_o such that $D \times I \subset U$ then if $y_o = (y'_o, t) \in (D \times I) \cap V$ is such that

$$B^{X\setminus U}f(y_o) < \infty$$

we have

$$B^{X\setminus U}f(y', t') < \infty$$

for any $y' \in D$ and any $t' < t$. Hence the set

$$\{y \in V \cap (D \times I) / \ B^{X\setminus U}f(y) = \infty\}$$

is a subset of a set of the form

$$\{y = (y', t') / \ t' = t'_o\},$$

where $t'_o := \sup\{t \in \mathbb{R} / \ (\exists) y' \in D,\ (y', t) \in V\}$. Obviously t'_o does not depend of f as above.

2. If $(X, {}^*\mathcal{H})$ is a Brelot harmonic space then any open subset of X is a Doob set.

Theorem 12. *Suppose hat U is a Doob set with respect to $\mathcal{S}$. Then there exists a fine neighbourhood V of z such that for any positive Borel function f o X we have*

$$\varepsilon_z^{X\setminus U}(f) < \infty \Longrightarrow \varepsilon_y^{X\setminus U}(f) < \infty \quad , \quad (\forall) y \in V.$$

Proof. Let F_z be the set of all positive Borel functions f on X with $\varepsilon_z^{X\setminus U}(f) < \infty$ and let for any $f \in F_z$, $A_f = \{y \in U / \ \varepsilon_y^{X\setminus U}(f) < \infty\}$. From Lemma 11 there exists $f_o \in F_z$ such that

$$A_{f_o} \subset \overline{A}^{fine} \quad , \quad (\forall) f \in F_z.$$

From this relation and from the fact that U is a Doob set it follows that there exists a fine open subset A_o of A_{f_o} such that $A_{f_o} \setminus A_o$ is totally thin and such that $B^{X\setminus U}f$ is finite on A_o whenever $f \in F_z$. The set $V_o = A_o \cup \{z\}$ is a fine neighbourhood of z for which we have

$$y \in V_o \, , \ \varepsilon_z^{X\setminus U}(f) < \infty \Longrightarrow \varepsilon_y^{X\setminus U}(f) < \infty.$$

Corollary 13. (Harnack inequality). *Suppose that U is a Doob set with respect to $\mathcal{S}$. Then there exists a fine neighbourhood V of z and a positive real function c on V such that*

$$y \in V \Longrightarrow \varepsilon_y^{X\setminus U} \leq c(y) \cdot \varepsilon_z^{X\setminus U}.$$

Proof. From the preceding theorem here exists a fine neighbourhood V of z such that for any positive Borel function f on X we have

$$y \in V \;,\; \varepsilon_z^{X\setminus U}(f) < \infty \Longrightarrow \varepsilon_y^{X\setminus U}(f) < \infty.$$

Hence we have

$$y \in V \;,\; \varepsilon_z^{X\setminus U}(f) = 0 \Longrightarrow \varepsilon_y^{X\setminus U}(f) = 0.$$

and therefore $\varepsilon_y^{X\setminus U}$ is absolutely continuous with respect to $\varepsilon_z^{X\setminus U}$ for any $y \in V$. Hence the Radon derivative g_y of $\varepsilon_y^{X\setminus U}$ with respect to $\varepsilon_z^{X\setminus U}$ is a positive bounded function. If we put, for any $y \in V$,

$$c(y) = \inf\{\alpha > 0 / \; g_y \leq \alpha, \; \varepsilon_z^{X\setminus U} - a.s.\}$$

then we have $\varepsilon_y^{X\setminus U} \leq c(y) \cdot \varepsilon_z^{X\setminus U}$.

References

1. H. Bauer :*Fine boundary limits of harmonic and caloric functions* . Bull. Sci. Math. **109**(1985), 337-361.
2. N. Boboc, Gh. Bucur and A. Cornea : *Order and Convexity in Potential Theory: H-Cones.* Lecture Notes in Math. 853, Springer-Verlag, 1981.
3. N. Boboc, Gh. Bucur : *Natural localization and natural sheaf property in standard H-cones of functions (I).* Rev. Roumaine Math. Pures Appl. **30**(1985), 1-26.
4. M. Brelot : *Sur les ensembles effilés.* Bull. Sci. Math. **68**(1944), 12-36.
5. M. Brelot :*On the behaviour of harmonic functions in the neighbourhood of an irregular point.* J. Anal. Math. **4**(1954-1955), 209-221.
6. W. Hansen :*Fine boundary limits of finely harmonic functions.* Math. Ann. **274**(1986), 643-647.
7. G. Mokobodzki : *Opérateur de subordination des résolvantes.* 1983 (Manuscrit non-publié).
8. I. Netuka :*Fine behaviour of solutions of the Dirichlet problem near an irregular point.* Bull. Sci. Math. **114**(1990), 1-22.
9. E. Smyrnelis : *Sur les limites fines des fonctions harmoniques et les suites maximales.* Bull. Sci. Math. **97**(1973), 161-175.

SOME RESULTS ABOUT SEQUENTIAL INTEGRATION ON WIENER SPACE

Nicolas Bouleau
Laboratoire de Mathématiques Appliquées
U.A. CNRS 1502
ENPC La Courtine
93167 Noisy-le-Grand cedex FRANCE

Abstract. Sequential integration is, in a large class of cases, the only available way for computing the expectation of random variables. These methods have been extensively studied in finite dimension. Their extension to infinite dimensional probability spaces would give useful tools for numerical computation of numerous probability models involving, for example, Brownian motion. In this paper we discuss aspects of this extension and give results concerning the case of Wiener space.

I. Presentation. We use the terms *sequential integration* for the domain of research which deals with numerical methods for computing the integral of a function f as limit of the average of its values along a sequence of points (ξ_n) taken in the space where f is defined:

$$\int f \, d\mu = \lim_{N \uparrow \infty} \frac{1}{N} \sum_{n=1}^{N} f(\xi_n).$$

The sequence (ξ_n) depends on the measure μ, not on f provided that f satisfies some hypotheses.

A. Finite dimensional case

When the function f is defined on an Euclidean space $\mathbf{R}^d$, the technics of *simulation algorithms* (cf e.g. Devroye (1986)) allow one to restrict the problem to the case where f is defined on a finite dimensional cube equipped with Lebesgue measure:

$$f : [0,1]^s \longrightarrow \mathbf{R} \quad , \quad \mu = \text{ Lebesgue measure}$$

(Note that the integer s is larger than d in general, and see the remark below).

In these cases the methods of sequential integration have been widely studied as well for the class of functions to which they apply or for the rate of convergence they give (for an introductory account see e.g. Bouleau (1993)). There are essentially three families of methods:

K. GowriSankaran et al. (eds.), Classical and Modern Potential Theory and Applications, 125–132.

1. The Monte Carlo method which, from a mathematical point of view, is based on the strong law of large numbers and apply (theoretically, see Bouleau-Lépingle (1993)) to any function in $\mathcal{L}^1([0,1]^s, dx)$.

2. The quasi-Monte Carlo methods in which the sequence (ξ_n) is no more an imitation of randomness but a deterministic sequence satisfying the only property of uniform distribution over the cube $[0,1]^s$. The function f has to be Riemann-integrable. The quality of such a sequence can be measured by its *discrepancy*,

3. The shift method which, from a mathematical point of view, is based on the pointwise ergodic theorem and apply (see remark below) to functions in $\mathcal{L}^1$.

Among a large literature we refer, for 1 and 2 to Niederreiter (1978), and for 1,2 and 3 to Bouleau-Lépingle (1993) and references in it.

Remark. The Monte Carlo method and the shift method apply in the more general case which can be called the case of *unbounded dimension* in which the dimension is finite but depends on the sample point, (see Bouleau-Lépingle (1993) chapter III). Many simulation algorithms (using the rejection method) yield an unbounded dimensional setting. Practically for the three methods (except particular cases) the function f has to be bounded.

B. Infinite dimensional case

In order to set the discussion of extending sequential integration to the infinite dimension on the most concrete basis, we limit the study to the case of Wiener space with which a significant class of models can be defined:

$$(W, \mathcal{W}, \mu) = (\mathcal{C}_0([0,1], \mathbf{R}^d), \mathcal{B}, m)$$

the measurable space is the space of continuous functions from $[0,1]$ into $\mathbf{R}^d$ vanishing at zero, equipped with its Borelian σ-field, and the probability measure is the Wiener measure.

Looking then to the situation in more details shows immediately some difficulties:

1. Most of interesting functionals (stochastic integrals, solutions of stochastic differential equations, etc) are classes for almost everywhere equality, so that for $\xi_n \in W$, $f(\xi_n)$ is not defined.

2. Useful and simply defined functionals can be very irregular so that their Borel versions are not Riemann-integrable. For example for $d \geq 2$, with obvious notation, if $f(\omega) = \int_0^1 B_s^1 dB_s^2$ then any Borel version $\tilde{f}$ of f is discontinuous at every point in W. (For a more detailed discussion of the irregularity of Wiener functionals see Bouleau-Lépingle (1993) chapter V).

Fortunately, the situation is not hopeless thanks to the *approximate continuity property* discovered for solutions of stochastic differential equations by Stroock and

Varadhan in 1972 (see Ikeda-Watanabe (1981) chapter 8). Let X_t be the solution of the s.d.e. in Stratonowich sense

$$\begin{cases} dX_t^i = \Sigma_{k=1}^r \sigma_k^i(X_t) \circ dB_t^k + b^i(X_t)dt \\ X_0 = x \end{cases}$$

where $i = 1, 2, \cdots, m$ and the coefficients σ and b are regular. Let now $\varphi \in \mathcal{C}^\infty([0,1], \mathbf{R}^r)$, $\varphi(0) = 0$ and let $Y_t = Y_t(\varphi)$ be the solution of the ordinary differential equation

$$\begin{cases} dY_t^i = \Sigma_{k=1}^r \sigma_k^i(Y_t)\dot{\varphi}(t)dt + b^i(Y_t)dt \\ Y_0 = x \end{cases}$$

Then, if $X_1(\omega)$ is taken to be $Y_1(\varphi)$ at the point φ, X_1 becomes approximately continuous at φ, in other words

$$\forall \varepsilon > 0 \quad m[|X_1 - Y_1(\varphi)| > \varepsilon | \ ||\omega - \varphi|| < \delta] \to 0$$

when $\delta \to 0$. The value of X_1 at the point φ satisfying this property is unique and is representative of the behaviour of X_1 in the neighborhood of φ. By this approximate continuity property one can reasonably expect that if φ_n are chosen regular along a sufficiently dense sequence in W the following formula be valid

$$\mathsf{E}X_1 = \lim_{N \uparrow \infty} \frac{1}{N} \sum_{n=1}^{N} X_1(\varphi_n)$$

where $X_1(\varphi_n)$ is defined by $Y_1(\varphi_n)$.

Other interesting functionals possess the same property. For example, for $h \in L^2([0,1])$ the random variable $\int_0^1 h(s)\, dB_s$ is approximately continuous at every φ in the Cameron-Martin space $CM = \{\psi : \psi(0) = 0, \dot{\psi} \in L^2([0,1], \mathbf{R}^r)\}$ when its value at φ is taken to be

$$\int_0^1 h(s)\dot{\varphi}(s)ds.$$

II. Continuous functions of functionals in the first chaos

We mention here an existence result concerning sequences in the Cameron-Martin space usable for integration of bounded continuous functions of functionals of the first chaos.

Theorem 1 *There is a sequence $(\varphi_n)_{n\geq 1}$ in CM such that*

$$\forall k \in \mathbf{N}^*, \forall F \in \mathcal{C}_b(\mathbf{R}^k, \mathbf{R}),\ \forall h_1, \cdots, h_k \in L^2([0,1], \mathbf{R}^r)$$

$$\lim_{N\to\infty} \frac{1}{N} \sum_{n=1}^{N} F\left(\int_0^\infty h_1(s)\dot{\varphi}_n(s)ds, \cdots, \int_0^\infty h_k(s)\dot{\varphi}_n(s)ds\right)$$
$$= \mathsf{E}\left[F\left(\int_0^\infty h_1(s)dB_s, \cdots, \int_0^\infty h_k(s)dB_s\right)\right].$$

Note that for general $h_1, \cdots, h_k$ and F the functional $F\left(\int_0^\infty h_1(s)dB_s, \cdots, \int_0^\infty h_k(s)dB_s\right)$ is not continuous. This theorem is a consequence of a property which can be stated for every separable Hilbert space and has its own interest:

Proposition 1 *Let H be a real separable Hilbert space with scalar product $(.,.)$ and norm $|.|$. There exists a sequence (a_n) in H such that for every $b \in H$ the real sequence (a_n, b) is distributed according to the Gaussian probability measure $N(0, |b|^2)$.*

Thus the sequence (a_n) is such that

$$\forall b \in H,\ \frac{1}{N} \sum_{n=1}^{N} \delta_{(b,a_n)} \to N(0, |b|^2) \text{ narrowly}$$

or equivalently

$$\forall b \in H,\ \forall u \in \mathbf{R},\ \frac{1}{N} \sum_{n=1}^{N} e^{iu(b,a_n)} \to e^{-\frac{u^2}{2}|b|^2}.$$

The proof is based on a linearization of this relation using the Fock space of H, and the result is then obtained by an argument of convergence almost everywhere of a Hilbert valued martingale (cf Bouleau (1992)). Thus the sequence (a_n) is not explicitly given here.

III. An effective result

Let $p_1, \cdots, p_n, \cdots$ be the sequence of prime numbers. For p prime and n an integer, let

$$n = a_0 + a_1 p + \cdots + a_k p^k \quad a_i \in \{0, \cdots, p-1\} \quad i = 1, \cdots, k,$$

be the representation of n in basis p. Let us define a function $\varphi_p : \mathbf{N} \to [0,1]$ by

$$\varphi_p(n) = \frac{a_0}{p} + \cdots + \frac{a_k}{p^{k+1}}.$$

It is a classical result of Halton (1960) (cf Hua-Wang (1981)) that the sequence

$$\xi_d(n) = (\varphi_{p1}(n), \cdots, \varphi_{pd}(n))$$

is uniformly distributed over $[0,1]^d$. Let us put

$$erf(x) = \frac{1}{\sqrt{2\pi}} \int_{-\infty}^{x} e^{-\frac{y^2}{2}} dy,$$

then the sequence

$$\eta_d(n) = \left(erf^{-1}(\varphi_{p1}(n)), \cdots, erf^{-1}(\varphi_{pd}(n))\right)$$

is $N(0, I_d)$-distributed on $\mathbf{R}^d$ i.e.

$$\frac{1}{N} \sum_{n=1}^{N} \delta_{\eta d(n)} \rightarrow N(0, I_d) \text{ narrowly.}$$

Here $N(0, I_d)$ is the centered Gaussian measure on $\mathbf{R}^d$ making coordinates independent and with unit variance.

Now, in infinite dimension, the sequence $\eta = \eta(n)$

$$\eta(n) = \left(erf^{-1}(\varphi_{p1}(n)), \cdots, erf^{-1}(\varphi_{pd}(n)), \cdots\right)$$

is a natural candidate to be distributed according to the Gaussian product measure on $\mathbf{R}^{\mathbf{N}}$. And the same is true for the truncated sequence $\tilde{\eta} = \tilde{\eta}(n)$ defined by

$$\begin{aligned} \tilde{\eta}_d(n) &= 0 & &\text{if } d \geq K(n) \\ \tilde{\eta}_d(n) &= \eta_d(n) & &\text{if } d < K(n) \end{aligned}$$

where K is an increasing function from $\mathbf{N}$ into $\mathbf{N}$ tending to infinity when n goes to infinity.

To put this on the Wiener space, let (χ_n) be a C.O.N.S. of $L^2[0,1]$ and let $\psi_n(t) = \int_0^t \chi_n(s)ds$. Then

$$B_t = \sum_{k=1}^{\infty} \psi_k(t) g_k ,$$

where the g_k are independent reduced Gaussian random variables, defines a Brownian motion (cf Bouleau-Hirsch (1991) chapter V example 1.3.2). Consequently the sequence h_n in CM

$$h_n(t) = \sum_{k=1}^{K(n)} \psi_k(t) \eta_k(n)$$

is a candidate to be distributed according to Wiener measure.

We have now to set precisely the topology with which our Wiener space is equipped. We shall see that it is possible to choose the function $K(n)$ such that the sequence $\tilde{\eta} = (\tilde{\eta}(n))$ be distributed according to Wiener measure for all Hölder topologies $C_\alpha, 0 < \alpha < \frac{1}{2}$ (and hence also for the uniform topology).

Let us consider the Haar basis

$$\chi_1(t) = 1$$

$$\chi_{2^n+k}(t) = \begin{cases} 2^{\frac{n}{2}} & \text{if } t \in \left[\frac{2k-2}{2^{n+1}}, \frac{2k-1}{2^{n+1}}\right] \\ -2^{\frac{n}{2}} & \text{if } t \in \left[\frac{2k-1}{2^{n+1}}, \frac{2k}{2^{n+1}}\right] \end{cases}$$

with $n \in \mathbf{N}^*$, $k = 1, \cdots, 2^n$. And let us consider the associated Schauder basis of $\mathcal{C}_0([0,1])$

$$\psi_m(t) = \int_0^t \chi_m(s)ds$$

so that any f in $\mathcal{C}_0([0,1])$ can be written uniquely

$$f = \sum f_m \psi_m.$$

Now for $0 < \alpha < 1$ let us consider the Hölder Banach spaces

$$C_\alpha = \left\{ f : [0,1] \to \mathbf{R},\ f(0) = 0 \quad \sup_{s,t} \frac{f(t) - f(s)}{|t-s|^\alpha} = \|f\|_{C_\alpha} < +\infty \right\}$$

and

$$C_\alpha^0 = \left\{ f \in C_\alpha : \quad \lim_{\delta \to 0} \sup_{|t-s|<\delta} \frac{f(t) - f(s)}{|t-s|^\alpha} = 0 \right\}$$

then we have the following theorem due to Ciesielski

Theorem 2 *Let*

$$c_{2^n+k}(\alpha) = 2^{n(\alpha+\frac{1}{2})+1-\alpha}$$

a) $f \in C_\alpha$ iff $\{f_m c_m(\alpha)\} \in l^\infty$ *and this defines an isomorphism of Banach spaces*
b) $f \in C_\alpha^0$ iff $\{f_m c_m(\alpha)\} \in c_0$ *and this defines an isomorphism of Banach spaces*
c) for $\alpha < \beta$ *every bounded set in* C_β *is relatively compact in* C_α^0.

With this result, to show that the measures

$$\mu_N = \frac{1}{N} \sum_{n=1}^{N} \delta_{\tilde{\eta}(n)}$$

converge to the Wiener measure narrowly for every C_α-topology $0 < \alpha < \frac{1}{2}$, we have to show that for all $\varepsilon > 0$ there is a bounded set in C_α which carries the μ_n's up to ε.

For this, the function $K(n)$ has to be chosen in relationship with the discrepancy of the Halton sequence. By definition this discrepancy is

$$D^*_\infty(\xi_d, N) = \sup_{x\in[0,1]^d} \left| \frac{1}{N} \sum_{n=1}^{N} 1_{[\![0,x]\!]}(\xi_d(n)) - \pi(x) \right|$$

where $\pi(x)$ is the distribution function of the Lebesgue measure, i.e. the volume of the rectangle $[\![0,x]\!]$. Explicit estimates of this discrepancy are known for the Halton sequence:

$$D^*_\infty(\xi_d, N) \leq \frac{1}{N} \prod_{i=1}^{d} \frac{p_i \log(p_i N)}{\log(p_i)} \leq 2\frac{1+\log N}{N} \prod_{i=1}^{d} p_i.$$

Thus, if we chose

$$K(N) = \sup\{d : \prod_{i=1}^{d} p_i \leq 2\sqrt{N}\ \}$$

we obtain that K is increasing to infinity and

$$D^*_\infty(\xi_{K(N)}, N) \to 0 \quad \text{as} \quad N \uparrow \infty$$

and this allows us to show the tightness property of the sequence (μ_N). We obtain

Theorem 3 *The sequence in CM*

$$f^{(n)} = \sum_{m=1}^{K(n)} \psi_m\ \eta_m(n)$$

is such that the measures

$$\frac{1}{N} \sum_{n=1}^{N} \delta_{f^{(n)}}$$

converge to the Wiener measure for every C_α topology, $0 < \alpha < \frac{1}{2}$.

Remark. The argument is not restricted to the Halton sequence and gives from any sequence cylindrically uniformly distributed on $[0,1]^{\mathbf{N}}$ a sequence in CM distributed according to Wiener measure for C_α topologies, $0 < \alpha < \frac{1}{2}$.

References

Bouleau N., Calculs d'espérance par simulation, Matapli (1993)

Bouleau N. and Lépingle D., *Numerical methods for stochastic processes*, Wiley (1993)

Bouleau N., Suite de points d'un espace de Hilbert distribuée selon la promesure gaussienne canonique, Note C. R. Acad. Sci. Paris t. 315, sI, 1297-1300, (1992)

Bouleau N. and Hirsch F., *Dirichlet forms and analysis on Wiener space*, de Gruyter (1991)

Devroye L., *Non-uniform random variate generation.* Springer (1986)

Hua L. K. and Wang Y., *Applications of number theory to numerical analysis*, Springer (1981)

Ikeda N. and S. Watanabe, *Stochastic differential equations and diffusion processes*, North Holland (1981)

Niederreiter H., Quasi-Monte Carlo methods and pseudo-random numbers, Bull. Amer. Math. Soc. 84, 957-1041, (1978)

SCHWARZ LEMMA TYPE INEQUALITIES FOR HARMONIC FUNCTIONS IN THE BALL

BERNHARD BURGETH
Mathematisches Institut
Universität Erlangen–Nürnberg
Bismarckstr. 1 1/2
D–91054 Erlangen
Germany

ABSTRACT. This paper is concerned with bounded functions h on the unit ball B^3 in $\mathbb{R}^3$ admitting a generalized Poisson type representation. Estimates for $h(x)$ in terms of $\|x\|$ and $h(0)$, as well as estimates for the radial derivative $\partial_r h(x)$ at $x \neq 0$ in terms of $\|x\|$ and $h(x)$ are presented. In both cases we determine the functions for which the inequalities become equalities.

AMS (1992) subject classification: primary 31B05, 31B10;
secondary 30C80

1. Subject

The well known Schwarz Lemma in classical complex analysis states that a holomorphic function h defined on $B^2 = \{z \in \mathbb{C} : |z| < 1\}$ with $|h| \leq 1$ satisfies for each $z \in B^2$

$$\left| \frac{h(z) - h(0)}{1 - h(z) \cdot \overline{h(0)}} \right| \leq |z|$$

and

$$|h'(z)| \leq \frac{1 - |h(z)|^2}{1 - |z|^2} .$$

Equality holds if and only if h belongs to the automorphism group of the open unit disk. Similar results in the two-dimensional case for harmonic functions seem to be known – see [Burckel] for results and further references.

We shall see that similar results hold for real-valued functions defined on the three-dimensional unit ball B^3 which are harmonic, hyperbolic-harmonic or, more general, admit a representation of the form $\mathcal{H}_{\alpha,\beta}$ introduced below. More precisely, we prove estimates for $h(x)$ in terms of $\|x\|$ and $h(0)$, as well as estimates for the radial derivative $\partial_r h(x)$ at a point $x \in B^3 \setminus \{0\}$ in terms of $\|x\|$ and $h(x)$. For a definition of ∂_r see below. The methods of proof are heavily based on an explicit solution of an extremal

K. GowriSankaran et al. (eds.), Classical and Modern Potential Theory and Applications, 133–147.

problem obtained in the fundamental lemma of this paper. Since linearity of ∂_r in h is essential for their applicability, we can't simply take $\|\nabla_x h(x)\|$ instead of $\partial_r h(x)$, as one might have expected.

Some Schwarz lemma type inequalities for harmonic functions can already be found in [Khavinson] and [Axler].

2. Notation

We set $B^3 = \{x \in \mathbb{R}^3 : \|x\| < 1\}$ and $S^2 = \{x \in \mathbb{R}^3 : \|x\| = 1\}$. If $x \in B^3 \setminus \{0\}$ we define $\bar{x} := x/\|x\| \in S^2$. $S(\bar{x}, \vartheta)$ denotes a relatively open spherical cap with center $\bar{x}$ and spherical angle $\vartheta \in [0, \pi]$:

$$S(\bar{x}, \vartheta) := \{y \in S^2 \mid \langle \bar{x}, y\rangle > \cos\vartheta\}.$$

σ denotes the Haar measure on S^2, normalized by $\sigma(S^2) = 1$. For positive Borel measures μ_1, μ_2 on S^2 with finite total mass, we define a partial order by

$$\mu_1 \leq \mu_2 \quad :\Longleftrightarrow \quad \mu_1(A) \leq \mu_2(A) \quad \text{for every Borel subset } A \text{ of } S^2.$$

Writing $\nabla_x = \left(\frac{\partial}{\partial x_1}, \frac{\partial}{\partial x_2}, \frac{\partial}{\partial x_3}\right)$ we define the radial derivative $\partial_r v$ of a continuously differentiable function v on B^3 by

$$\partial_r v(x) = \langle \bar{x}, \nabla_x v(x)\rangle, \quad x \in B^3 \setminus \{0\}.$$

Define for α, $\beta \in \mathbb{R}$, $\beta > 0$, the generalized Poisson kernel by

$$P_{\alpha,\beta} : \begin{cases} B^3 \times S^2 \to \mathbb{R} \\ (x, y) \mapsto \dfrac{(1 - \|x\|^2)^\alpha}{\|x - y\|^{2\beta}}. \end{cases} \tag{1}$$

Set

$$\mathcal{H}_{\alpha,\beta} := \Big\{h : B^3 \to \mathbb{R} \mid h(x) = \int_{S^2} P_{\alpha,\beta}(x, y)\mu(dy), \\ x \in B^3,\ \mu \text{ a Borel measure on } S^2 \text{ with } 0 \leq \mu \leq \sigma\Big\}. \tag{2}$$

Since σ plays a special role, we write

$$u_{\alpha,\beta}(x) := \int_{S^2} P_{\alpha,\beta}(x, y)\sigma(dy).$$

Introducing spherical coordinates, one can show that

$$u_{\alpha,\beta}(x) = \begin{cases} \dfrac{(1 - \|x\|^2)^\alpha}{2\|x\|} \ln\left(\dfrac{1 + \|x\|}{1 - \|x\|}\right) & \text{for } \beta = 1,\ x \neq 0; \\ \dfrac{(1 - \|x\|^2)^\alpha}{4(\beta - 1)\|x\|} \cdot \dfrac{1}{(1 - \|x\|)^{2\beta-2}} \left(1 - \left(\dfrac{1 - \|x\|}{1 + \|x\|}\right)^{2\beta-2}\right) & \text{for } \beta \neq 1,\ x \neq 0; \\ 1 & \text{for } x = 0. \end{cases}$$

Also define for $x \neq 0$, resp., $x = 0$

$$\begin{aligned} L_{\alpha,\beta}(\|x\|, h(0)) &:= \int\limits_{S(\bar{x},\gamma)} P_{\alpha,\beta}(x,y)\sigma(dy)\,, \\ L_{\alpha,\beta}(0, h(0)) &:= h(0) \end{aligned} \tag{3}$$

with $\cos\gamma = 1 - 2h(0)$.
Finally we shall use for $x \neq 0$

$$R_{\alpha,\beta}(\|x\|, h(x)) := \int\limits_{S(\bar{x},\tau)} \partial_r P_{\alpha,\beta}(x,y)\, d\sigma(y)\,, \tag{4}$$

with

$$\cos\tau = \begin{cases} \dfrac{1+\|x\|^2}{2\|x\|} - \dfrac{1}{2\|x\|}\left(\dfrac{1}{(1-\|x\|)^{2\beta-2}} - \dfrac{4(\beta-1)\|x\|h(x)}{(1-\|x\|^2)^{\alpha}}\right)^{\frac{1}{1-\beta}} & \text{for } \beta \neq 1; \\ \dfrac{1+\|x\|^2}{2\|x\|} - \dfrac{(1-\|x\|)^2}{2\|x\|}\exp\left(\dfrac{4\|x\|h(x)}{(1-\|x\|^2)^{\alpha}}\right) & \text{for } \beta = 1\,. \end{cases} \tag{5}$$

For later use we define the Möbius transformation $T_a : B^3 \to B^3$, $a \in B^3$ by

$$T_a(x) = \frac{(1-\|a\|^2)(x-a) - \|x-a\|^2 a}{[x,a]^2}\,,$$

$x \in B^3$, where $[x,a]^2 := 1 - 2\langle x,a\rangle + \|x\|^2\|a\|^2$ (see [Ahlfors]).

3. The sets $\mathcal{H}_{\alpha,\beta}$

Due to results in [Leutwiler] the sets $\mathcal{H}_{\alpha,\beta}$ are closely related to the solutions for $n = 3$ of the Weinstein equation

$$\Delta v(x) - \frac{2k}{1-\|x\|^2}\langle x, \nabla_x v(x)\rangle + \left(\frac{k(2-n-k)}{1-\|x\|^2} + \frac{4l}{(1-\|x\|^2)^2}\right) v(x) = 0 \tag{6}$$

where $k,\ l \in \mathbb{R}$, $(k-1)^2 - 4l \geq 0$, Δ denotes the usual Laplace Operator, $\nabla_x = (\frac{\partial}{\partial x_1}, \ldots, \frac{\partial}{\partial x_n})$ and v is a sufficiently smooth real function on B^n.

Theorem 3.2 in [Leutwiler] yields an order preserving bijection between the sets $\mathcal{H}_{\alpha_0,\beta_0}$ and

$$\{v : B^3 \to \mathbb{R} \mid v \text{ is solution of (6) for } n = 3 \text{ and } 0 \leq v(x) \leq u_{\alpha_0,\beta_0}(x) \text{ for all } x \in B^3\}$$

where $\alpha_0 := (1-k+s)/2$ and $\beta_0 := (2+s)/2$ with $s := \sqrt{(k-1)^2 - 4l}$.

We shall pay attention to the following special cases:

a) If we specify $k = 0$, $l = 0$, $s = 1$ (or equivalently $\alpha = 1$, $\beta = \frac{3}{2}$), equation (6) is just the usual Laplace equation $\Delta v = 0$, and $\mathcal{H}_{1,\frac{3}{2}}$ is the set of all positive harmonic functions on B^3 bounded by $u \equiv 1$.

b) Setting $k = -1$, $l = 0$, $s = 2$ (or equivalently $\alpha = 2$, $\beta = 2$), equation (6) turns out to be the Laplace–Beltrami equation

$$\Delta v(x) + \frac{2}{1 - \|x\|^2} \cdot \langle x, \nabla_x v(x) \rangle = 0$$

associated with the hyperbolic (or Poincaré) metric

$$ds^2 = \frac{4(dx_1^2 + dx_2^2 + dx_3^2)}{(1 - \|x\|^2)^2}$$

on B^3. $\mathcal{H}_{2,2}$ is the set of the so-called hyperbolic-harmonic functions on B^3, positive and bounded by $u \equiv 1$. For a detailed treatment see [Ahlfors].

c) We note that $\mathcal{H}_{0,\frac{1}{2}}$ just consists of the Newtonian potentials of positive Borel measures supported by S^2 and bounded by σ (with respect to the ordering defined above).

Finally we mention the so-called α–harmonic functions on B^3, with $0 < \alpha \leq 2$. As pointed out in [Hmissi] the α–harmonic functions on B^3 also admit a unique integral representation of the type considered in (2). In particular, from our formula for $u_{\alpha,\beta}$ one obtains $u_{\frac{\alpha}{2},\frac{3}{2}}(x) = (1 - \|x\|^2)^{\frac{\alpha}{2}-1}$ and consequently $\mathcal{H}_{\frac{\alpha}{2},\frac{3}{2}}$ is in one-to-one correspondence with the set

$$\{h : B^3 \to \mathbb{R} \mid h \ \alpha\text{–harmonic on } B^3,\ 0 \leq h(x) \leq (1 - \|x\|^2)^{\frac{\alpha}{2}-1}, x \in B^3\}.$$

4. Theorem 1

Theorem 1 provides estimates for functions $h \in \mathcal{H}_{\alpha,\beta}$ in terms of $\|x\|$ and $h(0)$.

Theorem 1 *Every function $h \in \mathcal{H}_{\alpha,\beta}$ satisfies for each $x \in B^3$ the sharp inequalities*

$$L_{\alpha,\beta}(-\|x\|, h(0)) \leq h(x) \leq L_{\alpha,\beta}(\|x\|, h(0)),$$

where $L_{\alpha,\beta}$ is defined by (3).
Equality for $x \in B^3 \setminus \{0\}$ in the upper, resp., lower estimate implies that h is the extremal function

$$z \mapsto \int_{S(\bar{x},\gamma)} P_{\alpha,\beta}(z, y)\sigma(dy), \ z \in B^3 \setminus \{0\}$$

where $\gamma \in [0, \pi]$ *is defined by*

$$\cos \gamma = 1 - 2h(0)\,, \tag{7}$$

resp.,

$$z \mapsto \int\limits_{S(-\bar{x},\gamma)} P_{\alpha,\beta}(z,y)\sigma(dy)\,, \;\; z \in B^3 \setminus \{0\}$$

with the same γ.

The key ingredient for the proof of this result is the following lemma stated in a much more general setting than is actually needed here.

Let Ω be a set and $\mathcal{A}$ a σ-algebra of subsets of Ω. Denote by $M^b(\Omega)$ the cone of all positive measures with finite total mass. Assume the corresponding vector space $M^b(\Omega) - M^b(\Omega)$ to be partially ordered by $\mu_1 \leq \mu_2 :\Longleftrightarrow \mu_1(A) \leq \mu_2(A)$ for every $A \in \mathcal{A}$. Suppose that $K, K_0 : \Omega \longrightarrow \mathbb{R}$ are two $\mathcal{A}$–measurable functions which are integrable with respect to both μ_1 and μ_2.

Lemma *Let* $\mu_1, \mu_2 \in M^b(\Omega) - M^b(\Omega)$ *be such that* $\mu_1 \leq \mu_2$.

1. *Define for every* $t \in \mathbb{R}$ *the measures*

$$\bar{\mu}_t := \mathbb{1}_{\{K > t \cdot K_0\}}\mu_2 + \mathbb{1}_{\{K < t \cdot K_0\}}\mu_1$$

and suppose that $c \in \operatorname{range}(t \mapsto \int_\Omega K_0\, d\bar{\mu}_t)$, *i.e., there exists* t^* *such that*

$$\int\limits_\Omega K_0 d\bar{\mu}_{t^*} \;=\; c.$$

Then every solution $\bar{\mu}$ *of the extremal problem*

$$\sup_{\mu_1 \leq \mu \leq \mu_2} \int\limits_\Omega K\, d\mu$$

subject to

$$\int\limits_\Omega K_0\, d\mu \;=\; c$$

is of the form

$$\bar{\mu} = \bar{\mu}_{t^*} + \mathbb{1}_{\{K = t^* K_0\}}\nu$$

where $\mu_1 \leq \nu \leq \mu_2$.

2. Define for every $t \in \mathbb{R}$ the measures

$$\underline{\mu}_t := \mathbb{1}_{\{K>t\cdot K_0\}}\mu_1 + \mathbb{1}_{\{K<t\cdot K_0\}}\mu_2$$

and suppose that $c \in \operatorname{range}(t \mapsto \int_\Omega K_0\, d\underline{\mu}_t)$, i.e., there exists t_* such that

$$\int_\Omega K_0 d\underline{\mu}_{t_*} = c.$$

Then every solution $\underline{\mu}$ of the extremal problem

$$\inf_{\mu_1 \le \mu \le \mu_2} \int_\Omega K\, d\mu$$

subject to

$$\int_\Omega K_0\, d\mu = c$$

is of the form

$$\underline{\mu} = \underline{\mu}_{t_*} + \mathbb{1}_{\{K=t_* K_0\}}\nu$$

where $\mu_1 \le \nu \le \mu_2$.

Proof: 1. For every admissible μ and arbitrary $t \in \mathbb{R}$ we have

$$\begin{aligned}
\int_\Omega K\, d\mu &= \int_{\{K>t\cdot K_0\}} K - t\cdot K_0\, d\mu + \int_{\{K<t\cdot K_0\}} K - t\cdot K_0\, d\mu + t\cdot \int_\Omega K_0\, d\mu \\
&\le \int_{\{K>t\cdot K_0\}} K - t\cdot K_0\, d\mu_2 + \int_{\{K<t\cdot K_0\}} K - t\cdot K_0\, d\mu_1 + t \int_\Omega K_0\, d\mu \\
&= \int_{\{K>t\cdot K_0\}} K\, d\mu_2 + \int_{\{K<t\cdot K_0\}} K\, d\mu_1 + t\cdot \Big(c - \Big(\int_{\{K>t\cdot K_0\}} K_0\, d\mu_2 + \int_{\{K<t\cdot K_0\}} K_0\, d\mu_1\Big)\Big) \\
&= \int_\Omega K\, d\bar{\mu}_t + t\Big(c - \int_\Omega K_0\, d\bar{\mu}_t\Big).
\end{aligned}$$

Choosing $t = t^*$, the last expression in the sum vanishes according to the definition of c and t^*, and we obtain

$$\int_\Omega K\, d\bar{\mu}_{t^*} \le \sup\Big\{ \int_\Omega K\, d\mu \ :\ \mu_1 \le \mu \le \mu_2\,,\ \int_\Omega K_0\, d\mu = c \Big\} \le \int_\Omega K\, d\bar{\mu}_{t^*}$$

and hence equality. Equality for some μ in the very first inequality of the proof can only appear if we have both

$$\int_{\{K>t\cdot K_0\}} K - t\cdot K_0\, d(\mu_2 - \mu) = 0$$

and

$$\int_{\{K<t\cdot K_0\}} K - t\cdot K_0 \, d(\mu - \mu_1) \;=\; 0\,.$$

Since the integrands are either strictly positive or strictly negative it follows that μ coincides with μ_2 on $\{K > t\cdot K_0\}$ and with μ_1 on the set $\{K < t\cdot K_0\}$. Up to the condition $\mu_1 \le \mu \le \mu_2$ we can prescribe μ on the set $\{K = t\cdot K_0\}$ arbitrarily.

2. The proof of the second part is established by interchanging the role of μ_1 and μ_2 and repeating the arguments of the first part of the proof. ■

Proof of the Theorem: We specify in the lemma $\Omega = S^2$, $\mathcal{A}$ the σ-algebra of Borel subsets of S^2 , $\mu_1 = 0$, $\mu_2 = \sigma$, as well as

$$K = P_{\alpha,\beta}(x,\cdot)\,, \quad \text{where } x \in B^3$$

and

$$K_0 = P_{\alpha,\beta}(0,\cdot) \equiv 1\,.$$

First we consider the estimate from above. According to the lemma we have to examine the sets

$$\{y\in S^2 \mid P_{\alpha,\beta}(x,y) > t\}\,, \quad t>0\,.$$

We observe that for any fixed $x \in B^3 \setminus \{0\}$

$$s \mapsto \frac{(1-\|x\|^2)^\alpha}{(1-2\|x\|s+\|x\|^2)^\beta}$$

is a strictly increasing function on $[-1,1]$. Therefore the sets in question are in fact spherical caps $S(\bar{x},\vartheta)$ (possibly degenerated to S^2 or the empty set $\emptyset$ for certain values of t) with center $\bar{x} \in S^2$ and spherical angle $\vartheta \in [0,\pi]$. ϑ is uniquely determined by t, but it is not necessary to know the exact relation between t and ϑ.
Following the lemma we set $c = h(0)$ and determine $\gamma \in [0,\pi]$ such that

$$\int_{S(\bar{x},\gamma)} P_{\alpha,\beta}(0,y)\sigma(dy) = \int_{S(\bar{x},\gamma)} \sigma(dy) = h(0)$$

holds. After introducing spherical coordinates, an evaluation of the integral leads to

$$\int_{S(\bar{x},\gamma)} \sigma(dy) = \frac{1}{2}(1-\cos\gamma) = h(0)\,,$$

and hence (7). From the lemma and by introducing spherical coordinates directed along $\bar{x}$ we obtain, with γ as in (7), the desired sharp inequality:

$$h(x) = \int_{S^2} P_{\alpha,\beta}(x,y)\mu(dy) \le \int_{S(\bar{x},\gamma)} P_{\alpha,\beta}(x,y)\sigma(dy) = L_{\alpha,\beta}(\|x\|, h(0))\,. \tag{8}$$

Since $\{y \in S^2 \mid P_{\alpha,\beta}(z,y) = t\}$ is a σ-null set for all $z \in B^3 \setminus \{0\}$ the lemma yields that the extremal function causing equality in (8) is unique and of the asserted form

$$z \mapsto \int_{S(\bar{x},\gamma)} P_{\alpha,\beta}(z,y)\sigma(dy)$$

with γ determined by (7).

The proof of the estimate from below is analogous: for every $x \in B^3 \setminus \{0\}$, $t > 0$, the sets

$$\{y \in S^2 \mid P_{\alpha,\beta}(x,y) < t\}$$

are spherical caps $S(-\bar{x}, \vartheta)$ centered at $-\bar{x}$. According to the lemma we have to solve

$$\int_{S(-\bar{x},\gamma)} P_{\alpha,\beta}(0,y)\sigma(dy) = h(0)$$

for $\cos\gamma$ which yields (7). We proceed as before with one minor change: we deal with $S(-\bar{x},\gamma)$ instead of $S(\bar{x},\gamma)$ and introduce spherical coordinates directed along $-\bar{x}$. This merely has the effect that one has to substitute $-\|x\|$ for $\|x\|$ in the upper bound in (8) to get the lower bound $L_{\alpha,\beta}(-\|x\|, h(0))$ of $h(x)$. ■

Calculation of $L_{\alpha,\beta}$

If $x \neq 0$ we introduce spherical coordinates directed along $\bar{x}$ to compute $L_{\alpha,\beta}$:

$$L_{\alpha,\beta}(\|x\|, h(0)) = \frac{1}{\mathrm{vol}(S^2)} \int_0^{\gamma}\int_0^{2\pi} \frac{(1-\|x\|^2)^\alpha}{(1-2\|x\|\cos\phi + \|x\|^2)^\beta} \cdot \sin\phi \; d\psi\, d\phi$$

$$= \begin{cases} \dfrac{(1-\|x\|^2)^\alpha}{4\|x\|} \cdot \Big[\ln(1-2\|x\|\cos\phi + \|x\|^2)\Big]_0^\gamma & \text{for } \beta = 1 \\ \dfrac{(1-\|x\|^2)^\alpha}{4(\beta-1)\|x\|} \cdot \left[\dfrac{-1}{(1-2\|x\|\cos\phi + \|x\|^2)^{\beta-1}}\right]_0^\gamma & \text{for } \beta \neq 1 \end{cases}$$

$$= \begin{cases} \dfrac{(1-\|x\|^2)^\alpha}{4\|x\|} \cdot \ln\left(1 + \dfrac{4\|x\|}{(1-\|x\|)^2} h(0)\right) & \text{for } \beta = 1\,; \\ \dfrac{(1-\|x\|^2)^\alpha}{4(\beta-1)\|x\|} \cdot \left(\dfrac{1}{(1-\|x\|)^{2\beta-2}} - \dfrac{1}{((1-\|x\|)^2 + 4\|x\|h(0))^{\beta-1}}\right) & \text{for } \beta \neq 1\,. \end{cases}$$

We substitute specific values for α and β to obtain

Corollary 1 *1) If $h : B^3 \to \mathbb{R}$ is a Newtonian potential (i.e. $\alpha = 0$, $\beta = 1/2$) of a measure μ with support in S^2 and $0 \leq \mu \leq \sigma$, then for all $x \in B^3$ the following holds:*

$$\|x\| \cdot h(x) \begin{cases} \geq \dfrac{1}{2}\left(1 + \|x\| - \sqrt{(1+\|x\|)^2 - 4\|x\|h(0)}\right); \\ \leq \dfrac{1}{2}\left(\sqrt{(1-\|x\|)^2 + 4\|x\|h(0)} + \|x\| - 1\right). \end{cases}$$

2) If $h : B^3 \to \mathbb{R}$ is a harmonic function (i.e. $\alpha = 1$, $\beta = 3/2$), with $0 \le h \le 1$, then

$$\|x\| \cdot h(x) \begin{cases} \geq \dfrac{1 - \|x\|}{2} \left(\dfrac{1 + \|x\|}{\sqrt{(1 + \|x\|)^2 - 4\|x\|h(0)}} - 1 \right) \\ \\ \leq \dfrac{1 + \|x\|}{2} \left(1 - \dfrac{1 - \|x\|}{\sqrt{(1 - \|x\|)^2 + 4\|x\|h(0)}} \right) \end{cases}$$

holds for all $x \in B^3$.

3) If $h : B^3 \to \mathbb{R}$ is a hyperbolic-harmonic function (i.e. $\alpha = 2$, $\beta = 2$) with $0 \le h \le 1$, then

$$h(x) \begin{cases} \geq (1 - \|x\|)^2 \cdot \dfrac{h(0)}{(1 + \|x\|)^2 - 4\|x\|h(0)} \\ \\ \leq (1 + \|x\|)^2 \cdot \dfrac{h(0)}{(1 - \|x\|)^2 + 4\|x\|h(0)} \end{cases} \tag{9}$$

holds for all $x \in B^3$.

In all three cases, equality for $x \in B^3 \setminus \{0\}$ in the upper, resp., lower estimate implies that h equals the function

$$g_{+\bar{x}} : z \mapsto \int_{S(\bar{x},\gamma)} P_{\alpha,\beta}(z, y)\sigma(dy), \quad z \in B^3,$$

with γ as in (7),
resp.,

$$g_{-\bar{x}} : z \mapsto \int_{S(-\bar{x},\gamma)} P_{\alpha,\beta}(z, y)\sigma(dy), \quad z \in B^3,$$

with γ as in (7). Here one has to substitute the appropriate values for α and β.

We can generalize the last result concerning hyperbolic-harmonic functions.

Corollary 2 *Let $h : B^3 \to \mathbb{R}$ be hyperbolic-harmonic with $0 \le h \le 1$. Then for any $x, a \in B^3$ the inequality*

$$\left| \frac{h(x) - h(a)}{h(x) + h(a) - 2h(x)h(a)} \right| \leq 2 \cdot \frac{[x, a] \cdot \|x - a\|}{[x, a]^2 + \|x - a\|^2} \tag{10}$$

holds. Equality for some $x, a \in B^3$ implies that h equals either $g_{+\bar{x}} \circ T_a$ or $g_{-\bar{x}} \circ T_a$.

Proof: A short calculation shows that (9) is equivalent to

$$\left|\frac{h(x)-h(0)}{h(x)+h(0)-2h(x)h(0)}\right| \leq 2\cdot\frac{\|x\|}{1+\|x\|^2} \tag{11}$$

where equality for $x \in B^3 \setminus \{0\}$ occurs only if

$$h = g_{+\bar{x}} \quad \text{or} \quad h = g_{-\bar{x}}\,. \tag{12}$$

In [Ahlfors], p. 55, it is shown that $h \circ T_a$ is hyperbolic-harmonic whenever h is. Furthermore for any $a \in B^3$ the mapping $v \mapsto v \circ T_a$ from $\mathcal{H}_{2,2}$ to $i\mathcal{H}_{2,2}$ is one-to-one and order preserving. We use $\|T_a(x)\| = \frac{\|x-a\|}{[x,a]}$ (see [Ahlfors], p. 27) and apply (11) to $h = v \circ T_a$ to obtain

$$\begin{aligned}\left|\frac{h(x)-h(a)}{h(x)+h(a)-2h(x)h(a)}\right| &= \left|\frac{v(T_a(x))-v(0)}{v(T_a(x))+v(0)-2v(T_a(x))v(0)}\right| \\ &\leq 2\cdot\frac{\|T_a(x)\|}{1+\|T_a(x)\|^2} = 2\cdot\frac{[x,a]\cdot\|x-a\|}{[x,a]^2+\|x-a\|^2}\end{aligned}$$

for every $v \in \mathcal{H}_{2,2}$, i.e. $h \in \mathcal{H}_{2,2}$. From (12) one can see that equality in (10) for $x \in B^3 \setminus \{0\}$ occurs only if

$$h = g_{+\bar{x}} \circ T_a \quad \text{or} \quad h = g_{-\bar{x}} \circ T_a\,.$$

■

5. Theorem 2

Theorem 2 provides estimates for the radial derivatives of functions $h \in \mathcal{H}_{\alpha,\beta}$ in terms of $\|x\|$ and $h(x)$.

Theorem 2 *If $h \in \mathcal{H}_{\alpha,\beta}$, then for every $x \in B^3 \setminus \{0\}$ the inequality*

$$\frac{1}{\|x\|}R_{\alpha,\beta}(-\|x\|, h(x)) \leq \partial_r h(x) \leq \frac{1}{\|x\|}R_{\alpha,\beta}(\|x\|, h(x)) \tag{13}$$

holds. Equality for $x \in B^3 \setminus \{0\}$ in the upper, resp., lower estimate implies that h equals the extremal function

$$z \mapsto \int_{S(\bar{x},\tau)} P_{\alpha,\beta}(z,y)\sigma(dy)\,, \quad z \in B^3\,, \tag{14}$$

where $\tau \in [0,\pi]$ is given by (5), resp., h equals the extremal function

$$z \mapsto \int_{S(-\bar{x},\tau)} P_{\alpha,\beta}(z,y)\sigma(dy)\,, \quad z \in B^3\,, \tag{15}$$

with $\tau \in [0,\pi]$ *given by*

$$\cos\tau = \begin{cases} \dfrac{1+\|x\|^2}{-2\|x\|} + \dfrac{1}{2\|x\|}\left(\dfrac{1}{(1+\|x\|)^{2\beta-2}} + \dfrac{4(\beta-1)\|x\|h(x)}{(1-\|x\|^2)^\alpha}\right)^{\frac{1}{1-\beta}} & \textit{for } \beta \neq 1; \\ \dfrac{1+\|x\|^2}{-2\|x\|} + \dfrac{(1+\|x\|)^2}{2\|x\|}\exp\left(-\dfrac{4\|x\|h(x)}{(1-\|x\|^2)^\alpha}\right) & \textit{for } \beta = 1. \end{cases} \tag{16}$$

Proof: Since

$$\|x\| \cdot \partial_r h(x) = \int_{S^2} \langle x, \nabla_x P_{\alpha,\beta}(x,y)\rangle \mu(dy), \quad x \in B^3,$$

whenever μ is the representing measure of h, we specify in the lemma as follows: $\Omega = S^2$, $\mathcal{A}$ is the σ-algebra of Borel subsets of S^2, $\mu_1 = 0$, $\mu_2 = \sigma$, as well as

$$K = \langle x, \nabla_x P_{\alpha,\beta}(x,\cdot)\rangle = \|x\| \cdot \partial_r P_{\alpha,\beta}(x,\cdot),$$

and

$$K_0 = P_{\alpha,\beta}(x,\cdot).$$

A short reckoning shows that

$$\langle x, \nabla_x P_{\alpha,\beta}(x,\cdot)\rangle = 2 \cdot P_{\alpha,\beta}(x,y)\left(\beta \cdot \frac{\langle y,x\rangle - \|x\|^2}{\|y-x\|^2} - \alpha \cdot \frac{\|x\|^2}{1-\|x\|^2}\right).$$

Hence for $x \in B^3 \setminus \{0\}$, $y \in S^2$, $t \in \mathbb{R}$,

$$\langle x, \nabla_x P_{\alpha,\beta}(x,y)\rangle > t \cdot P_{\alpha,\beta}(x,y)$$

is equivalent to

$$2\|x\|\left(\beta \cdot \frac{\langle y,\bar{x}\rangle - \|x\|}{1-2\|x\|\langle \bar{x},y\rangle + \|x\|^2} - \alpha \cdot \frac{\|x\|}{1-\|x\|^2}\right) > t.$$

The function

$$s \mapsto \beta \cdot \frac{s-\|x\|}{1-2\|x\|s+\|x\|^2} - \alpha \cdot \frac{\|x\|}{1-\|x\|^2}$$

is a strictly increasing function of $s \in [-1,1]$ for any α, $\beta \in \mathbb{R}$, $\beta > 0$, $x \in B^3 \setminus \{0\}$. This implies that the sets in question, namely

$$\{y \in S^2 \mid \langle x, \nabla_x P_{\alpha,\beta}(x,y)\rangle > t \cdot P_{\alpha,\beta}(x,y)\}, \quad t \in \mathbb{R}$$

are spherical caps (possibly degenerated to S^2 or $\emptyset$) with center $\bar{x}$ and spherical angle ϑ, where ϑ depends uniquely on t. Setting $c = h(x)$ in the lemma, we have to solve the following equation with respect to $\cos\tau$:

$$h(x) = \frac{1}{\mathrm{vol}(S^2)} \int_0^{\tau}\int_0^{2\pi} \frac{(1-\|x\|^2)^\alpha}{(1-2\|x\|\cos\phi+\|x\|^2)^\beta} \cdot \sin\phi \, d\psi \, d\phi$$

$$= \begin{cases} \dfrac{(1-\|x\|^2)^\alpha}{4\|x\|}\Big(\ln(1-2\|x\|\cos\tau+\|x\|^2) - 2\ln(1-\|x\|)\Big) & \text{for } \beta = 1\,; \\[2ex] \dfrac{(1-\|x\|^2)^\alpha}{4(\beta-1)\|x\|}\left(\dfrac{1}{(1-\|x\|)^{2\beta-2}} - \dfrac{1}{(1-2\|x\|\cos\tau+\|x\|^2)^{\beta-1}}\right) & \text{for } \beta \neq 1\,. \end{cases} \tag{17}$$

The integral is evaluated by introducing spherical coordinates directed along $\bar{x}$. (17) directly leads to (5). According to the lemma we conclude

$$\|x\| \cdot \partial_r h(x) \leq \int_{S(\bar{x},\tau)} \langle x, \nabla_x P_{\alpha,\beta}(x,y)\rangle \sigma(dy) = R_{\alpha,\beta}(\|x\|, h(x))\,. \tag{18}$$

Realizing that

$$\{y \in S^2 \mid \langle x, P_{\alpha,\beta}(x,y)\rangle = t \cdot P_{\alpha,\beta}(x,y)\}$$

is a σ–null set for every $t \in \mathbb{R}$ we can conclude as before that equality for $x \in B^3 \setminus \{0\}$ can only occur if h equals the extremal function

$$z \mapsto \int_{S(\bar{x},\tau)} P_{\alpha,\beta}(z,y)\sigma(dy)\,, \quad z \in B^3$$

where $\tau \in [0,\pi]$ is given by (5).

To get the lower estimate we consider the sets

$$\{y \in S^2 \mid \langle x, P_{\alpha,\beta}(x,y)\rangle < t \cdot P_{\alpha,\beta}(x,y)\}\,, \quad t \in \mathbb{R}\,.$$

Obviously these are spherical caps $S(-\bar{x},\vartheta)$ centered at $-\bar{x}$. We proceed as before, but with a substitution of $S(\bar{x},\tau)$ by $S(-\bar{x},\tau)$ in (14) to get (15). For the evaluation of the integral in (17) we now use spherical coordinates directed along $-\bar{x}$. This merely has the effect that one has to substitute $-\|x\|$ for $\|x\|$ in (17) (which yields (16)) and in $R_{\alpha,\beta}(\|x\|, h(x))$, the right hand side of (18), to get the lower bound $R_{\alpha,\beta}(-\|x\|, h(x))$ of $\|x\| \cdot \partial_r h(x)$. ■

Calculation of $R_{\alpha,\beta}$

Again we introduce spherical coordinates directed along $\bar{x}$ to compute $R_{\alpha,\beta}$:

$$\begin{aligned} R_{\alpha,\beta}(\|x\|, h(x)) &= \frac{1}{\mathrm{vol}(S^2)} \int_0^\tau \int_0^{2\pi} \frac{(1-\|x\|^2)^\alpha}{(1-2\|x\|\cos\phi + \|x\|^2)^\beta} \\ &\quad \cdot \left(2\beta \cdot \frac{\|x\| \cdot \cos\phi - \|x\|^2}{1-2\|x\|\cos\phi + \|x\|^2} - 2\alpha \cdot \frac{\|x\|^2}{1-\|x\|^2}\right) \cdot \sin\phi \, d\vartheta \, d\phi \\ &= \int_0^\tau \frac{(1-\|x\|^2)^\alpha}{(1-2\|x\|\cos\phi + \|x\|^2)^\beta} \cdot \|x\| \\ &\quad \cdot \left(\beta \cdot \frac{\cos\phi - \|x\|}{1-2\|x\|\cos\phi + \|x\|^2} - \alpha \cdot \frac{\|x\|}{1-\|x\|^2}\right) \cdot \sin\phi \, d\phi \, . \end{aligned}$$

In the case $\beta = 1$ we obtain with τ as in (5) for $x \in B^3 \setminus \{0\}$

$$\begin{aligned} & R_{\alpha,1}(\|x\|, h(x)) = \\ &= \left[-\frac{(1-\|x\|^2)^\alpha}{4\|x\|} \cdot \left(\left(1 + 2\alpha \cdot \frac{\|x\|^2}{1-\|x\|^2}\right) \right.\right. \\ &\quad \left.\left. \cdot \ln(1-2\|x\|\cos\phi + \|x\|^2) + \frac{1-\|x\|^2}{1-2\|x\|\cos\phi + \|x\|^2} \right) \right]_0^\tau \\ &= \frac{(1-\|x\|^2)^{\alpha-1}}{4\|x\|(1-\|x\|)^2} \left(1 - \exp\left(-\frac{4h(x)\|x\|}{(1-\|x\|^2)^\alpha}\right)\right) - h(x)\left(1 + 2\alpha \frac{\|x\|^2}{1-\|x\|^2}\right) . \qquad (19) \end{aligned}$$

In the case $\beta \neq 1$ we obtain with τ as in (5) for $x \in B^3 \setminus \{0\}$

$$\begin{aligned} & R_{\alpha,\beta}(\|x\|, h(x)) = \\ &= \left[\frac{1}{2} \cdot \frac{(1-\|x\|^2)^\alpha}{(1-2\|x\|\cos\phi + \|x\|^2)^{\beta-1}} \right. \\ &\quad \left. \cdot \left(\frac{\|x\| - \cos\phi}{1-2\|x\|\cos\phi + \|x\|^2} + \frac{1}{2\|x\|(\beta-1)} \left(1 + 2\alpha \frac{\|x\|^2}{1-\|x\|^2}\right) \right) \right]_0^\tau \\ &= \frac{(1-\|x\|^2)^\alpha}{4\|x\|} \cdot \left(\frac{2\alpha\|x\|^2 + \beta(1-\|x\|^2)}{(\beta-1)(1-\|x\|^2)} \cdot k - (1-\|x\|^2) k^{\frac{\beta}{\beta-1}} - \right. \\ &\quad \left. - \frac{(2\alpha - 2\beta + 1)\|x\|^2 - 2(\beta-1)\|x\| + 1}{(\beta-1) \cdot (1+\|x\|) \cdot (1-\|x\|)^{2\beta-1}} \right) , \qquad (20) \end{aligned}$$

where

$$k := \frac{1}{(1-\|x\|)^{2\beta-2}} - \frac{4(\beta-1)\|x\| \cdot h(x)}{(1-\|x\|^2)^\alpha} .$$

In each of the cases already considered in Corollary 1 the equations (19), (20) and hence (13) simplify remarkably.

Corollary 3 *1) If $h : B^3 \to \mathbb{R}$ is a Newtonian potential of a measure μ with support in S^2 (i.e. $\alpha = 0$, $\beta = 1/2$) and $0 \leq h \leq 1$, then*

$$-\frac{h(x) \cdot (1 - h(x))}{1 - \|x\|(2h(x) - 1)} \leq \partial_r h(x) \leq \frac{h(x) \cdot (1 - h(x))}{1 + \|x\|(2h(x) - 1)}$$

holds for all $x \in B^3 \setminus \{0\}$.

2) If $h : B^3 \to \mathbb{R}$ is a harmonic function (i.e. $\alpha = 1$, $\beta = 3/2$) with $0 \leq h \leq 1$, then

$$\frac{-3 - \|x\|(2h(x) - 1)}{1 - \|x\|^2} h(x) \cdot (1 - h(x)) \leq \partial_r h(x) \leq \frac{3 - \|x\|(2h(x) - 1)}{1 - \|x\|^2} h(x) \cdot (1 - h(x))$$

holds for all $x \in B^3 \setminus \{0\}$.

3) If $h : B^3 \to \mathbb{R}$ is a hyperbolic-harmonic function (i.e. $\alpha = 2$, $\beta = 2$) with $0 \leq h \leq 1$, then

$$-\frac{4}{1 - \|x\|^2} h(x) \cdot (1 - h(x)) \leq \partial_r h(x) \leq \frac{4}{1 - \|x\|^2} h(x) \cdot (1 - h(x))$$

holds for all $x \in B^3 \setminus \{0\}$.

The extreme functions in each case are obtained by substituting the specific values for α and β in equation (14) and (5), resp., (15) and (16).

Remark The result in Corollary 1, case 2), 3), can already be found in a slightly different form in [Burgeth]. Corollary 3, case 2), gives a sharper estimate of the radial derivative of harmonic functions than the one achieved by [Khavinson].

6. Acknowledgements

The author would like to thank Prof. H. Leutwiler and Prof. H. Bauer for their support and encouragement during the preparation of this paper, and Markus Denzer for drawing the author's attention to the paper of F. Hmissi.

The author would like to thank the Scientific Affairs Division of the NATO.

7. References

Ahlfors, L.V. *Möbius transformations in several dimensions.* Ordway Lectures in Mathematics, University of Minnesota (1981).

Axler, S., Bourdon, P. and Ramey, W. *Harmonic Function Theory.* Springer, New York (1992).

Burckel, R.B. *An Introduction to Classical Complex Analysis.* Volume 1, Birkhäuser, Basel (1979).

Burgeth, B. A Schwarz Lemma for harmonic and hyperbolic-harmonic functions in higher dimensions. Manuscripta Math. **77**, 283–291 (1992).

Hmissi, F. Fonctions harmoniques pour les potentiels de Riesz sur la boule unité. Preprint.

Khavinson, D. An extremal problem for harmonic functions in the ball. Canad. Math. Bull. **35** (2), 218–220 (1992).

Leutwiler, H. Best constants in the Harnack inequality for the Weinstein equation. Aequationes Mathematicae **34**, 304–315 (1987).

DUALITY OF H-CONES

S.-L. ERIKSSON-BIQUE
University of Joensuu
Department of Mathematics
P.O. Box 111
SF-80101 Joensuu
Finland

ABSTRACT. A potential-theoretic model of an H-cone is the cone of positive superharmonic functions on a harmonic space. We consider duals of H-cones. The dual of an H-cone is the set of extended real-valued additive and left order continuous functions finite on a dense set. Since the introduction of H-cones it has been unknown how well an H-cone can be embedded into its second dual. We have proved earlier that an H-cone is always specifically solid in its second dual and even solid and increasingly dense if a special unit exists. In this paper we shall improve these results so that they cover all potential-theoretic examples.

1. Introduction

An H-cone, studied notably by Boboc, Bucur and Cornea, is a subcone of a positive cone in a vector lattice that has two partial orders initial order and specific order denoted by $\leq$ and $\preceq$, respectively. Any pair of elements u and v of an H-cone has a reduced function

$$R(u-v) = \min \{w | w \geq u-v\}$$

and $R(u-v) \preceq u$. Moreover it is known that an H-cone may be extended to a vector lattice that is Dedekind complete with respect to the specific order ([2, Theorem 2.1.5]). Otherwise, it is not completely understood how specific order and initial order are related in H-cones. Our results show that mixed envelopes defined by Arsove and Leutwiler in algebraic potential theory ([1]) play a very important role when properties of initial and specific order are studied.

In our main theorem we show that an H-cone is solid and increasingly dense in its second dual if it possesses a generating set defined in terms of mixed envelopes. Then we verify that a cone of positive superharmonic functions on any harmonic space possesses a generating set. Essential concepts are defined in the next section.

2. Preliminaries

We use the following definition for an H-cone.

Definition 2.1. *Let E be an ordered vector space and S be a convex subcone of E^+ such that $E = S - S$. The cone S is called an H-cone in E if it possesses the following properties:*

K. GowriSankaran et al. (eds.), Classical and Modern Potential Theory and Applications, 149–157.

(A_1) any non-empty upward directed and dominated subset F of S has the least upper bound in E that belongs to S;

(A_2) any non-empty subset F of S has the least element in E that belongs to S;

(A_3) for any s and t in S the set $\{ u \in S | s - t \leq u \}$ has the least element denoted by $R(s-t)$ satisfying the condition $s - R(s-t) \in S$.

It is known that the preceding definition is equivalent with the original one (see, e.g., [8, Theorem 1.3]). For the reference to the theory of H-cones we refer to [2]. By [7, Theorem 2.5] the Axiom (A_2) follows from (A_1) and (A_3) as stated next.

Theorem 2.2. *Let E be an ordered vector space and S be a convex subcone of E^+ such that $E = S - S$. Then S is an H-cone in E if and only if Axioms (A_1) and (A_3) hold in S.*

We denote the least upper bound and the greatest lower bound of a set E by $\bigvee E$ and $\bigwedge E$, respectively. In addition of the initial order $\leq$ we consider also another partial order called *specific order* denoted by $\preceq$ and defined in an H-cone by

$$s \preceq t \Leftrightarrow t = s + s' \quad \text{for some } s' \in S.$$

If S is an H-cone in E, then by [2, Theorem 2.1.5] the set $E = S - S$ is a Dedekind complete with respect to the specific order.

Any pair of elements in an H-cone has unsymmetrical mixed envelopes defined by Arsove and Leutwiler in algebraic potential theory ([1] and other references there).

Theorem 2.3. *Let S be an H-cone. Then any pair of elements s and t in S admits a mixed lower envelope*

$$s \curlywedge t = \max\{ x \in S | x \preceq s, \ x \leq t \}$$

and a mixed upper envelope

$$s \curlyvee t = \min\{ x \in s | x \succeq s, \ x \geq t \}$$

satisfying the equality

$$s \curlywedge t + t \curlyvee s = s + t.$$

Proof. See [1, Theorem 2.5].

Let S be an H-cone. An element u of S is called a *generator*, if

$$x = \bigvee_{n \in \mathbf{N}} (nu) \curlywedge x$$

for any $x \in S$. Recall that a subset T of S is called *solid* if for all $s \in T$ the condition $t \leq s$ implies that $t \in T$. A subset T of S is called *order dense* in S if the condition

$$s = \bigvee \{ t \in T | t \leq s \}$$

holds for all s in S. A subset T of S is *increasingly dense* if it is dense and the set $\{t \in T | t \leq s\}$ is upward directed for any $s \in S$. Similarly we define the concepts *specifically solid* and *specifically order dense* with respect to the specific order.

Let $(S, \leq)$ be an H-cone. We recall that a function φ from S into a partially ordered set T is called *left order continuous*, if $\varphi(\bigvee F) = \bigvee_{f \in F} \varphi(f)$ for any non-empty upward directed bounded subset F of S. A left order continuous additive function from S into $\mathbf{R}^+ \cup \{+\infty\}$ is called a *hyperharmonic integral.* Hyperharmonic integrals may be extended to hyperharmonic morphisms defined in [6] If a hyperharmonic integral $\varphi : S \to \mathbf{R}^+ \cup \{+\infty\}$ is finite on a dense set of S it is called an *H-integral* (studied in [2]). The set of H-integrals on S is denoted by S^*. A partial ordering in S^* is defined by $\mu \leq \psi$ if $\mu(s) \leq \psi(s)$ for all $s \in S$.

Note that the set S^* is also an H-cone by [2, Theorem 2.2.6]. *The evaluation map* $x \mapsto \hat{x}$ from S into S^{**} is defined by

$$\hat{x}(\mu) = \mu(x)$$

for $x \in S$ and $\mu \in S^*$. The main result in [9, Theorem 2.11 and Theorem 2.14] states the following

Theorem 2.4. *Let S be an H-cone. The image of S through the evaluation map is specifically solid in S^{**}. Moreover, if S possesses a generator then the image of S through the evaluation map is solid and increasingly dense in S^{**}.*

3. Embedding theorems

The set of positive superharmonic functions on a harmonic space does not have necessarily a generator and therefore Theorem 2.4 cannot be applied in all potential-theoretic examples. Although Theorem 2.4 holds for the cone of positive superharmonic functions on P-harmonic spaces with a countable base by [9, Theorem 2.15], we are interested to improve the result so that it covers all cones of positive superharmonic functions on any harmonic spaces. We shall first generalize the idea of a generator. Recall that a mapping B from an H-cone S into S a balayage if it is additive and left order continuous.

Definition 3.1. *Let S be an H-cone. An element u in S is generating a balayage, if the mapping B_u from S into S defined by*

$$B_u(x) = \bigvee_{n \in \mathbf{N}} (nu) \curlywedge x$$

is a balayage. A set C is called a generating set if all elements of C are generating a balayage and the property

$$x = \bigvee_{\substack{n \in \mathbf{N} \\ u \in C}} (nu) \curlywedge x$$

holds for any element $x \in S$.

Elements generating balayages are characterized in [10] and [11]. We need the following result.

Lemma 3.2. *If an H-cone S has a generating set, it has an upward directed generating set.*

Proof. Let C be a generating set of an H-cone S and assume that $u \in S$ and v$\in S$. We shall show that the element $u+v$ generates a balayage. By [10, Proposition 2.5] the element $u+v$ generates a balayage if the condition

$$u+v = \bigvee_{\substack{n\in\mathbf{N}\\ f\in F}} (n(u+v)) \curlywedge f$$

holds for any upward directed family F with $\bigvee F = u$. Let F be an upward directed family with $\bigvee F = u$. Then we have $\bigvee_{f\in F} R(f-u) = v$ and $\bigvee_{f\in F} R(f-v) = u$ by [1, Lemma 4.5]. Moreover the inequality

$$f \geq R(f-u) + R(f-v) \tag{1}$$

follows from

$$f - R(f-v) \geq f - u.$$

Hence we obtain

$$\bigvee_{\substack{n\in\mathbf{N}\\ f\in F}} n(u+v) \curlywedge f \geq \bigvee_{\substack{n\in\mathbf{N}\\ f\in F}} nu \curlywedge R(f-v) + \bigvee_{\substack{n\in\mathbf{N}\\ f\in F}} nv \curlywedge R(f-u).$$

Since u and v generates a balayage we infer

$$u+v \geq \bigvee_{\substack{n\in\mathbf{N}\\ f\in F}} (u+v) \curlywedge f \geq u+v.$$

Hence the element $u+v$ generates a balayage and so the set $\left\{\sum_{n=1}^{k} u_k | u_k \in C, k \in \mathbf{N}\right\}$ is an upward directed generating set.

We define an equivalence relation $\approx$ in an $H-$cone S by

$$s \approx t \quad \text{if } \mu(s) = \mu(t) \text{ for all } \mu \in S^*.$$

Denote the equivalence classes with respect to the relation $\approx$ by $[S]$. Applying [9, Theorem 2.6] we infer that the set $[S]$ is an H-cone under the partial order, addition and multiplication by positive real numbers defined as follows

$$[s] \leq [t] \Leftrightarrow \mu(s) \leq \mu(t) \quad \text{for all } \mu \in S^*,$$
$$[s] + [t] = [s+t], \quad \alpha[s] = [\alpha s].$$

An $H-$cone $[S]$ has the following properties:

Lemma 3.3. *Let S be an $H-$cone. Then the H-cones S^* and $[S]^*$ are order isomorphic. Moreover for all s and t in S we have*

(a) $R([s]-[t]) = [R(s-t)]$,

(b) $[s] \mathbin{\prec\!\!\!\backslash} [t] = [s \mathbin{\prec\!\!\!\backslash} t]$,

(c) $[s] \bigvee [t] = [s \bigvee t]$,

(d) $[s] \bigwedge [t] = [s \bigwedge t]$,

(e) If $[F]$ is directed upwards in $[S]$ and $\bigvee [F] = [s]$, then there exists an upward directed family G with $s = \bigvee G$ and any element $[g]$ in $[G]$ is majorized by $[f]$ for some $f \in F$.

Proof. The sets S^* and $[S]^*$ are order isomorphic by [9, Proposition 2.8]. The property (a) follows from [3, Proposition 3]. The second one is obtained form (a), since the equality

$$[s \mathbin{\prec\!\!\!\backslash} t] + [R(s-t)] = [s] = [s] \mathbin{\prec\!\!\!\backslash} [t] + R([s]-[t])$$

holds by [1, (3.5)]. The properties (c), (d) follows from the proof of [9, Lemma 2.5]. The last assertions is valid by [9, Lemma 2.7].

Lemma 3.4. *Let S be an H-cone. If an element u in S generates a balayage then $[u]$ generates a balayage in $[S]$. Moreover, if C is a generating family in S then $[C]$ is a generating family in $[S]$ and $[C]$ is directed upwards provided that C is directed upwards.*

Proof. Assume that B_u is a balayage generated by $u \in S$. Suppose that $[F]$ is directed upwards and $[u] = \bigvee [F]$. Reviewing to [10, Proposition 2.5] we only have to prove that

$$\bigvee_{\substack{n \in \mathbf{N} \\ f \in F}} (n[u]) \mathbin{\prec\!\!\!\backslash} [f] = [u].$$

By Lemma 3.3 there exists an upward directed family G with $u = \bigvee G$ and any element $[g]$ is majorized by $[f]$ for some $f \in F$. Hence we have

$$[u] \geq \bigvee_{\substack{n \in \mathbf{N} \\ f \in F}} n[u] \mathbin{\prec\!\!\!\backslash} [f] \geq \bigvee_{\substack{n \in \mathbf{N} \\ g \in F}} n[u] \mathbin{\prec\!\!\!\backslash} [g] = \left[\bigvee_{\substack{g \in G \\ n \in \mathbf{N}}} (nu) \mathbin{\prec\!\!\!\backslash} g \right] = [u],$$

since u generates a balayage. Consequently also $[u]$ generates a balayage. The second assertion of the lemma follows from the first one.

Lemma 3.5. *Let S be an H-cone and u an element of S. Then the set*

$$S_u = \left\{ w \in S | \bigvee_{n \in \mathbf{N}} (nu) \mathbin{\prec\!\!\!\backslash} w = w \right\}$$

is an H-cone possessing a generator when addition, multiplication by positive real numbers and partial order are the same as in S. Moreover the image of S_u through the evaluation map is increasingly dense and solid in S_u^{**}.

Proof. Let F be an upward directed bounded set in S_u. Then we have

$$f = \bigvee_{\substack{n \in \mathbf{N} \\ f \in F}} nu \curlywedge f \leq \bigvee nu \curlywedge \left(\bigvee F\right) \leq \bigvee F$$

for all $f \in F$. Hence the element $\bigvee F$ belongs to S_u. We prove that S_u is a specifically solid subset of S. Assume that t belong to S_u and $s \preceq t$ for some $s \in S$. Since

$$(nu) \curlywedge s + (mu) \curlywedge (t-s) \leq (nu+mu) \curlywedge t \leq (nu+mu) \curlywedge s + (nu+mu) \curlywedge (t-s)$$

by [1, (3.4)] we obtain

$$\bigvee_{n \in \mathbf{N}} (nu) \curlywedge s + \bigvee_{n \in \mathbf{N}} (mu) \curlywedge (t-s) = t = s + (t-s)\,.$$

Thus the element s belongs to S_u and therefore the set S_u is specifically solid. Assume that v and w are arbitrary elements of S_u. Then from the inequality $R(v-w) \preceq v$ we infer that $R(v-w)$ belongs to S_u. Consequently, the set S_u is an H-cone. The last assertion follows directly from Theorem 2.4.

Theorem 3.6. *Let S be an H-cone possessing a generating set. Then the image of S through the evaluation map is increasingly dense and solid in S^{**}.*

Proof. Denote by C an upward directed generating set of an H-cone S. Assume first that the set S^* separates the points of S. Let $\mu \in S^{**}$ and suppose there exists s in S such that $\widehat{s} \geq \mu$. Let $u \in C$. Then the set

$$S_u = \left\{ w \in S| \bigvee_{n \in \mathbf{N}} (nu) \curlywedge w = w \right\}.$$

is an H-cone by Lemma 3.5 under the same addition, partial order and multiplication by positive real numbers as in S. We first show that the set

$$\mathcal{F} = \left\{ f_{|S_u} | f \in S^* \right\}$$

is solid and increasingly dense in S_u^*. Suppose that $g \in S_u^*$ and $g \leq f$ for some $f \in \mathcal{F}$. Then the mapping $g \circ B_u$ is a hyperharmonic integral from S into $\mathbf{R} \cup \{+\infty\}$ and it satisfies the condition

$$g \circ B_u \leq f \circ B_u \leq f.$$

Since f is cancellable, the function $g \circ B_u$ is cancellable and therefore an H-integral by [6, Theorem 6.24]. Since $(g \circ B_u)_{|S_u} = g$, the set $\mathcal{F}$ is solid in S_u^*. The set $\mathcal{F}$ is also increasingly dense, as for any $x \in S_u$ there exists $f \in S^*$ such that $f(x) > 0$ and so by [5, Proposition 4.2]

$$g = \bigvee_{f \in \mathcal{F}} g \wedge f.$$

Define $\mu_u : \mathcal{F} \to \mathbf{R} \cup \{+\infty\}$ by

$$\mu_u(f) = \mu(f \circ B_u).$$

Then μ_u is additive and left order continuous. In addition it satisfies the condition

$$\mu_u(f) = \mu(f \circ B_u) \leq \widehat{s}(f \circ B_u) = f\left(\bigvee_{n \in \mathbf{N}} (nu) \curlywedge s\right) = f(B_u(s))$$

for all $f \in \mathcal{F}$. Hence we obtain $\mu_u \leq \widehat{B_u(s)}$, which implies that the mapping μ_u may be extended to an H-integral on S_u^* denoted also by μ_u. Applying Lemma3.5 we find an element w_u in S_u such that $\mu_u = \widehat{w_u}$ on S_u^*. Since C is a generating set we infer that $f = \sup_{u \in C} f \circ B_u$. Consequently we have

$$\mu(f) = \sup_{u \in C} \mu(f \circ B_u) = \sup_{u \in C} \mu_u(f) = \sup_{u \in C} \widehat{w_u}(f)$$

for all $f \in S^*$. Applying Lemma 3.3 we may direct the set $V = \{v \in S \mid \widehat{v} \leq \mu, \widehat{v} \leq \widehat{s}\}$ upwards. Since S^* separates the points of S the condition $\widehat{v} \leq \widehat{s}$ implies that $v \leq s$. Hence the set V has the greatest element denoted by $\widehat{m}$ and $\mu = \widehat{m}$.

Assume now that μ is an arbitrary element in S^{**}. As S^* separates the point of S, we note that

$$\mu = \bigvee_{s \in S} \mu \wedge \widehat{s}$$

by [9, Proposition 4.2]. Hence $\widehat{S}$ is increasingly dense and solid in S^{**} provided that S^* separates the point of S. Since $[S]^*$ separates the points of $[S]$ the set $\widehat{[S]}$ is increasingly dense and solid in $\widehat{[S]}^{**}$. But this means that also $\widehat{S}$ is increasingly dense and solid in S^{**} completing the proof.

There is a close connection between Radon-measures and H-integrals on harmonic spaces. We use the same notations as in [4, p. 7].

Theorem 3.7. *Let X be a P-harmonic space in the sense of Constantinescu and Cornea [4, p. 30] and $\mathcal{S}$ be a cone of positive superharmonic functions on X. If μ is a Radon-measure on X, then the mapping $f \to \int^* f\,d\mu$ for $f \in \mathcal{S}$ is a hyperharmonic integral. Moreover, if μ is a hyperharmonic integral from S into $\mathbf{R}^+ \cup \{+\infty\}$then there exists an upward directed family $\mathcal{F}$ of Radon-measures such that $\mu(f) = \sup_{\nu \in \mathcal{F}} \int^* f d\nu$ for any $f \in \mathcal{S}$ and the mapping $f \to \int^* f\,d\nu$ is an H-integral for any $\nu \in \mathcal{F}$.*

Proof. Let $\mathcal{P}_c$ be a set of real continuous potential on X harmonic outside a compact set. Clearly the mapping $f \to \int^* f\,d\mu$ for $f \in \mathcal{S}$ is a hyperharmonic integral for any Radon measure μ.

For the proof of the second statement denote by ϵ_x the Dirac measure on x. Then the mapping $f \to \int^* f\,d\epsilon_x$ for $f \in \mathcal{S}$ is an H-integral on S, since by [4, Corollary 2.3.1] the set $\mathcal{P}_c$ is increasingly dense in $\mathcal{S}$. Assume that μ is a hyperharmonic integral from S into $\mathbf{R}^+ \cup \{+\infty\}$. Denote by G the set of Radon measures of the form $\sum_{k=1}^n \alpha_k \epsilon_{x_k}$ where $\alpha_k \geq 0$

and $x_k \in X$As a finite sum of H-integrals the mapping $\overline{\nu} : f \to \int^* f \, d\nu$ for $f \in \mathcal{S}$ is an H-integral for any $\nu \in \mathcal{G}$. Moreover, by [5, Proposition 4.2] we have

$$\mu = \bigvee_{\nu \in \mathcal{G}} \mu \wedge \overline{\nu} = \sup_{\nu \in \mathcal{G}} \mu \wedge \overline{\nu}.$$

Using [4, Lemma p.160 and Proposition 7.1.1] we note that for each $\nu \in \mathcal{G}$ there exists a unique Radon-measure $\widetilde{\nu}$ on X such that $\mu \wedge \overline{\nu}(f) = \int^* f \, d\widetilde{\nu}$ which implies the assertion.

We recall that an element f of an H-cone S is s-continuous for any $s \in S$ if for any upward directed non-empty set F with $f = \bigvee F$ and $\epsilon > 0$ there exists an element f_ϵ such that $f \leq f_\epsilon + \epsilon s$. Using this concept we can apply Theorem 3.6 in any harmonic spaces as stated next.

Theorem 3.8. *Let $\mathcal{S}$ be a cone of positive superharmonic functions on a harmonic space X in the sense of Constantinescu and Cornea. Then $\mathcal{S}$ is an H-cone admits a generating set and therefore the image of S through the evaluation map is increasingly dense and solid in S^{**}.*

Proof. By [2, p. 69] the set S is an H-cone. We verify that the set

$$\mathcal{F} = \{Rf \mid f \in \mathcal{K}^+(X), f \equiv 0 \text{ or } \operatorname{supp} f \subset \{x \in X | s(x) > 0\} \quad \text{for some } s \in S\}$$

is a generating set in S. If a non-zero function Rf belongs to $\mathcal{F}$ then there exists a superharmonic function s satisfies $s > \beta$ on supp f for some strictly positive real number β. Moreover there exists a strictly positive real number α such that $f \leq \alpha s$ for and so by [4, Proposition 2.2.3] the function Rf is continuous. We claim that Rf is s-continuous. Indeed, if $Rf = \sup_{g \in G} g$ and $G \subset \mathcal{S}$ is directed upwards then by Dini's theorem for any $\epsilon > 0$ there exists a function g in G so that $f \leq g + \beta\epsilon$. Hence we have $f \leq g + \epsilon s$ which implies $Rf \leq g + \epsilon s$ and Rf is s-continuous. Applying [10, Lemma 2.7] we note that the superharmonic function Rf generates a balayage. Since $u = \sup\{Rg \mid g \in \mathcal{F} \text{ and } g \leq u\}$ for any positive superharmonic functions u, the set $\mathcal{F}$ is a generating set. By virtue of Theorem 3.6 the image of the set $\mathcal{S}$ through the evaluation map is increasingly dense and solid in $\mathcal{S}^{**}$.

Acknowledgements. The author thanks the Department of Mathematics and Statistics of McGill University for the hospitality and support she enjoyed there while this work was done, and the Academy of Finland for the financial support that made it possible

References

[1] Arsove, M. and H. Leutwiler, *Algebraic potential theory.* Mem. Amer. Math. Soc. **226** (1980).

[2] Boboc, N., Gh. Bucur and A. Cornea, *Order and convexity in potential theory: H-cones.* Lecture notes in mathematics **853**, Springer-Verlag., Berlin-Heidelberg-New York, 1981.

[3] Boboc, N. and A. Cornea, *Cônes convexes ordonnés. H-cônes et adjoints de H-cônes.* C. R. Acad. Sci. Paris Sér. A-B **270** (1970), 595–599.

[4] Constantinescu, C. and A. Cornea, *Potential theory on harmonic spaces.* Springer-Verlag, Berlin-Heidelberg-New York,1972.

[5] Cornea, A. and S.-L. Eriksson, *Order continuity of the greatest lower bound of two functionals.* Analysis **7** (1987), 173–184.

[6] Eriksson, S.-L, *Hyperharmonic cones and hyperharmonic morphisms.* Ann. Acad. Sci. Fenn. Ser. AI Math. Dissertationes **49** (1984).

[7] Eriksson-Bique, S.-L., *Generalized Riesz spaces.* Rev. Roum. Pures. Appl. **36** (1991), 47–69.

[8] Eriksson-Bique, S.-L., *Real valued duals of H-cones.* Math. Scand. **71** (1992), 243-251.

[9] Eriksson-Bique, S.-L., *Duals of H-cones.* Analysis **13** (1993), 121-132.

[10] Eriksson-Bique, S.-L., *Characterizations of balayages.* To appear in Ann. Acad. Sci. Fenn.

[11] Eriksson-Bique, S.-L., *Elements generating balayages,* Submitted for publication. 1–12.

Régularité et intégrabilité des fonctionnelles de Wiener

D. FEYEL
Département de Mathématiques
Université Evry-Essonne
91025 EVRY
France

Abstract: We prove a kind of generalization of the clasical Ascoli theorem to infinite dimensional gaussian space.

Introduction. Le théorème classique d'Ascoli montre qu'une suite de fonctions dont les dérivées sont uniformément bornées sur [0,1] est relativement compacte en norme uniforme dès qu'elle est bornée en un point. Il y a des extensions bien connues de cette propriété aux espaces de Sobolev classiques sur $\mathbf{R}^m (m < \infty)$, que l'on peut trouver chez Schwartz [Sc, p.40].

On se propose ici d'étendre cela au cas des espaces de Sobolev gaussiens en dimension infinie. Dans le cas particulier des fonctions H-lipschitziennes, ces propriétés sont liées à l'intégrabilité exponentielle étudiée par Pisier [P], Kusuoka [K] et Ustünel [U1-U2].

Dans tout l'article, E est un espace localement convexe lusinien muni d'une mesure gaussienne centrée μ chargeant tous les ouverts. Si f est une fonction sur E, on pose $f_\theta(x,y) = f(x\cos\theta + y\sin\theta)$. L'espace de Sobolev gaussien $W^{1,p}(\mu)$ n'est autre que l'espace des f telles que $\theta \to f_\theta$ soit de classe $\mathcal{C}^1$ à valeurs dans $L^p(\mu\otimes\mu)$. C'est évidemment un sous-espace de $L^p(\mu)$, qui est un espace de Banach pour sa norme naturelle. De plus, la fonction $f' = \left[\dfrac{df_\theta}{d\theta}\right]_{\theta=0}$ est linéaire dans sa seconde variable (*cf.* [F2]).

Le carré du champ Γf est défini par $\Gamma f(x) = \int f'(x,y)^2 d\mu(y)$, de sorte que l'on a

$$N_p(\sqrt{\Gamma f}) = a_p N_p(f')$$

où $a_p^p = \int |t|^p e^{-t^2/2} dt/\sqrt{2\pi}$.

K. GowriSankaran et al. (eds.), Classical and Modern Potential Theory and Applications, 159–164.

1. LEMME: *soit une suite $f_n \in W^{1,p}(\mu)$ telle que la suite des $\sqrt{\Gamma f_n}$ soit bornée dans $L^p(\mu)$. Alors la suite des $f_n - \mu(f_n)$ est bornée dans $W^{1,p}(\mu)$.*

Démonstration: selon une idée de Pisier [P], on a pour presque tout $(x, y) \in E \times E$

$$f_n(y) - f_n(x) = \int_0^{\pi/2} f'_n(x\cos\theta + y\sin\theta, -x\sin\theta + y\cos\theta)d\theta.$$

Pour $g \in L^q(\mu)$ avec $q = p/(p-1)$, on obtient en tenant compte de l'invariance de $\mu \otimes \mu$

$$\int [f_n(y) - \mu(f_n)]g(y)d\mu(y) \leq \int_0^{\pi/2} N_p(f'_n)N_q(g)d\theta$$

et finalement $N_p(f_n - \mu(f_n)) \leq \frac{\pi}{2}N_p(f'_n) = \frac{\pi}{2}a_pN_p(\sqrt{\Gamma f_n})$.

Rappelons ([FLP1]) que la capacité $C_{1,p}$ est définie par

$$C_{1,p}(h) = Inf\{\|f\|_{W^{1,p}(\mu)} / \ \exists g \text{ s.c.}i.\ |h| \leq g \leq f\}.$$

L'espace $\mathcal{L}^1(C_{1,p})$ est l'adhérence en (semi-)norme $C_{1,p}$ du sous-espace des fonctions continues bornées. On a une injection linéaire continue canonique de $W^{1,p}(\mu)$ dans $\mathcal{L}^1(C_{1,p})$. L'image $\tilde{f} \in \mathcal{L}^1(C_{1,p})$ de $f \in W^{1,p}(\mu)$ est dite "modification quasi-continue canonique" de f. Les ensembles de $C_{1,p}$-capacité nulle sont dits "polaires" ou "$C_{1,p}$-polaires", et sont μ-négligeables. Une propriété vraie sauf sur un ensemble polaire est dite vraie "quasi-partout". Enfin le dual de $\mathcal{L}^1(C_{1,p})$ est un espace de mesures bornées négligeant les ensembles $C_{1,p}$-polaires.

2. PROPOSITION: *Soit $f_n \in W^{1,p}(\mu)$ telle que la suite $\sqrt{\Gamma f_n}$ soit bornée dans $L^p(\mu)$. On suppose que la fonction $\sup_n |\tilde{f}_n|$ est finie quasi-partout sur un ensemble A de capacité $C_{1,p}(A)$ non nulle. Alors la suite f_n est bornée dans $W^{1,p}(\mu)$.*

Démonstration: on prend une mesure $\nu \geq 0$ appartenant au dual de $\mathcal{L}^1(C_{1,p})$ portée par A et de masse $\nu(1) = 1$. On a

$$f_n = f_n - \mu(f_n) - \nu(\tilde{f}_n - \mu(f_n)) + \nu(\tilde{f}_n)$$

de sorte que f_n est la somme de trois termes tous bornés dans $W^{1,p}(\mu)$ (les deux derniers sont des suites de constantes).

3. THÉORÈME: *Soit f_n une suite croissante. On suppose la suite $\sqrt{\Gamma f_n}$ bornée dans $L^p(\mu)$. Si l'enveloppe supérieure $\sup_n \tilde{f}_n$ des modifications canoniques est finie sur un ensemble de capacité non nulle, alors la suite $\tilde{f}_n$ converge fortement dans $\mathcal{L}^1(C_{1,p})$.*

Démonstration: la proposition 3 montre que la suite f_n converge faiblement dans $W^{1,p}(\mu)$ vers $f = \sup_n f_n$. Alors la suite $\tilde{f}_n$ converge faiblement vers $\tilde{f}$ dans $\mathcal{L}^1(C_{1,p})$. La suite $\tilde{f} - \tilde{f}_n$ est décroissante de sorte que l'on peut appliquer le lemme de Dini à la boule unité positive du dual de $\mathcal{L}^1(C_{1,p})$ (cf. [F2]), et que $C_{1,p}(\tilde{f} - \tilde{f}_n)$ tend vers 0.

4. THÉORÈME: *Soit f_n une suite appartenant à $W^{1,p}(\mu)$. On note $\tilde{h}$ et $\tilde{g}$ les limites supérieures et inférieures de la suite des modifications canoniques $\tilde{f}_n$. On suppose que la suite $\sqrt{\Gamma f_n}$ est dominée dans $L^p(\mu)$, et que la fonction $\sup_n \tilde{f}_n$ est finie presque partout sur un ensemble de $C_{1,p}$-capacité non nulle.*

Alors $\tilde{h}$ et $\tilde{g}$ appartiennent à $W^{1,p}(\mu) \cap \mathcal{L}^1(C_{1,p})$. De plus, si la suite $\tilde{f}_n$ converge presque partout, elle converge quasi-partout et fortement dans $\mathcal{L}^1(C_{1,p})$.

Démonstration: on applique d'abord le théorème 3 à la suite $v_n = f_1 \vee f_2 \vee \cdots \vee f_n$, ce qui est loisible vu la relation $\Gamma v_n \leq \sup_n \Gamma f_n$, et la propriété de domination dans $L^p(\mu)$. Donc $\tilde{v} = \sup_n \tilde{f}_n$ appartient à $\mathcal{L}^1(C_{1,p})$. Le même raisonnement s'applique bien sûr aux deux suites $h_n = f_n \wedge f_{n+1} \wedge \cdots$ et $g_n = f_n \vee f_{n+1} \vee \cdots$, de sorte que $\tilde{h}$ appartient à $\mathcal{L}^1(C_{1,p})$, et que $\tilde{h}_n$ converge fortement vers $\tilde{h}$ dans $\mathcal{L}^1(C_{1,p})$. On a évidemment la propriété analogue pour $\tilde{g}$. Ensuite, si l'on a $\tilde{h} = \tilde{g}$ presque partout, on a nécessairement $\tilde{h} = \tilde{g}$ quasi-partout, et la suite $C_{1,p}(\tilde{h}_n - \tilde{g}_n)$ tend vers 0 grâce au lemme de Dini comme au théorème 3. On a enfin $C_{1,p}(\tilde{h} - \tilde{f}_n) \leq C_{1,p}(\tilde{h}_n - \tilde{g}_n)$, ce qui termine la démonstration.

5. EXEMPLES: *i) soit $E = \mathbf{R}^{\mathbf{N}}$ muni de sa gaussienne canonique μ. Si $f \in W^{1,p}(\mu)$, la suite des espérances conditionelles f_n sur les sous-tribus canoniques $\mathfrak{F}_n$ vérifie les hypothèses du théorème 4, car la suite $\sqrt{\Gamma f_n}$ forme une sous-martingale bornée dans $L^p(\mu)$. Alors les $\tilde{f}_n$ convergent quasi-partout et fortement dans $\mathcal{L}^1(C_{1,p})$.*

ii) soit P_t le semi-groupe d'Ornstein-Uhlenbeck, et soit $f \in W^{1,p}(\mu)$. Il résulte du théorème de ergodique de Stein [St] que $P_t f$ converge presque partout vers f quand t tend vers 0, et que les fonctions $\sqrt{\Gamma P_t f_n} \leq e^{-t} P_t \sqrt{\Gamma f_n}$ sont dominées dans $L^p(\mu)$ pour $t < 1$. Par ailleurs, $P_t f$ est quasi-continu pour $t > 0$, de sorte que $P_t f$ converge $C_{1,p}$-quasi-partout vers $\tilde{f}$.

6. REMARQUE: pour $p = 2$ (espace de Dirichlet), des résultats de ce type ont été obtenus par Bouleau et Hirsch ([BH]). Signalons aussi les résultats de L. Denis ([D]).

7. REMARQUE: on peut étendre le théorème 4 à une suite de fonctions f_n à valeurs banachiques. Il suffit de remarquer que l'on a toujours $\Gamma\|f_n\| \leq \Gamma f_n$.

Indiquons enfin une conséquence immédiate du théorème 4:

8. PROPOSITION: *si une suite $f_n \in W^{1,p}(\mu)$ converge en mesure, et si la suite $\sqrt{\Gamma f_n}$ est dominée dans $L^p(\mu)$, alors la suite $\tilde{f}_n$ converge dans $\mathcal{L}^1(C_{1,p})$.*

Fonctions H-lipschitziennes de rapport 1

Rappelons que l'espace H de Cameron-Martin est compactement inclus dans E, et est en dualité hilbertienne avec l'espace E' (dual de E) muni de la topologie induite par $L^2(\mu)$.

On s'intéresse maintenant aux fonctions f finies presque partout et telles que $N_\infty(\Gamma f) \leq 1$. (Cela entraîne que $f \in \cap_p W^{1,p}(\mu)$ d'après notre théorème 4 appliqué à la suite $n\mathrm{Arctg}(f/n)$).

On sait alors d'après Enchev et Stroock [ES] que f admet une modification $\tilde{f}$ qui est H-lipschitzienne de rapport 1, c'est à dire vérifie

$$\tilde{f}(x+u) \leq \tilde{f}(x) + |u|_H \tag{1}$$

pour tout couple $(x,u) \in E \times H$. Il est d'ailleurs facile d'obtenir $\tilde{f}$ en considérant une suite d'espérances conditionnelles comme dans l'exemple 5. Les fonctions $\tilde{f}_n$ sont continues, et leur limite supérieure $\tilde{f}$ vérifie (1). On notera que la fonction $\tilde{f}$ appartient à $\cap_p \mathcal{L}^1(C_{1,p})$.

9. THÉORÈME: *soient A et B deux ensembles boréliens complémentaires sur lesquels on a respectivement $f \leq c$ et $f \geq c$ presque partout. Alors pour tout $\lambda \geq 0$*

$$\mu(\{|f-c| > \lambda\}) \leq \int_{\lambda+\alpha}^{+\infty} e^{-t^2/2} dt/\sqrt{2\pi} + \int_{\lambda-\alpha}^{+\infty} e^{-t^2/2} dt/\sqrt{2\pi} \tag{2}$$

où α est défini par $\mu(A) = \mathrm{Er}f(\alpha) = \int_{-\infty}^{\alpha} e^{-t^2/2} dt/\sqrt{2\pi}$.

Démonstration: notons φ_A la H-distance à A définie par

$$\varphi_A(x) = \mathrm{Inf}\{|x-y|_H / y \in A\}$$

(cf. [FLP2]). On a $f-c \leq \varphi_A$ et $-f+c \leq \varphi_B$, donc $|f-c| \leq \varphi_A \vee \varphi_B$. La formule (2) découle alors de l'inégalité de Borell [B] selon laquelle $\mu(\{\varphi_A > \lambda\}) \leq 1 - \mathrm{Er}f(\lambda+\alpha)$ et $\mu(\{\varphi_B > \lambda\}) \leq 1 - \mathrm{Er}f(\lambda-\alpha)$ car $\mu(B) = 1-\mu(A) = 1-\mathrm{Er}f(\alpha) = \mathrm{Er}f(-\alpha)$.

10. REMARQUE: le second membre de (2) vaut aussi $2 - 2\mathrm{Er}f(\lambda) + \alpha^2 \xi e^{-\xi^2/2}/\sqrt{2\pi}$ où ξ est un nombre de l'intervalle $[\lambda-\alpha, \lambda+\alpha]$.

Disons que c est une *médiane* de f si l'on est dans les conditions du théorème 9 et si de plus A et B sont complémentaires. Une telle médiane existe toujours car μ est diffuse.

11. COROLLAIRE: *si c est une médiane de f, on a*

$$\mu(\{|f-c|>\lambda\}) \leq 2\int_{\lambda}^{+\infty} e^{-t^2/2}dt/\sqrt{2\pi} \tag{3}$$

et

$$\int \exp(\alpha|f-c|^2)d\mu \leq \frac{1}{\sqrt{1-2\alpha}} \tag{4}$$

Démonstration: en effet, dans ce cas $\mu(A)=\mu(B)=\frac{1}{2}$, donc $\alpha=0$. L'inégalité (4) se déduit de (3) par un calcul facile.

12. REMARQUE: la méthode d'Ustünel [U1-U2], utilisant la formule de Clark, convient au cas où la médiane c est remplacée par la *moyenne* $\mu(f)$.

On trouve alors la majoration (4') $\int \exp(\alpha|f-\mu(f)|^2) \leq 1/\sqrt{1-2\alpha}$, mais on ne trouve que la majoration (3') $\mu(\{|f-\mu(f)|>\lambda\}) \leq 2\exp(-\lambda^2/2)$ moins précise que (3). Cependant Kusuoka [K] cite la majoration (3) avec $\mu(f)$ au lieu de c.

13. COROLLAIRE: *la fonction* $\exp(\alpha f^2)$ *est intégrable pour* $2\alpha<1$.

14. REMARQUE: soit q une semi-norme généralisée et finie presque partout sur E. Il est bien facile de voir que q est majorée par une semi-norme analogue q_1 borélienne sur E. La restriction de q_1 à H est alors finie continue car H est inclus dans tout sous-espace de E portant μ (cf. [FLP3]). Il en est bien sûr de même de la restriction de q à H. Mais alors q/σ est H-lipschitzienne de rapport 1 pour σ convenable, et par suite $q \in \cap_p W^{1,p}(\mu)$, ce qui redémontre le théorème bien connu de Fernique. On sait aussi par ailleurs ([FLP1]) que $q \in \cap_p \mathcal{L}^1(C_{1,p})$ (et même mieux).

15. PROBLÈME: si f est H-lipschitzienne de rapport 1, elle a une modification $\tilde{f}$ qui est H-lipschitzienne de rapport 1 et $C_{1,p}$-quasi-continue pour tout $p>1$. Mais f elle-même est-elle $C_{1,p}$-quasi-continue? C'est bien sûr le cas en dimension finie.

BIBLIOGRAPHIE

[B] Ch. Borell, *The Brunn-Minkowski Inequality in Gauss Space* Inventiones Math. 30, 207-216, (1975).

[BH] N.Bouleau, F. Hirsch, *Dirichlet Forms and Analysis on Wiener Space* de Gruyter Studies in Math. 14, (1991).

[D] L. Denis, *Convergence quasi-partout pour les capacités définies par un semi-groupe sous-markovien* CRAS, Paris, série I, t.315, (1992).

[ES] O. Enchev, D.W. Stroock, *Rademacher's Theorem for Wiener Functionals* Ann. Probab. 21 no.1 (1993).

[F1] D. Feyel, *Espaces de Banach adaptés, quasi-topologie et balayage* Sém. Théorie du Potentiel, Paris, Lec. Notes in Math. 681, Springer, (1978).

[F2] D. Feyel, *Transformations de Hilbert-Riesz* CRAS, Paris, série I, t.310, 653-655, (1990).

[FLP1] D. Feyel, A. de La Pradelle *Capacités gaussiennes* Ann. Inst. Fourier, t.41, f.1, p. 49-76, (1991).

[FLP2] D. Feyel, A. de La Pradelle *Démonstration géom. d'une loi de tout ou rien* Publication de l'Université d'Evry-Val d'Essonne, (1993).

[FLP3] D. Feyel, A. de La Pradelle *Opérateurs linéaires gaussiens* A paraître in Potential Analysis (1992).

[K] S. Kusuoka, *Analysis on Wiener Spaces I* J. of Functional Analysis 98, 122-168, (1991).

[P] G. Pisier, *Probabilistic methods in the Geometry of Banach spaces* Probability and Analysis, Varenna-Como 1985, LN in Math. No. 1206, (1986).

[Sc] L. Schwartz, *Théorie des distributions, tome II* Paris, Hermann, ASI 1122, (1959).

[St] E.M. Stein, *Singular integrals and differentiability properties of functions* Princeton University Press, (1970).

[U1] A.S. Ustünel, *Intégrabilité exponentielle des fonctionnelles de Wiener* CRAS, série I, t.315, 997-1000, (1992).

[U2] A.S. Ustünel, *Exponential tightness of the Wiener Functionals* Preprint Series, University of Oslo, ISBN 82- 553-0854-7, (1993).

POINCARÉ INEQUALITIES IN L^1-NORM FOR THE SPHERE AND A STRONG ISOPERIMETRIC INEQUALITY IN $\mathbf{R}^n$

BENT FUGLEDE

University of Copenhagen, Mathematics Institute, Universitetsparken 5, 2100 Copenhagen, Denmark

ABSTRACT. For real-valued functions on the unit sphere in $\mathbf{R}^n$ with mean-value 0 the L^1-norm is estimated from above by a best possible constant times the Dirichlet norm. A dual version of this inequality involves the expansion of bounded functions in a series of spherical harmonics. A similar pair of dual inequalities leads to a lower estimate of the isoperimetric deficit of nearly spherical, convex domains K in $\mathbf{R}^n$ in terms of the asymmetry of K, which is essentially the volume of the symmetric difference of K and a suitable ball.

Introduction

The inspiration to the present work came from the inequalities (2) and (5) below, obtained recently by Hall, Hayman, and Weitsman [HH], [HHW]. We shall prove (or reprove) all the inequalities (1) through (5) and address the problem of obtaining higher dimensional versions of them. This extension requires a different approach from that of [HH] which was based on subordination theory from complex analysis.

If a_k, b_k are the Fourier coefficients of a measurable function f on $\mathbf{R}$ (mod 2π) with values between -1 and 1 then

$$\frac{1}{2}\sum_{k=1}^{\infty}\frac{a_k^2+b_k^2}{k^2}\leqslant\frac{\pi^2}{12}, \tag{1}$$

$$\frac{1}{2}\sum_{k=2}^{\infty}\frac{a_k^2+b_k^2}{k^2-1}\leqslant\frac{4}{\pi}-1. \tag{2}$$

Both inequalities are sharp. The equality sign prevails in (1) e.g. for $f(\theta)=\operatorname{sgn}(\cos\theta)$, and in (2) e.g. for $f(\theta)=\operatorname{sgn}(\cos 2\theta)$. (See also below.) As pointed out by the referee, (1) can be established by subordination, like (2) in [HH].

Each of the inequalities (1), (2) has a *dual* version in terms of 2π-periodic, absolutely continuous real-valued functions u:

$$\left(\int_{-\pi}^{\pi}|u(\theta)|\,\frac{d\theta}{2\pi}\right)^2\leqslant\frac{\pi^2}{12}\int_{-\pi}^{\pi}u'(\theta)^2\,\frac{d\theta}{2\pi}, \tag{3}$$

$$\left(\int_{-\pi}^{\pi}|u(\theta)|\,\frac{d\theta}{2\pi}\right)^2\leqslant\left(\frac{4}{\pi}-1\right)\int_{-\pi}^{\pi}\left(u'(\theta)^2-u(\theta)^2\right)\frac{d\theta}{2\pi}, \tag{4}$$

K. GowriSankaran et al. (eds.), Classical and Modern Potential Theory and Applications, 165–183.

where it is assumed in (3) that $\int_{-\pi}^{\pi} u(\theta)\, d\theta = 0$, and in (4) that the Fourier coefficients of u of orders $k = 0$ and $k = 1$ are 0.

The sign of equality in (3) occurs for

$$u(\theta) = \tfrac{1}{2}\left((\pi/2)^2 - \theta^2\right) \quad \text{for } -\pi/2 \leqslant \theta \leqslant \pi/2,$$

continued so as to be π-antiperiodic, i.e. $u(\theta + \pi) = -u(\theta)$. The sign of equality in (4) holds for

$$u(\theta) = \sqrt{2}\cos\theta - 1 \quad \text{for } -\pi/4 \leqslant \theta \leqslant \pi/4,$$

continued so as to be $(\pi/2)$-antiperiodic. In each of the inequalities (1) through (4) the equality sign holds *only* for the translates of the particular function f stated above, resp. only for the constant multiples of translates of the above particular function u (see Section 1 as to (1) and (3), and Section 2, or in detail [F3], as to (2) and (4)). In the case of (2) this uniqueness does not seem to follow from [HH].

Our primary aim in studying the inequalities (1) through (4), and especially their higher dimensional versions, is the determination of the best possible constants and of the functions for which the equality sign holds. If we would be content with bigger constants, the inequalities (1) through (4) would follow immediately from the corresponding, much simpler inequalities in which the L^2-norm of f and u enters in place of the L^∞-norm of f (implicit in (1) and (2)) and the L^1-norm of u (on the left of (3) and (4)). For example, the L^2 analogue of (3) is known as Wirtinger's inequality (with the best possible constant 1 in place of $\pi^2/12$ in (3)). And the L^2 analogue of the n-dimensional version of (3) is the Poincaré inequality for the unit sphere in $\mathbf{R}^n$. For other Poincaré type inequalities see [MPF, p. 96 ff.].

The inequality (4) is implicitly contained in [HHW], where the classical isoperimetric inequality for convex bodies K in the plane, with area $A(K) = A$ and perimeter L, was strengthened as follows by application of (2):

$$L^2 \geqslant 4\pi A\left(1 + \frac{\pi}{4-\pi}\alpha^2 + O(\alpha^3)\right). \tag{5}$$

Here α denotes the following measure of the "average" deviation of K from circular shape:

$$\alpha = \min_{x\in\mathbf{R}^2} \frac{A(K \setminus B(x,K))}{A(K)}, \tag{6}$$

where $B(x,K)$ is the disk centred at x and having the same area as K. This quantity α was proposed by L. E. Fraenkel and is known as the *asymmetry* of K. The constant $\pi/(4-\pi)$ in (5) was shown in [HHW] to be best possible, briefly because it is the reciprocal of the best possible constant $4/\pi - 1$ in (2) and (4).

In Section 1 of the present article we establish (1) and (3) as the case $n = 2$ of suitable n-dimensional versions of them, including the explicit determination of the best possible constant κ_n to replace $\pi^2/12$ if $n > 2$. In Section 2 we describe

n-dimensional versions of (2), (4), and (5), all involving one and the same best possible constant, now denoted by $\kappa(n)$, but we are unable to actually determine $\kappa(n)$ when $n > 2$. Instead we obtain good bounds for $\kappa(n)$, the lower bound being conjectured to actually equal $\kappa(n)$. For a detailed exposition of the contents of Section 2 see [F3].

Notation. Referring e.g. to [Sp, p. 188 f.], [M, p. 38 f.] we have on the unit sphere $\Sigma_n = \Sigma$ in $\mathbf{R}^n$ the Laplace-Beltrami operator Δ with the increasing sequence of eigenvalues

$$\lambda_k = k(k+n-2), \quad k = 0, 1, 2, \dots .$$

The eigenspace $\mathcal{H}_k$ corresponding to λ_k is the space of all spherical harmonics of degree k on Σ. For any (real-valued) L^2-function f on Σ let f_k denote the orthogonal projection of f on $\mathcal{H}_k$; then $\|f\|^2 = \sum_{k=0}^{\infty} \|f_k\|^2$ in terms of the L^2-norm w.r.t. normalized surface measure σ on Σ. Note that $f_0 = \int f \, d\sigma$, the mean-value of f. For $n = 2$ and $k \geqslant 1$ we have $f_k(\cos\theta, \sin\theta) = a_k \cos(k\theta) + b_k \sin(k\theta)$ in terms of the Fourier coefficients a_k, b_k of f, and hence $\|f_k\|^2 = \frac{1}{2}(a_k^2 + b_k^2)$.

For $1 \leqslant p \leqslant \infty$ the $L^p(\sigma)$-norm will be denoted by $\|\cdot\|_p$, but we continue to write simply $\|\cdot\|$ for $\|\cdot\|_2$.

For $m = 1$ or 2 we denote by $W^{m,p} = W^{m,p}(\Sigma)$ the Sobolev space of all real distributions u on Σ whose partial derivatives of order m (hence also of orders $\leqslant m$) in local coordinates on Σ are (locally) in $L^p(\sigma)$. For $u \in W^{1,2}$ we denote by ∇u the gradient of u in the sense of Riemannian geometry on Σ, and by $\|\nabla u\|$ the $L^2(\sigma)$-norm of the length $|\nabla u|$ of ∇u.

We denote by $\operatorname{dom}\Delta$ the domain of Δ as an operator in $L^2(\sigma)$ (acting in the distribution sense). It is known that $\operatorname{dom}\Delta = W^{2,2}$, cf. e.g. [Se, p. 685], or argue as in the last paragraph of the proof of Theorem 1.1 below.

1. Extension of (1) and (3) to dimension $n \geqslant 2$

These extensions amount to determining the following constants κ_n and c_n:

$$\kappa_n = \max\left\{ \sum_{k=1}^{\infty} \|f_k\|^2/\lambda_k \;\middle|\; f : \Sigma_n \to [-1,1] \ \sigma\text{-measurable} \right\}, \tag{7}$$

$$c_n = \min\left\{ \|\nabla u\|^2/\|u\|_1^2 \mid u \in W^{1,2}(\Sigma_n),\ u_0 = 0 \right\}. \tag{8}$$

The existence of the maximum κ_n and the minimum c_n is easy to prove by compactness, see the proof of Theorem 1.1 below. Note the following expansion in spherical harmonics

$$\|\nabla u\|^2 = -\int_{\Sigma} u \Delta u \, d\sigma = \sum_{k=1}^{\infty} \lambda_k \|u_k\|^2 \tag{9}$$

for functions $u \in \operatorname{dom}\Sigma = W^{2,2}$. Actually, $\|\nabla u\|^2 = \sum \lambda_k \|u_k\|^2$ holds even for $u \in W^{1,2}$, cf. e.g. [F3, Lemma 2]. In view of these observations it is not surprising that we have the following duality between the variational problems (7) and (8):

THEOREM 1.1. $c_n = 1/\kappa_n$.

Every maximizing function f for κ_n in (7) takes σ-almost everywhere the values ± 1 only, and the associated function

$$u = Tf := \sum_{k=1}^{\infty} \frac{1}{\lambda_k} f_k$$

is minimizing for c_n in (8), and $\|u\|_1 = \kappa_n$.

Conversely, any minimizing function u for c_n, normalized so that $\|u\|_1 = \kappa_n$, has the form $u = Tf$ for a unique maximizing function f for κ_n. We have $u(\xi) \neq 0$ σ-a.e., and $f = \operatorname{sgn} u$. Consequently u is in $\operatorname{dom} \Delta$ and satisfies (in the distribution sense) the Euler type equation

$$-\Delta u = f - f_0\,, \quad f = \operatorname{sgn} u\,. \tag{10}$$

In particular, u is C^1-smooth.

Proof: To see that there is an actual maximum in (7) we define for $f \in L^2(\sigma)$

$$Tf = \sum_{k=1}^{\infty} \frac{1}{\lambda_k} f_k \tag{11}$$

(convergent in the $L^2(\sigma)$-norm) in terms of the expansion $f = \sum_{k=0}^{\infty} f_k$, $f_k \in \mathcal{H}_k$. Then $T : L^2(\sigma) \to L^2(\sigma)$ is a compact self-adjoint operator, and the quantity

$$\Lambda(f) := \int_{\Sigma} (Tf)f \, d\sigma = \sum_{k=1}^{\infty} \|f_k\|^2/\lambda_k \tag{12}$$

is therefore continuous as a function of f in the weak topology on $L^2(\sigma)$, a fortiori in the weak* topology on $L^\infty(\sigma)$ viewed as the dual of $L^1(\sigma)$, and the unit ball in $L^\infty(\sigma)$ is weak* compact.

Consider any maximizing function f for κ_n in (7), and imagine that the set $F = \{\xi \in \Sigma \mid |f(\xi)| < 1 - \varepsilon\}$ has measure $\sigma(F) > 0$ for some ε, $0 < \varepsilon < 1$. Choose $g \in L^\infty(\sigma)$ with $\|g\|_\infty = 1$ so that $g = 0$ off F and $\int g\, d\sigma = \int (Tf) g \, d\sigma = 0$ with Tf from (11); this is possible since $L^\infty(F, \sigma)$ is infinite dimensional. Then $\|f + \varepsilon g\|_\infty \leqslant 1$ and

$$\int \left(T(f + \varepsilon g)\right)(f + \varepsilon g) \, d\sigma = \int (Tf) f \, d\sigma + \varepsilon^2 \int (Tg) g \, d\sigma > \kappa_n\,,$$

because $\int (Tg) g \, d\sigma = \sum_{k=1}^{\infty} \|g_k\|^2/\lambda_k > 0$. From this contradiction we see that actually $\sigma(F) = 0$ for any choice of ε, and so indeed $f(\xi) = \pm 1$ a.e.

Clearly $u = Tf$ is in $\operatorname{dom}\Delta = W^{2,2}$ (see the end of the introduction), and the differential equation in (10) holds. Since $u_0 = \int u\,d\sigma = 0$ we have from (9), (12)

$$\|\nabla u\|^2 = \int (f - f_0)u\,d\sigma = \int fu\,d\sigma = \Lambda(f) = \kappa_n\,.$$

It follows by (8) that

$$\kappa_n = \|\nabla u\|^2 = \int fu\,d\sigma \leqslant \|u\|_1 \leqslant \|\nabla u\|/\sqrt{c_n} = \sqrt{\kappa_n/c_n}$$

and consequently $c_n \leqslant 1/\kappa_n$. We show below that $c_n = 1/\kappa_n$, from which it will follow that the two inequalities here are equalities, and so $u = Tf$ is minimizing for c_n in (8), $\|u\|_1 = \kappa_n$, and $f(\xi) = \operatorname{sgn} u(\xi)$ for almost every $\xi \in \Sigma$ with $u(\xi) \neq 0$.

To obtain the remaining inequality $c_n \geqslant 1/\kappa_n$, consider any non-zero function $u \in W^{1,2}(\Sigma)$ with $u_0 = 0$, and write

$$f(\xi) = \begin{cases} \operatorname{sgn} u(\xi) & \text{if } u(\xi) \neq 0 \\ 0 & \text{if } u(\xi) = 0. \end{cases} \tag{13}$$

Then

$$\|u\|_1 = \int fu\,d\sigma = \sum_{k=1}^{\infty} \int \frac{f_k}{\sqrt{\lambda_k}}\, u_k\sqrt{\lambda_k}\,d\sigma\,, \tag{14}$$

$$\|u\|_1^2 \leqslant \sum_{k=1}^{\infty} \frac{\|f_k\|^2}{\lambda_k} \sum_{k=1}^{\infty} \lambda_k\|u_k\|^2 \leqslant \kappa_n\|\nabla u\|^2 \tag{15}$$

by (7), (9) and the Cauchy-Schwarz inequality applied to the vectors $\sum_{k=1}^{\infty} u_k\sqrt{\lambda_k}$ and $\sum_{k=1}^{\infty} f_k/\sqrt{\lambda_k}$ in the Hilbert space $L^2(\sigma) = \bigoplus_{k=0}^{\infty} \mathcal{H}_k$. This shows that $c_n \geqslant 1/\kappa_n$ in view of (8), and so actually $c_n = 1/\kappa_n$.

Suppose henceforth that u is minimizing for c_n and normalized so that

$$\|u\|_1 = 1/c_n = \kappa_n\,.$$

With f from (13) the relations (14) and (15) hold, the latter with equality because $\kappa_n = 1/c_n$. In view of (14) there is hence a constant $\gamma > 0$ such that

$$u_k\sqrt{\lambda_k} = \gamma f_k/\sqrt{\lambda_k}\,, \quad k \geqslant 1\,. \tag{16}$$

Since $u_0 = 0$ it follows from (11) and (9) that

$$u = \sum_{k=1}^{\infty} u_k = \gamma \sum_{k=1}^{\infty} f_k/\lambda_k = \gamma Tf,$$

$$c_n\|u\|_1^2 = \|\nabla u\|^2 = \sum_{k=1}^{\infty}\lambda_k\|u_k\|^2 = \gamma\sum_{k=1}^{\infty}\int f_k u_k\,d\sigma = \gamma\int fu\,d\sigma = \gamma\|u\|_1\,.$$

Consequently, $\gamma = c_n\|u\|_1 = 1$, whence $u = Tf$, and by (12)

$$\Lambda(f) = \int fu\,d\sigma = \|u\|_1 = \kappa_n\,,$$

so that f is maximizing for κ_n in (7). In particular, $f(\xi) = \pm 1$ a.e., as shown in the second paragraph of the proof, and so it follows from (13) that

$$u(\xi) \neq 0 \quad \sigma\text{-a.e. for } \xi \in \Sigma\,. \tag{17}$$

Since $f - f_0 \in L^\infty(\sigma)$, we have $u \in W^{2,p}$ for $p < \infty$. This follows from standard regularity theory for elliptic operators, cf. e.g. [Hö2, Theorem 17.1.1] which shows (when applied in local coordinates in Σ) that the Euler equation (10) locally has solutions of class $W^{2,p}$, and hence the relevant solution $u = Tf$ is likewise of this class (even globally on Σ, by compactness), the solutions of $\Delta v = 0$ being analytic. Finally, every function $u \in W^{2,p}$ with p sufficiently large is C^1-smooth according to the Sobolev embedding theorem, cf. e.g. [Hö1, Th. 4.5.13]. □

In terms of the elementary function φ_n defined for $n \geqslant 2$ by

$$\varphi_n(t) = \int_0^t (1-s^2)^{\frac{n-3}{2}}\,ds\,, \quad -1 \leqslant t \leqslant 1\,, \tag{18}$$

we have

$$\frac{\varphi_n(t)}{\varphi_n(1)} = \sigma(\{\xi = (\xi_1,\dots,\xi_n) \in \Sigma_n \mid -t \leqslant \xi_1 \leqslant t\}). \tag{19}$$

The solution of the extremum problems (7) and (8) above is given in the following theorem in view of the relation $c_n = 1/\kappa_n$ from Theorem 1.1.

THEOREM 1.2. *$\kappa_2 = \pi^2/12$, and for $n \geqslant 3$*

$$\kappa_n = \begin{cases} \dfrac{1}{n-2}\displaystyle\sum_{j=1}^{n-2}\frac{(-1)^{j-1}}{j}\,, & n \text{ even} \\[2ex] \dfrac{1}{n-2}\Big(2\ln 2 - \displaystyle\sum_{j=1}^{n-2}\frac{(-1)^{j-1}}{j}\Big)\,, & n \text{ odd}. \end{cases}$$

For every $a \in \Sigma_n$ the function $f(\xi) = \operatorname{sgn}(a\cdot\xi)$ is maximizing for κ_n in (7), and the function

$$u(\xi) = \int_0^{a\cdot\xi}(1-t^2)^{\frac{1-n}{2}}\big(\varphi_n(1) - \varphi_n(|t|)\big)\,dt\,, \quad \xi \in \Sigma_n\,, \tag{20}$$

is minimizing for c_n *in (8) and normalized so that* $\|u\|_1 = \kappa_n = 1/c_n$. *When* a *ranges over* Σ_n *we obtain in this way all maximizing and all thus normalized minimizing functions.*

Remark. For any $n \geqslant 3$ (even or odd) it follows that

$$\kappa_n = \frac{1}{n-2}\left(\ln 2 + \sum_{j=1}^{\infty} \frac{(-1)^j}{j+n-2}\right).$$

With n as a real variable this expression for κ_n has the limit

$$\frac{1}{1^2} - \frac{1}{2^2} + \frac{1}{3^2} - \frac{1}{4^2} + \cdots = \frac{\pi^2}{12} = \kappa_2$$

as $n \to 2$, cf. also (36) below. The recursion formula (38) for u (there denoted by u_n) in its dependence on n shows that u is an elementary function of $a \cdot \xi$ for every n because this is true for $n = 2,\ 3$, as we shall see below from (20), which also shows that u is *odd*: $u(-\xi) = -u(\xi)$.

Proof: The proof of Theorem 1.2 falls in two parts. First we show by a symmetrization argument that any minimizing function $u(\xi)$ for c_n depends only on $a \cdot \xi$, for a certain $a \in \Sigma$. Next we determine u by solving the Euler equation (10). In the process we show that $f_0 = 0$ (which is non-trivial).

First Part. Consider any minimizing function u for c_n in (8), normalized so that $\|u\|_1 = \kappa_n$ ($= 1/c_n$, by Theorem 1.1). We subject the level sets

$$E(\alpha) = \{\xi \in \Sigma \mid u(\xi) > \alpha\}, \quad \alpha \in u(\Sigma), \tag{21}$$

to *spherical symmetrization*, replacing $E(\alpha)$ by the cap

$$F(\alpha) = \{\xi \in \Sigma \mid \xi_1 > \beta\}, \tag{22}$$

where $\beta = \beta(\alpha)$ is such that $\sigma(F(\alpha)) = \sigma(E(\alpha))$, that is by (19)

$$\frac{1}{2}\left(1 - \frac{\varphi_n(\beta)}{\varphi_n(1)}\right) = \sigma(E(\alpha)). \tag{23}$$

For any $\alpha \in u(\Sigma)$ we have

$$\sigma(\{\xi \in \Sigma \mid u(\xi) = \alpha\}) = 0.$$

For $\alpha = 0$ this is (17) above. For $\alpha \neq 0$ use the continuity of u on Σ and the fact that u as a solution of (10) is analytic and non-constant in each domain of Σ where $u \gtrless 0$ (and so $f - f_0$ is constant). It follows that the set

$$C = \{u(\xi) \neq 0 \mid \nabla u(\xi) = 0\}$$

of non-zero critical values of u is countable and has no limit point $\neq 0$, cf. e.g. [K, p. 276]. The strictly decreasing function $\alpha \mapsto \sigma(E(\alpha))$ is a homeomorphism of the compact interval $u(\Sigma)$ onto $[-1,1]$. The function $\alpha \mapsto \beta = \beta(\alpha)$ defined by (23) is therefore a continuous and strictly increasing mapping of $u(\Sigma)$ onto $[-1,1]$, analytic with non-zero derivative for $\alpha \notin C \cup \{0\}$; we denote by $\alpha = \alpha(\beta)$ its inverse function defined for $-1 \leqslant \beta \leqslant 1$. The continuous function $v : \Sigma \to \mathbf{R}$ defined by

$$v(\xi) = \alpha(\xi_1) \tag{24}$$

is said to arise from the given function $u : \Sigma \to \mathbf{R}$ by spherical symmetrization. When α and β are related as above we clearly have from (22), (24)

$$F(\alpha) = \{\xi \in \Sigma \mid v(\xi) > \alpha\}, \tag{25}$$

and hence from $\sigma(F(\alpha)) = \sigma(E(\alpha))$ and $v_0 = \int v \, d\sigma$, etc.,

$$v_0 = u_0 = 0, \quad \|v\|_1 = \|u\|_1 = \kappa_n . \tag{26}$$

We proceed to show by a standard argument that

$$v \in W^{1,2}(\Sigma), \quad \|\nabla v\| \leqslant \|\nabla u\| . \tag{27}$$

Consider the decreasing functions

$$A(\alpha) = k_n \sigma(E(\alpha)), \quad D(\alpha) = k_n \int_{E(\alpha)} |\nabla u|^2 \, d\sigma$$

of $\alpha \in u(\Sigma)$, where k_n is the $(n-1)$-dimensional surface area of Σ. These functions are analytic for $\alpha \notin C \cup \{0\}$. If ds refers to the $(n-2)$-dimensional surface measure on the $(n-2)$-dimensional analytic manifold $\partial E(\alpha)$ (embedded in Σ), and if $S(\alpha)$ denotes the $(n-2)$-dimensional surface area of $\partial E(\alpha)$, then we have for $\alpha \notin C \cup \{0\}$

$$S(\alpha) = \int_{\partial E(\alpha)} ds , \quad -A'(\alpha) = \int_{\partial E(\alpha)} \frac{ds}{|\nabla u|}, \quad -D'(\alpha) = \int_{\partial E(\alpha)} |\nabla u| \, ds ,$$

whence by the Schwarz inequality and the isoperimetric theorem for subsets of Σ, cf. e.g. [BZ, p. 86],

$$-D'(\alpha) \geqslant S(\alpha)^2/(-A'(\alpha)) \geqslant \tilde{S}(\alpha)^2/(-(\tilde{A})'(\alpha)) , \tag{28}$$

where $\tilde{A}(\alpha)$ $(= A(\alpha))$ denotes the $(n-1)$-dimensional surface area of $F(\alpha)$, cf. (25), and $\tilde{S}(\alpha)$ $(\leqslant S(\alpha))$ the $(n-2)$-dimensional surface area of $\partial F(\alpha)$.

For any interval $[\alpha_1, \alpha_2] \subset u(\Sigma)$ not meeting $C \cup \{0\}$ we infer from (28) that

$$\begin{aligned} k_n \int_{E(\alpha_1)\setminus E(\alpha_2)} |\nabla u|^2 \, d\sigma &= D(\alpha_1) - D(\alpha_2) = \int_{\alpha_1}^{\alpha_2} -D'(\alpha) \, d\alpha \\ &\geqslant \int_{\alpha_1}^{\alpha_2} \frac{\tilde{S}(\alpha)^2}{-(\tilde{A})'(\alpha)} \, d\alpha = k_n \int_{F(\alpha_1)\setminus F(\alpha_2)} |\nabla v|^2 \, d\sigma \end{aligned} \tag{29}$$

because v is smooth in the closed spherical annulus $\overline{F(\alpha_1)} \setminus F(\alpha_2)$ and depends on ξ_1 only. Writing $\beta_1 = \beta(\alpha_1)$, $\beta_2 = \beta(\alpha_2)$, we obtain in view of (18), (19), (22), (24),

$$2\varphi_n(1) \int_{F(\alpha_1)\setminus F(\alpha_2)} |\nabla v|^2 \, d\sigma = \int_{\beta_1}^{\beta_2} (1 - \xi_1^2)\alpha'(\xi_1)^2 (1 - \xi_1^2)^{\frac{n-3}{2}} \, d\xi_1 \, ,$$

noting that

$$|\nabla v(\xi)| = \sqrt{1 - \xi^2}\, |\alpha'(\xi_1)|$$

for any $\xi \in \Sigma$ such that $\xi_1 \in [\beta_1, \beta_2]$, or more generally such that $\xi_1 \notin \beta(C \cup \{0\})$.

Adding up, the above inequality extends by continuity to the case where $[\alpha_1, \alpha_2] = u(\Sigma)$, and it follows easily in view of (29) and the Schwarz inequality that the continuous function α is locally absolutely continuous as a function of $\xi_1 \in]-1, 1[$ with locally square integrable derivative α'. Consequently, (29) holds with $F(\alpha_1)\setminus F(\alpha_2) = \Sigma\setminus\{-a\}$, where $a := (1, 0, \ldots, 0)$, cf. (22). It follows that $v \in W^{1,2}(\Sigma\setminus\{a, -a\})$, hence $v \in W^{1,2}(\Sigma)$ in view of [S, p. 58] in case $n > 2$. For $n = 2$ one uses the continuity of v as in the case of α above. This establishes (27).

Since u was chosen to be minimizing for c_n and normalized so that $\|u\|_1 = \kappa_n$, so is therefore v by (26), (27). It follows that $\|\nabla v\| = \|\nabla u\|$, and hence there is equality in (29) for every $[\alpha_1, \alpha_2] \subset u(\Sigma)$ not meeting $C \cup \{0\}$. This implies in turn that the sign of equality holds in both inequalities (28) for every $\alpha \in u(\Sigma)$ not in $C \cup \{0\}$. Hence $E(\alpha)$ from (21) must be a cap, by the isoperimetric theorem on the sphere Σ, and $|\nabla u|$ must be constant on $\partial E(\alpha)$. For each open interval $I \subset u(\Sigma) \setminus (C \cup \{0\})$ any two of the level "circles" $\partial E(\alpha)$, $\alpha \in I$, must therefore have constant distance, so these parallel "circles" $\partial E(\alpha)$ are concentric $(n-2)$-spheres. In other words, the $E(\alpha)$, $\alpha \in I$, are concentric caps, and so are therefore the $E(\alpha)$ for *all* $\alpha \in u(\Sigma)$, by the continuity of $\sigma(E_\alpha)$ as a function of $\alpha \in u(\Sigma)$. We may take this common centre of all $E(\alpha)$ to be the point $a = (1, 0, \ldots, 0)$ of Σ, and so

$$E(\alpha) = \{\xi \in \Sigma \mid \xi_1 > \beta(\alpha)\} = F(\alpha)$$

in view of (22) together with $\sigma(E(\alpha)) = \sigma(F(\alpha))$. It follows by (21), (25) that $u = v$, a continuous and strictly increasing function of $\xi_1 = t$ (viz. $t \mapsto \alpha(t)$, by (24)) taking both negative and positive values because $u_0 = 0$.

Second Part. Let $\tau \in]-1,1[$ denote the unique zero of this continuous and strictly increasing function $u = u(t)$ $(= \alpha(t))$. We show below that $\tau = 0$. Anyway,

$$f(t) = \operatorname{sgn} u(t) = \pm 1 \text{ for } t \gtrless \tau\,, \tag{30}$$

and so, by (19) and the fact that φ_n is odd,

$$f_0 = \int_\Sigma f\, d\sigma = \frac{1}{2\varphi_n(1)}\left(\int_\tau^1 \varphi_n'(t)\,dt - \int_{-1}^{\tau} \varphi_n'(t)\,dt\right) = -\frac{\varphi_n(\tau)}{\varphi_n(1)}\,.$$

The Euler equation (10), multiplied by $-(1-t^2)^{\frac{n-3}{2}} = -\varphi_n'(t)$, therefore reads

$$\frac{d}{dt}\left((1-t^2)^{\frac{n-1}{2}}\frac{du}{dt}\right) = -\left(\pm 1 + \frac{\varphi_n(\tau)}{\varphi_n(1)}\right)\varphi_n'(t) \tag{31}$$

for $-1 < t < 1$, where the upper sign refers to points t such that $u(t) > 0$, i.e., $t > \tau$, cf. (30). Because $u(t)$ is biggest for $t = 1$ and smallest for $t = -1$, and because $n \geqslant 2$, we have

$$(1-t^2)^{\frac{n-1}{2}}\,|du/dt| = (1-t^2)^{\frac{n-2}{2}}|\nabla u| \to 0 \quad \text{as } t \to \pm 1\,.$$

Since $u(\tau) = 0$ the relevant solution of (31) is therefore:

$$u(\xi_1) = \begin{cases} \left(1 + \dfrac{\varphi_n(\tau)}{\varphi_n(1)}\right)\displaystyle\int_\tau^{\xi_1} (1-t^2)^{\frac{1-n}{2}}(\varphi_n(1) - \varphi_n(t))\,dt\,, & \xi_1 > \tau\,, \\ \left(1 - \dfrac{\varphi_n(\tau)}{\varphi_n(1)}\right)\displaystyle\int_\tau^{\xi_1} (1-t^2)^{\frac{1-n}{2}}(\varphi_n(1) + \varphi_n(t))\,dt\,, & \xi_1 < \tau\,. \end{cases} \tag{32}$$

The mean-value $u_0 = 0$ of u over Σ equals, by (19),

$$u_0 = \frac{1}{2\varphi_n(1)}\int_{-1}^{1} u(\xi_1)\varphi_n'(\xi_1)\,d\xi_1 = \frac{\psi(\tau) - \psi(-\tau)}{2\varphi_n(1)} = 0\,, \tag{33}$$

where we define for $-1 \leqslant t \leqslant 1$ (and fixed $n \geqslant 2$) the C^1-function

$$\psi(t) = \left(1 + \frac{\varphi_n(t)}{\varphi_n(1)}\right)\int_t^1 (1-s^2)^{\frac{1-n}{2}}\,(\varphi_n(1) - \varphi_n(s))^2\,ds\,, \tag{34}$$

the integrand having the limit 0 as $s \to \pm 1$ because

$$(1-t^2)^{\frac{1-n}{2}}(\varphi_n(1) \mp \varphi_n(t)) \to \frac{1}{n-1} \quad \text{as } t \to \pm 1\,,$$

by L'Hospital's rule. To obtain the second equality in (33), insert the two expressions (32) and interchange the order of integration w.r.t. t and ξ_1. Also evaluate $\psi(-\tau)$ by substituting $-s$ for s in the integral in (34) while noting that φ_n is odd, cf. (18).

In order to deduce from (33) that $\tau = 0$ we show that

$$\psi(t) - \psi(-t) < 0 \quad \text{for } 0 < t < 1 .$$

Because $\psi(1) = \psi(-1) = 0$ it suffices to prove that the odd function $\psi(t) - \psi(-t)$, $-1 < t < 1$, is strictly *convex* for $0 < t < 1$ as a function of $\varphi_n(t)$ (which might serve as a new independent variable in place of t). From (34) we obtain

$$\frac{d}{dt}\left(\frac{1}{\varphi_n'(t)}\frac{d}{dt}(\psi(t) - \psi(-t))\right) = \frac{4t}{(1-t^2)^{n-1}}g(t) , \quad 0 < t < 1 ,$$

where we define for $0 < t \leqslant 1$

$$g(t) = \frac{2}{t}(1-t^2)^{\frac{n-1}{2}}\varphi_n(t) - (n-2)(\varphi_n(1)^2 - \varphi_n(t)^2).$$

To prove that $g(t) > 0$ for $0 < t < 1$ we note that $g(1) = 0$ and

$$g(0+) = 2\varphi_n'(0) - (n-2)\varphi_n(1)^2 = 2 - \frac{\pi}{2}\frac{\left(\Gamma(\frac{n-1}{2})\right)^2}{\Gamma(\frac{n}{2})\Gamma(\frac{n-2}{2})} > 2 - \frac{\pi}{2} > 0 ,$$

using the fact that $\ln\Gamma(x)$ is convex for $x > 0$. It suffices therefore to show that $g(t)$ is strictly *concave* as a function of the strictly increasing function

$$h(t) = -\int_t^1 (1-t^2)^{\frac{n-3}{2}}\frac{dt}{t^2} , \quad 0 < t < 1 ,$$

and this follows from

$$\frac{d}{dt}\frac{g'(t)}{h'(t)} = -2nt^2(1-t^2)^{\frac{n-3}{2}} < 0 .$$

Having thus shown that $\tau = 0$, and hence $f_0 = 0$, we obtain from (32)

$$u = u(\xi_1) = \int_0^{\xi_1}(1-t^2)^{\frac{1-n}{2}}\left(\varphi_n(1) - \varphi_n(|t|)\right)dt , \quad \xi \in \Sigma , \tag{35}$$

showing that the minimizing function u is unique up to isometries of Σ when subject to the normalization $\|u\|_1 = \kappa_n$. This amounts to (20) in the theorem with $a = (1, 0, \ldots, 0)$. By Theorem 1.1,

$$f = \operatorname{sgn} u = \operatorname{sgn} \xi_1$$

is therefore the unique maximizing function for κ_n up to isometries, and we find from (19) and (34) in the same way as (33),

$$\begin{aligned}\kappa_n &= \int_\Sigma |u|\,d\sigma = \frac{1}{\varphi_n(1)}\int_0^1 u(\xi_1)\varphi_n'(\xi_1)\,d\xi_1 \\ &= \frac{\psi(0)}{\varphi_n(1)} = \frac{1}{\varphi_n(1)}\int_0^1 (1-t^2)^{\frac{1-n}{2}}(\varphi_n(1)-\varphi_n(t))^2\,dt\,.\end{aligned} \tag{36}$$

In the rest of the proof we write $u_n = u_n(t)$ in place of $u = u(\xi_1)$ to indicate the dependence of u on the dimension n, whereby $t = \xi_1$ corresponds to $a \cdot \xi$ in (20).

For $n = 2$, where $\lambda_k = k^2$, we obtain $\varphi_2(t) = \arcsin t$, $\kappa_2 = \pi^2/12$, and

$$u_2(t) = \tfrac{1}{2}(\operatorname{sgn} t)\big((\pi/2)^2 - (\arccos|t|)^2\big), \quad -1 \leqslant t \leqslant 1\,.$$

With the substitution $t = \cos\theta$, $\pi \geqslant \theta \geqslant 0$, this leads in view of (7), (8) to (1), (3) and the expression for $u(\theta)$ stated in the introduction.

For $n = 3$, where $\lambda_k = k(k+1)$, we obtain $\varphi_3(t) = t$; $\kappa_3 = 2\ln 2 - 1$, and

$$u_3(t) = (\operatorname{sgn} t)\ln(1+|t|), \quad -1 \leqslant t \leqslant 1\,.$$

From (36), (35) we shall finally deduce the following recursion formulae for $n \geqslant 2$:

$$\kappa_{n+2} = \frac{n-2}{n}\kappa_n + \frac{1}{n^2(n-1)}\,, \tag{37}$$

$$u_{n+2}(\xi_1) = \frac{n-2}{n}u_n(\xi_1) + \frac{1}{n}\xi_1 u_n'(\xi_1)\,. \tag{38}$$

It is well known that

$$\varphi_{n+2}(t) = \frac{n-1}{n}\varphi_n(t) + \frac{1}{n}t(1-t^2)^{\frac{n-1}{2}}\,,$$

in particular $\varphi_{n+2}(1) = \frac{n-1}{n}\varphi_n(1)$. As to (37) we first evaluate

$$\begin{aligned}\big(\varphi_{n+2}(1)-\varphi_{n+2}(t)\big)^2 &= \left(\frac{n-1}{n}\right)^2\big(\varphi_n(1)-\varphi_n(t)\big)^2 \\ &\quad - 2\frac{n-1}{n^2}t(1-t^2)^{\frac{n-1}{2}}\big(\varphi_n(1)-\varphi_n(t)\big) \\ &\quad + \frac{1}{n^2}t^2(1-t^2)^{n-1}\,.\end{aligned}$$

In view of (36) this leads to

$$\begin{aligned}\kappa_{n+2} - \frac{n-1}{n}\kappa_n &= \frac{n-1}{n}\frac{1}{\varphi_n(1)}\int_0^1 (1-t^2)^{\frac{-1-n}{2}}t^2\big(\varphi_n(1)-\varphi_n(t)\big)^2 dt \\ &\quad - \frac{2}{n}\frac{1}{\varphi_n(1)}\int_0^1 \frac{t}{1-t^2}\big(\varphi_n(1)-\varphi_n(t)\big)\,dt \\ &\quad + \frac{1}{n(n-1)}\frac{1}{\varphi_n(1)}\int_0^1 (1-t^2)^{\frac{n-3}{2}}t^2\,dt\,.\end{aligned}$$

Here the last term on the right equals $\frac{1}{n^2(n-1)}$. The first term becomes after partial integration

$$-\frac{1}{n}\kappa_n + \frac{2}{n}\frac{1}{\varphi_n(1)}\int_0^1 \frac{t}{1-t^2}\left(\varphi_n(1)-\varphi_n(t)\right)dt$$

in view of the line following (34). This proves (37). As to (38) we obtain for $0 < \xi_1 < 1$ in view of (35)

$$\begin{aligned} u_{n+2}(\xi_1) - \frac{n-1}{n}u_n(\xi_1) =& \frac{n-1}{n}\int_0^{\xi_1}(1-t^2)^{\frac{-1-n}{2}}t^2\left(\varphi_n(1)-\varphi_n(t)\right)dt \\ & -\frac{1}{n}\int_0^{\xi_1}\frac{t}{1-t^2}\,dt \\ =& -\frac{1}{n}u_n(\xi_1) + \frac{1}{n}(1-\xi_1^2)^{\frac{1-n}{2}}\xi_1\left(\varphi_n(1)-\varphi_n(\xi_1)\right), \end{aligned}$$

as shown by partial integration in the former integral. This proves (38) because $(1-\xi_1^2)^{\frac{1-n}{2}}\left(\varphi_n(1)-\varphi_n(\xi_1)\right) = u_n'(\xi_1)$, by (35), and because u is even.

Starting from the evaluations of κ_n and u_n for $n = 2,\ 3$ (see after (36)) we obtain from (37) after multiplication by n the explicit expression for $\kappa(n)$ for $n \geqslant 3$ given in the theorem, and (38) shows that u is an elementary function of $a \cdot \xi = t$. □

2. Extension of (2), (4), and (5) to dimension $n \geqslant 2$

In order to extend the strong isoperimetric inequality (5) in the introduction we denote by ω_n the volume of the unit ball in $\mathbf{R}^n$; its surface area is then $n\omega_n$. The classical isoperimetric inequality for a convex body K in $\mathbf{R}^n$ is

$$\left(\frac{S}{n\omega_n}\right)^n \geqslant \left(\frac{V}{\omega_n}\right)^{n-1},$$

where S denotes the surface area and $V = V(K)$ the volume of K , cf. [BZ, p. 83]. The following result is based on [F1] and provides an n-dimensional version of (5) in the introduction, though without specifying the best possible constant $c(n)$ to replace $\pi/(4-\pi)$ when $n > 2$:

THEOREM 2.1. *For convex bodies K in $\mathbf{R}^n$ we have*

$$\left(\frac{S}{n\omega_n}\right)^n \geqslant \left(\frac{V}{\omega_n}\right)^{n-1}\left(1+\frac{2}{n}c(n)\beta^2+o(\beta^2)\right), \tag{39}$$

where $c(n) > n+1$, $c(n)$ depends on n only, and β denotes the barycentric asymmetry of K:

$$\beta = \frac{V(K\setminus B(b,K))}{V(K)},$$

$B(b, K)$ being the ball of the same volume as K and centred at the centre of gravity b of K endowed with Lebesgue measure.

Proofs of this and other results described in the present section are given in [F3], where it is also shown that $o(\beta^2)$ in (39) can be sharpened to $\beta^2 O(\beta)$ for $n = 2$ (compare with (5)), to $\beta^2 O(\beta\sqrt{\ln(1/\beta)})$ for $n = 3$, and to $\beta^2 O(\beta^{\frac{4}{n+1}})$ for $n \geqslant 4$. An inequality similar to (39), but without the remainder term $o(\beta^2)$ and without assuming K to approach a ball, was obtained in [F2], though with a very small constant $c(n)$ (unspecified, but calculable). For somewhat related results see [GS] and the survey article [G].

Clearly $\beta \geqslant \alpha$, where α is the n-dimensional version of the asymmetry α given by (6), and so (39) implies the corresponding inequality with β replaced by α, as in (5). On the other hand, α has the advantage in the planar situation that (5) extends to general open sets K in the plane, though with the smaller constant 1/6 (not best possible) in place of $4/(4-\pi)$, as shown in [HHW]. For non-convex measurable sets K in $\mathbf{R}^n$, $n > 2$, certain estimates in the spirit of (5) are obtained in [H], though generally with the term containing α^2 replaced by a constant times α^4.

In order to obtain a less non-linear version of (39) we let K have centre of gravity at 0 and volume equal to that of the unit ball in $\mathbf{R}^n$. Suppose, so to speak, that K is "infinitely close" to the unit ball. Then the remainder term $o(\beta^2)$ in (39) disappears, and in the representation of ∂K in polar coordinates R, ξ ($\xi \in \Sigma$) the radial function takes the form

$$R(\xi) = 1 + u(\xi), \quad \xi \in \Sigma,$$

with $u(\xi)$ "infinitesimal". The conditions that the volume and the centre of gravity of K be the same as for the unit ball translate into

$$u_0 = 0, \quad u_1 = 0 \tag{40}$$

in terms of the expansion $u = \sum u_k$ of u in spherical harmonics $u_k \in \mathcal{H}_k$. Moreover, β becomes the L^1-norm $\|u\|_1$ of u, and we are led to the following quadratic inequality, proved in [F3] and valid for functions u in the Sobolev space $W^{1,2}(\Sigma)$ satisfying (40):

THEOREM 2.2. *For all functions $u \in W^{1,2}(\Sigma)$ such that $u_0 = u_1 = 0$ we have*

$$\|\nabla u\|^2 - (n-1)\|u\|^2 \geqslant c(n)\|u\|_1^2$$

with the same best possible constant $c(n)$ as in (39) in Theorem 2.1.

Thus we may write

$$c(n) = \min\left\{ \frac{\|\nabla u\|^2 - (n-1)\|u\|^2}{\|u\|_1^2} \;\middle|\; u \in W^{1,2}(\Sigma)\backslash\{0\},\ u_0 = u_1 = 0 \right\}. \tag{41}$$

The existence of this minimum is easy to prove by compactness. The quantity

$$\Lambda(f) := \sum_{k=2}^{\infty} \frac{\|f_k\|^2}{\lambda_k - \lambda_1}, \tag{42}$$

defined for $f \in L^2(\sigma)$, generalizes the left hand member of (2) since $\lambda_k = k^2$ when $n = 2$. Now write

$$\kappa(n) = \max\{\Lambda(f) \mid f \in L^\infty(\sigma),\ -1 \leqslant f \leqslant 1\}. \tag{43}$$

This maximum exists by the compactness argument given in the proof of Theorem 1.1 above. Because $\lambda_2 - \lambda_1 = n + 1$ we have

$$\kappa(n) < \frac{\|f\|^2}{\lambda_2 - \lambda_1} < \frac{1}{n+1}. \tag{44}$$

The following duality is established in [F3] in the same way as Theorem 1.1:

THEOREM 2.3. $c(n) = 1/\kappa(n)$.
Every maximizing function f for $\kappa(n)$ in (43) takes σ-almost everywhere the values ± 1 only, and the associated function

$$u = Tf := \sum_{k=2}^{\infty} \frac{1}{\lambda_k - \lambda_1} f_k \tag{45}$$

(convergence in $L^2(\sigma)$-norm) is minimizing for $c(n)$ in (41), and $\|u\|_1 = \kappa(n)$.

Conversely, any minimizing function u for $c(n)$, normalized so that $\|u\|_1 = \kappa(n)$, has the form $u = Tf$ for a unique maximizing function f for $\kappa(n)$. We have $u(\xi) \neq 0$ σ-a.e., and $f = \operatorname{sgn} u$. Moreover, u is C^1-smooth.

From (44) and Theorem 2.3 we deduce the estimate $c(n) > n + 1$ stated in Theorem 2.1. For any function $f \in L^\infty(\sigma)$ the function $u = Tf$ as in (45) clearly satisfies (40) and the elliptic equation

$$-\Delta u - (n-1)u = f - f_0 - f_1. \tag{46}$$

The differential operator on the left of (46) governs the spherical potential theory developed by Berg [B] for a completely different purpose, namely the study of the first surface measure μ_1 of a convex body in $\mathbf{R}^n$. His results play an important role in the proof, given in [F3], of the following theorem:

THEOREM 2.4. *At least in dimension $n \leqslant 4$ every maximizing function f for $\kappa(n)$ is even: $f(-\xi) = f(\xi)$, and so is therefore every minimizing function u for $c(n)$.*

This innocent looking result has required a lot of effort, and I doubt if there is a simple proof. The theorem probably holds in any dimension. The proof for $n \leqslant 4$ falls in two parts: First to show that there *exists* an even, maximizing function, and next to prove that *every* maximizing function is even. This latter part seems to be new even in the planar case.

The proof of the existence of an even, maximizing function (when $n \leqslant 4$) is inspired by Hall and Hayman's proof of the same for $n = 2$ in [HH]: Consider an arbitrary maximizing function f. Associate with any point $a \in \Sigma$ the following even function f_a:

$$f_a(\xi) = \begin{cases} f(\xi) & \text{for } a \cdot \xi > 0 \\ f(-\xi) & \text{for } a \cdot \xi < 0, \end{cases}$$

and form the sum $\Lambda(f_a)$, cf. (42). Averaging this over Σ one hopes that

$$\int_\Sigma \Lambda(f_a)\, d\sigma(a) \geqslant \Lambda(f). \tag{47}$$

If this inequality holds (necessarily with the equality sign) then $\Lambda(f_a) = \kappa(n)$ a.e. for $a \in \Sigma$ (because $\Lambda(f_a) \leqslant \kappa(n) = \Lambda(f)$ for every $a \in \Sigma$). Thus the even function f_a is maximizing for almost every $a \in \Sigma$ (actually for *every* a because $\Lambda(f_a)$ is continuous in a).

In dimensions $n = 2$ and $n = 4$ it is relatively simple to verify the inequality (47), but for $n = 3$ the proof depends on quite delicate estimates of certain integrals involving Legendre polynomials. For $n > 4$ I am unable to prove the inequality (47), and it may well be false, so that a different method seems to be required to prove Theorem 2.4 in dimensions bigger than 4.

A consequence of Theorem 2.4 (whenever valid) is that the best possible constant $c(n)$ in Theorem 2.1 remains unchanged if the barycentric asymmetry β is replaced by the minimal asymmetry α used in [HHW] in the planar case.

With the aim of determining the maximum $\kappa(n)$ and the maximizing functions f in (43), one may begin by looking for *stationary functions* f and the corresponding *stationary values* $\Lambda(f)$ according to the following definition:

Definition. A function $f \in L^\infty(\sigma)$ is called a *stationary function*, and $\Lambda(f)$ from (42) a *stationary value*, if the associated function $u = Tf$ from (45) satisfies $u \neq 0$ σ-a.e. and $\operatorname{sgn} u = f$ σ-a.e.

Since $u = Tf$ always satisfies (46), the stationarity of f amounts to the validity of the Euler equation, i.e., equation (46) with $f = \operatorname{sgn} u$, including the requirement $u \neq 0$ a.e. The presence of the term $\operatorname{sgn} u$ makes the Euler equation non-linear. Note that a stationary function f takes the values ± 1 only.

In particular, every maximizing function f for $\kappa(n)$ is stationary, by Theorem 2.3. Hence $\kappa(n)$ is the biggest stationary value.

In dimension $n = 2$ it is not difficult, using the Euler equation, to find all *even* stationary functions (cf. Theorem 2.4). Up to rotation of the unit circle Σ there is one such for each even integer $p \geqslant 2$, namely (writing $\xi = (\cos\theta, \sin\theta)$ for $\xi \in \Sigma$):

$$f(\xi) = \operatorname{sgn}(\cos p\theta)\,, \quad \Lambda(f) = \frac{2p}{\pi}\tan\frac{\pi}{2p} - 1\,.$$

It is also easy to write down the associated stationary functions $u = Tf$ for the dual variational problem (41). The largest among the above stationary values $\Lambda(f)$ is clearly the one for $p = 2$, and so we recover Hall and Hayman's result that the function

$$f(\theta) = \operatorname{sgn}(\cos 2\theta) = \operatorname{sgn}(\xi_1^2 - \xi_2^2)$$

is maximizing for $\kappa(2) = \Lambda(f) = 4/\pi - 1$. In addition we have shown in view of Theorem 2.4 that the translates of f are the *only* maximizing functions for $\kappa(2)$. The minimizing function $u = Tf$ for $c(2)$, associated with the above f, is stated in the introduction in connection with (4). Note that $2p$ is the number of connectivity components (arcs) of the set where $f(\xi) \neq 0$ or equivalently $u(\xi) \neq 0$. And for $p = 2$ the function f gives rise to the smallest number of such components, namely 4.

Unfortunately, in dimension > 2, it seems beyond reach to find all stationary functions and values, but we shall discuss a few interesting examples.

For *even* dimension $n = 2m$ write

$$\begin{aligned} f(n,\xi) &= \operatorname{sgn}(\xi_1^2 + \cdots + \xi_m^2 - \tfrac{1}{2}) \\ &= \operatorname{sgn}\left(\sum_{i=1}^{m}\xi_i^2 - \sum_{i=m+1}^{2m}\xi_i^2\right) \end{aligned} \tag{48}$$

(noting that $\sum_{i=1}^{2m}\xi_i^2 = 1$). It can be shown that this even function $f(n,\cdot)$ is stationary, and the corresponding stationary value is

$$\Lambda(f(n,\cdot)) = \frac{1}{2m-1}\left(\frac{\Gamma(\frac{m+1}{2})}{\Gamma(\frac{1}{2})\Gamma(\frac{m}{2})}\frac{\Gamma(\frac{1}{4})\,\Gamma(\frac{m}{2}+\frac{1}{4})}{\Gamma(\frac{3}{4})\,\Gamma(\frac{m}{2}+\frac{3}{4})} - 1\right) \tag{49}$$

(which equals $4/\pi - 1$ for $m = 1$). This is obtained by first finding the explicit solution $u(n,\cdot) = Tf(n,\cdot)$ to the Euler equation, expressed in terms of hypergeometric functions, see [F3].

For *odd* $n = 2m+1$ write

$$f(n,\xi) = \operatorname{sgn}(\xi_1^2 + \cdots + \xi_m^2 - \varrho^2)\,. \tag{50}$$

It can be shown that this even function is stationary for precisely one value of $\varrho = \varrho(m)$ (> 0), the root in a certain transcendental equation which can be written down explicitly. It involves hypergeometric functions, as does the associated solution $u(n,\cdot) = Tf(n,\cdot)$ to the Euler equation.

For $n = 3$ we obtain

$$\ln\frac{1+\varrho}{1-\varrho} = \frac{2}{1+\varrho}, \quad \text{i.e., } \varrho \approx 0.5644,$$
$$\Lambda(f(3,\cdot)) = (1-\varrho)^2 \approx 0.1898. \tag{51}$$

Another interesting stationary function g for odd $n = 2m+1$ is the second expression for $f(n,\xi)$ in (48), which now is independent of the last coordinate ξ_{2m+1}. Here one can determine explicitly the projections g_k on $\mathcal{H}_k$ and thereby calculate $\Lambda(g)$, but it turns out that this stationary value is smaller than $\Lambda(f(n,\cdot))$. For example, for $n = 3$ we find $\Lambda(g) \approx 0.1763$, to be compared with (51).

Stirling's formula applied to (49) leads to the following asymptotic formula for the ratio between the lower bound $\Lambda(f(n,\cdot))$ (at least for even n) and the elementary upper bound in the estimate $\Lambda(f(n,\cdot)) \leqslant \kappa(n) < 1/(n+1)$, cf. (44):

$$\lim_{n\to\infty} (n+1)\Lambda(f(n,\cdot)) = \frac{1}{\sqrt{\pi}}\frac{\Gamma(1/4)}{\Gamma(3/4)} - 1 \approx 0.6692.$$

Conjecture. For any dimension $n \geqslant 2$, even or odd, the above stationary function $f(n,\cdot)$ from (48), resp. (50), is maximizing; and $f(n,\cdot)$ and $-f(n,\cdot)$ are the only maximizing functions up to isometries of the unit sphere Σ_n.

For $n = 3$ the same conjecture has been proposed by Richard Hall, who also calculated what corresponds to the stationary value (51). (Personal communication.) For even $n = 2m$ the function $-f(n,\cdot)$ arises from $f(n,\cdot)$ by composition with the isometry $\xi \mapsto (\xi_{m+1},\ldots,\xi_{2m},\xi_1,\ldots,\xi_m)$ of Σ_n.

The motivations for the above conjecture include the high degree of symmetry in the proposed stationary functions, together with the fact that the number of connectivity components of the set where $f(n,\xi) \neq 0$, or equivalently $u(n,\xi) \neq 0$, is as small as possible for an even function $u \not\equiv 0$ (cf. Theorem 2.4) with mean-value $u_0 = 0$, namely 4 components in dimension $n = 2$; 3 components for $n = 3$ (this uses the Jordan curve theorem); and 2 components in dimension $n \geqslant 4$. There is also further evidence supporting the conjecture, but no proof is in sight. In particular, there seems to be no procedure of symmetrization of the level sets (cf. the proof of Theorem 1.2) which works here, except in rather special situations.

REFERENCES

[B] C. Berg, *Corps convexes et potentiels sphériques*, Mat. Fys. Medd. Dan. Vid. Selsk. **37** 6 (1969).

[BZ] Yu. D. Burago and V. A. Zalgaller, "Geometric Inequalities," Springer-Verlag, Berlin, 1988.

[F1] B. Fuglede, *Stability in the isoperimetric problem for convex or nearly spherical domains in* $\mathbf{R}^n$, Trans. Amer. Math. Soc. **314** (1989), 619–638.

[F2] B. Fuglede, *Lower estimate of the isoperimetric deficit of convex domains in* $\mathbf{R}^n$ *in terms of asymmetry*, Geometriae Dedicata **47** (1993), 41–48.

[F3] B. Fuglede, *Lower estimates of the isoperimetric deficit of nearly spherical domains in* $\mathbf{R}^n$ *in terms of asymmetry* (Manuscript).

[G] H. Groemer, *Stability of geometric inequalities*, Chapter 1.4 in "Handbook of Convex Geometry I", Ed. P. M. Gruber and J. M. Wills, Elsevier Science Publishers, 1993.

[GS] H. Groemer and R. Schneider, *Stability estimates for some geometric inequalities*, Bull. London Math. Soc. **23** (1991), 67–74.

[H] R. R. Hall, *A quantitative isoperimetric inequality in n-dimensional space*, J. reine angew. Math. **428** (1992), 161–176.

[HH] R. R. Hall and W. K. Hayman, *A problem in the theory of subordination* (To appear in the Mandelbrojt volume).

[HHW] R. R. Hall, W. K. Hayman, and A. W. Weitsman, *On asymmetries and capacity*, J. d'Anal. Math. **56** (1991), 87–123.

[Hö1] L. Hörmander, "The Analysis of Linear Partial Differential Operators I," Springer-Verlag, Berlin, 1983, 1990.

[Hö2] L. Hörmander, "The Analysis of Linear Partial Differential Operators III," Springer-Verlag, Berlin, 1985.

[K] O. D. Kellogg, "Foundations of Potential Theory," Springer-Verlag, Berlin, 1929 (reprinted 1967).

[MPF] D. S. Mitrinović, J. E. Pečarić, and A. M. Fink, "Inequalities Involving Functions and Their Integrals and Derivatives," Kluwer Acad. Publ., Dordrecht, 1991.

[M] C. Müller, "Spherical Harmonics," Lecture Notes in Mathematics, 17, Springer, 1966.

[S] L. Schwartz, "Théorie des distributions I," Hermann, Paris, 1950.

[Se] R. T. Seeley, *Singular integrals on compact manifolds*, Amer. J. Math. **81** (1959), 658–690.

[Sp] M. Spivak, "A Comprehensive Introduction to Differential Geometry IV," Publish or Perish, Boston, 1975.

Keywords. asymmetry, convex body, isoperimetric inequality, Poincaré inequality, spherical harmonics, symmetrization

UNIFORM AND TANGENTIAL HARMONIC APPROXIMATION

STEPHEN J. GARDINER
Department of Mathematics,
University College,
Dublin 4,
Ireland.

ABSTRACT. Let Ω be an open set in $\mathbb{R}^n$ and E be a relatively closed subset of Ω. Which pairs (Ω,E) have the property that functions harmonic on (some neighbourhood of) E can be uniformly approximated by functions harmonic on Ω? Which pairs permit similar approximation of functions continuous on E and harmonic on E°? This paper reviews recently-obtained solutions to the above problems, and adds some new results concerning better - than - uniform approximation.

Key words: harmonic approximation, thin set, superharmonic function.

1. Introduction

If $A \subseteq \mathbb{C}$, let $Hol(A)$ denote the collection of all functions which are holomorphic on some open set containing A. Also, if Ω is an open subset of $\mathbb{C}$ (or $\mathbb{R}^n$), let Ω^* be the Alexandroff, or one-point, compactification of Ω. We state below two landmark results in the history of uniform holomorphic approximation. For the first result see [15] or [5, pp.198,201]; for the second see [1] or [2].

RUNGE'S THEOREM. *Let Ω be an open subset of $\mathbb{C}$ and K be a compact subset of Ω. The following are equivalent:*

K. GowriSankaran et al. (eds.), Classical and Modern Potential Theory and Applications, 185–198.

(a) for every f in Hol(K) and every positive number ε *there exists g in* $Hol(\Omega)$ *such that* $|g - f| < \varepsilon$ *on K;*
(b) $\Omega^*\backslash K$ *is connected.*

ARAKELYAN'S THEOREM. *Let* Ω *be an open subset of* $\mathbb{C}$ *and E be a relatively closed subset of* Ω. *The following are equivalent:*
(a) for every f in $C(E) \cap Hol(E^{\circ})$ *and every positive number* ε *there exists g in* $Hol(\Omega)$ *such that* $|g - f| < \varepsilon$ *on E;*
(b) $\Omega^*\backslash E$ *is connected and locally connected.*

We now consider the corresponding questions concerning harmonic approximation in $\mathbb{R}^n$ $(n \geq 2)$. If $A \subseteq \mathbb{R}^n$, let $\mathcal{H}(A)$ (resp. $\mathcal{S}(A)$) be the collection of all functions which are harmonic (resp. superharmonic) on an open set containing A. Let Ω be an open set in $\mathbb{R}^n$ and E be a relatively closed subset of Ω. Inspired by the above results we make the following definitions. We call (Ω,E) a *Runge* (resp. *Arakelyan*) *pair for harmonic functions* if, for every u in $\mathcal{H}(E)$ (resp. $C(E) \cap \mathcal{H}(E^{\circ})$) and every positive number ε, there exists v in $\mathcal{H}(\Omega)$ such that $|v - u| < \varepsilon$ on E. One of the earliest results concerning harmonic approximation is as follows (see [17, p.541]).

THEOREM A. *Let K be a compact subset of* $\mathbb{R}^n$. *If* $\mathbb{R}^n\backslash K$ *is connected, then* $(\mathbb{R}^n,K)$ *is a Runge pair for harmonic functions.*

Gauthier, Goldstein and Ow [11], [12] generalized this result to deal with approximation on non-compact sets by introducing a method "fusing together" two harmonic functions. Their result is stated below.

THEOREM B. *Let* Ω *be an open subset of* $\mathbb{R}^n$ *and E be a relatively closed subset of* Ω. *If* $\Omega^*\backslash E$ *is connected and locally connected, then* (Ω,E) *is a Runge pair for harmonic functions.*

However, unlike the case of holomorphic approximation, the connectedness hypotheses in Theorems A and B are not necessary. Below we present a precise characterization of both Runge pairs and Arakelyan pairs for harmonic functions.

2. Main results

From now on we use Ω to denote an open set in $\mathbb{R}^n$ and E to denote a relatively closed subset of Ω. A subset A of Ω will be called *Ω-bounded* if its closure $\overline{A}$ is a compact subset of Ω. We use $\hat{E}$ to denote the union of E with all the Ω-bounded (connected) components of $\Omega\backslash E$. Also, we will say that the pair (Ω,E) *satisfies the (K,L)-condition* if, for each compact subset K of Ω, there is a compact subset L of Ω which contains every Ω-bounded component of $\Omega\backslash(E\cup K)$ whose closure intersects K.

THEOREM 1. *The following are equivalent:*
(a) (Ω,E) is a Runge pair for harmonic functions;
(b) for each u in $\mathcal{H}(E)$ and each positive function s in $\mathcal{S}(E)$, there exists v in $\mathcal{H}(\Omega)$ such that $0 < v - u < s$ on E;
(c) $\Omega\backslash\hat{E}$ and $\Omega\backslash E$ are thin at the same points of E, and (Ω,E) satisfies the (K,L)-condition.

Theorem 1 is a refinement of recent work of the author: the equivalence of (a) and (c) appears in [9, Theorem 4], while in [10, Theorem 1] it was shown that (c) implies the following weaker form of (b):

(b′) for each u in $\mathcal{H}(E)$ and each positive function s in $\mathcal{S}(\hat{E})$, there exists v in $\mathcal{H}(\Omega)$ such that $0 < v - u < s$ on E.

As an indication of the approach used in [9], we outline in §3 the proof of "(c) $\Rightarrow$ (a)". This argument forms part of the proof of "(c) $\Rightarrow$ (b′)" in [10]. In §5.1 we show how to strengthen the implication "(c) $\Rightarrow$ (b′)" to obtain "(c) $\Rightarrow$ (b)", and hence Theorem 1 (since (b) clearly implies (a)). To demonstrate some of the implications of the thinness condition in (c) above, we record here a simple lemma. Its proof can be found in §3.1.

LEMMA 1. *Suppose that $\Omega\backslash\hat{E}$ and $\Omega\backslash E$ are thin at the same points of E. Then each component V of $\hat{E}\backslash E$ satisfies:*
(i) V is regular for the Dirichlet problem, and
(ii) $\partial V \subseteq \partial\hat{E}$.

However, when $n \geq 3$, the converse to Lemma 1 does not hold, as the following example shows.

EXAMPLE 1. Let $n \geq 3$, let $\{Y'_k : k \in \mathbb{N}\}$ be a dense subset of $[0,1]^{n-1}$, and define

$$u(X') = \sum_{k=1}^{\infty} 2^{-k} \phi_{n-1}(|X' - Y'_k|) \qquad (X' \in \mathbb{R}^{n-1}),$$

where $\phi_2(t) = \log(1/t)$, and $\phi_m(t) = t^{2-m}$ if $m \geq 3$. (We define $\phi_m(0) = +\infty$.) Further, let

$$E = \partial\Big([0,1]^{n-1} \times [-1,0]\Big) \cup \{(X',x_n) \in [0,1]^{n-1} \times (0,1]: u(X') \leq \phi_n(x_n)\}.$$

Then the only bounded component of $\mathbb{R}^n \backslash E$ is given by $V = (0,1)^{n-1} \times (-1,0)$, which is regular for the Dirichlet problem and satisfies $\partial V \subseteq \partial \hat{E}$. However, $\mathbb{R}^n \backslash \hat{E}$ is thin at each point of $(0,1)^{n-1} \times \{0\}$, whereas $\mathbb{R}^n \backslash E$ is not thin at any of these points. Thus $(\mathbb{R}^n, E)$ is not a Runge pair for harmonic functions, in view of Theorem 1.

A recent application of Runge-type harmonic approximation can be found in [8]. It makes use of a result of Armitage and Goldstein [3], stated as Theorem C in §4.2, and also ideas from the proof of Theorem 1.

THEOREM 2. *The following are equivalent:*
(a) for each u in $C(E) \cap \mathcal{H}(E^\circ)$ and each positive number ε, there exists v in $\mathcal{H}(E)$ such that $|v - u| < \varepsilon$ on E;
(b) for each u in $C(E) \cap \mathcal{H}(E^\circ)$ and each positive function s in $C(E) \cap \mathcal{S}(E^\circ)$, there exists v in $\mathcal{H}(E)$ such that $0 < v - u < s$ on E;
(c) $\Omega \backslash E$ and $\Omega \backslash E^\circ$ are thin at the same points of E.

The equivalence of (a) and (c) above is due to Keldyš [13] and Deny [6] under the additional assumption that E is compact; the general case is due to Labrèche [14]. The addition of condition (b) here is new. It improves Theorem C below (§4.2). The proof can be found in §4.

THEOREM 3. *The following are equivalent:*
(a) (Ω,E) *is an Arakelyan pair for harmonic functions;*
(b) for each u *in* $C(E) \cap \mathcal{H}(E^o)$ *and each positive function* s *in* $C(E) \cap \mathcal{S}(E^o)$, *there exists* v *in* $\mathcal{H}(\Omega)$ *such that* $0 < v - u < s$ *on* E;
(c) $\Omega\backslash\hat{E}$ *and* $\Omega\backslash E^o$ *are thin at the same points of* E, *and* (Ω,E) *satisfies the* (K,L)*-condition.*

Theorem 3 follows easily from Theorems 1 and 2 (see §5.2). Again, the addition of condition (b) is new: it improves [10, Theorem 2]. An earlier result identifying Arakelyan pairs which satisfy $E^o = \emptyset$ was given by Shaginyan [16].

EXAMPLE 2. Let $\Omega = \mathbb{R}^n$, and let E be a closed subset of $\mathbb{R}^{n-1}\times[0,+\infty)$ such that condition (c) of Theorem 3 holds. Then, for each u in $C(E) \cap \mathcal{H}(E^o)$, each positive number ε and each continuous function $f:\mathbb{R}^{n-1} \to (0,1]$, there exists v in $\mathcal{H}(\mathbb{R}^n)$ such that

$$0 < v(X) - u(X) < \varepsilon \frac{1 + x_n}{1 + |X|^n} \qquad (X \in E;\ x_n > 0)$$

and

$$0 < v(X',0) - u(X',0) < f(X') \qquad ((X',0) \in E).$$

Thus the error in the approximation can be arranged to decay arbitrarily quickly on the part of E which intersects the hyperplane boundary. To see this, let h_1 be the Poisson integral of f in ω, where $\omega = \mathbb{R}^{n-1}\times(0,+\infty)$, and let

$$h_2(X',x_n) = \frac{1 + x_n}{\{|X'|^2 + (1 + x_n)^2\}^{n/2}} . \qquad (X' \in \mathbb{R}^{n-1};\ x_n > -1).$$

Then h_1, h_2 are positive functions in $C(\overline{\omega}) \cap \mathcal{H}(\omega)$. It follows that the function $s(X) = \min\{h_1(X), \varepsilon h_2(X)\}$ is a positive member of $C(\overline{\omega}) \cap \mathcal{S}(\omega)$. Hence, by Theorem 3, we may achieve the desired approximation.

3. Proof of Theorem 1 (c) $\Rightarrow$ (a)

3.1. We begin with the proof of Lemma 1. Let V be a component of $\hat{E}\backslash E$ and suppose that there exists an irregular boundary point Y of V. Then $\Omega\backslash V$ is thin at Y, and so $\Omega\backslash\hat{E}$ is thin at Y. By hypothesis $\Omega\backslash E$ is thin at Y. Hence V is thin at Y. This leads to the contradictory conclusion that Ω (being the union of V and $\Omega\backslash V$) is thin at Y, so (i) must hold.

To prove (ii), let $A = \partial V\backslash\partial\hat{E}$. Then $A \subseteq (\hat{E})^{\circ}$, so $\Omega\backslash\hat{E}$ is certainly thin at each point of A. By hypothesis, $\Omega\backslash E$ is thin at each point of A, so V is thin at each point of A. Thus A is a relatively open subset of ∂V which has zero harmonic measure for each component of V (see [7, 1.XI.13]). Each point of A is therefore irregular for the Dirichlet problem on V, and it follows from (i) that A is empty. Thus (ii) is also established.

3.2. It will now be shown that (c) implies (a) in Theorem 1. (This material is based on [9].) Suppose that condition (c) of Theorem 1 holds, let $u \in \mathcal{H}(E)$ and $\varepsilon > 0$. It follows from Lemma 1 that each component V of $\hat{E}\backslash E$ is regular for the Dirichlet problem and satisfies $\partial V \subseteq \partial\hat{E}$. There is an open subset ω_1 of Ω such that u is harmonic on ω_1 and $E \subset \omega_1$. We denote by $\{V_k\colon k \in I\}$, where $I \subseteq \mathbb{N}$, the collection of Ω-bounded components of $\Omega\backslash E$ which are not subsets of ω_1, choose Y_k in V_k for each k in I, and define

$$\omega_2 \;=\; \omega_1 \cup \left(\bigcup_{k\in I} (V_k \setminus \{Y_k\}) \right).$$

(If there are no such components V_k, then the desired conclusion follows from Theorem B.)

Let K be an arbitrary compact subset of Ω, and define

$$S \;=\; \bigcup_{k\in J} V_k, \quad \text{where} \quad J \;=\; \{\, k \in I\colon \overline{V}_k \cap K \neq \varnothing \,\}.$$

It follows from the (K,L)-condition that S is Ω-bounded. Since

$$\operatorname{dist}(\overline{S}\cap E,\ \mathbb{R}^n\backslash\omega_1) \;>\; 0,$$

the set J must be finite. Thus $\{Y_k\colon k \in I\}$ has no limit point in Ω. It also follows that we can choose a collection $\{U_k\colon k \in I\}$ of Ω-bounded open sets such that $\bar{V}_k \subset U_k$ and $\bar{U}_k \subset \omega_2$ for each k, and such that any given compact subset of Ω intersects only finitely many of the sets $\bar{U}_k$.

Now fix k in I and let W_k be a V_k-bounded open set such that $Y_k \in W_k$ and $V_k \backslash W_k \subset \omega_1$. Let $G_k(\cdot,\cdot)$ denote the Green function for V_k. Since V_k is regular, $G_k(Y_k,\cdot)$ vanishes on ∂V_k. Now define

$$s_k(X) = a_k - b_k G_k(Y_k,X) \qquad (X \in V_k),$$

where

$$a_k = 1 + \sup\{ u(X): X \in \partial V_k \}$$

and b_k is a positive number chosen sufficiently large so that $s_k < u$ on ∂W_k. This ensures that the function defined by

$$u_k(X) = \begin{cases} s_k(X) & (X \in W_k) \\ \min\{s_k(X), u(X)\} & (X \in V_k \backslash W_k) \end{cases}$$

is superharmonic on $V_k \backslash \{Y_k\}$, and also that $u_k = u$ at all points of V_k which are sufficiently close to ∂V_k.

We carry out the above construction for each k in I, and define

$$u'(X) = \begin{cases} u_k(X) & (X \in V_k;\ k \in I) \\ u(X) & (X \in \omega \setminus (\cup_k V_k)). \end{cases}$$

Then u' is a superharmonic function on ω_2 such that $u' = u$ on an open set which contains E and such that, for each k in I, the function $u'(X) + b_k \phi_n(|X-Y_k|)$ has a superharmonic extension to $\omega_2 \cup \{Y_k\}$ for a suitable choice of positive constant b_k.

We again fix k temporarily. For each m in $\mathbb{N}$ let

$$A_{k,m} = \{ X \in \bar{U}_k: \operatorname{dist}(X,\hat{E}) \geq 1/m \},$$

and let $g_{k,m}$ be the Green function for $\Omega \backslash A_{k,m}$ with pole at X_k. (This must exist for all sufficiently large m, even if $n = 2$, because $\partial V_k \subseteq \partial\hat{E}$.) If we define $g_{k,m}(X) = 0$ when $X \in A_{k,m}$, then the upper regularization $\check{g}_{k,m}$ is subharmonic on $\Omega \backslash \{Y_k\}$. Thus the function $g_k = \lim_{m \to \infty} \check{g}_{k,m}$, being the limit of a decreasing sequence, is subharmonic on $\Omega\backslash\{Y_k\}$ and harmonic on $V_k\backslash\{Y_k\}$. Further, g_k vanishes on $U_k\backslash\hat{E}$, and so vanishes at each point of ∂V_k where $\Omega\backslash\hat{E}$ is non-thin. If X is a point of ∂V_k at which $\Omega\backslash\hat{E}$ is thin, then condition (c) shows that $\Omega\backslash E$, and hence V_k, are also thin at X. The set of all such points X therefore has zero harmonic measure for V_k. It follows that g_k coincides (on V_k) with the Green function for the regular set V_k with pole at Y_k. Thus, given a positive number ε, there is a compact subset K_k of V_k such that $Y_k \in K_k^{\circ}$ and $g_k < 2^{-k-3}b_k^{-1}\varepsilon$ on $V_k\backslash K_k^{\circ}$. Hence, by the monotonicity of the sequence $(\check{g}_{k,m})_{m\geq 1}$ and Dini's theorem, there exists m_k such that

$$\check{g}_{k,m_k}(X) \leq g_k(X) + 2^{-k-3}b_k^{-1}\varepsilon < 2^{-k-2}b_k^{-1}\varepsilon \qquad (X \in \partial K_k).$$

It follows that $\check{g}_{k,m_k}(X) \leq 2^{-k-2}b_k^{-1}\varepsilon$ on E.

Now let

$$W = \Omega \setminus \left(\bigcup_{k\in I} A_{k,m_k} \right).$$

This is an open set because only finitely many of the sets A_{k,m_k} intersect a given compact subset of Ω. Also, $\hat{E} \subseteq W$. Let $G_W(\cdot,\cdot)$ denote the Green function for W. Then

$$G_W(Y_k,X) \leq \check{g}_{k,m_k}(X) < 2^{-k-2}b_k^{-1}\varepsilon \qquad (X \in E). \tag{1}$$

We define

$$v_1(X) = \sum_{k\in I} b_k\, G_W(Y_k,X) \qquad (X \in W).$$

It follows from (1) that v_1 defines a potential on W, and that $v_1 < \varepsilon/4$ on E. The function $v_2 = u' + v_1 + \varepsilon/4$, suitably redefined on the set where it is the difference of two infinite values, is superharmonic on the open set $W \cap (\omega_1 \cup (\cup_k V_k))$, which contains $\hat{E}$. Also, $u < v_2 < u + \varepsilon/2$ on E.

Replacing u by $-u$ in the above argument, we obtain v_3 in $\mathcal{S}(\hat{E})$ such that $-u < v_3 < -u + \varepsilon/2$ on E, whence $-v_3 < v_2$ on E. It follows from the maximum principle that $-v_3 < v_2$ on an open set ω_3 which contains $\hat{E}$. The greatest harmonic minorant w of v_2 on ω_3 therefore satisfies $-v_3 \le w \le v_2$ on E, whence $|w - u| < \varepsilon/2$ on E. We observe that $w \in \mathcal{H}(\hat{E})$. Further, $\Omega^*\backslash\hat{E}$ is connected (obviously) and also locally connected (by the (K,L)-condition). Thus we may apply Theorem B to obtain v in $\mathcal{H}(\Omega)$ such that $|v - w| < \varepsilon/2$ on $\hat{E}$. Hence $|v - u| < \varepsilon$ on E, proving (a).

4. Proof of Theorem 2

4.1. We require the following lemma.

LEMMA 2. *Suppose that $\Omega\backslash E$ and $\Omega\backslash E^{\circ}$ are thin at the same points of E. If s is a positive function in $C(E) \cap \mathcal{S}(E^{\circ})$, then there exists a continuous function w in $\mathcal{S}(E)$ such that $0 < w < s$ on E.*

In proving Lemma 2 we make use of the fact (see [4, Théorème 3]) that, if E satisfies the stated thinness condition, then for each u in $C(E) \cap \mathcal{S}(E^{\circ})$ and each positive number ε, there exists a continuous function v in $\mathcal{S}(E)$ such that $|v - u| < \varepsilon$ on E. Thus the lemma is immediate in the case where s has a positive lower bound, a say, on E.

We can assume that $E \neq \Omega$, for otherwise the result is trivial. Let (K_m) be a sequence of compact subsets of Ω such that $K_1 \cap E = \varnothing$, such that $K_m \subset K^{\circ}_{m+1}$ for each m, and such that $\cup_m K_m = \Omega$. We define

$$\varepsilon_m = \inf\left(\{ s(X): X \in E \cap K_m \} \cup \{1\} \right) \qquad (m \in \mathbb{N}). \tag{2}$$

For each m we use the result recorded in the first paragraph to obtain a continuous function w_m in $\mathcal{S}(E)$ such that

$$\left(1 - \frac{1}{m}\right) s(X) < w_m(X) < \left(1 - \frac{1}{m}\right) s(X) + \frac{\varepsilon_m}{m(m+1)} \qquad (X \in E). \tag{3}$$

If $X \in E \cap K_m$, then

$$w_{m+1}(X) > \left(1 - \frac{1}{m+1}\right) s(X)$$

$$\geq \left(1 - \frac{1}{m}\right) s(X) + \frac{\varepsilon_m}{m(m+1)} > w_m(X). \tag{4}$$

Let U_m denote the open set where w_m is defined and positive, and let V_m be the subset of $U_m \cap U_{m+1}$ where $w_{m+1} > w_m$. Thus V_m is open and $E \cap K_m \subset V_m$. Let

$$W_m = \left(K^{\circ}_{m+2} \setminus K_m\right) \cap \left(\bigcap_{k=1}^{m+1} U_k\right) \cap V_{m+2} \qquad (m \geq 1).$$

Thus W_m is an open set which contains $E \cap (K^{\circ}_{m+2} \setminus K_m)$. It follows that the set W, given by $W = \cup_m W_m$, is open and contains E. We define a function on W by

$$w(X) = \min\{w_1(X), w_2(X), \ldots, w_{m+2}(X)\} \qquad (X \in W_m;\ m \in \mathbb{N}).$$

To see that w is well-defined, we note that $w_{m+3} > w_{m+2}$ on V_{m+2}, so

$$\min\{w_1(X), \ldots, w_{m+3}(X)\} = \min\{w_1(X), \ldots, w_{m+2}(X)\} \qquad (X \in W_m \cap W_{m+1}).$$

Clearly w is continuous, positive and superharmonic on W. Finally, if $X \in E \cap (K^{\circ}_{m+2} \setminus K_m)$, then it follows from (2) and (3) that

$$w(X) \leq w_{m+2}(X) < \left(1 - \frac{1}{m+3}\right) s(X) < s(X),$$

whence $w < s$ on E, as required.

4.2. In order to complete the proof of Theorem 2, we recall the following recent result of Armitage and Goldstein [3, Theorem 1.2].

THEOREM C. *Let Ω be a connected open set which possesses a Green function $G_{\Omega}(\cdot,\cdot)$, and let $P \in \Omega$. If $u \in C(E) \cap \mathcal{H}(E^{\circ})$ and ε is a positive number, then there exists v in $\mathcal{H}(E)$ such that*

$$|v(X) - u(X)| < \varepsilon \min\{1, G_{\Omega}(P,X)\} \qquad (X \in E).$$

As was pointed out in §2, the equivalence of (a) and (c) in Theorem 2 is known. Further, (b) clearly implies (a). It remains to prove that (c) implies (b). So suppose that (c) holds, let $u \in C(E) \cap \mathcal{H}(E^o)$ and let s be a positive function in $C(E) \cap \mathcal{S}(E^o)$. It follows from Lemma 2 that there is a positive superharmonic function w on an open set ω which contains E, such that $w < s$ on E. We can assume that ω possesses a Green function: the only case this excludes is where $E = \Omega$ and Ω does not possess a Green function, in which case (b) is trivially true. Let ω' be any component of ω, let B be a closed ball in ω' with centre P and choose a in the interval $[1,+\infty)$ such that

$$a \quad > \quad \sup \{ \, G_{\omega'}(P,X) \colon X \in \omega_0 \backslash B \, \}.$$

Let w' be the reduced function of w in ω' relative to some closed ball in $\omega' \backslash E$. Thus $w' \in \mathcal{H}(E \cap \omega')$. If we define

$$b \quad = \quad \inf \left\{ \frac{w'(X)}{\min\{a, \; G_{\omega'}(P,X)\}} : X \in B \right\},$$

then

$$w'(X) \quad \geq \quad b \, \min\{a, G_{\omega'}(P,X)\} \quad \geq \quad b \, \min\{1, G_{\omega'}(P,X)\} \qquad (X \in \Omega),$$

and it follows from Theorem C that there exists v in $\mathcal{H}(E \cap \omega')$ such that $|v - (u + w'/2)| < w'/2$, whence $0 < v - u < w' \leq w < s$, on $E \cap \omega'$. Repeating this argument for each component of ω, we obtain v in $\mathcal{H}(E)$ such that $0 < v - u < s$ on E, proving (b).

5. Proof of Theorems 1 and 3

5.1. We now complete the proof of Theorem 1 by showing that (c) implies (b), noting as in the paragraph following the theorem that we already know that (c) implies the weaker condition (b′). It is enough to prove the following lemma.

LEMMA 3. *Suppose that* (Ω, E) *is a Runge pair for harmonic functions. Then, for each positive function* s *in* $\mathcal{S}(E)$, *there exists a positive function* w *in* $\mathcal{S}(\hat{E})$ *such that* $w < s$ *on* E.

To prove this, let (Ω,E) be a Runge pair for harmonic functions, and let s be a positive superharmonic functions on some open set ω which contains E. On each component ω_k of ω which contains E we replace s by its reduced function relative to some closed ball in $\omega_k\backslash E$. This results in a function s' in $\mathcal{H}(E)$ such that $0 < s' \le s$ on ω. There is thus no loss of generality in assuming that $s \in \mathcal{H}(E)$. We may also assume that $E \neq \Omega$.

The argument now follows the pattern of §4.1. We choose an exhaustion (K_m) of Ω by compact sets with the following additional property: any Ω-bounded component of $\Omega\backslash E$ which intersects K_m is contained in K_{m+1}. This is possible since the (K,L)-condition holds, in view of the known equivalence of (a) and (c). We can also arrange that $K_1 \cap \hat{E} = \emptyset$. We now define ε_m as in (2), and use our hypothesis on (Ω,E) to obtain w_m in $\mathcal{H}(\Omega)$ such that (3) holds. Thus (4) holds for X in $E \cap K_m$, and it follows from the minimum principle and our construction of (K_m) that $w_{m+1} > w_m$ also on $\hat{E} \cap K_{m-1}$. It also follows from the minimum principle that $w_m > 0$ on an open set U_m which contains $\hat{E}$. Let V_m be the open subset of $U_m \cap U_{m+1}$ where $w_{m+1} > w_m$, so that $\hat{E} \cap K_{m-1} \subset V_m$, and define

$$W_m = \left(K^\circ_{m+2} \backslash K_m\right) \cap \left(\bigcap_{k=1}^{m+2} U_k\right) \cap V_{m+3} \qquad (m \ge 1).$$

Thus W_m is an open set which contains $\hat{E} \cap (K^\circ_{m+2}\backslash K_m)$. It follows, as in §4.1, that the function

$$w(X) = \min\{w_1(X), w_2(X), \ldots, w_{m+3}(X)\} \qquad (X \in W_m;\ m \in \mathbb{N})$$

is a well-defined positive superharmonic function on the open set $W = \cup_m W_m$, that $\hat{E} \subset W$, and that $w < s$ on E.

5.2. Turning now to the proof of Theorem 3, it is clear that (b) implies (a). Further, the implications "(a) $\Rightarrow$ (c)" of Theorems 1 and 2 can be combined to show that (a) implies (c) in this result also. Now suppose that (c) holds, let $u, s \in C(E) \cap \mathcal{S}(E^\circ)$ and suppose that $s > 0$. Since $\Omega\backslash\hat{E} \subset \Omega\backslash E \subset \Omega\backslash E^\circ$, we can apply Lemma 2 to obtain w in $\mathcal{S}(E)$ such that $0 < w < s$ on E. It now follows from Theorems 1 and 2 that (b) holds.

References

1. Arakeljan, N. U.: Uniform and tangential approximations by analytic functions, *Izv. Akad. Nauk Armjan. SSR Ser. Mat.* **3**, 273-286 (1968) (Russian); *Amer. Math. Soc. Transl. (2)* **122**, 85-97 (1984).

2. Arakeljan, N. U.: Approximation complexe et propriétés des fonctions analytiques, *Actes, Congrès intern. Math.* Tome **2**, 595-600 (1970).

3. Armitage, D. H. and M. Goldstein: Tangential harmonic approximation on relatively closed sets, *Proc. London Math. Soc.*, (to appear).

4. Bensouda, C. and P. M. Gauthier: Approximation surharmonique sur les fermés, *preprint*.

5. Conway, J. B.: *Functions of one complex variable*, Second Edition, Springer, New York, 1978.

6. Deny, J.: Systèmes totaux de fonctions harmoniques, *Ann. Inst. Fourier (Grenoble)* **1**, 103-113 (1949).

7. Doob, J. L.: *Classical potential theory and its probabilistic counterpart*, Springer, New York, 1983.

8. Gardiner, S. J.: The Dirichlet problem with non-compact boundary, *Math. Z.* **213**, 163-170 (1993).

9. Gardiner, S. J.: Superharmonic extension and harmonic approximation, *Ann. Inst. Fourier (Grenoble)* (to appear).

10. Gardiner, S. J.: Tangential harmonic approximation on relatively closed sets, *Illinois J. Math.* (to appear).

11. Gauthier, P. M., M. Goldstein and W. H. Ow: Uniform approximation on unbounded sets by harmonic functions with logarithmic singularities, *Trans. Amer. Math. Soc.* **261**, 169-183 (1980).

12. Gauthier, P. M., M. Goldstein and W. H. Ow: Uniform approximation on closed sets by harmonic functions with Newtonian singularities, *J. London Math. Soc. (2)* **28**, 71-82 (1983).

13. Keldyš, M. V.: On the solvability and stability of the Dirichlet problem, *Uspehi Mat. Nauk* **8**, 171-231 (1941) (Russian); *Amer. Math. Soc. Transl.* **51**, 1-73 (1966).

14. Labrèche, M.: *De l'approximation harmonique uniforme*, Doctoral thesis, Université de Montréal, 1982.

15. Runge, C.: Zur Theorie der eindeutigen analytischen Funktionen, *Acta*

Math. **6**, 228-244 (1885).

16. Shaginyan, A. A.: Uniform and tangential harmonic approximation of continuous functions on arbitrary sets, *Math. Notes* **9**, 78-84 (1971).

17. Walsh, J. L.: The approximation of harmonic functions by harmonic polynomials and by harmonic rational functions, *Bull. Amer. Math. Soc.* (2) **35**, 499-544 (1929).

Inversion and Reflecting Brownian Motion

JOSEPH GLOVER and MURALI RAO
Department of Mathematics
University of Florida
201 Walker Hall
PO Box 118000
Gainesville FL 32611–8000

ABSTRACT. For a class of domains $E \subset \mathbf{R}^p R^2$, we construct a function Φ and a time change ρ_t of two dimensional Brownian motion b_t such that $\Phi(b(\rho_t))$ is Brownian motion with normal reflection at the boundary of E.

1. Introduction

In this article, we construct a function Φ and a time change ρ_t of two dimensional Brownian motion b_t such that $\Phi(b(\rho_t))$ is Brownian motion with normal reflection at the boundary of a domain $E \subset \mathbf{R}^2$. Of course, we cannot provide such an explicit construction for an arbitrary domain, but the collection of domains which do admit this construction is suprisingly large. In general, there are several ways to approach the construction of reflecting Brownian motion. There is the analytic approach based on the infinitesimal generator (the Laplacian with Neumann boundary conditions), a construction via stochastic differential equations, and a Dirichlet space formulation. Our algorithm provides a concrete construction of this process from two dimensional Brownian motion which may provide more insight into the behavior of reflecting Brownian motion.

This article could be subtitled "A case study in Markov functions" since the ideas and techniques developed within spring directly from recent work on symmetry groups and Markov functions [4,5,9]. While these articles provided inspiration, the present article is largely self–contained.

Before describing the general construction, we discuss first the motivating example in which E is the complement in $\mathbf{R}^2$ of the open unit ball B centered at the origin. In this case, $\Phi(x) = x$ if $x \in E$, and $\Phi(x) = (x_1/\|x\|^2, x_2/\|x\|^2)$ if $x \in B - \{0\}$. Define $g(x)$ on $\mathbf{R}^2 - \{0\}$ by setting $g(x) = 1$ on E and $g(x) = |x|^{-4}$ on $B - \{0\}$. Let

$$\begin{aligned} a_t &= \int_0^t g(b_s)ds \\ \rho_t &= \inf\{s : a_s > t\} \end{aligned}$$

K. GowriSankaran et al. (eds.), Classical and Modern Potential Theory and Applications, 199–215.

Then $\Phi(b(\rho_t))$ is reflecting Brownian motion on E with normal reflection. This well known fact is the starting point of our investigation.

Let J be a finite or countably infinite index set, and let $(B_j)_{j\in J}$ be a collection of nonoverlapping open balls in $\mathbf{R}^2$. Their boundaries may or may not intersect. Our domains will be of the form $E = \mathbf{R}^2 - \bigcup_{j\in J} B_j$. Each ball B_j of center z_j and radius r_j generates an inversion in $\overline{\mathbf{R}^2} = \mathbf{R}^2 \bigcup\{\infty\}$ defined by

$$\begin{aligned} T_j(x) &= z_j + \frac{r_j^2}{|x-z_j|^2}(x-z_j) \text{ if } x \notin \{z_j, \infty\} \\ T_j(z_j) &= \infty \\ T_j(\infty) &= z_j \end{aligned}$$

These transformations of $\overline{\mathbf{R}^2}$ are connected to the famous Kelvin transforms of potential theory. Let $\mathcal{G}$ be the smallest group containing all of the inversions $(T_j)_{j\in J}$. Then $\mathcal{G}$ induces a splitting of $\overline{\mathbf{R}^2}$ into two sets H and H^c. The horizon $H = \{x \in \overline{\mathbf{R}^2} : Sx \notin E \text{ for every } S \in \mathcal{G}\}$, while $H^c = \{x \in \overline{\mathbf{R}^2} : Sx \in E \text{ for some } S \in \mathcal{G}\}$. If we consider $\mathcal{G}$ as a transformation group on H^c, then we can look at the orbit $[x]$ of a point $x \in H^c$ by setting $[x] = \{Sx : S \in \mathcal{G}\}$. Because $B_j \cap B_k = \emptyset$ for every $k \neq j$, it turns out that each such orbit $[x]$ contains exactly one point $z \in E$. We define $\varphi([x]) = z$. Our function $\Phi : H^c \to E$ is defined by setting $\Phi(x) = \varphi([x])$. The geometry of $\mathcal{G}$, H, H^c and Φ is discussed in section 2, which contains no probability theory. Probability enters in section 3 when we examine Brownian motion killed the first time it enters H. Call this killed process b_t. If we let $fc(H)$ denote the fine closure of H for Brownian motion, then the state space of b_t is $\mathbf{R}^2 - fc(H)$. Using the inversions, we construct a function g on H^c such that $g(z) = g(T_n z) r_n^4 |z - z_n|^{-4}$ for every inversion $(T_n)_{n\in J}$ and for every $z \in \mathbf{R}^2 - \{g = \infty\}$. We define

$$\begin{aligned} a_t &= \int_0^t g(b_s)ds \\ \rho_t &= \inf\{s : a_s > t\} \end{aligned}$$

and $Z_s = b(\rho_s)$. We assume there is a set $\Gamma \subset \mathbf{R}^2 - fc(H)$ which is polar for b_t such that $a_t < \infty$ on $[0, T_H)$ $a.s.$ P^x for each $x \in \mathbf{R}^2 - (fc(H) \cup \Gamma)$, so the state space of Z_s is $\mathbf{R}^2 - (fc(H) \cup \Gamma)$. This assumption is easily verified for a large class of examples. (In particular, if J is finite, it is true). In (3.4), we prove that $(S(Z_t),\ P^{S^{-1}(x)})$ is identical in law to (Z_t, P^x) for every $S \in \mathcal{G}$, for every $x \in \mathbf{R}^2 - (fc(H) \cup \Gamma)$. If ∂E has zero Lebesgue measure and if $\bigcap_{n\in J} \bigcup_{j\geq n} B_j$ is a polar set, then it follows that $\Phi(b(\rho_t))$ is a diffusion which is an extension (in the sense of Dirichlet spaces) of killed Brownian motion on the interior of E. If the process has infinite lifetime, then it is the maximal Markovian extension. Silverstein and Fukushima identify this process with the Dirichlet space $(\mathcal{E}, H^1(E))$, where

$$\mathcal{E}(u,v) = \frac{1}{2}\int \nabla u \nabla v dx$$

so $\Phi(b(\rho_t))$ is reflecting Brownian motion (see (3.8)). However, whether or not the lifetime of $b(\rho_t)$ is infinite depends on the geometry of E. We give an example to show it may be finite. At the end of section 3, we study conditions which imply that $b(\rho_t)$ has infinite lifetime (see (3.12)). In particular, if J is finite, then the lifetime is infinite, and our conditions are general enough to apply to certain geometries where J is infinite.

2. The Geometry of Inversions

In this article, J will be either a finite set $\{1,2,\ldots,K\}$ or the positive integers $\{1,2,\ldots\}$. We assume that J indexes a collection $B_n = B(z_n, r_n)$ of open balls in $\mathbf{R}^2$ with centers z_n and radii r_n such that $B_n \cap B_k = \emptyset$ for every $n \neq k$. The boundaries of two balls may or may not intersect at one point. For each $n \in J$, we define the inversion $T_n : \overline{\mathbf{R}^2} \to \overline{\mathbf{R}^2}$ with respect to B_n as follows. If $x \in \mathbf{R}^2 - \{z_n\}$, then

$$T_n(x) = z_n + \frac{r_n^2}{|x - z_n|^2}(x - z_n).$$

The definition is completed by setting $T_n(z_n) = \infty$ and $T_n(\infty) = z_n$. Various properties of inversions are collected in [6,7]. Each T_n is a continuous map of $\overline{\mathbf{R}^2}$ onto itself, and T_n is its own inverse: T_n^2 is the identity on $\overline{\mathbf{R}^2}$. If $B(q,r)$ is the ball with center q and radius r, then

$$T_n(B(q,r)) = B\left(z_n + \frac{r_n^2(q - z_n)}{|q - z_n|^2 - r^2}, \frac{r_n^2 r}{||q - z_n|^2 - r^2|}\right)$$

Let $\mathcal{G}$ denote the smallest group of transformations $S : \overline{\mathbf{R}^2} \to \overline{\mathbf{R}^2}$ containing all of the inversions $(T_n)_{n\in J}$. Then $\mathcal{G}$ is a countable group of invertible continuous transformations of $\overline{\mathbf{R}^2}$.

(2.1) Lemma. Let $S \in \mathcal{G}-\{\text{identity}\}$. Then $S = T_{n_1}T_{n_2}\cdots T_{n_k}$ for some finite collection $\{T_{n_1}, T_{n_2}, \cdots, T_{n_k}\}$ of inversions where $n_j \neq n_{j+1}$ for every $1 \leq j < k$. Moreover, this representation is unique.

Proof. That S can be written as $T_{n_1}T_{n_2}\cdots T_{n_k}$ with $n_j \neq n_{j+1}$ follows immediately from the definition of $\mathcal{G}$ and the fact that T_j^2 is the identity for each $j \in J$. Let us suppose the representation is not unique. That is, suppose there are inversions $\{T_{p_1}, \cdots, T_{p_m}\}$ with $p_j \neq p_{j+1}$ such that $S = T_{p_1}T_{p_2}\cdots T_{p_m} = T_{n_1}T_{n_2}\cdots T_{n_k}$. Consider the action of each representation on the point ∞. Since $n_j \neq n_{j+1}$ for each j, $(T_{n_1}\cdots T_{n_k})(\infty) \in B_{n_1}$. Since $p_j \neq p_{j+1}$ for each j, $(T_{p_1}\cdots T_{p_m})(\infty) \in B_{p_1}$.

But B_{n_1} and B_{p_1} intersect if and only if $n_1 = p_1$, and we conclude $T_{p_1} = T_{n_1}$, so $T_{p_2} \cdots T_{p_m} = T_{n_2} \ldots T_{n_k}$. Repeating this reasoning, we obtain $T_{p_2} = T_{n_2}$, etc. Finally, we end up with two possibilities. Either $m = k$ and $p_m = n_k$ (in which case we are done), or $m \neq k$. If $m \neq k$, then one is smaller (say $m < k$), and we have $T_{n_{m+1}} \cdots T_{n_k} =$ identity. Consider once again the action of this element on the point ∞. The identity element fixes this point, while $T_{n_{m+1}} \ldots T_{n_k}$ maps it into $B_{n_{m+1}}$, which is a contradiction. Thus $m = k$, and $T_{p_j} = T_{n_j}$ for each $1 \leq j \leq k$.

Q.E.D.

Whenever we write $S = T_{n_1} \cdots T_{n_k}$, we shall assume that $n_j \neq n_{j+1}$ for $1 \leq j < k$, so it is the unique representation of S.

If we equip $\mathcal{G}$ with the discrete topology, we may consider it as a topological transformation group on $\overline{\mathbf{R}^2}$. We shall denote by $[x]$ the <u>orbit</u> of x under $\mathcal{G}$. That is, $[x] = \{Sx : S \in \mathcal{G}\}$. The collection of orbits is denoted by $\overline{\mathbf{R}^2}/\mathcal{G}$. In general, the quotient topology on $\overline{\mathbf{R}^2}/\mathcal{G}$ is not Hausdorff and is useless to us without a more careful examination of the orbit structure.

(2.2) <u>Lemma</u>. Each orbit contains at most one point in $\overline{\mathbf{R}^2} - \bigcup_{n \in J} B_n$.

<u>Proof</u>. Suppose x and y are in $\overline{\mathbf{R}^2} - \bigcup_{n \in J} B_n$, and they are both in the same orbit. Then $[x] = [y]$, and there exists an element $S = T_{n_1} T_{n_2} \cdots T_{n_k} \in \mathcal{G}$ with $Sx = y$. Since $x \notin B_{n_k}$, $T_{n_k} x \in \overline{B_{n_k}}$. Since $B_{n_k} \cap B_{n_{k-1}} = \emptyset$, $T_{n_{k-1}} T_{n_k} \in \overline{B_{n_{k-1}}}$. Continuing this reasoning, we see that $Sx = y \in \overline{B_{n_1}}$. Since $y \notin B_{n_1}$, y must be in the boundary of B_{n_1} (denoted ∂B_{n_1}). But T_{n_1} fixes points in ∂B_{n_1}, so $T_{n_2} \cdots T_{n_k} x = T_{n_1} y = y$. The argument is now repeated: y must be in $\overline{B_{n_2}}$. Since $y \notin B_{n_2}, y$ must be in ∂B_{n_2}. But then $T_{n_3} \cdots T_{n_k} x = T_{n_2} y = y$. Continuing in this manner, we arrive at $T_{n_k} x = y$. Since $x \notin B_{n_k}$, $y \in \overline{B_{n_k}}$. Since $y \notin B_{n_k}$, $y \in \partial B_{n_k}$, and we obtain $x = T_{n_k} y = y$.

Q.E.D.

In general, then, there are two types of orbits, namely those containing a point in $\overline{\mathbf{R}^2} - \bigcup_{n \in J} B_n$ and those containing no such point. To simplify our notation, let

$$E = \overline{\mathbf{R}^2} - \bigcup_{n \in J} B_n$$

and define

$$H = \{x \in \overline{\mathbf{R}^2} : [x] \cap E = \emptyset\}.$$

H is called the <u>horizon</u>. If we let $E_S = SE = \{Sx : x \in E\}$, then

$$H^c = \bigcup_{S \in \mathcal{G}} E_S$$

$$H = \bigcap_{S \in \mathcal{G}} S \left(\bigcup_{j \in J} B_j \right)$$

That is, $y \in H^c$ if and only if y can be "reached" from a point $x \in E$ through the action of $\mathcal{G}$. The horizon H is a countable intersection of open sets and H^c is a union of closed sets.

Note that $S : H \to H$ and $S : H^c \to H^c$ whenever $S \in \mathcal{G}$. Also, $H^c/\mathcal{G}$ is precisely the collection of orbits containing a point in E, while $H/\mathcal{G}$ is the collection of orbits containing no such point. Unfortunately, the quotient topologies on $H^c/\mathcal{G}$ and $H/\mathcal{G}$ may still fail to be Hausdorff.

Our work in this article depends on a very orderly construction of H^c. By virtue of Lemma (2.1), each $S \in \mathcal{G}-\{\text{identity}\}$ can be written uniquely as $T_{n_1} \cdots T_{n_k}$ with $n_j \neq n_{j+1}$ for $1 \leq j < k$. Thus we can assign S the number k (namely, the number of inversions needed in the representation). We name this function χ, so $\chi(S) = k$. If we define $\chi(\text{identity}) = 0$, then $\chi : \mathcal{G} \to \mathbf{Z}$. Define

$$\begin{aligned}
\mathcal{G}_k &= \{S \in \mathcal{G} : \chi(S) = k\} \\
\mathcal{H}_k &= \{S \in \mathcal{G} : \chi(S) \leq k\} \\
F_k &= \bigcup_{S \in \mathcal{G}_k} E_S \\
E_k &= \bigcup_{p \leq k} F_p = \bigcup_{S \in \mathcal{H}_k} E_S
\end{aligned}$$

In figure 1, we have indicated F_1 and F_2 in the case where $J = \{1, 2, 3\}$ and B_1, B_2, and B_3 are three unit balls arranged so that each ball is tangent to the other two. (See also the discussion in (2.5) below).

Note that

$$\begin{aligned}
F_1 &= \bigcup_{k_1 \in J} T_{k_1}\left(\overline{\mathbf{R}^2} - \bigcup_{j \in J} B_j\right) = \bigcup_{k_1 \in J}\left(B_{k_1} - T_{k_1}\Big(\bigcup_{j \neq k_1} B_j\Big)\right) \\
&= \bigcup_{k_1 \in J} B_{k_1} - \bigcup_{k_1 \in J} T_{k_1}\left(\bigcup_{j \neq k_1} B_j\right) \\
F_2 &= \bigcup_{k_2 \in J} \bigcup_{k_1 \neq k_2} T_{k_2} T_{k_1}\left(\overline{\mathbf{R}^2} - \bigcup_{j \in J} B_j\right) \\
&= \bigcup_{k_2 \in J} T_{k_2}\left(\bigcup_{k_1 \neq k_2}\left(B_{k_1} - T_{k_1}\Big(\bigcup_{j \neq k_1} B_j\Big)\right)\right) \\
&= \bigcup_{k_2 \in J} T_{k_2}\left(\bigcup_{k_1 \neq k_2} B_{k_1}\right) - \bigcup_{k_2 \in J} T_{k_2} \bigcup_{k_1 \neq k_2} T_{k_1}\left(\bigcup_{j \neq k_1} B_j\right)
\end{aligned}$$

and in general,

$$
\begin{aligned}
F_n &= \bigcup_{k_n \in J} \bigcup_{k_{n-1} \neq k_n} \cdots \bigcup_{k_1 \neq k_2} T_{k_n} \cdots T_{k_1} \left(\overline{\mathbf{R}^2} - \bigcup_{j \in J} B_j \right) \\
&= \bigcup_{k_n \in J} T_{k_n} \bigcup_{k_{n-1} \neq k_n} T_{k_{n-1}} \cdots \bigcup_{k_2 \neq k_3} T_{k_2} \left(\bigcup_{k_1 \neq k_2} B_{k_1} \right) \\
&\quad - \bigcup_{k_n \in J} T_{k_n} \bigcup_{k_{n-1} \neq k_n} T_{k_{n-1}} \cdots \bigcup_{k_1 \neq k_2} T_{k_1} \left(\bigcup_{j \neq k_1} B_j \right).
\end{aligned}
$$

Before continuing our general investigation of this geometry, let us examine some examples which will be pursued in the next section.

(2.3) Example 1. Let $J = \{1\}$, and let B_1 be the ball of radius 1 centered at the origin. Then $\mathcal{G}$ consists of only the identity and the inversion T_1, and $H = \emptyset$.

(2.4) Example 2. Let $J = \{1, 2\}$, and let B_1 and B_2 be two disjoint open balls. In this case $\mathcal{G}$ consists of

$$
\{\text{identity}, \ T_1(T_2T_1)^k, \ T_2(T_1T_2)^k, (T_1T_2)^k, (T_2T_1)^k, \ k \geq 0\}
$$

and

$$
H = \left(\bigcap_{k \geq 0} T_1(T_2T_1)^k B_2 \right) \cup \left(\bigcap_{k \geq 0} T_2(T_1T_2)^k B_1 \right)
$$

If $\overline{B_1} \cap \overline{B_2} = \emptyset$, then H consists of 2 points, one in B_1 and one in B_2. If $\overline{B}_1 \cap \overline{B}_2 = z \in \partial B_1 \cap \partial B_2$, then $H = \emptyset$ since

$$
\bigcap_{k \geq 0} T_1(T_2T_1)^k \overline{B}_2 = \bigcap_{k \geq 0} T_2(T_1T_2)^k \overline{B}_1 = z \in E.
$$

(2.5) Example 3. Let $J = \{1, 2, 3\}$, and let B_1, B_2 and B_3 be three unit balls arranged so that each ball is tangent to the other two. (Think of three pennies pushed together on a flat surface). For $i \neq j$, let $z_{ij} = \partial B_i \cap \partial B_j$. Then

$$
H_1 = \bigcap_{S \in \mathcal{H}_1} S\Big(\bigcup_{j \in J} B_j \Big)
$$

is a chain of six open balls $(B_k^1)_{1 \leq k \leq 6}$ inside $B_1 \cup B_2 \cup B_3$ as illustrated on the left in figure 2. Note that you can draw a closed arc contained in $\bigcup \overline{B}_k^1$ which intersects each B_k^1. The set $H_2 = \bigcap_{S \in \mathcal{H}_2} S\Big(\bigcup_{j \in J} B_j \Big)$ is a chain of 12 balls $(B_k^2)_{1 \leq k \leq 12}$ inside these 6 as illustrated on the right in figure 2. Note that you can draw a closed arc contained in $\bigcup \overline{B}_k^2$ which intersects each B_k^2. If $\overline{H}_n = \bigcap_{S \in \mathcal{H}_n} S\Big(\bigcup_{j \in J} \overline{B}_j \Big)$, then $\overline{H}_n$

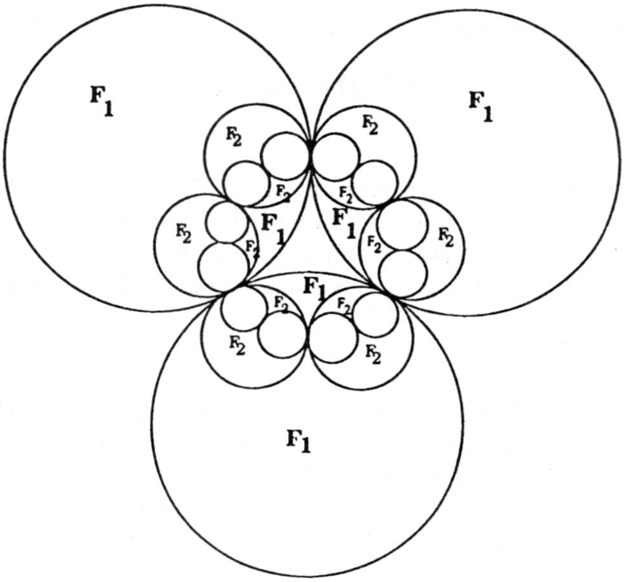

FIGURE 1

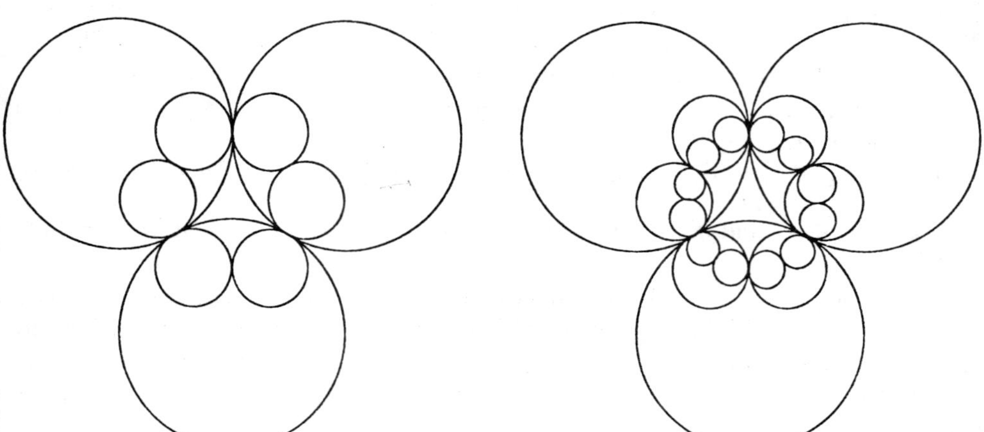

FIGURE 2

decreases to a closed arc $\overline{H}$ connecting the points z_{ij}. In fact, $\overline{H}$ is the closure of the horizon H, and

$$\overline{H} - H = \bigcup_{S \in \mathcal{G}} S\{z_{12}, z_{23}, z_{13}\}$$

is a countable set.

(2.6) Example 4. Consider the sequence of vertical lines $(L_n = \{(x,y) : x = 2^{-n}\})_{n \geq 2}$ in $\mathbf{R}^2$. On the line L_n, consider all balls with centers $(2^{-n},\ 2k \cdot 2^{-3n})_{k \in Z}$ and radius 2^{-3n}. Each ball is tangent to the one next to it and balls on different lines do not intersect. Then each line L_n is contained in $\overline{H}$, and $L_n - H$ is contained in

$\bigcup_{S \in \mathcal{G}} S\Big(\bigcup_{i \neq j} \partial B_i \cap \partial B_j\Big)$.

For each orbit $[x] \in H^c/\mathcal{G}$, define $\varphi([x])$ to be the unique point $y \in [x]$ such that $y \in E$. Then φ is an invertible map from $H^c/\mathcal{G}$ onto E and identifies $H^c/\mathcal{G}$ with E. We also define a map $\Phi : H^c \longrightarrow E$ by setting $\Phi(x) = \varphi([x])$. In the next section, we show that $\Phi(b(\rho_t))$ is a Markov process, where $b(\rho_t)$ is an appropriate time change of Brownian motion b_t. Therefore, we need to examine the continuity properties of Φ. While Φ is not a continuous function, in general, it is "almost" continuous, as we make precise below.

(2.7) Definition. For each $x \in H^c$, define $\eta(x)$ to be the cardinality of the set $\{S : x \in E_S\}$.

(2.8) Lemma. If $[x] = [z]$, then $\eta(x) = \eta(z)$.

Proof. If $[x] = [z]$, then $x = Uz$ for some $U \in \mathcal{G}$. Therefore, $x \in E_S$ if and only if $z \in E_{U^{-1}S}$.

Q.E.D.

(2.9) Proposition. η takes values in the set $\{1, 2, \infty\}$.

Proof. Fix $x \in H^c$, and let $y = \Phi(x) \in E$. By (2.8), $\eta(y) = \eta(x)$. Assume $\eta(y) > 1$. Then there are two different elements T and S in $\mathcal{G}$ with $y \in E_S \cap E_T$. That is, $Sz = y = Tq$ for two points z and q in E. But [y] contains exactly one point in E, so $z = q = y$, and $S^{-1}y = T^{-1}y$ or $TS^{-1}y = y$. Let us write TS^{-1} in its unique representation as $T_{n_1} \cdots T_{n_k}$.

First, if $y \notin \bigcup_{j \in J} \partial B_j$, then, as we have noted before, $TS^{-1}y \in B_{n_1}$. Since $TS^{-1}y = y \in E$ and $B_{n_1} \cap E = \emptyset$, we have a contradiction, and we conclude $\eta(y) = 1$.

Second, suppose y is on the boundary of exactly one ball, say B_j. If $n_k \neq j$, then $TS^{-1}y \in B_{n_1}$. But $TS^{-1}y = y \in \partial B_j$, and $\partial B_{n_1} \cap \partial B_{n_j} = \emptyset$ by our supposition. So n_k must be j, and we obtain $T_{n_1} \cdots T_{n_{k-1}} y = T_{n_k} y = y$. Continuing this argument, we see that $T_{n_1} = T_{n_2} = \ldots = T_{n_k} = T_j$, and $\{S : y \in E_S\} = \{\text{identity}, T_j\}$, so $\eta(y) = 2$.

Finally, suppose $y \in \partial B_n \bigcap \partial B_j$. Then $\{S : y \in E_S\} \supset \{(T_j T_n)^k : k \geq 1\}$, so $\eta(y) = \infty$.

Q.E.D.

(2.10) Corollary. (i) $\{\eta = 1\} \cap H^c = \{x \in H^c : \Phi(x) \in \overline{\mathbf{R}^2} - \bigcup_{j \in J} \overline{B}_j\}$

(ii) $\{\eta = 2\} \cap H^c = \{x : \Phi(x)$ is on the boundary of exactly one ball$\}$

(iii) $\{\eta = \infty\} \cap H^c = \{x :$ for some $n \neq j$, $\Phi(x) \in \partial B_n \cap \partial B_j\}$.

Define

$$\begin{aligned} D &= E - \left(\bigcap_{n \in J} \overline{\bigcup_{j \geq n} B_j} \right) \\ D_S &= \{SD : S \in \mathcal{G}\} \\ I &= \bigcup D_S. \end{aligned}$$

(2.11) Proposition. The restriction of $\Phi(x)$ to $I \cap \{x : \eta(x) = 1 \text{ or } 2\}$ is a continuous function.

Before beginning the proof of this proposition, let us look at some examples to see that Φ may be discontinuous elsewhere.

Example 2 (continued). Recall the case of two tangent balls with $z = \overline{B}_1 \cap \overline{B}_2$. Then $\eta(z) = \infty$. Consider the centers (c_n) of the sequence of balls $T_1(T_2T_1)^n B_2$: $\lim_{n \to \infty} c_n = z$. Since $\Phi(z) = z$ and $\Phi(c_n) = \infty$ for every n, Φ is not continuous at z.

Example 5. This is a new example. Let $q_n = (0, 2^{-n})$ be a sequence of centers of non-intersecting balls B_n with radii r_n chosen so that $\overline{B}_n \cap \overline{B}_{n+1} = z_n$ for every $n \geq 1$; i.e. the closed balls touch. Then $\lim_{n \to \infty} q_n = (0,0) \in E$, $\eta(0,0) = 1$, and $(0,0) \in \bigcap_{n \in J} \overline{\bigcup_{j \geq n} B_j}$. Again $\Phi(0,0) = (0,0)$, while $\Phi(q_n) = \infty$ for every n, so Φ is not continuous at $(0,0)$.

Proof of (2.11). First, suppose $x \in I$ and $\eta(x) = 1$. Then there is a unique $S \in \mathcal{G}$ with $x = S\Phi(x)$. Since $x \in D_S$, $x \notin S\left(\bigcap_{n \in J} \overline{\bigcup_{j \geq n} B_j}\right)$, so $\Phi(x) \notin \bigcap_{n \in J} \overline{\bigcup_{j \geq n} B_j}$. Since $\eta(x) = 1$, $\Phi(x) \notin \bigcup_{j \in J} \partial B_j$, so $\Phi(x)$ must be in E^o. Therefore, $x \in S(E)^o$, and $\Phi(w) = S^{-1}w$ on $S(E)^o$, and this is a continuous function.

Second, suppose $x \in I$ and $\eta(x) = 2$. By (2.10 ii), $\Phi(x)$ is on the boundary of exactly one ball, say ∂B_j, and there exists an $S \in \mathcal{G}$ such that $x = S\Phi(x)$. Since $\Phi(x)$ is not in the closed set $\bigcap_{n \in J} \overline{\bigcup_{j \geq n} B_j}$, there is a neighborhood N of $\Phi(x)$ such that

$$N \subset E \cup \left(B_j - T_j\left(\bigcup_{k \neq j} B_k \right) \right)$$

On E, Φ is the identity. On $B_j - T_j\left(\bigcup_{k\neq j} B_k\right)$, Φ is T_j. These two functions are continuous and coincide on ∂B_j. Therefore, on $S(E)$, $\Phi = S^{-1}$, and on $S\left(B_j - T_j\left(\bigcup_{k\neq j} B_k\right)\right)$, $\Phi = T^{-1}S^{-1}$. These two functions are continuous and coincide on $S(\partial B_j)$. Therefore, Φ is continuous on the open set $S(N)$ containing x.

Q.E.D.

3. Brownian Motion and $\mathcal{G}$

We begin this section by constructing a function $g : \mathbf{R}^2 \to [0, \infty]$ with several properties: $\{x : g(x) = \infty\}$ is countable, $g > 0$ on H^c, $g = 0$ on H, and

$$g(z) = \frac{g(T_n z) r_n^4}{|z - z_n|^4} \tag{3.1}$$

for every inversion T_n and for every $z \in \mathbf{R}^2 - \{g = \infty\}$.

Recall that $\chi(S) = k$ if $S = T_{n_1} \cdots T_{n_k}$, $\mathcal{G}_k = \{S \in \mathcal{G} : \chi(S) = k\}$, $F_k = \bigcup_{S\in\mathcal{G}_k} E_S$ and $E_k = \bigcup_{p\leq k} F_p$.

Define $g(z) = 1$ on $E_0 = E$. We extend the definition of g from E_0 to $E_1 = E_0 \cup F_1$ as follows. For each $k \in J$, set

$$g(z) = g(T_k z) r_k^4 |z - z_k|^{-4} \quad \text{if } z \in \Lambda_k^1 = T_k\left(B_k^c - \bigcup_{j\neq k} B_j\right). \tag{3.2}$$

We need to check that this extension of g is well-defined, for a point z may be in $\Lambda_k^1 \cap \Lambda_p^1$ or in $\Lambda_k^1 \cap E$, in which case $g(z)$ is subject to two possibly different definitions. If $z \in \Lambda_k^1 \cap \Lambda_p^1 = E_{T_k} \cap E_{T_p}$, then $z \in \partial B_k \cap \partial B_p$, and

$$g(T_k z) r_k^4 |z - z_k|^{-4} = 1 = g(T_p z) r_p^4 |z - z_p|^{-4}.$$

If $z \in \Lambda_k^1 \cap E = E_{T_k} \cap E$, then $z \in \partial B_k$, and $g(T_k z) r_k^4 |z - z_k|^{-4} = 1$. Thus, this extension of g is well-defined. Since $F_1 = \bigcup_{k\in J} \Lambda_k^1$, we have extended the definition of g to E_1. This extension is infinite at only countably many points in E_1 (namely, at the points $(z_j)_{j\in J}$), and it is positive on E_1. Let us check that (3.1) holds on E_1. If $z \in F_1$, then (3.1) holds by definition. What if $z \in E_0$? If we let $w = T_k z$, then (3.2) becomes $g(T_k w) = g(w) r_k^4 |T_k w - z_k|^{-4}$. Since $|T_k w - z_k||w - z_k| = r_k^2$, we obtain $g(w) = g(T_k w) r_k^4 |w - z_k|^{-4}$, so (3.1) holds if $z \in E_0$.

For the sake of clarity, we sketch the second step before proceeding to the general induction step. For each $k \in J$, we next extend g to the set

$$\Lambda_{k_2}^2 = T_{k_2}\left(\bigcup_{k_1\neq k_2} B_{k_1} - \bigcup_{k_1\neq k_2} T_{k_1}\left(\bigcup_{j\neq k_1} B_j\right)\right)$$

by defining $g(z) = g(T_{k_2}z)r_{k_2}^4|z - z_{k_2}|^{-4}$, just as before. Note that $T_{k_2}z \in E_1$, so $g(T_{k_2}z)$ is defined. It is simple to check that this extension is well-defined, it is infinite at only countably many points, and it is positive on $E_2 = E_1 \cup F_2$ since $F_2 = \bigcup_{k_2 \in J} \Lambda_{k_2}^2$. Furthermore, it satisfies (3.1) on E_2.

Now we are ready for the induction step. Having defined the desired function g on E_{n-1}, for each $k_n \in J$, we extend g to the set

$$\begin{aligned}\Lambda_{k_n}^n &= T_{k_n} \bigcup_{k_{n-1} \neq k_n} T_{k_{n-1}} \cdots \bigcup_{k_2 \neq k_3} T_{k_2} \bigcup_{k_1 \neq k_2} B_{k_1} \\ &\quad - T_{k_n} \bigcup_{k_{n-1} \neq k_n} T_{k_{n-1}} \cdots \bigcup_{k_1 \neq k_2} T_{k_1} \left(\bigcup_{j \neq k_1} B_j \right)\end{aligned}$$

by defining $g(z) = g(T_{k_n}z)r_{k_n}^4|z - z_{k_n}|^{-4}$. This extension is well-defined, it is infinite at only countably many points, and it is positive. It also satisfies (3.1) on $E_n = E_{n-1} \cup F_n$, since $F_n = \bigcup_{k \in J} \Lambda_k^n$. Since E_n increases to H^c, we have defined g on all of H^c. We finish defining g on $\overline{\mathbf{R}^2}$ by setting $g = 0$ on H. Since each $T_n : H \to H$ and $T_n : H^c \to H^c$, g has all of the properties advertised in the first paragraph of this section.

Let $\overline{b}_t = (\overline{b}_t^1, \overline{b}_t^2)$ be Brownian motion in $\mathbf{R}^2$. More precisely, let $\overline{b} = (\Omega, \mathfrak{F}, \mathfrak{F}_t, \overline{b}_t, \Theta_t, P^x)$ be a canonical realization of two dimensional Brownian motion starting at x with filtration $\mathfrak{F}_t$ and translation operators Θ_t.

Define

$$T_H = \inf\{t \geq 0 : b_t \in H\}$$

and let $b_t = (b_t^1, b_t^2)$ be $\overline{b}_t$ killed at T_H. Let $fc(H)$ be the fine closure of H: $fc(H)$ is the union of H and its regular points H^r. The state space of the killed Brownian motion b_t is $\mathbf{R}^2 - fc(H)$, and if $\zeta = \inf\{t : b_t = \Delta\}$, then $b_{\zeta-} \in H^r$ *a.s.* on $\{\zeta > 0\}$. Introduce an additive functional a_t of b_t by defining

$$a_t = \int_0^t g(b_s)ds$$

and define its right continuous inverse ρ_t by setting

$$\rho_t = \inf\{s : a_s > t\}$$

In order for ρ_t to function as a legitimate time change of b_t, we must have $a_t < \infty$ on $[0, T_H)$. We assume there is a set $\Gamma \subset \mathbf{R}^2 - fc(H)$ which is polar for b_t such that $a_t < \infty$ on $[0, T_H)$ *a.s.* P^x for each $x \in \mathbf{R}^2 - (fc(H) \cup \Gamma)$. In general, Γ will contain the countable set $\{S(\infty) : S \in \mathcal{G}\}$, but may contain other points as well. We may assume, in fact, that Γ is invariant under $\mathcal{G}$ by replacing Γ with the polar set

$\bigcup_{S\in\mathcal{G}} S\Gamma$, if necessary. Checking the existence of this polar set Γ can be easily done in many examples, due to the fact that the restriction of g to $I \cap \{x : \eta(x) = 1 \text{ or } 2\}$ is continuous. To see this, recall the proof of (2.11). If $x \in I$ and $\eta(x) = 1$, then $x \in S(E)^o$, and g is continuous on $S(E)^o$. If $x \in I$ and $\eta(x) = 2$, then there is exactly one j such that $x \in S(\partial B_j)$. There is a neighborhood $N \subset E \cup (B_j - T_j(\bigcup_{k\neq j} B_k))$ such that $x \in S(N)$ and g is continuous on $S(N)$. Thus, if $H^c - I \cap \{\eta = 1 \text{ or } 2\}$ is polar, then the additive functional a_t is finite on $[0, T_H)$ since the integrand $g(b_s)$ is continuous on $[0, T_H)$. The set $H^c - I \cap \{\eta = 1 \text{ or } 2\}$ will be polar in examples where

$$\bigcap_{n\in J} \overline{\bigcup_{j\geq n} B_j}$$

is polar.

What can we say about $a(T_H)$? This random variable can be either finite or infinite. The simplest example is when H is polar (see examples 1 and 2), for then $a(T_H) = \infty$. On the other hand, in example 4, $a(T_H) = T_H < \infty$ *a.s* P^x for each x in the left half–plane. This example is interesting in part because $T_H = T_{\{y-axis\}}$, while the y–axis is actually contained in E. However, it often occurs that $T_H < \infty$ and $a(T_H) = \infty$ since g tends to "blow up" near H. See (3.12) for some sufficient conditions. Our interest in this article centers on the case where $a(T_H) = \infty$.

If we set $X_t = b^1(\rho_t)$ and $Y_t = b^2(\rho_t)$, then $Z_t = (X_t, Y_t)$ is a time changed diffusion with state space $\mathbf{R}^2 - (fc(H) \cup \Gamma)$. It has finite lifetime if $a(T_H) < \infty$ and infinite lifetime if $a(T_H) = \infty$.

(3.3) Lemma. $S : H^r \to H^r$ for every $S \in \mathcal{G}$.

Proof. S is a finely continuous invertible map taking H to H. Therefore, it maps the fine closure $H \cup H^r$ to $H \cup H^r$, and we conclude that $S : H^r \to H^r$.

Q.E.D.

Thanks to this lemma and the fact that $S\Gamma = \Gamma$, we know that S maps the state space of b_t onto itself whenever $S \in \mathcal{G}$.

(3.4) Proposition. For each $S \in \mathcal{G}$ and for each $x \in \mathbf{R}^2 - (fc(H) \cup \Gamma)$, $\left(S(Z_t), P^{S^{-1}(x)}\right)$ is identical in law to (Z_t, P^x).

Proof. If suffices to prove the proposition for $S = T_n$. Let $(V_t, W_t) = T_n(Z_t)$, and let $(x_n, y_n) = z_n$, so

$$\begin{aligned} V_t &= x_n + \frac{r_n^2}{|Z_t - z_n|^2}(X_t - x_n) \\ W_t &= y_n + \frac{r_n^2}{|Z_t - z_n|^2}(Y_t - y_n) \end{aligned}$$

For $t < a(T_H)$, Itô's lemma yields

$$\begin{aligned}
V_t - V_0 &= r_n^2 \int_0^t \frac{(Y_s - y_n)^2 - (X_s - x_n)^2}{|Z_s - z_n|^4} dX_s \\
&\quad -2r_n^2 \int_0^t \frac{(X_s - x_n)(Y_s - y_n)}{|Z_s - z_n|^4} dY_s \\
&\quad +\frac{r_n^2}{2} \int_0^t \frac{2(X_s - x_n)^2 - 6(X_s - x_n)(Y_s - y_n)^2}{|Z_s - z_n|^6} d < X >_s \\
&\quad +\frac{r_n^2}{2} \int_0^t \frac{6(X_s - x_n)(Y_s - y_n)^2 - 2(X_s - x_n)^2}{|Z_s - z_n|^6} d < Y >_s
\end{aligned}$$

Since $d < X >_s = d < Y >_s$, the last two integrals cancel, and V_t is a martingale. Similarly,

$$\begin{aligned}
W_t - W_0 &= r_n^2 \int_0^t \frac{(X_s - x_n)^2 - (Y_s - y_n)^2}{|Z_s - z_n|^4} dY_s \\
&\quad -2r_n^2 \int_0^t \frac{(X_s - x_n)(Y_s - y_n)}{|Z_s - z_n|^4} dX_s
\end{aligned}$$

An elementary calculation shows that $V_t - V_0$ and $W_t - W_0$ are orthogonal. In calculating their increasing processes, straightforward algebra yields

$$< V >_t = \int_0^t r_n^4 |Z_s - z_n|^{-4} d < X >_s = \int_0^t r_n^4 |Z_s - z_n|^{-4} d < Y >_s = < W >_t$$

Note that

$$< V >_t = \int_0^t r_n^{-4} \left((V_t - x_n)^2 + (W_t - y_n)^2\right)^2 d < X >_s$$

Since $X_t = b^1(\rho_t)$, $< X >_t = \rho_t$.

(3.5) Lemma. $\rho_t = \int_0^t g^{-1}(X_s, Y_s)ds$.

Proof. Let

$$\gamma_t = \inf\{u : C_u = \int_0^u g^{-1}(X_s, Y_s)ds > t\}.$$

If h is any bounded measurable function on $\mathbf{R}$ with compact support, then by Lebesgue's time change lemma, we have

$$\begin{aligned}
\int_0^{C_\infty} h(\rho(\gamma_t))dt &= \int_0^{a(T_H)} h(\rho_t) g^{-1}(X_t, Y_t)dt \\
&= \int_0^{T_H} h(t) g^{-1}(b_t) g(b_t) dt = \int_0^{T_H} h(t)dt.
\end{aligned}$$

Therefore, $\rho(\gamma_t) = t$, so

$$\rho_t = \int_0^t g^{-1}(X_s, Y_s)ds$$

Q.E.D.

Since $(X_s, Y_s) = T_n(V_s, W_s)$, $d\rho(s) = g^{-1}(T_n(V_s, W_s))$, so

$$\begin{aligned} < V >_t &= \int_0^t r_n^{-4}\left((V_s - x_n)^2 + (W_s - y_n)^2\right)^2 g^{-1}\Big(T_n(V_s, W_s)\Big) ds \\ &= < W >_t . \end{aligned}$$

By (3.1), $< V >_t = < W >_t = \int_0^t g^{-1}(V_s, W_s) ds$. If we let ξ_t be the inverse of $< V >_t$, then $(V(\xi_t), W(\xi_t))$ is a Brownian motion killed the first time it hits H. If we define

$$\begin{aligned} c_t &= \int_0^t g(V(\xi_s), W(\xi_s)) ds \\ \mu_t &= \inf\{u : c_u > t\}, \end{aligned}$$

then $\Big(V(\xi(\mu_t)), W(\xi(\mu_t))\Big)$ is identical in law to (X_t, Y_t). But by the reasoning in (3.5), $\xi(\mu_t) = t$.

Q.E.D.

The proper state space of Z_t is $\mathbf{R}^2 - (fc(H) \cup \Gamma)$. We can consider it to be $\mathbf{R}^2 - H$, if we permit Z_t to branch directly to Δ, starting from $x \in H^r \cup \Gamma$. This quibble would permit us to regard the state space of $B_t = \Phi(Z_t)$ as E rather than $\Phi(\mathbf{R}^2 - (fc(H) \cup \Gamma)) \subset E$.

(3.6) Theorem. $B_t = \Phi(Z_t)$ is a strong Markov process on E.

Proof. If $x \in E \cap (fc(H) \cup \Gamma)$, then B_t dies immediately, and there is nothing to prove. So let $x \in \Phi(\mathbf{R}^2 - (fc(H) \cup \Gamma))$, let h be a bounded measurable function on E, and let P_t be the semigroup of Z_t. Since $\Phi \circ S = \Phi$ for every $S \in \mathcal{G}$,

$$P_t(h \circ \Phi)(Sx) = E^{Sx}[h \circ \Phi(Z_t)] = E^{Sx}[h \circ \Phi(S^{-1} Z_t)].$$

By (3.4), this is $P_t(h \circ \Phi)(x)$, so $P_t(h \circ \Phi)$ can be written as $v \circ \Phi$ for some function v on E. By a well-known theorem of Dynkin (see [1,9]), $B_t = \Phi(Z_t)$ is a strong Markov process.

Q.E.D.

Let us recall the results from section 2 concerning the continuity of Φ. The restriction of Φ to $I \cap \{x : \eta(x) = 1 \text{ or } 2\}$ is continuous, and examples show that it can be discontinuous elsewhere. The set $\{\eta = \infty\}$ causes no difficulty since it is countable, therefore polar, and can be deleted from the state space if desired. The set $I^c \subset \partial E$ seems to be polar for Z_t in the examples we have examined, but we do not know how to prove it is polar in general, so we state the following result, which follows directly from the continuity properties of Φ.

(3.7) Proposition. (i) The sample paths of B_t are continuous on the random set $\{t > 0 : B_t \notin I^c\}$ a.s. P^x for every $x \in E$. (ii) B_t is continuous at $t = 0$ a.s. (P^x) for every $x \in I \cap \{\eta = 1 \text{ or } 2\}$.

(3.8) Theorem. Assume the Lebesgue measure of ∂E is zero and I^c is polar. If $a(T_H) = \infty$ a.s. (P^x) for every $x \in \mathbf{R}^2 - (fc(H) \cup \Gamma)$, then B_t is reflecting Brownian motion on E.

Proof. Consider the process B_t killed when it leaves the interior of E: this process is simply killed Brownian motion with Dirichlet form $\left(\mathcal{E}, H_0^1(\mathbf{R}^2)\right)$, where the bilinear form

$$\mathcal{E}(u,v) = \frac{1}{2}\int \nabla u \cdot \nabla v dx$$

has domain $H_0^1(\mathbf{R}^2)$. Now B_t is a symmetric local process spending zero time on the boundary and having infinite lifetime. Therefore, it must be the maximal Markovian extension of killed Brownian motion ([8], Thm. 20.5). Silverstein [8] and Fukushima ([3], Thm. 2.3.1) identify this process as Brownian motion with normal reflection; namely, the process with quadratic form $\mathcal{E}$ above and domain $H^1(\mathbf{R}^2)$.

Q.E.D.

We suspect that the hypothesis "I^c is polar" in (3.8) above is unnecessary, since we suspect that Φ is finely continuous for the process Z_t under the other hypotheses of (3.8).

Let us now examine criteria guaranteeing that $a(T_H) = \infty$ a.s. (P^x) for every $x \in \mathbf{R}^2 - (fc(H) \cup \Gamma)$. Even though we constructed g explicitly, it can be difficult to estimate $a(T_H)$. We consider a mixed geometry–probability approach to the problem. Recall

$$\begin{aligned} H_n &= \bigcap_{S \in \mathcal{H}_n} S(\bigcup_{j \in J} B_j) \\ H &= \lim_{n\to\infty} H_n. \end{aligned}$$

Let $\tau_n = \inf\{t : b_t \in H_n\}$, $\tau = \inf\{t : b_t \in H\}$. In our discussion, we shall assume:

(3.9) Hypothesis. For each $x \in \mathbf{R}^2 - (fc(H) \cup \Gamma)$, there is an N_x such that $\tau_{N_x} < \tau_{N_x+1} < \tau_{N_x+2} < \ldots < \tau$ a.s. (P^x).

This hypothesis is easily verified, for example, if J is finite. It is not true for example 4. Now $b(\tau-)$ can be in either H or $H^r \cap H^c$. We shall also assume:

(3.10) Hypothesis. $H^r \cap H^c$ is polar.

Thus, we may assume that $b(\tau-) = d \in H$. For some $p \in J, d \in B_p$. For each n, there is a sequence $(k_n^j)_{1\le j\le n} \subset J$ such that

$$d \in T_p T_{k_n^1} T_{k_n^2} \cdots T_{k_{n-1}^{n-1}} B_{k_n^n}.$$

Call this set Q_n. Since $b(\tau-) = d$, there is a sequence (t_n) of times increasing to τ such that $b(t_n) \in \partial Q_n$. Let us check that $k_n^n \neq k_{n+1}^{n+1}$. Since $Q_{n+1} \subset Q_n$, we have $k_{n+1}^1 = k_n^1$, and we get

$$T_{k_{n+1}^2} \dots T_{k_{n+1}^n} B_{k_{n+1}^{n+1}} \subset T_{k_n^2} \dots T_{k_n^{n-1}} B_{k_n^n}.$$

Continuing this reasoning, we obtain $T_{k_{n+1}^n} B_{k_{n+1}^{n+1}} \subset B_{k_n^n}$. But then $k_{n+1}^n = k_n^n$, and $k_{n+1}^{n+1} \neq k_n^n$. So $\Phi(b(t_n)) \in \partial B_{k_n^n}$, where $k_{n+1}^{n+1} \neq k_n^n$. There are two possibilities for the sequence $\Phi(b(t_n))$. First, for some N, k_n^n takes values in a two point set $\{\ell, m\} \subset J$ for every $n \geq N$. In this case, $\lim_{n\to\infty} \Phi(b(t_n)) \in \partial B_\ell \cap \partial B_m$. In general, B_t is a subMarkovian extension of killed Brownian motion, so it has a limit at its lifetime ζ. Thus, $B_{\zeta-} \in \partial B_\ell \cap \partial B_m$. Since $\bigcup_{\ell\in J} \bigcup_{m\neq\ell} \partial B_\ell \cap \partial B_m$ is a polar set, we can discount this event. The second possibility is that k_n^n could take an infinite number of different values, in which case $B_{\zeta-}$ is a limit point of $(\partial B_j)_{j\in J}$. Let L be the set of limit points of $(\partial B_j)_{j\in J}$ which can be written as $\lim_{n\to\infty} p(j_n)$, where the sequence (j_n) contains infinitely many different values of J.

(3.11) Hypothesis. L is polar.

(3.12) Theorem. If hypotheses (3.9), (3.10) and (3.11) are satisfied, then $a(T_H) = \infty$ a.s. (P^x) for every $x \in \mathbf{R}^2 - (fc(H) \cup \Gamma)$.

It is easy to verify the three hypotheses if J is finite.

(3.13) Corollary. If J is finite, then $a(T_H) = \infty$ a.s. (P^x) for every $x \in \mathbf{R}^2 - (fc(H) \cup \Gamma)$.

The three hypotheses can also be verified in Example 5, in which J is infinite.

ACKNOWLEDGEMENT. Research of the first author supported in part by NSA Grant MDA904–92–H–3034.

REFERENCES

1. E.B. Dynkin, Markov Processes, Vol. 1, Springer-Verlag, Berlin–Heidelberg–New York 1965.

2. M. Fukushima, A construction of reflecting barrier Brownian motions for bounded domains, Osaka J. Math. vol.4 1967, 183–215.

3. M. Fukushima, Dirichlet Forms and Markov Processes, North-Holland, Amsterdam–Oxford–New York 1980.

4. J. Glover and J.B. Mitro, Symmetries and functions of Markov processes, Ann. of Probab. vol.18 1990, 655–668.

5. J. Glover, Symmetry groups and translation invariant representations of Markov processes, Ann. of Probab. vol. 19 1991, 562–586.

6. N.S. Landkof, Foundations of Modern Potential Theory, Springer-Verlag, New York–Heidelberg–Berlin 1972.

7. M. Rao, Brownian Motion and Classical Potential Theory, Aarhus Lecture Notes, vol. No. 47 1977.

8. M. Silverstein, Symmetric Markov Processes, Springer–Verlag Lecture Notes in Math., Berlin–Heidelberg–New York, vol. 426, 1974.

9. J. Watkins, Diffusion processes and their distributional symmetries, Prob. Th. Rel. Fields, (to appear).

Γ-Potentials

Joseph Glover, Murali Rao, Hrvoje Šikić*
Department of Mathematics, University of Florida
Gainesville, FL 32611

Renming Song
Department of Mathematics, Northwestern University
Evanston, IL 60208

September 28, 1993

Introduction

While much is known about solutions of the equation $\Delta u = 0$ on a domain D, very little seems to be known about solutions of the equation $(-\Delta)^\alpha u = 0$ for $0 < \alpha < 1$. Since $(-\Delta)^\alpha$ is just a fractional power of the Laplacian, one might expect the analysis of these two equations to be quite similar, but standard techniques yielding insight into the first equation often yield little or no information about the second. In fact, this is because $(-\Delta)^\alpha$ is no longer a differential operator, but is instead an integro-differential operator. We approach $(-\Delta)^\alpha$ as the infinitesimal generator of a Markov process on a domain D obtained by subordinating Brownian motion. From the probabilistic point of view, new difficulties are introduced because the subordinated process does not have continuous trajectories like Brownian motion. Based on this characterization and the associated potential theory, we characterize the solutions of $(-\Delta)^\alpha u = 0$ as a certain class of "harmonic" functions, and we give conditions under which the solution is continuous.

In order to make these results accessible to both probabilists and analysts, we discuss the necessary potential theory in section 1. Probabilists

*The third author is also a member of the Department of Mathematics, University of Zagreb, 41 000 Zagreb, Croatia

K. GowriSankaran et al. (eds.), Classical and Modern Potential Theory and Applications, 217–232.

will recognize the facts in this section as statements about subordinations couched in analytic terms. We apply these facts in section 2 to establish an integral representation which provides an explicit one-to-one correspondence between the Δ-harmonic functions associated with Brownian motion on D and the $(-\Delta)^\alpha$-harmonic functions associated with the subordinated process on D. In addition, continuity of these harmonic functions is established.

1 Definitions and General Results

Let D be a domain in R^n, i.e., D is an open and connected subset of R^n. Let $(X_t; t \geq 0)$ be a Brownian motion killed upon exit from D. By $\tau = \tau_D$ we denote the first exit time from D.

For an additive functional A we define its Γ_α-potential by

$$\Gamma_{\alpha,A}(x) = \frac{1}{\Gamma(\alpha)} E^x[\int_0^\infty t^{\alpha-1} dA_t], \quad x \in D, 0 < \alpha < \infty. \tag{1}$$

An easy application of the Markov property and Fubini's theorem shows that the following formula is valid: for any $\beta > 0$,

$$\frac{1}{\Gamma(\beta)} E^x[\int_0^\infty s^{\beta-1} \Gamma_{\alpha,A}(X_s) ds] = \Gamma_{\alpha+\beta,A}(x). \tag{2}$$

The next proposition, which is a consequence of (2), will be used in the sequel several times.

Proposition 1 *Suppose* $\Gamma_{\alpha,A}$ *is not identically infinite. Then* $\Gamma_{\beta,A}$ *is locally integrable for all* $0 < \beta \leq \alpha$.

Proof. If $\beta < \alpha$, then, by (2), $\Gamma_{\beta,A}$ is also not identically infinite. Therefore it is enough to prove the statement for $\beta = \alpha$. Notice that in the case $\alpha = 1$, we have an ordinary potential, thus the statement of the proposition is correct.

Consider the case $0 < \alpha < 1$. We claim that, for every $t > 0$, the following inequality is valid:

$$\alpha \int_0^t e^{-s}(t-s)^{\alpha-1} ds \leq t^{\alpha-1}. \tag{3}$$

For $t \leq 1$, the left hand side is smaller than

$$\alpha \int_0^t (t-s)^{\alpha-1} ds = t^\alpha \leq t^{\alpha-1}.$$

To prove (3) for $t \geq 1$ now, it is enough to prove that the following function

$$t \mapsto e^t t^{\alpha-1} - \alpha \int_0^t e^s s^{\alpha-1} ds$$

is increasing on $[1, \infty)$. This follows immediately by differentiation.

Consider now the resolvent operator G_1 for the process (X_t). Using (3) we obtain

$$\begin{aligned}
(G_1 \Gamma_{\alpha,A})(x) &= E^x[\int_0^\infty e^{-t} \Gamma_{\alpha,A}(X_t) dt] \\
&= \frac{1}{\Gamma(\alpha)} E^x[\int_0^\infty e^{-t} \int_t^\infty (s-t)^{\alpha-1} dA_s dt] \\
&= \frac{1}{\Gamma(\alpha)} E^x[\int_0^\infty dA_s \int_0^s e^{-t}(s-t)^{\alpha-1} dt] \\
&\leq \frac{1}{\Gamma(\alpha)} \frac{1}{\alpha} E^x[\int_0^\infty s^{\alpha-1} dA_s] = \frac{1}{\alpha} \Gamma_{\alpha,A}(x),
\end{aligned}$$

where in the first line above we used the Markov property, in the second line we used Fubini's theorem and in the third line we used (3). Since (X_t) is the Brownian motion killed upon exit from D, G_1 has a density. Therefore, a standard proof (see, for example, page 267 of [1]) shows that $\Gamma_{\alpha,A}$ is locally integrable, since it is not identically infinite.

It remains to prove the statement for $\alpha > 1$. Applying (2) to 1 and $\alpha - 1 > 0$, we obtain

$$\Gamma_{\alpha,A}(x) = E^x[\int_0^\infty \Gamma_{\alpha-1,A}(X_s) ds].$$

Hence $\Gamma_{\alpha,A}$ is an ordinary potential, and, since it is not identically infinite, it must be locally integrable.

Q.E.D.

The next important property that we expect for Γ-potentials is that they uniquely determine additive functionals. In proving the uniqueness theorem for Γ-potentials we will use the following, easily derived, formula:

$$\Gamma_{\alpha,A}(x) = \begin{cases} \frac{1-\alpha}{\Gamma(\alpha)} \int_0^\infty t^{\alpha-2} E^x[A_t] dt; & 0 < \alpha < 1 \\ E^x[A_\infty]; & \alpha = 1 \\ \frac{\alpha-1}{\Gamma(\alpha)} \int_0^\infty t^{\alpha-2} E^x[A_\infty - A_t] dt; & \alpha > 1. \end{cases} \tag{4}$$

Theorem 1 *If, for some $\alpha \in (0, \infty)$, $\Gamma_{\alpha,A}$ and $\Gamma_{\alpha,B}$ are not identically infinite, and $\Gamma_{\alpha,A} = \Gamma_{\alpha,B}$, then $A = B$.*

Proof. Notice that, by Proposition 1, $\Gamma_{\alpha,A}$ and $\Gamma_{\alpha,B}$ are finite almost everywhere. Hence, if $\alpha = 1$, then $A = B$, by (4) and the uniqueness theorem for ordinary potentials ([1], Chapter IV, Theorem 2.13).

Consider the case $0 < \alpha < 1$. It follows, by (4), that for almost every x and for almost every t, $E^x[A_t]$ and $E^x[B_t]$ are finite. Since they are nondecreasing functions in t, we conclude that for almost every x, $E^x[A_t]$ and $E^x[B_t]$ are finite for every t. Therefore, for almost every x, $t \mapsto E^x[A_t]$ and $t \mapsto E^x[B_t]$ determine σ-finite measures on $[0, \infty)$. Since $\Gamma_{\alpha,A} = \Gamma_{\alpha,B}$, it follows that $E^x[\Gamma_{\alpha,A}(X_t)] = E^x[\Gamma_{\alpha,B}(X_t)]$. Thus by Fubini's theorem we have

$$E^x[\int_t^\infty (s-t)^{\alpha-1} dA_s] = E^x[\int_t^\infty (s-t)^{\alpha-1} dB_s].$$

Since $s \mapsto (s-t)^{\alpha-1}$ is deterministic, we obtain that for almost every x, and, for every t,

$$\int_t^\infty (s-t)^{\alpha-1} dE^x[A_s] = \int_t^\infty (s-t)^{\alpha-1} dE^x[B_s].$$

Recall (see, for example, [4], page 3.31) that $s \mapsto (s-t)^{\alpha-1}$, $0 < \alpha < 1$, is a potential kernel on $[0, \infty)$ and determines measures uniquely. Therefore, for almost every x,

$$E^x[A_t] = E^x[B_t] < \infty, \quad \text{for every } t.$$

It follows (see [1], page 159) that $A = B$.

Assume now that the statement is proved for every $0 < \alpha \leq n$, where n is a positive integer. We will show that then it must be valid for every $n < \alpha \leq n+1$. Since $\alpha > 1$, it follows, by (4) and Proposition 1, that for almost every x, $E^x[A_\infty]$ and $E^x[B_\infty]$ are finite. Since $\Gamma_{\alpha,A} = \Gamma_{\alpha,B}$, it follows, by (2),

$$E^x[\int_0^\infty s^{(\alpha-1)-1} \Gamma_{1,A}(X_s) ds] = E^x[\int_0^\infty s^{(\alpha-1)-1} \Gamma_{1,B}(X_s) ds],$$

and both sides are finite for almost every x. Notice that these are $\Gamma_{\alpha-1}$ potentials of $\int_0^t \Gamma_{1,A}(X_s)ds$ and $\int_0^t \Gamma_{1,B}(X_s)ds$, respectively. Since $\alpha - 1 \leq n$, it follows that the two additive functionals are equivalent, by the inductive assumption. Since $\Gamma_{1,A}(x) = E^x[A_\infty]$, it follows that, for almost every x,

$$E^x[A_\infty] = E^x[B_\infty] < \infty,$$

which proves the theorem.

Q.E.D

Notice that one of the consequences of the proof of Proposition 1 (case $\alpha > 1$) is that for $\alpha > 1$,

$$u(x) = \Gamma_{\alpha,A}(x) \tag{5}$$

is superharmonic. For this superharmonic function the following is true:

$$\int_0^s t^{\alpha-1} dA_t + u(X_s) \leq E^x[\int_0^\infty t^{\alpha-1} dA_t | \mathcal{F}_s]. \tag{6}$$

In fact,

$$\begin{aligned} E^x[\int_0^\infty t^{\alpha-1} dA_t | \mathcal{F}_s] &= \int_0^s t^{\alpha-1} dA_t + E^x[\int_s^\infty t^{\alpha-1} dA_t | \mathcal{F}_s] \\ &= \int_0^s t^{\alpha-1} dA_t + E^x[\int_0^\infty (t+s)^{\alpha-1} dA_t(\theta_s) | \mathcal{F}_s] \\ &= \int_0^s t^{\alpha-1} dA_t + E^{X_s}[\int_0^\infty (t+s)^{\alpha-1} dA_t] \\ &\geq \int_0^s t^{\alpha-1} dA_t + u(X_s). \end{aligned}$$

From (6) we can get that for any $\lambda > 0$,

$$\begin{aligned} \lambda P^x[u^* > \lambda] &\leq \lambda P^x[M^* > \lambda] \\ &\leq E^x[\int_0^\infty t^{\alpha-1} dA_t], \end{aligned}$$

where M is the martingale

$$M_s = E^x[\int_0^\infty t^{\alpha-1} dA_t | \mathcal{F}_s],$$

and * is the usual notation for the supremum.

Remark 1. The definition of the Γ-potential and the results of this section can be obtained for much more general Markov processes (X_t). The assumptions needed are that the resolvent of (X_t) has a density with the usual properties (see, for example, [1], Chapter VI) and the additive functionals are natural (in the sense of [1], Chapter IV).

However, the main results of this paper (contained in the following sections) are obtained only for Brownian motion killed upon exit from D. For this reason we did not introduce Γ- potentials in the most general setting.

2 Representations of Harmonic Functions

It is possible to represent positive harmonic functions as Γ_α-potentials for $0 < \alpha < 1$. Indeed, let $s = Uf$, where U is the potential kernel of Brownian motion killed upon exit from D and $f \geq 0$. Let us denote the $\Gamma_{\alpha,A}$- potential, with respect to the additive functional

$$A_t = \int_0^t f(X_s)ds,$$

by $V^\alpha f$, i.e.,

$$V^\alpha f(x) = \frac{1}{\Gamma(\alpha)} E^x[\int_0^\infty t^{\alpha-1} f(X_t)dt]. \tag{7}$$

By (2), we get, for every $0 < \alpha < 1$,

$$V^\alpha V^{1-\alpha} f(x) = Uf(x). \tag{8}$$

As usual, we use P_t to denote the transition semigroup of (X_t). Then we obtain

$$\begin{aligned}
\Gamma(1-\alpha)V^{1-\alpha}f(x) &= E^x \int_0^\infty t^{-\alpha} f(X_t)dt \\
&= \int_0^\infty t^{-\alpha} P_t f(x)dt \\
&= \int_0^\infty P_t f(x)dt \int_0^\infty \frac{1}{\Gamma(\alpha)} \lambda^{-(1-\alpha)} e^{-\lambda t} d\lambda \\
&= \frac{1}{\Gamma(\alpha)} \int_0^\infty \lambda^{\alpha-1} d\lambda \int_0^\infty e^{-\lambda t} P_t f(x)dt \\
&= \frac{1}{\Gamma(\alpha)} \int_0^\infty \lambda^{\alpha-1} s^\lambda d\lambda,
\end{aligned}$$

where $s^\lambda = U^\lambda f$ satisfies

$$s = s^\lambda + \lambda U^\lambda s. \tag{9}$$

Thus we can rewrite (8) as

$$Uf = V^\alpha g, \quad g = \frac{1}{\Gamma(\alpha)\Gamma(1-\alpha)} \int_0^\infty \lambda^{\alpha-1} s^\lambda d\lambda. \tag{10}$$

Now we are going to extend (10) to more general excessive functions. For any excessive function s, it is easy to show (see, for example, [5]) that for each $\lambda > 0$, there is a unique λ-excessive function s^λ such that

$$s = s^\lambda + \lambda U^\lambda s. \tag{11}$$

And if $s_n = Uf_n$ increases to s, $s_n^\lambda = U^\lambda f_n$ tends to s^λ almost everywhere. Applying Fatou's lemma and (10) to this fact, we obtain that, for every $0 < \alpha < 1$, and, for any excessive function s,

$$V^\alpha[\frac{1}{\Gamma(\alpha)\Gamma(1-\alpha)}\int_0^\infty \lambda^{\alpha-1}s^\lambda d\lambda] \leq s. \tag{12}$$

We can show that for a special class of excessive functions (12) becomes an equality.

An excessive function s is called purely excessive if

$$\lim_{t\to\infty} P_t s = 0,$$

almost everywhere. If s is purely excesive, (11) can be expanded to

$$s = s^\lambda + \lambda U^\lambda s = s^\lambda + \lambda U s^\lambda, \tag{13}$$

and for any $\lambda, \mu \geq 0$,

$$U^\lambda s^\mu = U^\mu s^\lambda. \tag{14}$$

From (14) we know that $U^1 s$ is the potential of s^1, i.e., $U^1 s = U s^1$. Therefore

$$(U^1 s)^\lambda = U^\lambda s^1 = U^1 s^\lambda. \tag{15}$$

For potentials of functions, (10) is true. Thus from (15)

$$\begin{aligned} U^1 s = U s^1 &= V^\alpha[\frac{1}{\Gamma(\alpha)\Gamma(1-\alpha)}\int_0^\infty \lambda^{\alpha-1}U^1 s^\lambda d\lambda] \\ &= U^1[V^\alpha\{\frac{1}{\Gamma(\alpha)\Gamma(1-\alpha)}\int_0^\infty \lambda^{\alpha-1}s^\lambda d\lambda\}]. \end{aligned}$$

Applications of (12) now shows that we just proved the following result.

Proposition 2 *If s is purely excessive, then, for every $0 < \alpha < 1$,*

$$s = V^\alpha g, \; g = \frac{1}{\Gamma(\alpha)\Gamma(1-\alpha)}\int_0^\infty \lambda^{\alpha-1}s^\lambda d\lambda. \tag{16}$$

Remark 2. If D is a bounded C^2-domain in R^n, every positive harmonic function is purely excessive. Hence, Proposition 2 represents every positive harmonic function on a nice domain as Γ_α-potentials of functions.

It is possible to represent g in (16) in terms of s directly. Suppose that s is purely excessive. By (13), we get

$$s^\lambda = \lambda \int_0^\infty (s - P_t s)e^{-\lambda t}dt.$$

Thus

$$\begin{aligned}\int_0^\infty \lambda^{\alpha-1}s^\lambda d\lambda &= \int_0^\infty \lambda^\alpha d\lambda \int_0^\infty (s - P_t s)e^{-\lambda t}dt \\ &= \int_0^\infty (s - P_t s)dt \int_0^\infty \lambda^\alpha e^{-\lambda t}d\lambda \\ &= \Gamma(\alpha+1)\int_0^\infty t^{-\alpha}(s - P_t s)\frac{dt}{t}.\end{aligned}$$

Hence, if s is purely excessive, then $s = V^\alpha g$, where

$$g = \frac{\alpha}{\Gamma(1-\alpha)}\int_0^\infty t^{-\alpha}(s - P_t s)\frac{dt}{t}. \tag{17}$$

Let us now apply (16) and (17) to some special excessive functions and compute g explicitly.

Example 1. Let D be a bounded domain. Then $s \equiv 1$ is purely excessive, and we can apply (16). Since

$$s^\lambda(x) = E^x[e^{-\lambda\tau}]$$

(recall that τ is the first exit time from D), we get

$$g(x) = \frac{1}{\Gamma(1-\alpha)}E^x[\tau^{-\alpha}]. \tag{18}$$

Also, since $1 = V^\alpha g$, we get

$$\begin{aligned}\int_D dx &= \int_D V^\alpha g(x)dx = \int_D g(x)V^\alpha 1(x)dx \\ &= \int_D g(x)\frac{1}{\alpha\Gamma(\alpha)}E^x[\tau^\alpha]dx.\end{aligned}$$

Thus for each $0 < \alpha < 1$,

$$\int_D E^x[\tau^{-\alpha}]E^x[\tau^\alpha]dx = \alpha\Gamma(\alpha)\Gamma(1-\alpha)\int_D dx. \tag{19}$$

Example 2. The assumptions are the same as in the previous example. Let us now compute g for $s(x) = E^x[\tau^m]$, where m is a positive integer.

We will use the following formula, which can be easily proved by induction: for every positive integer m, and for every real non-integer a,

$$\sum_{k=0}^{m} \binom{m}{k} \frac{(-1)^{m-k}}{a-(m-k)} = \frac{-m!}{\prod_{k=0}^{m}(k-a)} = \frac{(-1)^m}{a \cdot \binom{a-1}{m}}. \tag{20}$$

By the Markov property, we get

$$\begin{aligned} P_t s(x) &= E^x[E^{X(t)}(\tau^m); t < \tau] \\ &= E^x[\tau^m \circ \theta_t; t < \tau] \\ &= E^x[(\tau - t)^m; t < \tau] \\ &= \sum_{k=0}^{m} \binom{m}{k} (-t)^{m-k} E^x[\tau^k; t < \tau]. \end{aligned}$$

Therefore

$$\frac{(s - P_t s)(x)}{t} = \frac{E^x[\tau^m; \tau \leq t]}{t} - \sum_{k=0}^{m-1} \binom{m}{k} (-1)^{m-k} t^{m-k-1} E^x[\tau^k; t < \tau].$$

Applying Fubini's theorem we get

$$\int_0^\infty t^{-\alpha} \frac{E^x[\tau^m; \tau \leq t]}{t} dt = E^x[\tau^m \int_0^\infty t^{-\alpha-1} 1_{(\tau \leq t)} dt] = \frac{E^x[\tau^{m-\alpha}]}{\alpha},$$

and, similarly, for $k = 0, 1, \cdots, m-1$,

$$\int_0^\infty t^{-\alpha} t^{m-k-1} E^x[\tau^k; t < \tau] dt = \frac{E^x[\tau^{m-\alpha}]}{(m-k)-\alpha}.$$

Finally, using (17) and applying (20), we obtain

$$g(x) = \frac{(-1)^m}{\Gamma(1-\alpha) \binom{\alpha - 1}{m}} E^x[\tau^{m-\alpha}]. \tag{21}$$

Now let us restrict our attention to the case of a domain D such that all excessive functions on D are purely excessive (see Remark 2). The main result of this section is the following representation theorem:

Theorem 2 *If $0 < \alpha < 1$ and if s is U-excessive on D, then there is a $V^{1-\alpha}$-excessive function g such that*

$$s = V^{\alpha} g. \tag{22}$$

Furthermore, if s is U-harmonic, then g is $V^{1-\alpha}$- harmonic.

Conversely, if g is $V^{1-\alpha}$-excessive and $s = V^{\alpha} g$ is not identically infinite, then s is U-excessive. If g is $V^{1-\alpha}$- harmonic and s is not identically infinite, then s is U-harmonic.

Before we give the proof of this theorem, a lemma is needed.

Lemma 1 *Let V be a potential kernel and suppose that $Vg_n \uparrow Vg$ where g_n are V-excessive. Then g is equal to an excessive function almost everywhere.*

Proof. We have

$$Vg_n = V^{\lambda} g_n + \lambda V^{\lambda} V g_n, \tag{23}$$

where V^{λ} is the resolvent of V. By the monotone convergence theorem we know that $\lim_{n\to\infty} V^{\lambda} V g_n = V^{\lambda} V g$. It follows, by (23), that

$$\lim_{n\to\infty} V^{\lambda} g_n(x) = V^{\lambda} g(x), \tag{24}$$

if $Vg(x) < \infty$. Since g_n is V-excessive, $\lambda V^{\lambda} g_n$ is increasing in λ. From (24) we conclude that if $Vg(x) < \infty$, then $\lambda V^{\lambda} g(x)$ is increasing in λ. Define

$$\tilde{g}(x) = \sup_{\lambda > 0} \lambda V^{\lambda} g(x).$$

Then, for all x such that $Vg(x) < \infty$,

$$\tilde{g}(x) = \lim_{n\to\infty} n V^{n} g(x),$$

so, by monotone convergence, we have that for any x such that $Vg(x) < \infty$,

$$\begin{aligned} \mu V^{\mu} \tilde{g}(x) &= \lim_{n} \mu V^{\mu} n V^{n} g(x) \\ &= \mu \lim_{n} n V^{\mu} V^{n} g(x) \\ &= \mu \lim_{n} n [\frac{V^{\mu} g(x) - V^{n} g(x)}{n - \mu}] \\ &= \mu V^{\mu} g(x) \leq \tilde{g}(x). \end{aligned}$$

Thus, $\mu V^{\mu}\tilde{g}(x) \leq \tilde{g}(x)$ almost everywhere, and, by the excessivity of both sides, everywhere. Thus $\tilde{g}$ is supermedian and the first equality shows that $\tilde{g} = g$ almost everywhere. Hence g is equal to an excessive function almost everywhere.

Q.E.D.

Proof of the Theorem. Notice that, by (8), if g is $V^{1-\alpha}$-excessive, then g is the monotone limit of $V^{1-\alpha}\phi_n$, and

$$V^{\alpha}g = \lim_{n\to\infty} V^{\alpha}V^{1-\alpha}\phi_n = \lim_{n\to\infty} U\phi_n,$$

i.e., $V^{\alpha}g$ is either identically infinite or U-excessive.

Let s be U-excessive. Then s is the increasing limit of potentials Uf_n, i.e.,

$$s = \lim_n Uf_n = \lim_n V^{\alpha}g_n$$

where $g_n = V^{1-\alpha}f_n$ are $V^{1-\alpha}$-excessive. From Proposition 2 we know that $s = V^{\alpha}g$, since s is purely excessive. From Lemma 1 we conclude that g is equal to a $V^{1-\alpha}$-excessive function almost everywhere.

Let s be U-harmonic and $s = V^{\alpha}g$, where g is $V^{1-\alpha}$-excessive. Then

$$g = V^{1-\alpha}\mu + \tilde{g},$$

where $\tilde{g}$ is $V^{1-\alpha}$-harmonic (and, of course, $V^{1-\alpha}$-excessive). We have, by (8),

$$s = V^{\alpha}g = V^{\alpha}V^{1-\alpha}\mu + V^{\alpha}\tilde{g} = U\mu + V^{\alpha}\tilde{g}.$$

Since $V^{\alpha}\tilde{g}$ is U-excessive and s is U-harmonic, the potential part must be zero, i.e., $s = V^{\alpha}\tilde{g}$. It follows that $g = \tilde{g}$.

Finally, suppose that g is $V^{1-\alpha}$-harmonic and $s = V^{\alpha}g$ is not identically infinite. Then s is U-excessive and therefore

$$s = U\mu + h,$$

where h is U-harmonic. By (8) and the proof above, there is a $V^{1-\alpha}$-harmonic function $\tilde{g}$ such that

$$V^{\alpha}g = s = V^{\alpha}(V^{1-\alpha}\mu + \tilde{g}).$$

It follows that $g = V^{1-\alpha}\mu + \tilde{g}$, and, since g is $V^{1-\alpha}$-harmonic, that $g = \tilde{g}$, i.e., $s = h$.

Q.E.D.

As in the previous theorem, Let D be a domaim such that all excessive functions on D are purely excessive. Assume that $h > 0$ is U-harmonic on D. We have proved that, for every $0 < \alpha < 1$, $h = V^{\alpha}g$, where

$$g = \frac{\alpha}{\Gamma(1-\alpha)} \int_0^{\infty} t^{-\alpha} \frac{(h - P_t h)}{t} dt \tag{25}$$

is $V^{1-\alpha}$-harmonic. Since h is continuous, it is natural to ask if g is also continuous. The following theorem answers this question.

Theorem 3 *Every positive $V^{1-\alpha}$-harmonic function g, such that $V^{\alpha}g$ is not identically infinite, is continuous.*

Proof. Since Theorem 2 establishes an one-to-one correspondence between positive U-harmonic functions and positive $V^{1-\alpha}$-harmonic functions, it is enough to prove the statement of the theorem for g, which is of the form (25).

For every $\epsilon > 0$,

$$y \mapsto \int_{\epsilon}^{\infty} t^{-\alpha} \frac{(h - P_t h)(y)}{t} dt$$

is continuous, by the dominated convergence theorem, since h and $P_t h$ are continuous (see [3]) and

$$\int_{\epsilon}^{\infty} t^{-\alpha-1} dt = \frac{1}{\alpha\epsilon^{\alpha}} < \infty.$$

Hence, we need only to prove that

$$y \mapsto \int_0^{\epsilon} t^{-\alpha} \frac{(h - P_t h)(y)}{t} dt$$

is continuous at every $y \in D$. Let$x \in K \subset L^{\circ}$, where K and L are compact subsets of D and as usual, L° is the interior of L. It is enough to prove that

$$\lim_{\epsilon \downarrow 0} \int_0^{\epsilon} t^{-\alpha} \frac{(h - P_t h)}{t} dt = 0$$

uniformly on K.

Let τ_L be the first exit time from L. Then, we have

$$(h - P_t h)(x) = h(x) - E^x[h(X_t); t < \tau_L] - E^x[h(X_t); \tau_L \leq t < \tau].$$

Recall that $(h(X_t)1_{\{t<\tau\}})$ is a positive martingale, since h is harmonic on D. Since τ_L is a stopping time, we get

$$\begin{aligned}
&|(h - P_t h)(x)| \\
&\le |E^x[h(X_{\tau_L})] - E^x[h(X_{\tau_L}); t < \tau_L]| + |E^x[h(X_{\tau_L}); \tau_L \le t < \tau]| \\
&= |E^x[h(X_{\tau_L}); \tau_L \le t]| + |E^x[h(X_{\tau_L}); \tau_L \le t < \tau]|.
\end{aligned}$$

Since (X_t) is continuous and L is compact, it follows that $X_{\tau_L} \in \partial L$ and that there exists a constant $M = M(L,h) > 0$ such that $h(X_{\tau_L}) \le M$. Therefore, we get

$$\begin{aligned}
|(h - P_t h)(x)| &\le M[P^x(\tau_L \le t) + P^x(\tau_L \le t < \tau)] \\
&\le 2MP^x(\tau_L \le t).
\end{aligned}$$

Since $K \subset L^\circ$ is compact, the distance d between K and $\overline{L^c}$ is strictly positive. Thus, for every $x \in K$, we have

$$\{\tau_L \le t\} \subset \{\sup_{0\le s\le t} |X_s - x| \ge d\}$$

almost surely with respect to P^x. Using the fact that

$$|X_s - x|^2 = \sum_{i=1}^{n} |X_s^i - x^i|^2$$

is a P^x-submartingale, we obtain (by submartingale inequality)

$$P^x(\tau_L \le t) \le P^x(\sup_{0\le s\le t} |X_s - x|^2 \ge d^2) \le \frac{E^x[|X_t - x|^2]}{d^2} \le \frac{nt}{d^2},$$

since (X_t) is the Brownian motion killed upon exit from D. This inequality completes the proof, since, for every $x \in K$,

$$\left|\int_0^\epsilon t^{-\alpha}\frac{(h - P_t h)(x)}{t}dt\right| \le \int_0^\epsilon t^{-\alpha}\frac{2Mnt}{td^2}dt = \frac{2Mn}{(1-\alpha)d^2}\epsilon^{1-\alpha}.$$

Q.E.D.

3 Fractional Powers of the Laplace Operator

Let D be a domain in R^n, such that all excessive functions on D are purely excessive. Recall that the Laplace operator Δ on D can be expressed in terms of the infinitessimal generator of the Markov semigroup of (X_t), more precisely, the infinitesimal generator of (X_t) is $\frac{\Delta}{2}$. For $0 < \alpha < 1$ we can define, analytically, fractional power $(-\Delta)^\alpha$ on D (see, for example, section 2.6 in [2]). From the probabilistic point of view, we consider α-subordinated Markov process, say (X_t^α), with respect to the Brownian motion killed upon exit from D. The potential of (X_t^α) is exactly the V^α-potential analyzed in the previous section. We claim that (X_t^α) and V^α are natural choices to treat $(-\Delta)^\alpha$ on D. The consequence is that the results of section 2 apply to the fractional power of $(-\Delta)$.

As usual, we use $C_0^\infty = C_0^\infty(D)$ to denote the family of infinitely differentiable functions with compact support in D.

Lemma 2 *If u is positive and $V^{1-\alpha}u$ is not identically infinite, then for every $\phi \in C_0^\infty$,*

$$\int_D u(x)|[(-\Delta)^\alpha \phi](x)|dx < \infty. \tag{26}$$

Proof. Using formulae (6.9) and (6.15) in [2], pages 70-72, we obtain

$$\begin{aligned}
\int_D u|(-\Delta)^\alpha \phi|dx &= \int_D u|(-\Delta)^\alpha V^\alpha V^{1-\alpha}\Delta\phi|dx \\
&= \int_D u|V^{1-\alpha}\Delta\phi|dx \\
&\leq \int_D uV^{1-\alpha}|\Delta\phi|dx \\
&= \int_D (V^{1-\alpha}u)|\Delta\phi|dx < \infty,
\end{aligned}$$

since $|\Delta\phi| > 0$ only on a compact set and $V^{1-\alpha}u$ is locally integrable, by Proposition 1.

Q.E.D.

The immediate consequence of this lemma is that, whenever the conditions of Lemma 2 are satisfied, we can perform the same computation as in the proof of Lemma 2 but without absolute values, to obtain the formula

$$\int_D u(x)[(-\Delta)^\alpha \phi](x)dx = \int_D (V^{1-\alpha}u)(x)(\Delta\phi)(x)dx. \tag{27}$$

The following theorem describes the relations between V^α-harmonic functions and the solutions (in the sense of distributions) of the equation

$$(-\Delta)^\alpha u = 0.$$

Theorem 4 *Let g be a positive V^α-harmonic function such that $V^{1-\alpha}g$ is not identically infinite. Then, for every $\phi \in C_0^\infty$,*

$$\int_D g \cdot [(-\Delta)^\alpha \phi] dx = 0. \tag{28}$$

Conversely, suppose that g is a positive function such that $V^{1-\alpha}g$ is not identically infinite and such that (28) is satisfied for all $\phi \in C_0^\infty$. Then g is V^α-harmonic.

Proof. Suppose that g is a positive V^α-harmonic function and that $h = V^{1-\alpha}g$ is not identically infinite. It follows, by Theorem 2, that h is a positive, locally integrable U-harmonic function. Thus

$$\int_D h(x)(\Delta\phi)(x) dx = 0, \tag{29}$$

for every $\phi \in C_0^\infty$ (see, for example, [4]). Since g satisfies the conditions of Lemma 2, (28) follows immediately, by (27).

Suppose now that g is a positive function such that (28) is true and such that $V^{1-\alpha}g$ is not identically infinite. Then g satisfies the conditions of Lemma 2 and (28). Applying (27) we obtain that h satisfies (29), i.e., h is a postive U-harmonic function. By Theorem 2, g is a V^α-harmonic function.

Q.E.D.

We will finish this paper with the following straightforward consequence of Theorem 3 and Theorem 4.

Corollary 1 *Suppose that u is positive, $V^{1-\alpha}u$ is not identically infinite, and that, for every $\phi \in C_0^\infty$,*

$$\int_D u[(-\Delta)^\alpha \phi] dx = 0. \tag{30}$$

Then u is continuous on D.

References

[1] Blumenthal, R. M.; Getoor, R. K. *Markov Processes and Potential Theory*, Academic Press, New York, 1968.

[2] Pazy, A. *Semigroups of Linear Operators and Applications to Partial Differential Equations*, Springer, New York, 1983.

[3] Pop-Stojanović, Z.R. *Behavior of excessive functions of certain diffusions under the action of the transition semigroup*, Sem. on Stoch. Proc. 1988, Birkhauser, 213-220.

[4] Rao, M. *Brownian Motion and Classical Potential Theory*, Aarhus Lecture Notes Series No. 47, 1977

[5] Rao, M. *Representations of excessive functions*, Math. Scand. **51**(1982), 367-381.

FATOU-DOOB LIMITS AND THE BEST FILTERS.

Kohur GowriSankaran[1]
Department of Mathematics & Statistics
McGill University
805 Sherbrooke Street West
Montréal, Québec H3A 2K6 CANADA

In a series of articles including one in this Proceedings, J. Bliedtner and P.A. Loeb [BL1, BL2, BL3, BL4] have given far reaching generalisations of the well known Fatou and fine limit theorems on the quotient of positive harmonic functions. They also make comments regarding the extension of their results to the superharmonic case [BL2, §3]. In this note we give a substantially different and complete treatment of the 'superharmonic' case. Accordingly, let Ω be a non-void set and $\mathbf{H}$ and $\mathbf{P}$ two non-empty classes of extended positive valued functions. We assume that both $\mathbf{H}$ and $\mathbf{P}$ are convex cones and that elements of $\mathbf{H}$ take only real values and further that if $u_1,\ u_2 \in \mathbf{H}$, with $u_1(x) \leq u_2(x)$ for all $x \in \Omega$, then we assume that $(u_1 - u_2) \in \mathbf{H}$. This property is equivalent to say that the natural order on $\mathbf{H}$ is the same as the order induced by $\mathbf{H}$. Further, let $\mathbf{S}$ be the class of functions on Ω which can be expressed in the form $u + p$ where $u \in \mathbf{H}$ and $p \in \mathbf{P}$. Lastly, we assume that $\mathbf{S}$ has the following properties; (1) $\forall v_1, v_2 \in \mathbf{S}$, the pointwise infimum function, $\inf(v_1, v_2)$ belongs to $\mathbf{S}$ and (2) $\forall v_1, v_2 \in \mathbf{S},\ v_j = u_j + p_j$, the condition $v_1 \leq v_2$ on Ω implies that $u_1 \leq u_2$ on Ω. It is easy to deduce from the last condition that every $v \in \mathbf{S}$ is written uniquely in the form $u + p$ with $u \in \mathbf{H}$ and $p \in \mathbf{P}$. [G1]

Now, we introduce the conditions to get the 'minimal boundary'. We suppose that there is a topology τ on $\mathbf{H}$ which is finer than the simple convergence topology as functions on Ω, such that $\mathbf{H}$ has a τ-metrizable compact base Δ. By Choquet integral representation theorem, each $u \in \mathbf{H}$ corresponds a finite Radon measure μ_u on Δ, carried by the set of extremal (or minimal) elements Δ_1, such that $u = \int h\mu_u(dh)$. We assume that the above representing measure is unique

Definition 1. Suppose $\mathcal{G} = \{\mathcal{G}_h : h \in \Delta_1\}$ is a family of filters on Ω. This family $\mathcal{G}$ is said to have the Fatou-Doob property relative to a $u \in \mathbf{H}$ if for every $v \in \mathbf{S},\ v = w + p$ with $w \in \mathbf{H},\ p \in P$, the limit of $v(x)/u(x)$, x following $\mathcal{G}_h$, exists

[1] Part of this research was supported by a NSERC operating grant.

K. GowriSankaran et al. (eds.), Classical and Modern Potential Theory and Applications, 233–236.

for μ_u almost every $h \in \Delta_1$; and the limit is a Radon-Nikodym derivative of the absolutely continuous part μ_w relative to μ_u.

With some simple restrictions on the 'reduced function' we can easily show that the limit function obtained in the above process *is in fact* the Radon-Nikodym derivative as specified in the above definition.

Let us now fix a $u \in \mathbf{H}$ with $u(x) \neq 0$ for any x. Suppose we assign a set $Z(p, A, u) \subset \Delta_1$ corresponding to each $p \in \mathbf{P}$ and each Borel set $A \subset \Delta_1$ with $\mu_u(A) > 0$. This is the "zero set" or the set of all points where we expect p/u_A to have "limit" zero where u_A is the element of $\mathbf{H}$ with μ restricted to A as the representing measure. Suppose this correspondence verifies the following fundamental conditions.

(a) $Z(0, A, u) = A$ for all A

(b) for all $\alpha > 0$ and all such A, $Z(\alpha p, A, u) = Z(p, A, u)$

(c) $Z(p_1, A, u) \cap Z(p_2, A, u) = Z(p_1 + p_2, A, u)$

(d) if $p_1, p_2 \in \mathbf{P}$ with $p_1(x) \leq p_2(x)$ for all x then $Z(p_1, A, u) \supset Z(p_2, A, u)$

(e) for Borel sets $A \subset B \subset \Delta_1$, with $\mu_u(A) > 0$, $\forall p \in \mathbf{P}$ $Z(p, B, u) \supset Z(p, A, u)$ and

(f) for all $p \in \mathbf{P}$, all such A, $\mu_u[A \backslash Z(p, A, u)] = 0$

We shall call a choice of sets as above, *a choice of u-zero sets* for $\mathbf{P}$. In the familiar case when Ω is a Brelot harmonic space with a positive potential and countable base for open sets, we take $Z(p, A, u)$ to be the set of all $h \in \Delta_1$ where p/u_A has the fine limit zero. We now have,

Theorem 2. Let $(p, A) \mapsto Z(p, A, u)$ be a choice of u-zero sets for a fixed $u \in \mathbf{H}$ as set out above. Define a family of filter $\{\mathcal{G}_h : h \in \Delta_1\} = \mathcal{G}$, as follows. For each $h \in \Delta_1$, $\mathcal{G}_h$ has the (filter) base consisting of the sets $\{x \in \Omega : p(x) < u_A(x)\}$ for all $p \in \mathbf{P}$ and all Borel sets A with $\mu_u(A) > 0$. Here $u_A = \int_A h \mu_u(dh)$.

Then $\mathcal{G}$ has the Fatou-Doob property relative to u. Further, if $\mathcal{G}' = \{\mathcal{G}'_h : h \in \Delta_1\}$ is another family of filters with the Fatou-Doob property relative to u such that for all $p \in \mathbf{P}$, $A \subset \Delta_1$ Borel with $\mu_u(A) > 0$,

$$Z(p, A, u) \subset \left\{ h : \lim_{\mathcal{G}_{h'}} \frac{p(x)}{u_A(x)} = 0 \right\} .$$

Then, $\mathcal{G}'_h$ is finer than $\mathcal{G}_h$ for all $h \in \Delta_1$. It follows that $\mathcal{G}$ constructed above is the smallest family of filters possessing the Fatou-Doob property relative u for the given u-zero sets.

Definition 3. Suppose for each $u \in \mathbf{H}$ is assigned *a choice of u-zero* sets. We shall say that this *choice of zero sets* is consistent if $\forall$ $u_1, u_2 \in \mathbf{H}$ with $u_1 \leq u_2$, then

$Z(p, A, u_2) \subset Z(p, A, u_1)$ for all $p \in \mathbf{P}$ and $Z(p, \Delta_1, u_A) = Z(p, A, u)$ for all p, u and applicable A.

Note that such a consistent choice automatically satisfies the condition (e) in the definition of u-zero sets conditions. We now have,

Theorem 4. Let us suppose we are given a consistent choice of u-zero sets for all $u \in \mathbf{H}$. Then, there exists a smallest family $\mathcal{F}^1 = \{\mathcal{F}'_h : h \in \Delta_1\}$ of filters having the Fatou-Doob property relative all $u \in \mathbf{H}$ subject to the condition that $\forall\, p \in \mathbf{P},\ Z(p, \Delta_1, u) \subset \left\{ h \in \Delta_1 : \lim_{\mathcal{F}'_h} \cdot \frac{p}{u} = 0 \right\}$

Remark 5. Under the hypothesis of our situation, for every h in Δ_1 there corresponds the 'fine' filter $\mathcal{F}_h$ defined in the usual way [G1]. The smallest filter family with the Fatou-Doob property constructed in Theorem 4 is finer than the family of fine filter for all $h \in \Delta_1$.

Remark 6. Suppose Ω is a Brelot harmonic space with a potential > 0 and a countable base for open sets. Let $\mathbf{H}$ be the set of all positive (≥ 0) harmonic functions and $\mathbf{P}$ the class of potentials on Ω. Let Δ_1 be the set of minimal elements in a convenient compact base of $\mathbf{H}$. Let $\mathcal{F} = \{\mathcal{F}_h : h \in \Delta_1\}$ consists of the collection of fine filters, for all $h \in \Delta_1$. It is known [G2] that for all $p \in \mathbf{P}$ and each $u \in \mathbf{H}$ (with μ_u the canonical representing measure on Δ_1), the $\lim \cdot \dfrac{p(x)}{u(x)}$ as x tends to h following $\mathcal{F}_h$, is zero except for a Borel set $E_{p,u}$ of μ_u measure zero. Set $Z(p, \Delta_1, u) = \Delta_1 \backslash E_{p,u}$ for all $p \in \mathbf{P}$ and all $u \in \mathbf{H}$ (note this gives a definition $Z(p, A, u)$ for all $A \subset \Delta_1$, A Borel by taking u_A).

It is simple to verify that this gives a *choice of u-zero sets* for $\mathbf{P}$ which is consistent. We can then deduce that the collection $\mathcal{F}$, of all fine filters $\mathcal{F}_h, h \in \Delta_1$, is the *smallest* one with the Fatou-Doob property relative to all $u \in \mathbf{H}$.

The detailed proofs of the above results as well as important applications to the case of n-superharmonic functions and a conjecture of Doob [D1] will appear elsewhere.

BIBLIOGRAPHY

[BL1] J. BLIEDTNER & P.A. LOEB, "A Measure Theoretic Boundary Limit Theorem" Archiv der Math. 43 (1984), 373-376.

[BL2] J. BLIEDTNER & P.A. LOEB, "A reduction technique for limit theorems in analysis and probability theory" Archiv für Mat. 30 (1992) 25-43.

[BL3] J. BLIEDTNER & P.A. LOEB, "Best filters for the general Fatou boundary limit theorem" (to appear) Proc. Amer. Math. Soc.

[BL4] J. BLIEDTNER & P.A. LOEB, "The best approach for boundary limits" (This Proceedings)

[D1] J.L. DOOB, "A non-probabilistic proof of the relative Fatou theorem" Ann. Inst. Fourier IX (1959), 293-300.

[G1] K. GOWRISANKARAN, "Extreme harmonic functions and boundary value problems" Ann. Inst. Fourier XIII, 2 (1963), 307-356.

[G2] K. GOWRISANKARAN, "Fatou-Naim-Doob limits in the axiomatic system of Brelot" Ann. Inst. Fourier XVi (1966).

GAUSSIAN UPPER BOUNDS FOR THE HEAT KERNEL AND FOR ITS DERIVATIVES ON A RIEMANNIAN MANIFOLD

A.GRIGOR'YAN
Department of Mathematics
Harvard University
Cambridge, MA 02138
USA
grigor@math.harvard.edu

ABSTRACT. Gaussian upper bounds of the heat kernel and of its time derivatives on a complete non-compact Riemannian manifold are shown to follow from certain diagonal estimates

1. Main results

Given a Riemannian manifold M we consider the heat kernel $p(x,y,t)$ being by definition the smallest positive fundamental solution to the heat equation $u_t - \Delta u = 0$ on $M \times (0, +\infty)$ where Δ is the Laplace operator associated with the Riemannian metric. The question to be discussed here is how to get Gaussian upper estimates of $p(x,y,t)$ and of its time derivatives $\frac{\partial^m p}{\partial t^m}(x,y,t)$ provided we know a priori for all $t > 0$ an on-diagonal upper bound

$$p(x,x,t) \leqslant \frac{1}{f(t)} \tag{1.1}$$

where $f(t)$ is an increasing function, x is a given point on M.

In the simplest case when the manifolds M is a Euclidean space $\mathbf{R}^n$ we have

$$p(x,y,t) = \frac{1}{(4\pi t)^{n/2}} \exp\left(-\frac{r^2}{4t}\right) \tag{1.2}$$

where $r = \mathrm{dist}(x,y)$, and (1.1) holds with $f(t) = \frac{\text{const}}{t^{n/2}}$. A plain computation yields that the $m-$ th time derivative $\frac{\partial^m p}{\partial t^m}$ has in this case the sign $(-1)^m$ and its absolute value is estimated as follows

$$\left|\frac{\partial^m p}{\partial t^m}\right|(x,y,t) \asymp \frac{1}{t^{n/2+m}}\left(1 + \frac{r^2}{t}\right)^m \exp\left(-\frac{r^2}{4t}\right) \tag{1.3}$$

where the sign $\asymp$ means that the ratio of the left- and right-hand sides in (1.3) is bounded from above and below by constants depending only on n and m.

K. GowriSankaran et al. (eds.), Classical and Modern Potential Theory and Applications, 237–252.

Similar inequalities can be obtained in a more general situation as will be shown below. From now on we assume that the manifold in question is non-compact and complete. The most interesting aspects of what follows are connected to behaviour of the heat kernel $p(x,y,t)$ and its derivatives as $t \to \infty$ and as $r \to \infty$ where r denotes a geodesic distance between the points x and y.

In order to formulate the main result let us introduce the following notation. Let us fix some constant $D > 2$ and put

$$E^D(x,t) = \int_M p^2(x,y,t) \exp\left(\frac{r^2}{Dt}\right) dy \tag{1.4}$$

where $r = \text{dist}(x,y)$ (dist stands for a geodesic distance and dy is the Riemannian volume element).

Theorem 1.1 *Let $f(t)$ be an increasing function on $(0,\infty)$ of at most a polynomial growth in the sense that*

$$f(2t) \leqslant Af(t) \tag{1.5}$$

for all $t > 0$ and for some constant A. Suppose that the heat kernel on a manifold M satisfies the on-diagonal estimate (1.1) for some (fixed) point $x \in M$ and for all $t > 0$. Then for any $D > 2$ and for all $t > 0$ we have

$$E^D(x,t) \leqslant \frac{\text{const}_A}{f(\delta t)} \tag{1.6}$$

where $\delta = \min(D-2, 1)$.

Of course, applying (1.5) one sees that $f(\delta t) \geqslant \text{const}_\delta f(t)$ and the factor δ can be absorbed into the constant in the numerator of (1.6) but sometimes the dependence on D in (1.6) is essential and the inequality (1.6) enables one to catch it. Note that the hypothesis $D > 2$ can not be relaxed: if $D = 2$ then in the Euclidean space we have $E^D = \infty$.

Let us also observe that the assumption that (1.1) is true for *all* $t > 0$ does not restrict applicability of the theorem to the case when (1.1) is known only for $t < T$. Indeed, the function $p(x,x,t)$ is decreasing in t, therefore, if (1.1) holds for $t < T$ then the function $f(t)$ can be extended for $t \geqslant T$ simply as the constant $f(T)$ and (1.1) will be valid for all $t > 0$.

Theorem 1.1 will be proved in Section 3 below. In Section 2, we shall outline other related results and methods. Now, let us explain how to get pointwise upper Gaussian estimations similar to (1.3) applying this theorem. As was proved in [9] the following inequality holds always irrespective of geometry of a manifold

$$p(x,y,t) \leqslant \sqrt{E^D(x,t/2)E^D(y,t/2)} \exp\left(-\frac{r^2}{2Dt}\right) . \tag{1.7}$$

Therefore, if we are given that the on-diagonal estimate (1.1) holds for a point x as well as for a point y then by (1.6) and (1.7)

$$p(x,y,t) \leqslant \frac{\text{const}}{f(\frac{\delta}{2}t)} \exp\left(-\frac{r^2}{2Dt}\right) . \tag{1.8}$$

In order to get inequalities involving derivatives of the heat kernel let us define the powers of the gradient ∇ as follows: ∇^m means Δ^m if m is even and $\nabla\Delta^{\frac{m-1}{2}}$ if m is odd. Let us fix some number $D > 2$ and introduce the following series of functions $E_m(x,t)$:

$$E_m(x,t) = \int_M |\nabla^m p|^2 (x,y,t) \exp\left(\frac{r^2}{Dt}\right) dy, \; m = 0,1,2,... \tag{1.9}$$

where the operator ∇ relates to the variable y and $r = \mathrm{dist}(x,y)$. Obviously, E^D is the same as E_0 in the new notation. Of course, it would be correct to write E^D_m in place of E_m but we skip the superscript D in order to simplify notations when the constant D is fixed.

The following two theorem were proved in [11] .

Theorem 1.2 *If $D > 2$ then for any $x \in M$ and any integer $m \geqslant 0$ the function $E_m(x,t)$ is finite and decreasing in t. Besides, for all $x, y \in M$ and any $t > 0$*

$$\left|\frac{\partial^m p}{\partial t^m}\right| (x,y,t) \leqslant \sqrt{E_{2m}(x,t/2)E_0(y,t/2)} \exp\left(-\frac{r^2}{2Dt}\right) \tag{1.10}$$

In fact, the inequality (1.10) as well as (1.7) above are derived from the semigroup identity

$$p(x,y,t) = \int_M p(x,z,t-s)p(z,y,s)dy \tag{1.11}$$

upon differentiation in t and a proper application of Cauchy-Schwarz inequality using the fact that $\frac{\partial^m p}{\partial t^m} = \Delta^m p = \nabla^{2m} p$.

Theorem 1.2 reduces the question of finding upper bounds of $\left|\frac{\partial^m p}{\partial t^m}\right|$ to that of E_m . It turns out that E_m for $m > 0$ can be estimated directly via E_0 as follows.

Theorem 1.3 *Suppose that for some point $x \in M$ and all $t > 0$*

$$E_0(x,t) \leqslant \frac{1}{\varphi(t)}$$

where $\varphi(t)$ is a positive increasing function on $(0,\infty)$ then for any integer $m \geqslant 1$ and for all $t > 0$

$$E_m(x,t) \leqslant \frac{C^m}{\varphi_m(t)} \tag{1.12}$$

where $C = \frac{D/2+8}{D-2}$ and $\varphi_m(t)$ denotes the $m-$ th integral of the function $\varphi(t)$, that is to say, $\varphi_0 = \varphi$ and for $m > 0$

$$\varphi_m(t) = \int_0^t \varphi_{m-1}(\tau)d\tau \; .$$

Let us observe that the inequality (1.12) can be rewritten in the following form

$$E_m(x,t) \leqslant C^m \left(\int_0^t \frac{(t-\tau)^{m-1}}{(m-1)!} \frac{d\tau}{E_0(x,\tau)} \right)^{-1}$$

if we put in Theorem 1.3 $\varphi(t) = 1/E_0(x,t)$.

The following statement is a straightforward consequence of the theorems 1.1-1.3.

Corollary 1.1 *Let $f(t)$ and $g(t)$ be increasing in t functions either satisfying the hypothesis (1.5) of at most a polynomial growth. Assume that for two points $x, y \in M$ and for all $t > 0$*

$$p(x,x,t) \leqslant \frac{1}{f(t)}, \quad p(y,y,t) \leqslant \frac{1}{g(t)}$$

then for any $D > 2$, any integer $m \geqslant 0$ and for all $t > 0$

$$\left| \frac{\partial^m p}{\partial t^m} \right| (x,y,t) \leqslant \frac{\mathrm{const}_{A,D,m}}{t^m \sqrt{f(\frac{\delta}{2}t) g(\frac{\delta}{2}t)}} \exp\left(-\frac{r^2}{2Dt} \right) \tag{1.13}$$

where

$$\mathrm{const}_{A,D,m} = \mathrm{const}_A \delta^{-m} 20^m \sqrt{(2m)!} \tag{1.14}$$

and $\delta = \min(D-2, 1)$ is the same as in Theorem 1.1.

Moreover, for some $\nu = \nu(A)$

$$\left| \frac{\partial^m p}{\partial t^m} \right| (x,y,t) \leqslant \frac{\mathrm{const}_{A,m}}{t^m \sqrt{f(t) g(t)}} \left(1 + \frac{r^2}{t} \right)^{\nu+m} \exp\left(-\frac{r^2}{4t} \right). \tag{1.15}$$

In fact, the exponent ν is exactly the number for which the inequality is valid:

$$\frac{f(t_2)}{f(t_1)} \leqslant \mathrm{const} \left(\frac{t_2}{t_1} \right)^{\nu} \quad \forall t_2 > t_1 > 0 \tag{1.16}$$

and the same for the function g. It is plain that (1.16) follows from (1.5) but sometimes it is useful to postulate (1.16) separately in order to have a better value of ν.

2. An outline of previous results and methods

Let us compare the announced Theorem 1.1 and Corollary 1.1 with the previous results of this kind. A numerous works are devoted to heat kernel's estimations - the size of this note is not enough to mention even a small part of them. We consider below several types of the known theorems from our standpoint - what Gaussian bounds of the heat kernel can be derived from an on-diagonal estimate. Let us note that the basic results in this direction are due to Davies [4] ; [5] and Varopoulos [16] .

We discuss more detailed approaches to estimation of the time derivatives of the heat kernel which are due to Porper [13], Cheng, Li, Yau [3] , Varopoulos [16], [18] , Davies [8] and and Kovalenko, Semenov [12] based on various ideas. The common achievement of these works is that upper bounds of the time derivatives follows from upper bounds of the heat kernel itself without any additional geometric assumptions. Of course, the same is stated also by Corollary 1.1 in the most flexible and sharp form.

2.1. Bounds Which Are Uniform In x. These are theorems which yield the off-diagonal upper bound (1.8) under the hypothesis that the on-diagonal bound (1.1) holds for *all* points x. Such a statement for the case of a polynomial function

$$f(t) = \text{const} \begin{cases} t^{\alpha} & , t < 1 \\ t^{\beta} & , t \geqslant 1 \end{cases}$$

was first proved by Davies [4] using a log-Sobolev inequality as an intermediate step between on-diagonal and off-diagonal upper bounds of the heat kernel. Another approach was offered earlier by Ushakov [15] for the setting of parabolic equations in unbounded domains in $\mathbf{R}^n$ but without sharp exponent $2D$ in (1.8) . We apply an improved version of the latter method in the proof of Theorem 1.1. Let us emphasize is this connection that when applying Corollary 1.1 the on-diagonal upper bound need be checked as a hypothesis only at *two* points x, y rather than for all points.

2.2. Bounds Which Are Non-Uniform In x. These are non-homogeneous estimations when one assumes that for any x

$$p(x,x,t) \leqslant \varphi(x,t)$$

where behaviour in t might be different for different x. A theorem of Davies [5] states that

$$p(x,y,t) \leqslant \text{const}\sqrt{\varphi(x,t)\varphi(y,t)}\exp\left(-\frac{r^2}{2Dt}\right) \tag{2.1}$$

provided the function $\psi(x,t) = \sqrt{\varphi(x,t)}$ satisfies the conditions

$$|\psi_t| \leqslant \text{const}\,\frac{\psi}{t} \tag{2.2}$$

and

$$\Delta\psi \leqslant \text{const}\,\frac{\psi}{t}\ . \tag{2.3}$$

In the view of Corollary 1.1, the most restrictive second condition (2.3) is actually superfluous: indeed, (2.2) implies that the function $f_x(t) = \frac{1}{\varphi(x,t)}$ satisfies the condition (1.5) of a polynomial growth in t for any point x. Therefore, by Corollary 1.1 (case $m = 0$) we get (2.1) .

2.3. Gaussian Estimates Of $p(x,y,t)$ With A Polynomial Correction Term. The first results are due to Varopoulos [16] , [17] and in the sharpest form to Davies, Pang [6] . Following [16] , such estimations as (1.15) containing the factor $1+\frac{r^2}{t}$ to some power are derived from (1.13) upon optimization with respect to D provided one knows an explicit dependence on D of other constants. Our estimate (1.15) (case m=0) gives the same power of this factor as in [6] provided $f(t)=g(t)=\mathrm{const}\, t^{\nu}$. Again, the advantage of Corollary 1.1 is that it needs the initial on-diagonal estimate only at two points x,y in contrast to all the previous results.

2.4. Superpolynomial Decay Of The Heat Kernel. By the hypotheses of Theorem 1.1 and Corollary 1.1, the function $f(t)$ can not increase faster than polynomially. There exists, in fact, only one related result which catches the opposite situation. This is a consequence of theorems 2.2 and 4.2 from [9] , and it states that if the inequality (1.1) is true for any $x\in M$ and for all $t>0$ then, again, for any $x\in M$ and $t>0$

$$E_0(x,t)\leqslant \frac{\mathrm{const}}{f(\delta t)}$$

where $\delta=\delta(D)$ and $D>2$ is arbitrary, provided the function $f(t)$ satisfies certain regularity conditions. Without going into details of these conditions, let us only mention that they admit also a superpolynomial function $f(t)$, for example, $f(t)=\exp t^{\nu}$, for large t where $0<\nu\leqslant 1$. On the other hand, in order to run this theorem, one must have the hypothesis (1.1) be true at once for *all* x whereas in Theorem 1.1, one needs the same at a *single* point x. Hence, in this sense, Theorem 1.1 is more flexible.

2.5. Estimates Of The Time Derivatives By The Method Of Porper. This method was developed by Porper [13] , [14] for the setting of a uniform parabolic equation in $\mathbf{R}^n$, and it goes through on a manifold as well. The starting point is the assumption that for some points x,y we are given that for all $t>0$

$$p(x,x,t)\leqslant \frac{\mathrm{const}}{t^{\nu}},\quad p(y,y,t)\leqslant \frac{\mathrm{const}}{t^{\nu}} \tag{2.4}$$

and

$$p(x,y,t)\leqslant \frac{\mathrm{const}}{t^{\nu}}\exp\left(-\frac{r^2}{\sigma t}\right) \tag{2.5}$$

and the objective is to obtain similar bounds for the time derivatives of the heat kernel.

Let us introduce the notation $W_k(x,t)$ similar to $E_k(x,t)$ but without the Gaussian weight:

$$W_k(x,t)=\int_M \left|\nabla^k p\right|^2(x,y,t)dy$$

in particular, we have by (1.11)

$$W_0(x,t)=p(x,x,2t)\ . \tag{2.6}$$

By differentiation in t one can show that

$$\frac{\partial^k W_0}{\partial t^k}(x,t) = (-2)^k W_k(x,t)$$

which implies, in particular, that the function $W_0(x,t)$ is convex and decreasing in t whence it follows that

$$W_1(x,t) = -\frac{1}{2}\frac{\partial W_0}{\partial t}(x,t) \leqslant \frac{1}{2}\frac{W_0(x,t/2) - W_0(x,t)}{t/2} \leqslant \frac{1}{t} W_0(x,t/2) \; . \qquad (2.7)$$

Similarly, for any integer k

$$W_k(x,t) \leqslant \frac{1}{t} W_{k-1}(x,t/2) \leqslant \frac{1}{t^k} W_0(x,t/2^k) \; . \qquad (2.8)$$

Besides, the semigroup identity (1.11) enables one to get the following initial estimate

$$\left|\frac{\partial^k p}{\partial t^k}\right|(x,y,t) \leqslant \sqrt{W_{2k}(x,t/2) W_0(y,t)} \; . \qquad (2.9)$$

Let now x, y be the points for which (2.4) and (2.5) hold, then by substituting into (2.9) successively (2.8) , (2.6) and (2.4) we get that

$$\left|\frac{\partial^k p}{\partial t^k}\right|(x,y,t) \leqslant \frac{\text{const}}{t^{\nu+k}} \; . \qquad (2.10)$$

In order to involve the Gaussian factor one estimates first $\frac{\partial p}{\partial t}(x,y,\theta)$ at a mean point $\theta \in (t, t+\delta)$ (for some $\delta > 0$) by the mean-value theorem

$$\left|\frac{\partial p}{\partial t}\right|(x,y,\theta) \leqslant \frac{1}{\delta}\, |p(x,y,t+\delta) - p(x,y,t)| \leqslant \frac{p(x,y,t+\delta) + p(x,y,t)}{\delta} \; . \qquad (2.11)$$

Applying again the mean-value theorem to the function $\frac{\partial p}{\partial t}$ we have

$$\left|\frac{\partial p}{\partial t}\right|(x,y,t) \leqslant \left|\frac{\partial p}{\partial t}\right|(x,y,\theta) + \delta \sup_{\tau \in (t,t+\delta)} \left|\frac{\partial^2 p}{\partial t^2}\right|(x,y,\tau) \; . \qquad (2.12)$$

Next we substitute into (2.12) the upper bound of $\left|\frac{\partial p}{\partial t}\right|(x,y,\theta)$ from (2.11) , the upper bound of $\left|\frac{\partial^2 p}{\partial t^2}\right|(x,y,\tau)$ obtained by (2.10) and the upper bounds of $p(x,y,t+\delta), p(x,y,t)$ according to (2.5) which yields

$$\left|\frac{\partial p}{\partial t}\right|(x,y,t) \leqslant \frac{1}{\delta}\frac{\text{const}}{t^\nu} \exp\left(-\frac{r^2}{\sigma(t+\delta)}\right) + \delta \frac{\text{const}}{t^{\nu+2}}$$

whence by choosing an optimal value of δ which is to be $\delta = \varepsilon t \exp\left(-\frac{r^2}{2\sigma t}\right) \leqslant \varepsilon t$ for a small $\varepsilon > 0$ it follows finally

$$\left|\frac{\partial p}{\partial t}\right|(x,y,t) \leqslant \frac{\text{const}}{t^{\nu+1}} \exp\left(-\frac{r^2}{\sigma_1 t}\right) \tag{2.13}$$

with $\sigma_1 = 2\frac{1+\varepsilon}{1-\varepsilon}\sigma$.

The advantage of this method is that in order to get an upper bound of $\left|\frac{\partial p}{\partial t}\right|(x,y,t)$ one need only be given the upper bound of the heat kernel itself at the same points x, y. On the other hand, it does not yield the sharp Gaussian exponent - we have in (2.13) under the exponential the coefficient $\sigma_1 > 2\sigma$ instead of expected σ.

2.6. Integral Estimates Of The Derivatives According To Cheng-Li-Yau. The original purpose of the method to be outlined below was to get pointwise upper bounds of the space derivatives of the heat kernel under the assumption that the curvature of the manifold is bounded. We have extracted a part of arguments of Cheng, Li, Yau [3] which involve no curvature and which enable one to get crucial integral estimations of $\nabla^k p$.

Let us consider side by side with the quantities $W_k(x,t)$ the integrals of $\nabla^k p$ over an exterior of a ball

$$W_k^R(x,t) = \int_{r \geqslant R} \left|\nabla^k p\right|^2 (x,y,t)dy \ .$$

where $r = \text{dist}(x,y)$. We shall concentrate on obtaining a Gaussian estimation of W_k of the following kind:

$$W_k^R(x,t) \leqslant \frac{\text{const}}{t^{\nu+k}} \exp\left(-\frac{r^2}{D_k t}\right) \tag{2.14}$$

(which implies in turn pointwise Gaussian upper bounds of the time derivatives of the heat kernel upon application of the semigroup identity) under assumption that (2.14) is known to hold for $k = 0$:

$$W_0^R(x,t) \leqslant \frac{\text{const}}{t^{\nu}} \exp\left(-\frac{r^2}{Dt}\right) \tag{2.15}$$

for a given point x and for all $t > 0$.

First, one proves (2.8) by means of the eigenvalue expansion. We skip this part of the arguments and pass directly to Gaussian bounds. Obviously, we have by (2.15) and (2.8)

$$W_k^R(x,t) \leqslant W_k(x,t) \leqslant \frac{1}{t^k} W_0(x, t/2^k) \leqslant \frac{\text{const}}{t^{\nu+k}} \ . \tag{2.16}$$

The following inequality is true for any smooth functions $u(y)$, $\varphi(y)$ provided $\varphi(y)$ is finitely supported:

$$\int_M \left|\nabla^{k+1}u\right|^2 \varphi^2 \leqslant 4\int_M \left|\nabla^k u\right|^2 |\nabla\varphi|^2 + 2\left(\int_M \left|\nabla^k u\right|^2 \varphi^2 \int_M \left|\nabla^{k+2}u\right|^2 \varphi^2\right)^{\frac{1}{2}}$$

which is proved by a standard technique of integration by parts. By an appropriate choice of φ and by putting $u(y) = p(x,y,t)$ one deduces from it for any $\gamma < 1$

$$W_{k+1}^R \leqslant \frac{\text{const}}{R^2} W_k^{\gamma R} + 2\left(W_k^{\gamma R} W_{k+2}^{\gamma R}\right)^{\frac{1}{2}} \leqslant \text{const}\sqrt{W_k^{\gamma R}}\left(\frac{1}{R^2}\sqrt{W_k} + \sqrt{W_{k+2}}\right) .$$

all functions W being taken at the point (x,t).

Let us estimate W_k and W_{k+2} by means of (2.16) and suppose that (2.14) is true by the inductive hypothesis, then the formula above yields

$$W_{k+1}^R(x,t) \leqslant \frac{\text{const}}{t^{\nu+k+1}} \exp\left(-\frac{\gamma^2 R^2}{2D_k t}\right)\left(\frac{t}{R^2} + 1\right) .$$

If $R^2/t > 1$ then the last factor on the right-hand side can be absorbed into the Gaussian term and we get

$$W_{k+1}^R(x,t) \leqslant \frac{\text{const}}{t^{\nu+k+1}} \exp\left(-\frac{r^2}{D_{k+1}t}\right) \tag{2.17}$$

where $D_{k+1} > 2D_k$ as in the above method of Porper. If $R^2/t \leqslant 1$ then (2.17) follows directly from (2.16) .

We see that this methods gives no sharp Gaussian term either. On the other side, an important advantage of the method is that one need not know that p is a solution to the heat equation in order to deduce (2.14) from (2.15) and (2.16) . Also, the hypotheses (2.15) and (2.16) need hold only at a single point x.

2.7. Methods Based Upon The Semigroup Theory. The following approach arose independently in the papers of Varopoulos [16] and Kovalenko, Semenov [12] and is based upon the fact that the heat semigroup $e^{t\Delta}$ in $L^2(M)$ is holomorphic. Since this semigroup is also submarkovian then by general properties of holomorphic semigroups we have

$$\left\|\Delta e^{t\Delta}\right\|_{2\to 2} \leqslant \frac{\text{const}}{t} \tag{2.18}$$

where $p \to q$ means that one considers an operator acting from $L^p(M)$ to $L^q(M)$. By a standard interpolation we have

$$\left\|\Delta e^{3t\Delta}\right\|_{1\to\infty} \leqslant \left\|e^{t\Delta}\right\|_{1\to 2}\left\|\Delta e^{t\Delta}\right\|_{2\to 2}\left\|e^{t\Delta}\right\|_{2\to\infty} . \tag{2.19}$$

Since either norm $\|e^{t\Delta}\|_{1\to 2}$ and $\|e^{t\Delta}\|_{2\to\infty}$ is equal to $\sup_x \sqrt{p(x,x,2t)}$ and the operator $\Delta e^{t\Delta}$ has the kernel $\frac{\partial p}{\partial t}(x,y,t)$ then (2.19) implies that

$$\sup_{x,y}\left|\frac{\partial p}{\partial t}\right|(x,y,3t) \leqslant \frac{\text{const}}{t}\sup_x p(x,x,2t)\ . \tag{2.20}$$

The higher order derivatives are estimated similarly. The Gaussian bounds are obtained using a perturbed semigroup in the spirit of Davies [7] . We do not go into further details because these arguments are presented very well in the literature. We shall only mention that the sharpest Gaussian term in estimations of the time derivative of the heat kernel is obtained by Davies [8] by a modification of this approach - in place of using (2.18) he considered the heat kernel for complex values of the time and applied the Cauchy integral formula in order to estimate $\frac{\partial p}{\partial t}$.

Although this method gives a sharp Gaussian factor and a sharp polynomial correction term it has also a drawback - its applicability depends very heavily on the initial on-diagonal upper bound of the heat kernel which must be valid uniformly for *all* x as it is seen from (2.20) .

Comparing the methods above with one based upon Theorems 1.1-1.3 we should like to mention that our approach possesses all their advantages - it yields the sharp Gaussian factor as well as it is not tied to a uniform in x behaviour of $p(x,x,t)$.

3. Proof of Theorem 1.1 and Corollary 1.1

Let us start with proof of Theorem 1.1. In the course of the proof, we would have to deal with different unproper integrals of the heat kernel which could a priori be equal to ∞. In order to avoid such difficulties we replace the heat kernel p by that of a bounded region. Indeed, by definition the heat kernel p is constructed as a limit of the local heat kernels p_Ω :

$$p(x,y,t) = \lim_{\Omega\to M} p_\Omega(x,y,t)$$

where Ω is a bounded region with a smooth boundary, p_Ω is a heat kernel in Ω with the vanishing Dirichlet boundary values and $\Omega \to M$ means an exhaustion of M by an expanding sequence of such regions (see [2] for details of this definition).

Since by the maximum principle $p_\Omega \leqslant p$ the hypothesis (1.1) holds for the local heat kernel p_Ω too. We fix from the very beginning a region Ω and shall perform all the proof for the heat kernel in this region obtaining the inequality (1.6) for E^D_Ω (with an obvious definition of E^D_Ω) instead of E^D.

If we have proved already that $E^D_\Omega(x,t) \leqslant \mathcal{R}$ where $\mathcal{R}$ stands for the right-hand side of (1.6) then for any compact $K \subset M$ and for any Ω containing K we have

$$\int_K p^2_\Omega(x,y,t)\exp\left(\frac{r^2}{Dt}\right)dy \leqslant \mathcal{R}$$

whence it follows upon passage to the limit as $\Omega \to M$ that the same inequality is true for p in place of p_Ω. Since K is arbitrary it follows that in this inequality K can be replaced by M which yields $E^D \leqslant \mathcal{R}$.

Thus, from now on we shall deal with the local heat kernel p_Ω and in order to simplify notations we shall suppress the subscript Ω. We suppose that Ω is wide enough so that the point x for which (1.1) holds belongs to Ω. Also, we extend the function $p(x, y, t) \equiv p_\Omega(x, y, t)$ to all $y \in M$ so that it vanishes if y lands outside Ω.

The main ingredient of the proof is the following lemma which is proved by a version of the method invented by Ushakov [15] for parabolic equations in $\mathbf{R}^n$. Let us denote by B_R^x a geodesic ball of radius R centred at the point x.

Lemma 3.1 *Suppose that $f(t)$ is an increasing function which is at most polynomial in the sense of (1.5) and let $D \geqslant D_0$ where D_0 is a large absolute constant (say, $D_0 = 200$). Then the on-diagonal bound (1.1) implies that for any $R > 0$*

$$\int_{M \setminus B_R^x} p^2(x, y, t)dy \leqslant \frac{\mathrm{const}_A}{f(t)} \exp\left(-\frac{R^2}{Dt}\right) \tag{3.1}$$

Proof. Let us introduce the function:

$$d(y) = \begin{cases} R - r & , r \leqslant R \\ 0 & , r > R \end{cases}$$

where $r = \mathrm{dist}(x, y)$ and for some $T > 0$ consider the function

$$\xi(y, t) = \frac{d(y)^2}{2(t - T)}, \quad 0 < t < T,$$

then we have obviously

$$\frac{\partial \xi}{\partial t} + \frac{1}{2} |\nabla \xi|^2 \leqslant 0 \, . \tag{3.2}$$

We shall apply the integral maximum principle which is stated in the following lemma.

Lemma 3.2 *If a function $u(y, t)$ satisfies the heat equation $u_t - \Delta u = 0$ in $\Omega \times (t_1, t_2)$ with Dirichlet boundary values $u|_{\partial\Omega} = 0$ then the integral*

$$\int_\Omega u^2(y, t) e^{\xi(y,t)} dy$$

is a decreasing function of $t \in (t_1, t_2)$ provided the function ξ satisfies (3.2) in $\Omega \times (t_1, t_2)$.

This property of solutions to parabolic equations was discovered by Aronson [1] . The proof for the setting of manifolds is found in [3] and in [10] . Let us note also that the proof of the first part of Theorem 1.2 is based on this lemma as well.

Applying Lemma 3.2 to the function $u(y,t) = p(x,y,t)$ we see for any $\tau < t < T$

$$\int_M p^2(x,y,t)e^{\xi(y,t)}\,dy \leqslant \int_M p^2(x,y,\tau)e^{\xi(y,\tau)}\,dy \tag{3.3}$$

whence, using the specific form of the function d, it follows that

$$\int_{M\setminus B_R^x} p^2(x,y,t)dy \leqslant \int_{B_\rho^x} p^2(x,y,\tau)\exp\left(-\frac{d(y)^2}{2(T-\tau)}\right)dy + \int_{M\setminus B_\rho^x} p^2(x,y,\tau)dy$$

where $\rho < R$. Observing here that $d(y)$ under the exponential is at least as much as $R-\rho$ and letting $T \to t+$ we obtain

$$\int_{M\setminus B_R^x} p^2(x,y,t)dy \leqslant \exp\left(-\frac{(R-\rho)^2}{2(t-\tau)}\right)\int_M p^2(x,y,\tau)dy + \int_{M\setminus B_\rho^x} p^2(x,y,\tau)dy$$

Since by the semigroup property (1.11) and (1.1)

$$\int_M p^2(x,y,\tau)dy \equiv p(x,x,2\tau) \leqslant \frac{1}{f(2\tau)} \leqslant \frac{1}{f(\tau)} \tag{3.4}$$

it follows that

$$\int_{M\setminus B_R^x} p^2(x,y,t)dy \leqslant \frac{1}{f(\tau)}\exp\left(-\frac{(R-\rho)^2}{2(t-\tau)}\right) + \int_{M\setminus B_\rho^x} p^2(x,y,\tau)dy\ . \tag{3.5}$$

Let us arrange now two decreasing sequences $t_k = t2^{-k}$ and $R_k = (\frac{1}{2} + \frac{1}{k+2})R$ where $k = 0,1,2,\ldots$. Obviously, $t_k \to 0$ as $k \to \infty$ and $\frac{1}{2}R < R_k \leqslant R$ for any k. We apply the inequality (3.5) for pairs t_k, t_{k-1} and R_k, R_{k-1} in place of τ, t and, respectively, ρ, R and sum up all such inequalities. Since

$$\int_{M\setminus B_{R_k}^x} p^2(x,y,t_k)dy \leqslant \int_{M\setminus B_{R/2}^x} p^2(x,y,t_k)dy$$

and the right-hand side integral approaches to 0 as $t_k \to 0$ (which follows from the fact that $p(x,y,t) \to 0$ as $t \to 0$ uniformly in $y \in \Omega \setminus B_{R/2}^x$) we obtain from (3.5)

$$\int_{M\setminus B_R^x} p^2(x,y,t)dy \leqslant \sum_{k=0}^{\infty}\frac{1}{f(t_k)}\exp\left(-\frac{(R_{k-1}-R_k)^2}{2(t_{k-1}-t_k)}\right)$$

or, applying (1.16) in the form $f(t) \leqslant \mathrm{const}_A 2^{k\nu} f(t_k)$ in order to estimate $f(t_k)$ via $f(t)$

$$\int_{M\setminus B_R^x} p^2(x,y,t)dy \leqslant \frac{\mathrm{const}_A}{f(t)}\sum_{k=0}^{\infty}\exp\left(ck - \frac{2^{k-1}}{(k+2)^4}\frac{R^2}{t}\right) \tag{3.6}$$

r some positive absolute constants c_1, c_2 the following inequality

$$\frac{2^{k-1}}{(k+2)^4} > c_1 k + c_2$$

-ugn then c_2 can be taken 0.006). Therefore, putting $X = R^2/t$

$$ck - \frac{2^{k-1}}{(k+2)^4} X < (c - c_1 X)k - c_2 X \leqslant -ck - c_2 X$$

provided

$$X \geqslant 2\frac{c}{c_1}\ . \tag{3.7}$$

Hence, under assumption (3.7) the sum on the right-hand side in (3.6) does not exceed the geometric series

$$\sum_{k=0}^{\infty} \exp(-ck - c_2 X) = \frac{e^{-c_2 X}}{1 - e^{-c}}$$

and (3.6) acquires the from

$$\int_{M \setminus B_R^x} p^2(x,y,t)dy \leqslant \frac{\mathrm{const}_A}{f(t)} \exp\left(-c_2 \frac{R^2}{t}\right)$$

which was to be proved.

Finally, let $R^2/t < 2\frac{c}{c_1}$, then

$$\int_{M \setminus B_R^x} p^2(x,y,t)dy \leqslant \int_M p^2(x,y,t)dy \leqslant \frac{1}{f(t)} \leqslant \frac{\mathrm{const}_A}{f(t)} \exp\left(-c_2 \frac{R^2}{t}\right)$$

where we have applied (3.4) and boundedness of R^2/t.

Thus, we have proved (3.1) with $D = D_0 \equiv 1/c_2$, and the more so is valid for $D > D_0$. □

Next, we proceed with the proof of the Theorem 1.1. There are two points on which we shall focus our attention. First, how to pass from the integral (3.1) to E^D and second, how to diminish the constant D so that it is arbitrarily close to 2.

First we prove (1.6) for a large D. To that end, let us observe that the integral E^{5D} is decomposed into the sum of the integrals

$$\int_{\{2^k R \leqslant r \leqslant 2^{k+1} R\}} p^2(x,y,t) \exp\left(\frac{r^2}{5Dt}\right) dy \tag{3.8}$$

for $k = 0, 1, 2, \ldots$ and of the integral

$$\int_{B_R^x} p^2(x,y,t) \exp\left(\frac{r^2}{5Dt}\right) dy \tag{3.9}$$

where R is an arbitrary positive number. We estimate the integral (3.9) from above by

$$\frac{1}{f(t)}\exp\left(\frac{R^2}{5Dt}\right)$$

which follows from (3.4) . The integral (3.8) does not exceed the following

$$\exp\left(\frac{4^{k+1}R^2}{5Dt}\right)\int_{\{r\geqslant 2^k R\}} p^2(x,y,t)dy \leqslant \frac{\mathrm{const}_A}{f(t)}\exp\left(-\frac{4^kR^2}{Dt}+\frac{4^{k+1}R^2}{5Dt}\right)$$
$$=\frac{\mathrm{const}_A}{f(t)}\exp\left(-\frac{4^kR^2}{5Dt}\right)$$

where we have applied Lemma 3.1 assuming that $D \geqslant D_0$. Adding all these inequalities we obtain

$$E^{5D}(x,t)\leqslant\frac{1}{f(t)}\exp\left(\frac{R^2}{5Dt}\right)+\frac{\mathrm{const}_A}{f(t)}\sum_{k=0}^{\infty}\exp\left(-\frac{4^kR^2}{5Dt}\right)$$

Taking here $R=\sqrt{Dt}$ and denoting $5D$ by D we get that for any $D\geqslant D_1\equiv 5D_0$

$$E^D\leqslant\frac{\mathrm{const}_A}{f(t)} \tag{3.10}$$

which coincides with (1.6) .

Finally, let us prove (1.6) for any D such that $2<D<D_1$. We apply again Lemma 3.2 and following from it the inequality (3.3) , on this occasion with the weight function $\xi(y,t)=\frac{r^2}{2(t+T)}$ for a positive T. Given t, we choose consecutively T and τ so that $2(t+T)=Dt$ i.e. $T=(D-2)t/2$ and $2(\tau+T)=D_1\tau$ i.e. $\tau=2T/(D_1-2)=\frac{D-2}{D_1-2}t<t$. Hence, (3.3) takes the form

$$E^D(x,t)\leqslant E^{D_1}(x,\tau)$$

whence, applying (3.10) to estimate the right-hand side we get

$$E^D(x,t)\leqslant\frac{\mathrm{const}_A}{f(\tau)}\ .$$

Since we have by (1.16) $f(\tau)\geqslant \mathrm{const}_A f((D-2)t)$ the desired estimate (1.6) follows from the inequality above. □

Proof of Corollary 1.1. Theorem 1.1 applies under the conditions of this corollary and gives (1.6) , or, using the notation E_0 instead of E^D

$$E_0(x,t)\leqslant\frac{\mathrm{const}_A}{f(\delta t)}\ .$$

By Theorem 1.2 we obtain that for any $m \geqslant 0$

$$E_m(x,t) \leqslant \frac{\text{const}_A C^m \delta^m}{f_m(\delta t)}$$

where δ^m in the numerator comes from integration of $f(\delta t)$ in t. Let us note that for $m \geqslant 1$

$$f_m(t) = \int_0^t \frac{(t-\tau)^{m-1}}{(m-1)!} f(\tau)d\tau > \int_{t/2}^t \frac{(t-\tau)^{m-1}}{(m-1)!} f(\tau) \geqslant f(t/2)\frac{(t/2)^m}{m!}$$

whence it follows that

$$f_m(t) \geqslant \frac{t^m}{A2^m m!} f(t)$$

and

$$E_m(x,t) \leqslant \frac{\text{const}_A (2C)^m m!}{t^m f(\delta t)} .$$

The analogous inequality holds for the point y. Applying finally Theorem 1.3 we obtain (1.13) . The form of the coefficient (1.14) comes from a remark that $2C\delta = \frac{D+16}{D-2}\min(D-2,1) < 20$ for any $D > 2$ and thereby $2C < 20\delta^{-1}$.

In order to prove (1.15) we apply (1.16) which implies $f(\delta t/2) \geqslant \text{const}_A \delta^\nu f(t)$ and (1.13) is transformed to

$$\left|\frac{\partial^m p}{\partial t^m}\right|(x,y,t) \leqslant \frac{\text{const}_{A,m}}{\delta^{\nu+m} t^m \sqrt{f(t)g(t)}} \exp\left(-\frac{r^2}{2Dt}\right) .$$

Let us put here $D = 2 + \min(1, \frac{t}{r^2})$. Since $D-2 \leqslant 1$ it follows that $\delta = D - 2 \leqslant \frac{t}{r^2}$ and we have evidently

$$\frac{r^2}{4t} - \frac{r^2}{2Dt} = \frac{\delta}{4D}\frac{r^2}{t} \leqslant \frac{t}{r^2}\frac{r^2}{4Dt} = \frac{1}{4D} \leqslant \frac{1}{8}$$

and

$$\frac{1}{\delta} = \max(1, \frac{r^2}{t}) < 1 + \frac{r^2}{t}$$

whence (1.15) follows.

REFERENCES

[1] **Aronson D.G.**, *Bounds for the fundamental solution of a parabolic equation*, Bull. of AMS, **73** (1967) pp. 890-896.

[2] **Chavel I.**, "Eigenvalues in Riemannian geometry", Academic Press, New York, 1984.

[3] **Cheng S.Y., Li P., Yau S.-T.**, *On the upper estimate of the heat kernel of a complete Riemannian manifold*, Amer. J. Math., **103** (1981) no.5, pp. 1021-1063.

[4] **Davies E.B.**, *Explicit constants for Gaussian upper bounds on heat kernels*, Amer. J. Math., **109** (1987) pp. 319-334.

[5] **Davies E.B.**, *Gaussian upper bounds for the heat kernel of some second-order operators on Riemannian manifolds*, J. Funct. Anal., **80** (1988) pp. 16-32.

[6] **Davies E.B., Pang M.M.H.**, *Sharp heat kernel bounds for some Laplace operators*, Qurt. J. Math., **40** (1989) pp. 281-290.

[7] **Davies E.B.**, "Heat kernels and spectral theory", Cambridge: Cambridge University Press, 1989.

[8] **Davies E.B.**, *Pointwise bounds on the space and time derivatives of the heat kernel*, J. Operator Theory, **21** (1989) pp. 367-378.

[9] **Grigor'yan A.**, *Heat kernel upper bounds on a complete non-compact manifold*, Revista Mathemática Iberoamericana, to appear.

[10] **Grigor'yan A.**, *Integral maximum principle and its applications*, Proc. of Edinburgh Royal Society, to appear.

[11] **Grigor'yan A.**, *Upper bounds of derivatives of the heat kernel on an arbitrary complete manifold*, preprint.

[12] **Kovalenko V.P., Semenov U.A.**, *Semigroups generated by an elliptic operator of second order*, (in Russian) in: Methods of Functional Analysis in Problems of Mathematical Physics, Kiev Inst. of Math., Ukranian Acad. of Sciences, 1987. pp. 17-36.

[13] **Porper F.O.**, *Estimates of the derivatives of the fundamental solution of a stationary parabolic divergence equation with constants that do not depend on the smoothness of the coefficients*, Soviet Math. Dokl., **18** (1977) pp. 1092-1096.

[14] **Porper F.O., Eidel'man S.D.**, *Two-side estimates of fundamental solutions of second-order parabolic equations and some applications*, Russian Math. Surveys, **39** (1984) no.3, pp. 119-178.

[15] **Ushakov V.I.**, *Stabilization of solutions of the third mixed problem for a second order parabolic equation in a non-cylindric domain*, Math. USSR Sb., **39** (1981) pp. 87-105.

[16] **Varopoulos N.Th.**, *Semi-groupes d'opérateurs sur les espaces L^p*, C. R. Acad. Sci. Paris Sér. I Math., **301** (1985) pp. 865-868.

[17] **Varopoulos N.Th.**, *Analysis on Lie groups*, J. Funct. Anal., **76** (1988) pp. 346-410.

[18] **Varopoulos N.Th.**, *Small time Gaussian estimates of heat diffusion kernel.I. The semigroup technique*, Bull. Sci. Math.(2), **113** (1989) no.3, pp. 253-277.

Integrals of analytic functions along 2 curves

R.R. HALL and W.K. HAYMAN
Department of Mathematics
University of York
York YO1 5DD
England

1. Introduction Suppose that C_0 is a circle and that C is a rectifiable Jordan curve whose interior D contains the interior of C_0, so that $C_0 \subset \overline{D} = C \cup D$. Let $\mathcal{F}$ be a class of functions defined in $\overline{D}$. We ask for the smallest constant $K = K(\mathcal{F}, C)$ such that

$$\int_{C_0} u(z)|dz| \leq K \int_C u(z)|dz| \tag{1.1}$$

whenever $u \in \mathcal{F}$. If no such K exists we set $K(\mathcal{F}, C) = \infty$.

Harold Shapiro conjectured in a lecture in York that if

$$\mathcal{P} = \{u : u(z) = |P(z)|, \text{ where } P \text{ is a polynomial}\} \tag{1.2}$$

then $K(\mathcal{P}, C) \leq K_0$ where K_0 is an absolute constant.

In fact, Garnett, Gehring and Jones [1983] at about the same time proved a stronger form of Shapiro's conjecture in which C_0 is replaced by the intersection of an arbitrary circle with D. More recently Fernandez and Hamilton [1987] have obtained a corresponding result where C_0 is the intersection with $\overline{D}$ of an arbitrary chord-arc curve and Bishop and Jones [1990] have extended this further to the case when C_0 is merely regular in the sense of Ahlfors-David. In the latter two cases K depends on the chord-arc constant of C_0 or the constant of regularity respectively. However none of the above results contain explicit constants.

In this paper we prove Shapiro's original conjecture with

$$K_0 \leq e(3 + 4/\pi) = 11.6158....;$$

our method, which relies on the Hardy-Littlewood maximal theorem with an explicit constant and a simple projection lemma has, we hope, some independent interest.

We define

$$\mathcal{S}^+ = \{u : u \text{ is subharmonic (s.h.), upper semicontinuous (u.s.c.) and } u \geq 0 \text{ in } D\} \tag{1.3}$$

and

$$\mathcal{A} = \{v : v = e^u, u \in \mathcal{S}^+\}. \tag{1.4}$$

K. GowriSankaran et al. (eds.), Classical and Modern Potential Theory and Applications, 253–265.

In a companion paper (Hayman [1993]) a characterization of $K(\mathcal{S}^+, C)$ is given and consequently necessary and sufficient conditions for $K(\mathcal{S}^+, C)$ to be finite. In particular if C has an interior angle of opening greater than π at any point then $K(\mathcal{S}^+, C) = \infty$. On the other hand we also prove in Hayman [1993]

THEOREM A. *We have for every rectifiable curve C*

$$K(\mathcal{P}, C) = K(\mathcal{A}, C). \tag{1.5}$$

The results of the present paper are based on the Hardy-Littlewood Maximal Theorem which we require in the following form [Hardy and Littlewood, 1930].

THEOREM B. *Suppose that S_0 is a closed subset of $\{z : |z| \leq 1\}$, which lies except possibly for $z = 1$ in $\{z : |z| < 1\}$ and in some angle $\{z : |\arg(1-z)| < \frac{\pi}{2} - \delta\}$, where $\delta > 0$. We define*

$$S_\phi = \{z : ze^{-i\phi} \in S_0\}. \tag{1.6}$$

Suppose that u is s.h., continuous and $u \geq 0$ in $\{z : |z| \leq 1\}$ and that $p > 1$. We define

$$U(\phi) = \sup\{u(z) : z \in S_\phi\}. \tag{1.7}$$

Then

$$\int_0^{2\pi} U(\phi)^p d\phi \leq A(p, S_0) \int_0^{2\pi} u(e^{i\phi})^p d\phi, \tag{1.8}$$

where the constant $A(p, S_0)$ depends on p, S_0 only.

In our case we are particularly interested in the domain

$$S_0 = \left\{ z : z = re^{i\theta}, |\theta| \leq \frac{\pi}{2}, \ 0 \leq r \leq \tan\left(\frac{\pi}{4} - \frac{|\theta|}{2}\right)\right\}. \tag{1.9}$$

For a general S_0 the best constant $A(p, S_0)$ decreases with increasing p. For if $q > p$ and u is s.h. and $u \geq 0$ then so is $u^{q/p}$ and we apply (1.8) with u replaced by $u^{q/p}$, to obtain

$$A(q, S_0) \leq A(p, S_0).$$

It follows that the limit

$$A(\infty, S_0) = \lim_{p \to \infty} A(p, S_0) \tag{1.10}$$

always exists.

We may now state our main result.

THEOREM 1. *We have*

$$K(\mathcal{P}, C) = K(\mathcal{A}, C) \leq K_0 \leq 2A(\infty, S_0) \leq e\left(3 + \frac{4}{\pi}\right) = 11.6158.... \tag{1.11}$$

where S_0 is the set (1.9).

Thus Shapiro's conjecture is verified with $K_0 < 12$. The rest of the paper will be divided in two parts. In the first part we establish the inequality

$$K(\mathcal{P}, C) \leq 2A(\infty, S_0). \tag{1.12}$$

In the second part we develop numerical estimates for $A(p, S_0)$ in Theorem B which yield in particular the bound

$$A(\infty, S_0) \leq e\left(\frac{3}{2} + \frac{2}{\pi}\right), \tag{1.13}$$

with S_0 as in (1.9).

2. A reformulation of the problem

Suppose that $Q(z)$ is a polynomial and $P(z) = Q'(z)$. We also assume initially that C is an analytic Jordan curve, containing C_0 in its interior D. Then

$$l = l(C) = \int_C |P(z)|\ |dz| = \int_C |Q'(z)|\ |dz| \tag{2.1}$$

is the length of the path γ which is the image of C by the polynomial map $\xi = Q(z)$. Similarly

$$l_0 = l(C_0) = \int_{C_0} |P(z)|\ |dz| = \int_{C_0} |Q'(z)|\ |dz| \tag{2.2}$$

is the length of the image γ_0 of C_0 by $Q(z)$. In this form the problem is conformally invariant. We make a conformal mapping

$$z = \psi(w) \text{ of } \{w : |w < 1\} \text{ onto } D, \text{ s.t. } \psi(0) = 0. \tag{2.3}$$

Since C is an analytic Jordan curve the map has an analytic extension to $\{w : |w| \leq 1\}$ which is a homeomorphism from $\{w : |w| \leq 1\}$ to $\overline{D}$. In particular

$$\xi = \phi(w) = Q(\psi(w)) \tag{2.4}$$

gives a map of $c = \{w : |w| = 1\}$ onto γ and a map of an analytic Jordan curve c_0 in the interior $\{w : |w| < 1\}$ onto γ_0. These maps yield reparametrizations of γ, γ_0 respectively which do not alter the lengths l, l_0. It is thus sufficient to compare the integrals

$$l_0 = \int_{c_0} |\phi'(w)|\ |dw| \text{ and } l = \int_c |\phi'(w)|\ |dw|. \tag{2.5}$$

Next we note that, by a theorem of Jorgensen [1956], c_0 is a hyperbolically convex curve w.r.t. the disc $\{w : |w| < 1\}$. This means that if D_0 is the interior of c_0 and $\overline{D} = D_0 \cup c_0$ then if w_1, w_2 are two points of $\overline{D}_0$ the hyperbolic geodesic arc joining w_1 to w_2 lies in $\overline{D}_0$. In fact Jorgensen showed that if a circle C_0 lies in a simply-connected domain D then C_0 is convex w.r.t. the hyperbolic metric in D. This property is invariant under conformal mapping since the hyperbolic metric is conformally invariant.

In order to prove Theorem 1 we need to compare the two integrals in (2.5). We recall the following facts about the hyperbolic metric. (See for example Hayman [1989], pp. 682-685.)

If w_1, w_2 are two points in $\{w : |w| < 1\}$ then the hyperbolic distance w.r.t. this disc is

$$d(w_1, w_2) = \inf \int_\gamma \frac{|dw|}{1 - |w|^2},$$

where the infimum is taken over all rectifiable paths $\gamma \subset \{w : |w| < 1\}$ joing w_1 to w_2. The unique geodesic, for which this infimum is attained, is an arc of a circle orthogonal to the unit circle. This is obvious when $w_1 = 0$ and γ is the segment $[0, w_2]$, and in the general case the result is proved by means of a bilinear map from the unit disc onto itself, which maps geodesics onto geodesics.

To obtain our conclusion we shall make use of a "conformal projection" $W = \pi(w)$ from c_0 to c. The relevant properties of $\pi(w)$ are contained in

LEMMA 1. *Let c_0 be a hyperbolically convex closed curve lying in $\{w : |w| < 1\}$ and surrounding the origin. There exists a conformal projection $W = \pi(w)$ from c_0 to the unit circle $W = e^{i\phi}$ with the following properties.*

(i) As w moves along c_0 in the positive sense, $W = \pi(w)$ moves around the unit circle in the positive sense, having jump discontinuities only at points where the tangent to c_0 has discontinuities. Further if w_1, w_2 are the endpoints of an arc γ of c_0 described in the positive sense and if $W_1 = \pi(w_1)$, $W_2 = \pi(w_2)$ then the lengths l, L of γ and the arc W_1W_2 of the unit circle satisfy $l \leq 2L$.

(ii) If $W = \pi(w) = e^{i\phi}$ then w lies in S_ϕ given by (1.6) and (1.9).

Let w be a point of c_0. By hypothesis there exists a geodesic through w and joining c_0 to the unit circle, orthogonal to the unit circle and to c_0, except at the countably many points where c_0 has a corner. Let this unique geodesic be $c_1(w)$. It is a circular arc, not meeting the interior of c_0. If $c_1(w)$ meets the unit circle at $W = e^{i\phi}$ then we define $\pi(w) = W$. We proceed to show that (i) and (ii) hold. We first prove (i). A hyperbolically convex curve can be approximated from the inside by a hyperbolic polygon, i.e. a Jordan curve consisting of a finite number of geodesic arcs and such that the interior angle at any corner is less than π. Thus it is enough to prove (i) when c_0 is such a polygon. Evidently, at a corner of c_0, ϕ has a positive jump, since the exterior angle is greater than π. We may therefore assume that w_1 and w_2 are points on the same geodesic γ. In this case it is enough to show that if $W = \pi(w)$ is the conformal projection from γ to the unit circle then

$$\left|\frac{dW}{dw}\right| > \frac{1}{2}. \tag{2.6}$$

Suppose that the geodesic γ (extended if necessary) meets the unit circle at $a = e^{i\theta_1}$ and $b = e^{i\theta_2}$. We make the transformation

$$z = \frac{w-a}{b-w}, \qquad Z = \frac{W-a}{b-W}.$$

Since γ separates W from the origin we have $\theta_1 < \phi < \theta_2 < \theta_1 + \pi$. As W moves round the unit circle from a to b, Z describes a ray,

$$Z = re^{-i\lambda},\ 0 < r < \infty,\ \lambda = \frac{1}{2}(\theta_2 - \theta_1) \in \left(0, \frac{\pi}{2}\right),$$

and γ maps onto the ray

$$z = re^{i(\frac{\pi}{2}-\lambda)},\ 0 < r < 1.$$

Also $c_1(w)$ transforms to a circular arc $|z| = t$ joining, and orthogonal to, these rays. Plainly when $W = \pi(w)$ we have $z = iZ$ that is

$$\frac{w-a}{b-w} = \frac{i(W-a)}{b-W},$$

whence

$$\frac{dw}{(b-w)^2} = \frac{idW}{(b-W)^2},$$

$$\frac{dW}{dw} = \frac{(b-W)^2}{i(b-w)^2} = \frac{(1+z)^2}{i(1+Z)^2} = \frac{(1+iZ)^2}{i(1+Z)^2}.$$

Thus

$$\left|\frac{dW}{dw}\right| = \frac{1+2r\sin\lambda + r^2}{1+2r\cos\lambda + r^2} > \frac{1+r^2}{(1+r)^2} \geq \frac{1}{2},$$

which is (2.6), and (i) is proved.

To prove (ii), suppose that $w = r$ is a point on the real axis and that $\pi(w) = e^{i\phi}$. Since the origin lies inside c_0 so does the segment $[0, r]$. Hence the geodesic $c_1(w)$ makes an angle between $\mp\frac{\pi}{2}$ with the segment $[r, 1]$. The extreme cases $\mp\frac{\pi}{2}$ correspond to the extreme values of ϕ, and $c_1(w)$ is an arc of the circle with $[r, 1/r]$ as diameter, meeting the unit circle orthogonally at $e^{i\phi}$. We see from the geometry that in this case $\tan\left(\frac{\pi}{4} - \frac{|\phi|}{2}\right) = r$ so that generally $\pi(r) = e^{i\phi}$ implies $\tan\left(\frac{\pi}{4} - \frac{|\phi|}{2}\right) \geq r$ and by rotation $\pi(re^{i\theta}) = e^{i\phi}$ implies $\tan\left(\frac{\pi}{4} - \frac{1}{2}|\phi - \theta|\right) \geq r$. This proves Lemma 1.

3. Extension of a Theorem of Fejér and F. Riesz

We shall deduce Theorem 1 from

THEOREM 2. *Suppose that $u(z)$ is s.h. and non-negative in the disc $\{z : |z| < 1\}$ and continuous in the closed disc $\{z : |z| \leq 1\}$, and that (1.7) holds, where $p > 1$. Let γ be a hyperbolically convex curve lying in $\{z : |z| < 1\}$ and surrounding the origin. Then*

$$\int_\gamma u(z)^p |dz| \leq 2A(p, S_0) \int_0^{2\pi} u(e^{i\phi})^p d\theta \tag{3.1}$$

where S_0 is the domain given by (1.9) and $A(p, S_0)$ is the constant in (1.8).

Let $\pi(z) : Z = e^{i\phi}$ be the projection from γ to the unit circle whose existence is asserted in Lemma 1, and let $U(\phi)$ be the maximal function given by (1.7). It follows from Lemma 1 (ii) that $z \in S_\phi$ and hence that

$$u(z) \leq U(\phi).$$

Moreover from Lemma 1 (i)

$$|dz| \leq 2|dZ| = 2d\phi.$$

Also the mapping $Z = \pi(z)$ is one-to-one from γ to a subset of the unit circle. Thus by (1.8)

$$\int_\gamma u(z)^p |dz| \leq 2 \int_0^{2\pi} U(\phi)^p d\phi \leq 2A(p, S_0) \int_0^{2\pi} u(e^{i\phi})^p d\phi.$$

This proves Theorem 2. We deduce

THEOREM 3. *Suppose that $f(z)$ is regular for $|z| < 1$ and continuous for $|z| \leq 1$, that γ is a hyperbolically convex curve lying in $\{z : |z| < 1\}$, and that $\delta > 0$. Then*

$$\int_\gamma |f(z)|^\delta |dz| \leq K_0 \int_{|z|=1} |f(z)|^\delta |dz|, \tag{3.2}$$

where

$$K_0 = 2A(\infty, S_0).$$

We set $u(z) = |f(z)|^{\delta/p}$ where $p > 1$. Then $u(z)$ satisfies the hypothesis of Theorem 2 and (3.1) yields

$$\int_\gamma |f(z)|^\delta |dz| \leq 2A(p, S_0) \int_{|z|=1} |f(z)|^\delta |dz|.$$

Letting p tend to ∞ we obtain (3.2). In the limiting case when γ reduces to an interval $[-r, r]$ described in each direction, and letting r tend to 1 we obtain

$$\int_{-1}^{1} |f(r)|^\delta dr \leq \frac{1}{2} K_0 \int_{|z|=1} |f(z)|^\delta |dz|.$$

This inequality was obtained by Fejér and Riesz [1921] but with the sharp constant $\frac{1}{2}$ instead of $\frac{1}{2}K_0$.

We should also mention a generalization due to F. Carlson [1943] in which the right hand side of (3.2) is replaced by

$$\frac{1}{\pi}\int_{|z|=1} |f(z)|^{\delta} V(z)|dz|,$$

where

$$V(z) = \int_{\gamma} \left| d_{\xi} \arg(z-\xi) \right|.$$

This result neither contains nor is contained in Theorem 3 since on a hyperbolically convex curve γ the oscillation $V(z)$ may be uniformly large for $|z| = 1$. On the other hand curves for which $V(z)$ is bounded need not be hyperbolically convex, and the constant $1/\pi$ is sharp in Carlson's Theorem.

To prove Theorem 1 we suppose first that C is analytic and that C_0 lies in the interior of C. Then the map $z = \psi(w)$ in (2.3) is analytic for $|w| \leq 1$ and so are $\phi(w)$ and $\phi'(w)$, given by (2.4). We apply (3.2) with $\phi'(w)$ instead of $f(z)$ and $\delta = 1$ and deduce that

$$l_0 \leq K_0 l$$

where l, l_0 are given by (2.1), (2.2) or (2.5). Since P is an arbitrary polynomial we obtain (1.11), apart from the bound for $A(\infty, S_0)$.

If C is an arbitrary rectifiable curve with interior D, we can approximate C from inside D by an analytic curve C' whose length is less than that of C. (See e.g. Collingwood and Lohwater [1966], p.50.) Then if $P(z)$ is continuous in $\overline{D}$ and in particular if $P(z)$ is a polynomial,

$$\int_{C'} |P(z)|\,|dz| \text{ approximates } \int_C |P(z)|\,|dz|.$$

At the same time, we replace C_0 by a concentric smaller circle C_0'. By what we have just proved we obtain

$$\int_{C_0'} |P(z)|\,|dz| \leq K_0 \int_{C'} |P(z)|\,|dz|,$$

and letting C_0' tend to C_0 and C' tend to C we obtain

$$\int_{C_0} |P(z)|\,|dz| \leq K_0 \int_C |P(z)|\,|dz|.$$

A similar argument yields

THEOREM 4. *Suppose that C is a Jordan curve with interior D and that $\overline{D}$ contains a circle C_0. Suppose that $f(z)$ is continuous in $\overline{D}$ and regular in D and maps C onto a path of finite length l. Then the image of C_0 has length l_0 where $l_0 \leq K_0 l$.*

We again approximate C, C_0 from the inside by C', C_0' where C_0' is concentric with C_0 and C' is now the image of $|w| = r$ when $z = \phi(w)$ maps $\{w : |w| < 1\}$ onto the interior of C'. Then $f(z)$ is regular inside and on C' and so Theorem 3 yields

$$\int_{C_0'} |f'(z)|\,|dz| \le K_0 \int_{C'} |f'(z)|\,|dz| = K_0 \int_0^{2\pi} \left|\frac{\partial}{\partial\theta} f\{\phi(re^{i\theta})\}\right| r\,d\theta.$$

The right hand side does not exceed, but approaches as $r \to 1$, the length of the image $|w| = 1$ by $g(w) = f\{\phi(w)\}$, (the argument of Collingwood and Lohwater [loc. cit.] extends to this case). A similar conclusion holds for the left hand side. This yields Theorem 4.

An example is given in Hayman [1993] to show that we cannot choose $K_0 = 1$ in Theorem 1 even if $|f(z)|^\delta = |z+\sqrt{2}|$ and C is convex. Another shows that we cannot replace $|P(z)|$ by $u(z)^p$ where $u(z)$ is s.h. and non-negative in D and continuous in $\overline{D}, p \le 2$, and K_0 is replaced by any finite constant. The case $p > 2$ remains open. On the other hand Theorem A shows that we can replace $P(z)$ by $\exp\{u(z)\}$ where $u(z)$ is s.h. in D and u.s.c. in $\overline{D}$.

4. Inequalities for subharmonic and maximal functions

In this section we give an upper bound for the non-negative subharmonic function $u(z)$ in $|z| < 1$ in terms of a suitable maximal function of its boundary values. We assume that $u(z)$ is continuous in the closed disc $\{z : |z| \le 1\}$ and we define the maximal function

$$M(\phi) = \sup\left\{\frac{1}{|I|}\int_I u(e^{i\gamma})d\tau : \phi \in I\right\}, \tag{4.1}$$

where the supremum is taken over intervals I containing ϕ and $|I|$ denotes the length of I. No restriction is imposed on I in (4.1) other than $|I| > 0$; however it is clear that we may assume $|I| \le 2\pi$. Our object is to show that, with $U(\phi)$ as in (1.7) subject to (1.9), we have

$$U(\phi) \le \left(\frac{3}{4} + \frac{1}{\pi}\right) M(\phi) \tag{4.2}$$

and we assume, as we may by rotation, that $\phi = 0$. We begin with an upper bound for $u(z)$ valid throughout the open disc: we write $z = re^{i\theta}$ where $0 \le r < 1$ and $|\theta| \le \pi$, also $M = M(0)$.

LEMMA 2. *In the circumstances described above we have*

$$u(z) \leq M(1 + g(r,\theta)) \tag{4.3}$$

where

$$g(r,\theta) = \frac{|\theta|}{2\pi} \cdot \frac{1+r}{1-r} - \frac{1}{\pi} \arctan\left(\frac{1+r}{1-r} \tan\frac{|\theta|}{2}\right), \tag{4.4}$$

(the right hand side being interpreted as $r/(1-r)$ when $|\theta| = \pi$). Provided that $0 < |\theta| < \pi$ we may write this in the equivalent form

$$g(r,\theta) = \frac{1}{\pi}\left\{\frac{|\theta|/2}{\tan(|\theta|/2)} \tan\left(\psi + \frac{|\theta|}{2}\right) - \psi - \frac{|\theta|}{2}\right\}, \tag{4.5}$$

where ψ denotes the angle ($\psi \geq 0$ always) between the lines joining 1 to the origin and to the point z.

We may assume that $\theta \geq 0$ by symmetry, indeed that $0 < \theta < \pi$, leaving the cases $\theta = 0$, π as simple extensions of our argument. The equivalence of (4.4) and (4.7) then is immediate from the sine rule; we have

$$\frac{1+r}{1-r} = \frac{\sin(\psi+\theta) + \sin\psi}{\sin(\psi+\theta) - \sin\psi} = \frac{\tan(\psi + \frac{\theta}{2})}{\tan\frac{\theta}{2}}\,. \tag{4.6}$$

Let $P(r,\tau)$ denote the Poisson kernel. Then

$$\begin{aligned} u(re^{i\theta}) &\leq \frac{1}{2\pi}\int_0^{2\pi} P(r,\theta-\tau)u(e^{i\tau})d\tau \\ &= \frac{1}{2\pi}\int_0^{R_0}\int_{E(R)} u(e^{i\tau})d\tau dR, \end{aligned} \tag{4.7}$$

where

$$R_0 = \max P(r,\theta-\tau) = P(r,0),$$

and

$$E(R) = \{\tau : P(r,\theta-\tau) \geq R\}/2\pi\mathbf{Z}. \tag{4.8}$$

We take $E(R)$ to be an interval centred at θ. It contains the point 0 if and only if $R \leq R_1 = P(r,\theta)$, and in this case we deduce from (4.1) that

$$\int_{E(R)} u(e^{i\tau})d\tau \leq M|E(R)|. \tag{4.9}$$

When $R > R_1$ we put $F(R) = E(R)U[0,\xi]$ where

$$P(r,\theta-\xi) = R, \quad 0 < \xi \leq \theta. \tag{4.10}$$

Then $F(R)$ is an interval containing 0 and (4.1) yields

$$\int_{E(R)} u(e^{i\tau})d\tau \leq \int_{F(R)} u(e^{i\tau})d\tau \leq M|F(R)|. \tag{4.11}$$

From (4.7), (4.9) and (4.11) we obtain

$$\begin{aligned} u(re^{i\theta}) &\leq \frac{M}{2\pi}\left\{\int_0^{R_1} |E(R)|dR + \int_{R_1}^{R_0} |F(R)|dR\right\} \\ &= \frac{M}{2\pi}\left\{\int_0^{R_0} |E(R)|dR + \int_{R_1}^{R_0} \xi(R)dR\right\}, \end{aligned} \tag{4.12}$$

where $\xi = \xi(\mathbf{R})$ is implicit in (4.10).

Since $P(r, \theta - \tau)/2\pi$ is a probability mass the first integral in the latter curly brackets equals 2π. In the second integral we substitute $R = P(r, \theta - \xi)$ to obtain

$$u(re^{i\theta}) \leq M\left\{1 + \frac{1}{2\pi}\int_0^{\theta} \xi \frac{d}{d\xi} P(r, \theta - \xi)d\xi\right\} \tag{4.13}$$

and the second term inside the brackets is

$$\frac{\theta}{2\pi} P(r, 0) - \frac{1}{2\pi}\int_0^{\theta} P(r, \theta - \xi)d\xi = g(r, \theta), \tag{4.14}$$

which yields (4.3) as required. This proves Lemma 2.

We require the supremum of $g(r, \theta)$ for $z = re^{i\theta} \in S_0$. We denote the function on the right hand side of (4.5) by $h(\psi, \theta)$ and we begin by showing that h is an increasing function of ψ: by (4.6) this means that g is an increasing function of r. We have (assuming $\theta > 0$)

$$\pi \frac{\partial h}{\partial \psi} = \frac{\theta/2}{\tan(\theta/2)} \sec^2\left(\psi + \frac{\theta}{2}\right) - 1. \tag{4.15}$$

From (4.6), $\psi + \frac{\theta}{2} < \frac{\pi}{2}$ whence

$$\cos^2\left(\psi + \frac{\theta}{2}\right) < \cos^2 \frac{\theta}{2} < \frac{\theta/2}{\tan(\theta/2)} \tag{4.16}$$

because $\sin\theta < \theta$. Thus the right hand side of (4.15) is positive as required. Next, from (1.9) we have

$$r = \frac{\sin\psi}{\sin(\psi + |\theta|)} \leq \tan\left(\frac{\pi}{4} - \frac{|\theta|}{2}\right) \tag{4.17}$$

in S_0; equivalently $\psi \leq \frac{\pi}{4} - \frac{|\theta|}{2}$. Hence in S_0,

$$h(\psi, \theta) \leq h\left(\frac{\pi}{4} - \frac{|\theta|}{2}, \theta\right) = \frac{1}{\pi}\left(\frac{|\theta|/2}{\tan(|\theta|/2)} - \frac{\pi}{4}\right) \leq \frac{1}{\pi} - \frac{1}{4}. \tag{4.18}$$

We insert this upper bound into (4.3) to obtain (4.2).

5. Powers of functions and Maximal functions

We can now complete the proof of Theorem 1 by means of

THEOREM 5. *Suppose that $u(e^{i\phi}) = f(\phi)$ is a non-negative continuous function on $[0, 2\pi]$, that $u(z)$ is the harmonic extension of $u(e^{i\phi})$ into $|z| < 1$ and that $U(\phi)$ is defined by (1.7). If $C = C(S_0)$ is a constant such that for each positive α and every f we have*

$$\Big|\{\phi : U(\phi) > \alpha\}\Big| \le \frac{C}{\alpha}\int_0^{2\pi} f(\phi)d\phi \tag{5.1}$$

then for $1 < p < \infty$ we have

$$\int_0^{2\pi} U(\phi)^p d\phi \le C\left(\frac{p}{p-1}\right)^p \int_0^{2\pi} u(e^{i\phi})^p d\phi. \tag{5.2}$$

Theorem 5 and its proof are based on ideas from the interpolation of operators and were kindly supplied to us by Professor Robert Kaufman. We define the operator

$$U(\phi) = T(f(\phi))$$

on non-negative continuous functions f on $[0, 2\pi]$ and not that it has the following properties:

$$T(f_1 + f_2) \le T(f_1) + T(f_2), \tag{5.3}$$

$$T(f) \le T(g) \text{ if } f \le g, \tag{5.4}$$

$$T(f) = f \text{ if } f \text{ is constant.} \tag{5.5}$$

We also note that if S_0 is given by (1.9) then (5.1) holds with

$$C = \frac{3}{2} + \frac{2}{\pi}. \tag{5.6}$$

In fact if $M(\phi)$ is given by (4.1) we have for every positive α

$$|\{\phi : M(\phi) > \alpha\}| \le \frac{2}{\alpha}\int_0^{2\pi} u(e^{i\phi})d\phi. \tag{5.7}$$

This result is known but we include the proof for completeness. Let E be a closed subset of $\{M(\phi) > \alpha\}$. Each point of E lies in an open interval I, such that

$$\int_I u(e^{i\phi})d\phi > \alpha|I|,$$

and by Heine-Borel Theorem a finite set of such intervals $I_\nu : (\alpha_\nu, \beta_\nu)$ covers E. We assume that the I_ν are arranged in order of increasing α_ν and that none of them lies in the union of the others. Then $\alpha_\nu < \alpha_{\nu+1}$, $\beta_\nu < \beta_{\nu+1}$ and $\alpha_{\nu+2} \geq \beta_\nu$, since otherwise $I_{\nu+1} \subset I_\nu \cup I_{\nu+2}$. We deduce that no point is covered more than twice by the I_ν. Hence

$$|E| \leq \sum |I_\nu| < \frac{1}{\alpha} \sum \int_{I_\nu} u(e^{i\phi}) d\phi \leq \frac{2}{\alpha} \int_0^{2\pi} u(e^{i\phi}) d\phi.$$

This proves (5.7).

Suppose now that $\alpha > 0$ and that S_0 is given by (1.9). Then by (4.2), (5.7) we have

$$|\{\phi : U(\phi) > \alpha\}| \leq \left| \left\{ \phi : M(\phi) > \frac{2\alpha}{C} \right\} \right| \leq \frac{C}{\alpha} \int_0^{2\pi} u(e^{i\phi}) d\phi,$$

where C is given by (5.6). Thus (5.1) holds with this value of C. Assuming (5.2) we obtain Theorem B with

$$A(p, S_0) \leq \left(\frac{3}{2} + \frac{2}{\pi}\right) \left(\frac{p}{p-1}\right)^p, \quad A(\infty, S_0) \leq e \left(\frac{3}{2} + \frac{2}{\pi}\right)$$

which is (1.13). This proves Theorem 1.

It remains to deduce (5.2) from (5.1). We suppose that $1 < p < \infty$, define

$$b = 1 - 1/p, \tag{5.8}$$

and, given $f(\theta)$ and $0 < t < \infty$, we define

$$\begin{aligned} f^{bt}(\theta) &= \max\{f(\theta) - bt, 0\}, \\ f_{bt}(\theta) &= f(\theta) - f^{bt}(\theta). \end{aligned}$$

Then by (5.3)

$$T(f) \leq T(f^{bt}) + T(f_{bt}) \leq T(f^{bt}) + bt$$

since $f_{bt} \leq bt$, so that $T(f_{bt}) \leq bt$ by (5.4) and (5.5). Thus

$$\begin{aligned} |\{T(f) > t\}| &\leq |\{T(f^{bt}) > (1-b)t\}| \\ &\leq \frac{C}{(1-b)t} \int_0^{2\pi} f^{bt}(\theta) d\theta = \frac{C}{(1-b)t} \int_0^{2\pi} (f(\theta) - bt)^+ d\theta \end{aligned} \tag{5.9}$$

by (5.1) applied to f^{bt}, with $\alpha = (1-b)t$. Now (5.9) yields

$$\begin{aligned} \int_0^{2\pi} U(\phi)^p d\phi &= \int_0^{2\pi} T(f)^p d\phi = \int_0^\infty |\{T(f) > t\}| p t^{p-1} dt \\ &\leq \int_0^\infty p t^{p-1} dt \frac{C}{(1-b)t} \int_{f(\theta) \geq bt} (f(\theta) - bt) d\theta \\ &= \frac{Cp}{1-b} \int_0^{2\pi} d\theta \int_0^{f(\theta)/b} t^{p-2} (f(\theta) - bt) dt \\ &= \frac{C}{(1-b) b^{p-1} (p-1)} \int_0^{2\pi} f(\theta)^p d\theta. \end{aligned}$$

Using (5.8) we obtain (5.2).

We should like to express our gratitude to Harold Shapiro for suggesting this problem, to Bob Kaufman and Jangmei Wu for discussion and help and particularly for Bob's Theorem 5, and to David Hamilton for acquainting us with recent literature.

References

1. C.J. Bishop and P.W. Jones, Harmonic measure and arc length, *Annals of Mathematics (2)* **132** (1990), 511-547. M R 92c 30026.

2. F. Carlson, Quelques inégalités concernant les fonctions analytiques, *Arkiv för mat. astronomi och fysik* 29B 11 (1943), 6pp.

3. E.F. Collingwood and A.J. Lohwater, *The theory of cluster sets*, Cambridge University Press, 1966.

4. L. Fejér and F. Riesz, Über einige funktionentheoretische Ungleichungen, *Math. Zeit.* **11** (1921), 305-314.

5. J.L. Fernandez and D.H. Hamilton, Length of curves under conformal mappings, *Comment, Math. Helv.* **62** (1987), 122-134. M R 88i 30011.

6. J.B. Garnett, F.W. Gehring and P.W. Jones, Conformally invariant length sums, *Indiana University Math. J.* **32** (1983), 366-403.

7. G.H. Hardy and J.E. Littlewood, A maximal theorem with function-theoretic applications, *Acta Math.* **54** (1930), 81-116.

8. W.K. Hayman, *Subharmonic Functions II*, Academic Press, London, 1989.

9. W.K. Hayman, Integrals of subharmonic functions along 2 curves, *Indagationes*, 1993.

10. V. Jorgensen, On an inequality for the hyperbolic measure and its applications to the theory of functions, *Math. Scand.* **4** (1956), 113-124.

ON THE RESTRICTED MEAN VALUE PROPERTY FOR MEASURABLE FUNCTIONS

W. HANSEN, N. NADIRASHVILI
Fakultät für Mathematik
Universität Bielefeld
Universitätsstraße
D - 33615 Bielefeld
Germany

Let U be a domain in $\mathbb{R}^d$, $d \geq 1$, and $r > 0$ a real function on U such that $B(x, r(x)) := \{y \in \mathbb{R}^d : \|x - y\| < r(x)\} \subset U$ for every $x \in U$. A Lebesgue measurable function f on U which is bounded by some harmonic function $h \geq 0$ on U is called r-median if

$$f(x) = \int f \, d\lambda_{B(x,r(x))}$$

for every $x \in U$ (λ Lebesgue measure on $\mathbb{R}^d$, $\lambda_B = \lambda(B)^{-1} 1_B \lambda$ for every ball B).

In [4,5] we showed that every continuous r-median function on U is harmonic provided $U \neq \mathbb{R}^d$ and, if $d = 2$, $\complement U$ is not polar (for the case $U = \mathbb{R}^d$, $d \geq 2$, see [7]). The crucial step was an inequality $f \geq H\overline{f}_\alpha$ for lower h-bounded r-supermedian functions f (and the result on r-median functions followed looking at f and $-f$). It was obtained using the Schrödinger equation $\Delta u - \delta 1_A \rho^{-2} u = 0$ on U (δ small, A a suitable subset of U, $\rho = \text{dist}(\cdot, \complement U)$) and a transfinite sweeping generated by the kernel $T : (x, E) \mapsto \lambda_{B(x,r(x))}(E)$ (we could assume r to be measurable). It was the use of limits in this procedure which forced us to assume that the r-supermedian function f be l.s.c.

Nevertheless, a careful application of Lusin's theorem then allowed us to deduce a corresponding result for measurable functions assuming that $\inf r(K) > 0$ for every compact subset K of U. Unfortunately, however, for unbounded sets U this method led to an additional boundedness condition for r: We needed that for every compact K in U there exists a compact set L in U such that $B(x, r(x)) \cap K = \emptyset$ for every $x \in U \setminus L$.

The purpose of this note is to present an approach where the two cases

a) f l.s.c., no condition on r (except $B(x, r(x)) \subset U$),

b) r locally bounded away from zero, no continuity condition on f

are treated simultaneously. This is possible if we use the simple fact that already the sequence $(T^m)_{m \in \mathbb{N}}$ is sweeping all mass out of (compact subsets of) U if r is locally bounded away from zero, and hence no transfinite sweeping is needed in this case.

In the following let U denote a domain in $\mathbb{R}^d$ and assume that $\complement U$ is non-polar if $d \leq 2$. Let $\mathcal{P}(U)$ denote the convex cone of all continuous real potentials on U.

K. GowriSankaran et al. (eds.), Classical and Modern Potential Theory and Applications, 267–271.

Lemma 1. *Let S be a kernel on U and $\gamma, r : U \to]0,\infty[$ such that $B(y, r(y)) \subset U$, $S(y,\cdot) \geq \gamma(y)\lambda_{B(y,r(y))}$ for every $y \in U$ and the functions r, r^{-1}, γ^{-1} are locally bounded on U. Assume that $Sp \leq p$ for every $p \in \mathcal{P}(U)$. Then $\lim_{m\to\infty} S^m p = 0$ for every $p \in \mathcal{P}(U)$.*

Proof. Obviously, $S1 \leq 1$ and hence $\gamma \leq 1$. Fix $p \in \mathcal{P}(U)$. Then there exists a strictly positive $q \in \mathcal{P}(U)$ such that $f := p/q$ tends to zero at infinity with respect to U (i.e., $\{f \geq \alpha\}$ is a compact subset of U for every real $\alpha > 0$). We may assume without loss of generality that $B(y, 2r(y)) \subset U$ for every $y \in U$ (replace r by $r/2$ and γ by $\gamma 2^{-d}$). Define

$$T(y,\cdot) = \frac{q}{q(y)} S(y,\cdot) \quad \text{and} \quad \delta(y) = \frac{\inf q(B(y, r(y)))}{q(y)} \gamma(y).$$

Then $T1 = \frac{1}{q} Sq \leq 1$ and $T^m f = \frac{1}{q} S^m p$ for every $m \in \mathbb{N}$. Moreover, $T(y,\cdot) \geq \delta(y)\lambda_{B(y,r(y))}$ for every $y \in U$, $\delta \leq 1$ and δ^{-1} is bounded: Indeed, fix a compact subset K of U and let $L = \overline{\bigcup_{x\in K} B(x, r(x))}$. Then L is a compact set since r is locally bounded. If $U \neq \mathbb{R}^d$ then L is contained in the closed subset $\{y \in \mathbb{R}^d : \operatorname{dist}(y, K) \leq \operatorname{dist}(y, \complement U)\}$ of U. So L is a compact subset of U,

$$\inf \delta(K) \geq \frac{\inf q(L)}{\sup q(K)} \inf \gamma(K) > 0.$$

Since $Sp \leq p$, the sequence $(S^m p)$ is decreasing. So $(T^m f)$ is decreasing to a function $g \geq 0$. Clearly $Tg = g$ and we have to show that $g = 0$.

Assume the contrary and let

$$\alpha = \sup g(U), \qquad A = \{x \in U : \limsup_{y\to x} g(y) = \alpha\}.$$

Then $\alpha > 0$ and A is a non-empty compact subset of U since $g \leq f$ and $\{f \geq \alpha/2\}$ is compact. Fix $x \in A$ having maximal distance from the origin. Choose $x_n \in U$ such that $\lim_{n\to\infty} x_n = x$ and $\lim_{n\to\infty} g(x_n) = \alpha$. We may assume without loss of generality that the sequences $(\delta(x_n)), (r(x_n))$ converge, and then $\delta := \lim_{n\to\infty} \delta(x_n) > 0$, $r := \lim_{n\to\infty} r(x_n) > 0$, and $B(x, r(x)) \subset U$. Moreover, for every $n \in \mathbb{N}$,

$$\begin{aligned} g(x_n) = Tg(x_n) &= \delta(x_n)\lambda_{B(x_n,r(x_n))}(g) + (T(x_n,\cdot) - \delta(x_n)\lambda_{B(x_n,r(x_n))})(g) \\ &\leq \delta(x_n)\lambda_{B(x_n,r(x_n))})(g) + (1 - \delta(x_n))\alpha \end{aligned}$$

since $g \leq \alpha$, $T1 \leq 1$, and $\delta(x_n)\lambda_{B(x_n,r(x_n))} \leq T(x_n,\cdot)$. Therefore,

$$\alpha \leq \delta\lambda_{B(x,r(x))}(g) + (1-\delta)\alpha.$$

This implies that $\alpha \leq \lambda_{B(x,r(x))}(g)$ and hence $g = \alpha$ λ-a.e. on $B(x, r(x))$ contradicting our choice of g. □

Now fix a harmonic function $h \geq 1$ on U and take Y and $\mathcal{P}$ as in [4] (Y is a metrizable compactification of U, $\mathcal{P} \subset \mathcal{C}^+(Y)$ is a min-stable convex cone separating the points of Y, and $\mathcal{P}|_U$ is generated by 1 and a sequence (q_n/h) with $q_n \in \mathcal{P}(U)$). If S is a kernel on Y such that $Sp \leq p$ for every $p \in \mathcal{P}$ then the sequence $(S^m p)$ is decreasing for every $p \in \mathcal{P}$ and we may define a kernel S^∞ by

$$S^\infty p := \lim_{m\to\infty} S^m p \qquad (m \in \mathbb{N}).$$

Given a kernel S on U and a Borel subset A of U we define a kernel S_A on Y by

$$S_A(y,\cdot) = \begin{cases} S(y,\cdot), & y \in U \setminus A, \\ \varepsilon_y, & y \in A \cup (Y \setminus U). \end{cases}$$

Lemma 2. *Let A be a subset of U and S a kernel on U such that $Sp \leq p$ for every $p \in \mathcal{P}(U)$ and $\lim_{m\to\infty} S^m\varphi = 0$ for every continuous real function φ on U having compact support. Then $((S_A)^m(y,\cdot)|_A)_{m\in\mathbb{N}}$ is increasing to $(S_A)^\infty(y,\cdot)|_A$ for every $y \in Y$ and $(S_A)^\infty 1_{U\setminus A} = 0$. In particular, $S_A(S_A)^\infty = (S_A)^\infty$.*

Proof. The definition of S_A obviously implies that $1_A(S_A)^m(y,\cdot) \leq 1_A(S_A)^{m+1}(y,\cdot)$ for every $y \in Y$ and $m \in \mathbb{N}$. Define a kernel S'_A by

$$S'_A(y,\cdot) = \begin{cases} 1_{\complement A} S(y,\cdot), & y \in U \setminus A, \\ \varepsilon_y, & y \in A \cup (Y \setminus U). \end{cases}$$

By induction it is easily verified that, for every $y \in U \setminus A$ and $m \in \mathbb{N}$,

$$(S'_A)^m(y,\cdot) \leq S^m(y,\cdot)$$

and

$$(S_A)^m(y,B) = (S'_A)^m(y,B)$$

for every Borel subset B of $U \setminus A$. Fix $y \in U \setminus A$ and a continuous real function $\varphi \geq 0$ on U with compact support. Since $\lim_{m\to\infty} S^m\varphi(y) = 0$, we obtain that $\lim_{m\to\infty}(S'_A)^m\varphi = 0$ and hence

$$\lim_{m\to\infty} (S_A)^m(\varphi 1_{U\setminus A}) = 0.$$

Thus

$$(S_A)^\infty\varphi(y) = \lim_{m\to\infty} (S_A)^m(\varphi 1_A)(y).$$

This proves the claim for $y \in U \setminus A$. For $y \in A \cup (Y \setminus U)$ the statement of the lemma is trivial. □

Lemma 3. *Let A and S be as in Lemma 2 and let $f \geq 0$ be a Borel function on U such that $Sf \leq f$. Extend f by zero to a function on Y. Then $(S_A)^\infty f \leq f$.*

Proof. Clearly, $S_A f \leq f$, hence $(S_A)^m f \leq f$ for every $m \in \mathbb{N}$. So we conclude from Lemma 2 that

$$(S_A)^\infty f = \lim_{m\to\infty} (S_A)^m (f|_A) \leq \lim_{m\to\infty} (S_A)^m (f) \leq f.$$

□

From now on suppose that $T : (x, E) \mapsto \mu_x(E)$ is a kernel on U such that for some real numbers $0 < \eta < 1$, $\gamma > 0$, and $a > 0$ the following holds for every $x \in U$:

(i) $\mu_x(s) \leq s(x)$ for every superharmonic function $s \geq 0$ on U, $\mu_x(h) = h(x)$,

(ii) there exists $0 < r(x) \leq \operatorname{dist}(x, \complement U) =: \rho(x)$ such that $\mu_x \geq \gamma \lambda_{B(x,r(x))}$ and $(G^{\varepsilon_x} - G^{\mu_x}) 1_{\complement B(x,\eta r(x))} \lambda \leq a\rho^2 \mu_x$.

If $d \leq 2$ assume in addition that

(iii) $\sup_{x\in U} r(x)^{-2} \int_{B(x,r(x))} (G_U^{\varepsilon_x} - G_U^{\mu_x})\, d\lambda < \infty$

(this holds in particular if, for some $c \in \mathbb{R}^+$ and all $x \in U$, $\mu_x(\complement B(x, cr(x)) = 0$ (cf. [5])).

Recall that a lower h-bounded measurable function f on U is μ-median (μ-median resp.) provided $\mu_x(f) = f(x)$ ($\mu_x(f) \leq f(x)$ resp.) for every $x \in U$.

Given $0 < \alpha \leq 1/2$ and a lower h-bounded measurable function f on U we define $\overline{f}_\alpha$ on the Martin compactification $U \cup \partial^M U$ as in [5]. The representation of h by the Martin kernel,

$$h = \int K(z, \cdot)\, \chi(dz)$$

(where χ is a measure on the set $\partial_1^M U$ of minimal points in $\partial^M U$) leads to a kernel $H : \varphi \mapsto \int K(z, \cdot)\varphi\, \chi(dz)$.

Theorem. *Let f be a lower h-bounded μ-supermedian function on U. If f is l.s.c. or r is locally bounded away from zero then $f \geq H\overline{f}_\alpha$.*

Proof. For lower semi-continuous f this is the statement of Theorem 2.1 in [5]. In the proof of that theorem a suitable subset A was used to define a kernel

$$P_A(y, \cdot) = \begin{cases} \frac{h}{h(y)}\mu_y, & y \in U \setminus A \\ \varepsilon_y, & y \in A \cup (Y \setminus U). \end{cases}$$

Given $x \in U$, $\mathcal{M}_{P_A}(x)$ denotes the smallest w^*-closed, P_A-stable subset of $\mathcal{M}_{P_A}(x)$ containing ε_x and σ_x is the unique P_A-invariant element in $\mathcal{M}_{P_A}(x)$. The lower semi-continuity of f was only used to obtain that $\sigma_x(f/h) \leq (f/h)(x)$. Consequently, in order to prove the theorem above it suffices to show that $\sigma_x(f/h) \leq (f/h)(x)$ holds as well if r is locally bounded away from zero.

So assume that r^{-1} is locally bounded on U. We conclude from Lemma 1 that $P(y,\cdot) := \frac{h}{h(y)}\mu_y$ satisfies $\lim_{m\to\infty} P^m p = 0$ for every $p \in \mathcal{P}(U)$, hence $\lim_{m\to\infty} P^m\varphi = 0$ for every continuous real function φ on U having compact support. Since f is μ-supermedian, we know that $P(f/h) \le f/h$. For every $x \in U$, we conclude by Lemma 2 and Lemma 3 that $\mathcal{M}_{P_A}(x) = \{P_A^m : m = 0,1,2,\dots,\infty\}$ and $\sigma_x(f/h) \le f/h(x)$. □

Corollary. *Let f be a μ-median h-bounded function on U. If f is continuous or r is locally bounded away from zero then $f = H\overline{f}_a$. In particular, f is harmonic.*

Remark. The preceding arguments are valid also for $U = \mathbb{R}^d$, $d \ge 3$. They lead to the following result (for the details see [7]).

Theorem. *Let $(\mu_x)_{x\in\mathbb{R}^d}$ be a kernel on $\mathbb{R}^d$, $d \ge 3$, having the properties (i) and (ii) with $r \le \|\cdot\| + M_0$ for some $M_0 \in \mathbb{R}^+$. Let f be a μ-median bounded function on $\mathbb{R}^d$ and assume that f is continuous or that r is locally bounded away from zero. Then f is constant.*

Corollary. *Let $r > 0$ be a real function on $\mathbb{R}^d$ such that $r \le \|\cdot\| + M_0$ for some $M_0 \in \mathbb{R}^+$ and let f be a Lebesgue measurable bounded function on U which is r-median. Assume that f is continuous or that r is locally bounded away from zero. Then f is constant.*

REFERENCES

1. Baxter, J.R., *Restricted mean values and harmonic functions*, Trans. Amer. Math. Soc. **167** (1972), 451–463.
2. ———, *Harmonic functions and mass cancellation*, Trans. Amer. Math. Soc. **245** (1978), 375–384.
3. Fenton, P.C., *On sufficient conditions for harmonicity*, Trans. Amer. Math. Soc. **253** (1979), 139–147.
4. Hansen, W., Nadirashvili, N., *A converse to the mean value theorem for harmonic functions*, to appear in Acta Math..
5. ———, *Mean values and harmonic functions*, to appear in Math. Ann..
6. ———, *Littlewood's one circle problem*, to appear in J. London Math. Soc..
7. ———, *Liouville's theorem and the restricted mean value property*, Preprint 1993.
8. Heath, D., *Functions possessing restricted mean value properties*, Proc. Amer. Math. Soc. **41** (1973), 588–595.
9. Kellogg, O.D., *Converses of Gauss's theorem on the arithmetic mean*, Trans. Amer. Math. Soc. **36** (1934), 227–242.
10. Littlewood, J.L., *Some Problems in Real and Complex Analysis*, Hath. Math. Monographs, Massachusetts, 1968.
11. Netuka, I., *Harmonic functions and mean value theorems*, in Czech, Časopis Pěst. Mat. **100** (1975), 391–409.
12. Susanka, L., *Potential theoretic and stopping time methods for converse mean value theorems*, Thesis 1991.
13. Veech, W.A., *A zero–one law for a class of random walks and a converse to Gauss' mean value theorem*, Ann. of Math. **97** (1973), 189–216.
14. ———, *A converse to the mean value theorem for harmonic functions*, Amer. J. Math. **97** (1975), 1007–1027.

A CONSTRUCTIVE METHOD FOR UNIVALENT LOGHARMONIC MAPPINGS

WALTER HENGARTNER and JÉRÉMIE ROSTAND
Département de Mathématiques
Université Laval
Québec, P.Q., Canada
G1K 7P4

ABSTRACT. Univalent logharmonic mappings are connected with minimal surfaces whose Gauss maps are periodic. Using an extremality property we give a constructive method for some of them.

1. Introduction and motivation

Let Ω be a domain of the complex plane $\mathbf{C}$ and let $\mathbf{S}$ be a nonparametric minimal surface lying over Ω. Then $\mathbf{S}$ can be represented by a function $s = G(u, v)$, $w = u + iv \in \Omega$, and there is a univalent orientation-preserving harmonic mapping $w = F(z)$ from an appropriate domain D of $\mathbf{C}$ onto Ω which determines $\mathbf{S}$ in the following sense. The mapping F is a solution of the system of linear elliptic partial differential equation

$$\overline{F_{\bar{z}}} = AF_z \tag{1}$$

where A belongs to the linear space $H(D)$ of all analytic functions defined on D and, since we assumed that F is orientation-preserving, we have $|A| < 1$ on D. We shall call A the second dilatation function of F. The value $\frac{1+|A(z)|}{1-|A(z)|}$ is the quotient of the maximum value and the minimum value of the differential $|dF(z)|$ when dz varies on the unit circle. The representation of the minimal surface $\mathbf{S}$ is then given by the three real-valued harmonic functions (see e.g. [10] and [14]).

$$u(z) = Re\ F(z), v(z) = Im\ F(z), s(z) = Im \int^z \sqrt{A} F_z dz. \tag{2}$$

Since $(s_z)^2 = -\overline{F_{\bar{z}}} F_z = -A(F_z)^2$ on D, it follows that $\sqrt{A}$ belongs to $H(D)$. In particular, each zero of A is of even order.

Since the Riemannien metric of $\mathbf{S}$ is $ds^2 = |F_z|^2(1 + |A|)^2|dz|^2$, it follows that $x = Re\ z$ and $y = Im\ z$ are isothermal parameters for $\mathbf{S}$. Moreover, the exterior unit normal vector $\vec{\mathbf{n}}(z) = (\mathbf{n}_1(z), \mathbf{n}_2(z), \mathbf{n}_3(z))$, $\mathbf{n}_3(z) \geq 0$, to the minimal surface

K. GowriSankaran et al. (eds.), Classical and Modern Potential Theory and Applications, 273–291.

S (known as the Gauss mapping) depends only on the second dilatation function $A(z)$ of F. More precisely, we have

$$\vec{\mathbf{n}} = \left(2Im\,\sqrt{A}, 2Re\,\sqrt{A}, (1-|A|)\right)/(1+|A|). \tag{3}$$

The inverse of the stereographic projection of the Gauss mapping $\vec{\mathbf{n}}$, $\dfrac{i}{\sqrt{A(z)}}$, is called the Weierstrass parameter.

The following question arises: What are the domains D ? If ϕ is univalent and analytic and if F is univalent and harmonic, then the composition $F \circ \phi$ (whenever it makes sense) is a univalent harmonic mapping but $\phi \circ F$ is in general not harmonic. Hence, if F represents a minimal surface over Ω (in the sense of relation (2)), then $F(\phi)$ represents the same minimal surface but in other isothermal parameters. We say that two univalent harmonic mappings F_1 and F_2 are equivalent, if $F_2^{-1} \circ F_1$ defines a conformal mapping.

Suppose that Ω is a proper simply connected domain of **C**. Then, we may choose for D any proper simply connected domain of **C**. In particular $D = U$ or $D = \Omega$ are appropriate choices.

Next, let Ω be the pointed unit disk $U\backslash\{0\}$. Then, D is a doubly connected domain of **C** and at least one boundary component is a continuum. Indeed, if not, then F is an affine transformation and F(D) is not the pointed disk. Applying a univalent conformal pre-mapping, we may assume that D is the annulus $A(r,1) = \{z; r < |z| < 1\}$ for some $r \in [0,1)$ and that F maps the exterior boundary $\{z; |z| = 1\}$ onto the unit circle. If $r = 0$, then F extends to a univalent harmonic mapping defined on the unit disk U. If $r > 0$, then $F(z) = \frac{z - \frac{r^2}{\bar{z}}}{1-r^2}$ is an example of a univalent harmonic orientation-preserving mapping from $A(r,1)$ onto the pointed unit disk $A(0,1)$. Observe that there are several other harmonic mappings which have the same properties (For more details see e.g. [7] and [9]). Hence, one cannot simply choose $D = \Omega$ in order to characterize a nonparametric minimial surface by a univalent harmonic map F from D onto Ω.

Consider now the left half-plane $D = \{z; Rez < 0\}$ and let F be a univalent harmonic and orientation-preserving map defined on D which satisfies the relation:

$$F(z + i\alpha) = F(z) + i\beta \; for \; all \; z \in D$$

where α and β are real constants. Applying the transformation $\frac{2\pi}{\beta}.F(\frac{z\alpha}{2\pi})$ we may assume without loss of generality that $\alpha = \beta = 2\pi$, i.e. that

$$F(z + 2\pi i) = F(z) + 2\pi i \; for \; all \; z \in D \tag{4}$$

Whenever the limit $\lim\limits_{x->-\infty} ReF(z) = c$ exists for some $c \in [-\infty, \infty)$, we shall write $ReF(-\infty) = c$. By the same way, $A(-\infty) = c$ means that the limit $\lim\limits_{x->-\infty} A(z) = c$ exists.

Let UHP (univalent harmonic periodic) denote the class of all univalent harmonic orientation-preseving mappings defined on the left-hand plane $D = \{z; Re z < 0\}$ which satisfy

$$F(z + 2\pi i) = F(z) + 2\pi i \ for \ all \ z \in D$$

and

$$ReF(-\infty) = -\infty.$$

It follows then that the second dilatation function A is periodic, i.e. that $A(z + 2\pi i) = A(z)$ on D and therefore the Gauss map is also periodic. Observe that $A(-\infty)$ exists. Furthermore, it was shown in [3] that mappings in the class UHP admit the representation

$$F(z) = z + 2\beta x + H(z) + \overline{G(z)} \tag{5}$$

where:

(1) H and G belong to $H(D)$ such that

(1a) $G(-\infty) = 0$ and $H(-\infty)$ exists in $\mathbb{C}$ and

(1b) $H(z + 2\pi i) = H(z)$ and $G(z + 2\pi i) = G(z)$ on D.

(2) $$\left|\frac{G'(z) + \overline{\beta}}{1 + \beta + H'(z)}\right| < 1 \text{ on D}$$

(3) $\beta = \frac{\overline{A(-\infty)}(1+A(-\infty))}{1-|A(-\infty)|^2}$ and hence, $Re\beta > -\frac{1}{2}$.

Define

$$f(z) = e^{F(\log(z))}, \quad z \in U. \tag{6}$$

Then f is univalent on U and is a solution of the nonlinear elliptic partial differential equation

$$\overline{f_{\overline{z}}} = a\frac{\overline{f}}{f} f_z \tag{7}$$

where $a(z) = A\log(z)$ and hence $a \in H(U)$, $|a| < 1$ on U. We call a solution of (7) a logharmonic map. Observe that the family of all univalent logharmonic and orientation preserving mappings f satisfying $f(0) = 0$ is isomorphic to the class UHP. As we shall see, in many cases it is easier to work with logharmonic mappings even if the differential equation becomes nonlinear. In this paper we give a method to construct univalent logharmonic mappings onto a strictly starlike domain. Section 2 contains a brief survey on univalent logharmonic mappings. In section 3 we develop a constructive method and in section 4, we give an explicit example.

2. Univalent logharmonic mappings

DEFINITION 1. *Let D be a domain of the complex plane $\mathbf{C}$. We say that a function f defined on D is a* **logharmonic mapping** *if f is a solution of the nonlinear elliptic partial differential equation*

$$\overline{f_{\bar{z}}} = a \frac{\bar{f}}{f} f_z \tag{8}$$

where $|a(z)| < 1$ on D and where a belongs to $H(D)$.

Suppose that f is a nonconstant logharmonic mapping defined on D and fix a point $z_0 \in D$. By Lemma 2.2 in [1], there is a nonnegative integer m and a complex number γ, $Re\,\gamma > -1/2$, such that f admits in a certain neighborhood $V(z_0)$ of z_0 the following representation:

$$f(z) = (z - z_0)^m |z - z_0|^{2m\gamma} h_{z_0}(z) \overline{g_{z_0}(z)}. \tag{9}$$

The functions h_{z_0} and g_{z_0} are analytic and nonvanishing on $V(z_0)$. Note that $f(z_0) \neq 0$ if and only if $m = 0$ and that a univalent logharmonic mapping vanishes at z_0 if and only if $m = 1$.

Let f be a univalent logharmonic mapping defined on the unit disk U satisfying $f(0) = 0$. Then f can be represented in the form

$$f(z) = z|z|^{2\beta} h(z) \overline{g(z)} \tag{10}$$

where h and g are nonvanishing analytic functions on U such that $g(0) = 1$,
$\left| \dfrac{g' + \overline{\beta}}{1 + \beta + h'} \right| < 1$ on U and
$\beta = \dfrac{\overline{a(0)}(1 + a(0))}{1 - |a(0)|^2}$ and hence $Re\,\beta > -\frac{1}{2}$.
We always take the branch $1^\beta = 1$. The case $a \equiv 0$ implies that $g \equiv 1$ and that f is analytic. Hence, conformal mappings are also logharmonic. We may multiply f with any complex constant without changing the dilatation function; but no translations are allowed. Furthermore, we may apply the postmapping $\psi(w) = w|w|^{2\tau}$, $Re\,\tau > -\frac{1}{2}$ in order to get a new univalent logharmonic mapping which is a solution of (8) with respect to another a(z). The same holds if we apply a conformal premapping.

DEFINITION 2. *Let Ω be a domain of $\mathbf{C}$ containing the origin.*
a) We say that Ω is **starlike** *(with respect to the origin) if $w \in \Omega$ implies that $tw \in \Omega$ for all $t \in [0,1]$.*
b) We say that Ω is **strictly starlike** *if each radial line hits the boundary $\partial\Omega$ of Ω in exactly one finite point.*
c) Let f be a univalent analytic mapping from U onto Ω normalized by $f(0) = 0$ and $f'(0) = 1$. Then f belongs to the class S^* if Ω is starlike.

REMARK 1: If $f \in S^*$ then $f_r(z) = \frac{1}{r}f(rz) \in S^*$ and $f_r(U)$ is a strictly starlike domain.

We start with the following mapping Theorem. The existence part stated here is a weak version of a result given in [1] and the uniqueness was shown for mappings onto strictly starlike domains in [3].

THEOREM 1. *Given $a \in H(U)$, $|a| \leq k < 1$ on U and let Ω be a simply connected Jordan domain of $\mathbf{C}$ containing the origin. Then there is a univalent solution $f(z) = z|z|^{2\beta}h(z)\overline{g(z)}$ of (8) such that $f(U) = \Omega$, $g(0) = 1$ and $h(0) > 0$. If, in addition, Ω is strictly starlike, then f is uniquely determined.*

The proof of the existence uses the general nonconstructive approach due to Boyarski [4]. It is hence desirable to obtain a constructive method in order to approximate (locally uniformly) solutions having the properties of Theorem 1. The proof of uniqueness is based on the following subordination principle. Since it is an essential part of what comes later, we repeat the proof of it.

THEOREM 2. *(Subordination principle) Let Ω be a strictly starlike domain and let $a \in H(U)$, $|a| \leq k < 1$ on U be given. Let $F = z|z|^{2\beta}H \cdot \overline{G}$ be a univalent solution of (8) such that $F(U) = \Omega$, $H(0) > 0$ and $G(0) = 1$. Finally, let $f = z|z|^{2\beta}h \cdot \overline{g}$ be any other solution of (8), with respect to the given a and assume that $f(U) \subset \Omega$, $g(0) = 1$ and $h(0) > 0$. Then we have $h(0) \leq H(0)$ and equality holds if and only if $f = F$.*

Proof: Suppose that $h(0) > H(0)$ and define

$$L(z) = \frac{2H(0)}{H(0) + h(0)} \cdot \frac{f(z)}{F(z)}$$

Then L is a solution of (8) with respect to the same dilatation function $a(z)$ and

$$L(0) = \frac{2h(0)}{H(0) + h(0)}$$

is a finite real value > 1. Since $\frac{2H(0)}{H(0)+h(0)} < 1$, the function $f_1 = \frac{2H(0)}{H(0)+h(0)} f$ maps U onto a precompact domain of Ω. Therefore, there is an $r_0 \in (0,1)$ such that for all z, $r_0 < |z| < 1$, L omits the interval $(1, \infty)$ of the real axis. But this is a contradiction to the argument principle which states that

$$\frac{1}{2\pi} \oint_{|z|=r} d \arg(L(z) - L(0)) > 0$$

for all $r \in (0,1)$.

Suppose now that $h(0) = H(0)$ and that $f \neq F$. For $\epsilon \in (0,1)$, put $f_\epsilon = (1-\epsilon)\frac{f}{F}$. Then, for each $\epsilon > 0$, there is (as above) an $r_\epsilon \in (0,1)$ such that

$$J_\epsilon = \frac{1}{2\pi}\oint_{|z|=r} d\,\arg(f_\epsilon - 1) = 0 \quad \forall r \in (r_\epsilon, 1)$$

By the argument principle, we conclude that $J_\epsilon = 0$ for all $r \in (0,1)$. On the other hand, for each $r \in (0,1)$, J_ϵ converges to J_0 as ϵ tends to 0, which leads to a contradiction. ∎

There are domains Ω for which we can explicitly determine the set of all univalent logharmonic mappings f from U onto Ω with $f(0) = 0$. One such special case is $\Omega = U$ which was studied in [2]. Surprisingly, the set of all univalent logharmonic automorphisms defined on U can be identified by the set $\{Re\,\beta > -\frac{1}{2}\} \times S^*$.

THEOREM 3. *A function $f = z|z|^{2\beta}h\overline{g}$ is a univalent logharmonic (and hence orientation-preserving) mapping from U onto U if and only if $Re\,\beta > -\frac{1}{2}$, $g(z) = \frac{1}{h(z)}$ and $h(z) = \sqrt{\frac{\phi(z)}{z}}$ for some $\phi \in S^*$.*

EXAMPLES: For $\phi(z) \equiv z$ we get $f(z) = T(z) = z|z|^{2\beta}$ and for $\phi(z) = \frac{z}{(1-z)^2}$ we have $f(z) = K(z) = \frac{z(1-\overline{z})}{1-z}$ which maps $\partial U \setminus \{1\}$ onto the point $\{-1\}$ and the cluster set at $z = 1$ is ∂U.

Denote by S^*_{LH} the set of all univalent logharmonic mappings $f(z) = z|z|^{2\beta}h\cdot\overline{g}$ such that $g(0) = h(0) = 1$ and where $f(U)$ is a starlike domain. The following Theorem has been shown in [2].

THEOREM 4.
a) If $f = z|z|^{2\beta}h\cdot\overline{g} \in S^*_{LH}$ then $Re\,\beta > -\frac{1}{2}$ and $\phi = z\frac{h}{g} \in S^*$.
b) If $\phi \in S^*$ and $a \in H(U)$, $|a| < 1$, are given, then there is a unique $f = z|z|^{2\beta}h\cdot\overline{g} \in S^*_{LH}$ such that $\phi = z\frac{h}{g}$.

As an immediate consequence of Theorem 4 and Remark 1 we have

COROLLARY: If $f \in S^*_{LH}$, then $f_r(z) = \frac{1}{r}f(rz) \in S^*_{LH}$ and $f_r(U)$ is a strictly starlike domain.

REMARK 2: There is no analogous result for univalent harmonic mappings.

Let f be a univalent logharmonic mapping defined on U and let a be its (second) dilatation function. Then $f_1 = f|f|^{2\tau}$, $\tau = \frac{-\beta}{1+\beta+\overline{\beta}}$, is again a univalent logharmonic map defined on U and its dilatation function is $a_1(z) = \frac{a(z)-a(0)}{1-\overline{a(0)}a(z)}$. In particular, we have $a_1(0) = 0$ and hence, $f_1(z) = zh_1(z)\overline{g_1(z)}$.

The next result gives a sharp coefficient estimate and was proved in [3].

THEOREM 5. *Let $f = zh(z)\overline{g(z)} \in S^*_{LH}$ (hence $a(0) = 0$) and put*

$$h(z) = \exp\left(\sum_{k=1}^{\infty} a_k z^k\right) \quad \text{and} \quad g(z) = \exp\left(\sum_{k=1}^{\infty} b_k z^k\right).$$

Then we have the sharp estimates

$$|a_n| \leq 2 + \frac{1}{n} \quad \text{and} \quad |b_n| \leq 2 - \frac{1}{n}$$

for all $n \in \mathbb{N}$. Equality holds for the mapping

$$f(z) = z\frac{1 - \overline{ze^{i\alpha}}}{1 - ze^{i\alpha}} e^{\frac{4ze^{i\alpha}}{1-ze^{i\alpha}}}, \quad \alpha \in (0, 2\pi).$$

We terminate this section with some remarks on univalent logharmonic mappings defined on the exterior of the unit disk.

Let f be a univalent logharmonic mapping defined on $\Delta = \mathbb{C} \setminus \overline{U} = \{z; |z| > 1\}$ which keeps infinity fixed. Such a function is called a univalent logharmonic exterior mapping. Suppose that f vanishes at a point p. Applying an appropriate rotation in the preimage we may assume that $p > 1$. Then, f admits the representation

$$f(z) = z|z|^{2\beta} \frac{z-p}{1-pz} \left|\frac{z-p}{1-pz}\right|^{2\gamma} h \cdot \overline{g} \tag{11}$$

where $Re\ \beta > -\frac{1}{2}$, $Re\ \gamma > -\frac{1}{2}$ and where h and g are two nonvanishing analytic functions on $\Delta \cup \{\infty\}$.

Note that not every mapping of the form (11) is univalent and logharmonic. If f does not vanish on Δ, then $f_1 = \frac{1}{f(\frac{1}{z})}$ is a univalent logharmonic mapping defined on U. In this case, by putting $p = 1$, f still can be represented in the form (11). In [4], the following characterisation of the one-point mappings has been given.

THEOREM 6. *A function f is a univalent logharmonic exterior mapping from Δ onto $\mathbb{C} \setminus \{1\}$ vanishing at a point $p > 1$ if and only if f is of the form*

$$f(z) = \overline{z}|z|^{2\beta} \frac{z-p}{1-p\overline{z}} \left|\frac{z-p}{1-pz}\right|^{2\gamma} \tag{12}$$

where $Re\ \gamma > -\frac{1}{2}$ and

$$\left|\frac{\beta(1+\overline{\gamma}) - \gamma(1+\overline{\beta})}{1+\gamma+\overline{\gamma}} - \frac{1}{p^2-1}\right| \leq \frac{p}{p^2-1} \tag{13}$$

In particular, it follows that

$$f(z) = \overline{z}\left(\frac{z-p}{1-p\overline{z}}\right) \tag{14}$$

satisfies the assertion of Theorem 6.

We get the following interesting application of Theorem 6 (see [4]). Let K be a compact continuum of $\mathbf{C}$, which satisfies the following two conditions:

(a) $\mathbf{C}\setminus K$ is connected

(b) $z \in K$ implies that $z + 2\pi in \notin K$ for all $n \in \mathbf{Z}\setminus\{\infty\}$.

Define

$$\begin{aligned} K_n &= \{z = \zeta + 2\pi in;\ \zeta \in K\} \\ D^* &= \mathbf{C}\setminus \bigcup_{n\in\mathbf{Z}} K_n \\ \text{and}\quad \Omega^* &= \mathbf{C}\setminus \bigcup_{n\in\mathbf{Z}} \{2\pi in\} \end{aligned}$$

We show that there is a univalent harmonic and orientation-preserving mapping F from D^* onto Ω which keeps infinity fixed. Define $s = f_1(z) = e^z$. Then, we have $f_1(D^*) = \mathbf{C}\setminus(E\cup\{0\})$, where $E = f_1(K) = f_1(K_n)$ is a compact set which does not contain the origin for all $n \in \mathbf{Z}$. Furthermore, E and $\mathbf{C}\setminus E$ are connected sets. Define $D_1^* = f_1(D^*)\cup\{0\}$ and let ϕ be the conformal mapping from D_1^* onto $\Delta = \mathbf{C}\setminus\overline{U}$, such that $\phi(\infty) = \infty$, $\phi(0) > 1$. Then, by Theorem 6, $t = f_2(s) = \overline{\phi(s)}\left(\frac{\phi(s)-\phi(0)}{1-\phi(0)\overline{\phi(s)}}\right)$ maps D_1^* onto $\mathbf{C}\setminus\{1\}$ satisfying $f_2(\infty) = \infty$ and $f_2(0) = 0$. Finally $\hat{F}(z) = \log f_2 \circ f_1(z)$ is a univalent harmonic (orientation-preserving) mapping from D^* onto Ω^*.

There are several numerical methods to get the conformal mapping ϕ. Hence, the mapping $\hat{F}$ can be approximated by constructive methods.

3. A constructive method

In this section we introduce a constructive method for univalent logharmonic mappings from the unit disk U onto a strictly starlike domain Ω which has been successfully applied to conformal mappings (see [11],[12] and [13]) as well as for univalent harmonic mappings (see [6] and [8]).

Let Ω be a strictly starlike domain of $\mathbf{C}$. Then $\partial\Omega$ can be expressed in the parametric form

$$w(t) = R(t)e^{it}, \quad 0 \leq t \leq 2\pi \tag{15}$$

where $R(t)$ is a positive continuous function on $[0, 2\pi]$. We shall use also the following notations:

$$\begin{aligned} ||f||_\infty &= \sup\{|f(z)|;\ z \in U\} \\ ||\Omega||_\infty &= \sup\{|w|;\ w \in \Omega\} \\ d(\partial\Omega) &= \text{distance from the origin to } \partial\Omega \end{aligned}$$

Define for all $w \in \mathbf{C}$:

$$\lambda_\Omega(w) = \begin{cases} \frac{|w|}{R(t)} & \text{if } 0 \neq w = |w|e^{it} \\ 0 & \text{if } w = 0\,. \end{cases} \tag{16}$$

Then, we have

$$\begin{aligned} \lambda_\Omega(w) &< 1 &\iff& \quad w \in \Omega \\ \lambda_\Omega(w) &= 1 &\iff& \quad w \in \partial\Omega \\ \lambda_\Omega(w) &> 1 &\iff& \quad w \in \mathbf{C}\setminus\overline{\Omega} \\ \lambda_\Omega(w) &= 0 &\iff& \quad w = 0\,. \end{aligned}$$

Let f be any complex valued function defined on U. Then, we define

$$\mu_\Omega(f) = \sup\{\lambda_\Omega(w);\ w \in f(U)\}\,. \tag{17}$$

The following properties have been shown in [6].

LEMMA 1. *We have*

$$\begin{aligned} &(a) \quad \mu_\Omega(f) \leq 1 \iff f(U) \subset \Omega \\ &(b) \quad \mu_\Omega(tf) = t\mu_\Omega(f)\ \ \forall t \geq 0 \\ &(c) \quad \mu_\Omega(f) \leq \frac{||f||_\infty}{d(\partial\Omega)} \\ &(d) \quad ||f||_\infty \leq \mu_\Omega(f)||\Omega||_\infty \\ &(e) \quad \mu_\Omega(f_1 + f_2) \leq (\mu_\Omega(f_1) + \mu_\Omega(f_2))\frac{||\Omega||_\infty}{d(\partial\Omega)}\,. \end{aligned}$$

The next lemma shows that μ_Ω is lower semi-continuous with respect to the pointwise convergence.

LEMMA 2. *Let Ω be a strictly starlike domain of* $\mathbf{C}$ *and let f_n be a sequence of mappings from U into* $\mathbf{C}$ *which converges pointwise to f. Then we have* $\liminf_{n\to\infty} \mu_\Omega(f_n) \geq \mu_\Omega(f)$. *Strict inequality can hold even in the case of locally uniform convergence.*

Proof:

a) Suppose $\mu_\Omega(f) < \infty$. For any given $\epsilon > 0$, there is a $z_\epsilon \in U$ such that

$$\mu_\Omega(f) \leq \lambda_\Omega(f(z_\epsilon)) + \epsilon$$

Define $w_n = f_n(z_\epsilon)$. Then we have $\mu_\Omega(f_n) \geq \lambda_\Omega(w_n)$. Since $w_n \to \hat{w} = f(z_\epsilon)$ and since $\frac{1}{R(t)}$ is continuous on $[0, 2\pi]$, we conclude that

$$\lim_{n\to\infty} \lambda_\Omega(w_n) = \lambda_\Omega(f(z_\epsilon)) \geq \mu_\Omega(f) - \epsilon$$

Hence, we get $\mu_\Omega(f_n) \geq \mu_\Omega(f) - \epsilon$ for all $\epsilon > 0$ which shows the lower semi-continuity of the functional μ_Ω for the case $\mu_\Omega(f) < \infty$.

b) If $\mu_\Omega(f) = \infty$, then there is for each fixed $M > 0$ a $z_M \in U$ such that $\lambda_\Omega(f(z_M)) \geq M$. By repeating the same arguments as above we get $\liminf_{n\to\infty} \mu_\Omega \geq M$ for all $M > 0$.

c) The following example has been given in [5] which shows that the inequlity can be strict even for the case of locally uniform convergence. Define $\Omega = U$, $E_n = \left\{w = e^{i\theta};\ |\theta| \leq \pi - \frac{1}{2n}\right\}$ and $G_n = \{|w| < 2\} \setminus \{[1, 2] \cup E_n\}$. Let f_n be the univalent conformal mapping from U onto G_n such that $f_n(0) = 0$ and $f_n'(0) > 0$. Then, by the kernel theorem of Carathéodory, $\{f_n\}$ converges locally uniformly to the mapping $f(z) = z$. Here we have $\mu_\Omega(f_n) = 2$ but $\mu_\Omega(f) = 1$. ∎

The next lemma states that μ_Ω is continuous on the Banach space $[C(\overline{U}), ||f||_\infty]$.

LEMMA 3. *Let Ω be a strictly starlike domain of* $\mathbf{C}$ *and let $\{f_n\}$ be a sequence in $C(\overline{U})$ which converges uniformly on $\overline{U}$ to a function f. Then we have* $\lim_{n\to\infty} \mu_\Omega(f_n) = \mu_\Omega(f)$.

Proof: The proof given for Lemma 5 in [6] can be applied without any change. ∎

From now on, let Ω be a fixed strictly starlike domain of $\mathbf{C}$ and let $a \in H(U)$, $a(0) = 0$, $|a| \leq k < 1$ be a given (second) dilatation function. Denote by N the set of all solutions $f(z) = zh(z)\overline{g(z)}$ of (8) with respect to the given dilatation function which are normalized by $g(0) = h(0) = 1$. Observe that $\beta = 0$ since we assume that $a(0) = 0$. Our constructive method is based on the following optimization result:

THEOREM 7. *Let Ω be a strictly starlike domain of* $\mathbf{C}$ *and let $a \in H(U)$, $a(0) = 0$ and $|a| \leq k < 1$ be given. Denote by $F(z) = zH(z)\overline{G(z)}$ the univalent solution of (8)* *which satisfies $F(U) = \Omega$, $G(0) = 1$ and $H(0) > 0$. (The existence and uniqueness of F has been discussed in Theorem 1). Then there exists a unique $f^* \in N$ such that $\mu_\Omega(f^*) \le \mu_\Omega(f)$ for all $f \in N$ and furthermore, we have $f^* = \frac{F}{H(0)}$.*

Proof: Since $F_1 = \frac{F}{H(0)} \in N$ and since, by Lemma 1 (a and b),

$$\mu_\Omega(F_1) = \mu_\Omega\left(\frac{F}{H(0)}\right) = \frac{1}{H(0)} < \infty ,$$

it is sufficient to consider the problem

$$\min_{f \in N} \{\mu_\Omega(f);\ \mu_\Omega(f) < \infty\} .$$

Fix $f \in N$ and define $f_1 = \frac{f}{\mu_\Omega(f)}$. Then, by Lemma 1a, we have $\mu_\Omega(f_1) = 1$ and, by Lemma 1b, we conclude that $f_1(U) \subset \Omega$. Furthermore, we have the representation $f_1 = zh_1\overline{g_1}$, where $h_1(0) = \frac{1}{\mu_\Omega(f)}$. By the subordination principle, Theorem 2, we get

$$h_1(0) = \frac{1}{\mu_\Omega(f)} \le H(0) = \frac{1}{\mu_\Omega(F_1)} \tag{18}$$

and hence, $\mu_\Omega(F_1) \le \mu_\Omega(f)$. Since, by Theorem 1, equality holds in (18) only for $f = F_1$, Theorem 7 has been proved. ∎

Theorem 7 allows us to solve the following mathematical program:

$$\min M;\ \lambda(f(z)) \le M \quad \forall z \in U, \forall f \in N \tag{19}$$

For $f \in N$, we have $f(z) = zh(z)\overline{g(z)}$ where $h(z) = \exp\left(\sum_{k=1}^{\infty} a_k z^k\right)$ and $g(z) = \exp\left(\int_0^z \frac{a(s)}{s} ds + \sum_{k=1}^{\infty} k a_k \int_0^z a(s)s^{k-1}\, ds\right)$. Furthermore, each $f \in N$ is an open mapping. In what follows, we apply the Ritz-Ansatz:

$$\begin{aligned} h(z) &= \exp\left(\sum_{k=1}^{n} a_k z^k\right) \quad \text{and} \\ g(z) &= \exp\left(\int_0^z \frac{a(s)}{s} ds + \sum_{k=1}^{n} k a_k \int_0^z a(s)s^{k-1}\, ds\right) \end{aligned} \tag{20}$$

Hence, we are lead to the semi-infinite program

$$\min\ M$$

$$\lambda\left[e^{it}\cdot\exp\left(\overline{\int_0^{it}\frac{a(s)}{s}\,ds}+\sum_{k=1}^{n}\left(a_k e^{ikt}+k\overline{a_k}\overline{\int_0^{it}a(s)s^{k-1}\,ds}\right)\right)\right]\ \le\ M \tag{21}$$
$$\text{for all}\ \ t\in[0,2\pi]$$

Denote by V_n the set of all mappings $f\in N$ which are of the form

$$f(z)=z\exp\left(\overline{\int_0^{z}\frac{a(s)}{s}\,ds}+\sum_{k=1}^{n}\left(a_k z^k+k\overline{a_k}\overline{\int_0^{z}a(s)s^{k-1}\,ds}\right)\right)$$

and by f_n^* any solution of the optimization problem

$$\min\ \{\mu_\Omega(f);\ f\in V_n\}\ .$$

We do not claim that f_n^* is unique nor that it is univalent. However, we want to show that f_n^* converges to $f^*=\frac{F}{H(0)}$. Unfortunately, we cannot prove such a statement in generality.

LEMMA 4. *Let a be a polynomial of z such that $|a|\le k<1$ on U. Then for any $f\in N\cap C(\overline{U})$, there is a sequence $\hat{f}_n\in V_n$ such that $\hat{f}_n$ converge uniformly to f on $\overline{U}$.*

Proof: Define

$$f_1(z)=\log\left(\frac{f}{z}\right)-\overline{\int_0^{z}\frac{a(s)}{s}\,ds}+(1-a_1)z+\overline{(1-a_1)\int_0^{z}a(s)\,ds}$$

Then f_1 is harmonic on U, continuous on $\overline{U}$ and is a solution of the partial differential equation $\overline{(f_1)_{\bar z}}=a(f_1)_z$. Furthermore, we have

$$f_1(z)=z+\sum_{k=2}^{\infty}a_kz^k+\overline{\sum_{k=2}^{\infty}b_kz^k}$$

and hence, we can apply Lemma 6 in [6] which states that there is a sequence of polynomials $p_n(z)=z+\sum_{k=2}^{n}c_{nk}z^k$ such that

$$\Pi_n(z)=p_n(z)+\overline{\int_0^{z}a(s)p_n'(s)\,ds}$$

converges uniformly to f_1 on $\overline{U}$. Define

$$\hat{f}_n = z\exp\left\{\Pi_n(z) + \overline{\int_0^z \frac{a(s)}{s}\,ds} + (a_1-1)z + \overline{(a_1-1)\int_0^z a(s)\,ds}\right\}$$

Then $\hat{f}_n \in V_n$ and converges uniformly to f on $\overline{U}$. ■

Now we proceed as in [6]. Let f_n^* be any solution of the problem

$$\min\ \mu_\Omega(f); \quad f \in V_n$$

and let F be as in Theorem 7. Since Ω is a Jordan domain and F is quasiconformal, it follows that f^* is continuous on $\overline{U}$. Define $M_n = \mu_\Omega(f_n^*)$ and $M = \mu_\Omega(f^*)$. Then we have $M_n \geq M_{n+1} \geq M$ for all $n \in \mathbb{N}$. Next, by Lemma 1d, we get

$$||f_n^*||_\infty \leq M_n||\Omega||_\infty \leq M_1||\Omega||_\infty.$$

Since the mappings f_n^* are quasiregular, it follows that there exists a subsequence $f_{n_k}^*$ which converges locally uniformly to f^*. By Lemma 3 and 4, we have $\mu_\Omega(\hat{f}_{n_k}) \geq \mu_\Omega(f_{n_k}^*) \geq M$ and $\lim_{k\to\infty}\mu_\Omega(\hat{f}_{n_k}) = \mu_\Omega(f^*)$. Hence, we conclude that $\lim_{k\to\infty}\mu_\Omega(\hat{f}_{n_k}) = M$. This holds for all normally convergent subsequences of f_n^* which leads to $\lim_{n\to\infty}\mu_\Omega(f_n^*) = M = \mu_\Omega(f^*)$. Finally, by the uniqueness part in Theorem 1, we conclude that the sequence f_n^* converges locally uniformly to f^*. Summarizing, we have shown

THEOREM 8. *Let $a(z)$ be a polynomial such that $||a||_\infty \leq k < 1$ on U. Then the sequence f_n^* of solutions of*

$$\min\ \mu_\Omega(f); \quad f \in V_n$$

converges locally uniformly to the univalent solution f^* of

$$\min\ \mu_\Omega(f); \quad f \in N.$$

It remains the question how big n should be. We have seen in Theorem 5, that $|a_n| \leq 2 + \frac{1}{n}$ and $|b_n| \leq 2 - \frac{1}{n}$. Suppose that Ω is a Jordan domain whose boundary $\partial\Omega$ is rectifiable and piecewise smooth. Then we get the following additional estimate for the coefficients.

THEOREM 9. *Let*

$$F(z) = z|z|^{2\beta}\exp\left\{\sum_{k=1}^{\infty} a_k z^k + \overline{\sum_{k=1}^{\infty} b_k z^k}\right\}$$

be a univalent logharmonic mapping from U onto Ω and let $L(r)$ be the length of $F(|z|=r)$, $0<r<1$. Then we have

$$|a_n| \le \frac{\liminf_{r\to 1} L(r)}{2\pi d(\partial\Omega)n}$$
$$|b_n| \le \frac{\liminf_{r\to 1} L(r)}{2\pi d(\partial\Omega)n}.$$

Equality holds for the case $\Omega = U$ and $f(z) = \frac{z(1-\bar{z})}{1-z}$.

Proof: Fix $r \in (0,1)$. Then we have

$$\begin{aligned} n|a_n| &= \left| \frac{r^{2Re\beta}}{2\pi} \oint_{|z|=r} \frac{zf_z - \bar{z}f_{\bar{z}}}{f} \cdot \frac{dz}{z^{n+1}} \right| \\ &\le \frac{r^{2Re\beta}}{2\pi d(\partial\Omega)} \oint_{|z|=r} |zf_z - \bar{z}f_{\bar{z}}| \left| \frac{dz}{z^{n+1}} \right| \\ &\le \frac{r^{2Re\beta}}{2\pi d(\partial\Omega) r^n} \int_0^{2\pi} \left| \frac{\partial f}{\partial t} \right| dt \\ &= \frac{L(r) r^{2Re\beta}}{2\pi d(\partial\Omega) r^n} \end{aligned}$$

Letting r tend to 1, we get

$$n|a_n| \le \frac{\liminf_{r\to 1} L(r)}{2\pi d(\partial\Omega)}$$

The case for b_n is done analogously

$$n|b_n| = \left| \frac{r^{2Re\beta}}{2\pi} \oint_{|z|=r} \frac{zf_z - \bar{z}f_{\bar{z}}}{f} \cdot \overline{\left(\frac{dz}{z^{n+1}} \right)} \right|$$

Finally, for $f(z) = \frac{z(1-\bar{z})}{1-z}$, we have

$$L(r) = 2\pi r, \quad d(\partial\Omega) = 1 \quad a_n = \frac{1}{n} \text{ and } b_n = -\frac{1}{n} \quad \blacksquare$$

REMARK 3:
1) The coefficients a_n and b_n do not change if F is multiplied by a positive factor.
2) The factor $\liminf_{r\to 1} L(r)$ cannot be replaced by

$$L(1) = \oint_{|z|=1} |zf_z - \bar{z}f_{\bar{z}}| \left| \frac{dz}{z^{n+1}} \right|$$

Indeed, for the extremal function $f(z) = \frac{z(1-\bar{z})}{1-z}$ we get $L(1) = 0$.

4. A simple example

In most cases, the mathematical program of the given constructive method will be non-linear. But in some cases of Ω it can be reduced to a linear program.

Let consider the strictly starlike domain bounded by the closed curve $R(t)e^{it}$ where

$$R(t) = \begin{cases} e^{it+(\pi-t)\cot\theta} & \text{if } 0 \le t \le \pi \\ e^{it+(\pi+t)\cot\theta} & \text{if } -\pi \le t < 0 \end{cases}$$

and where θ is a fixed constant parameter in $(0, \frac{\pi}{2})$. Keeping in mind that we can work either with the logharmonic mapping f or with the mapping $F \in UHP$, we realise that the problem becomes simpler if we look at the harmonic (periodic) situation. Let

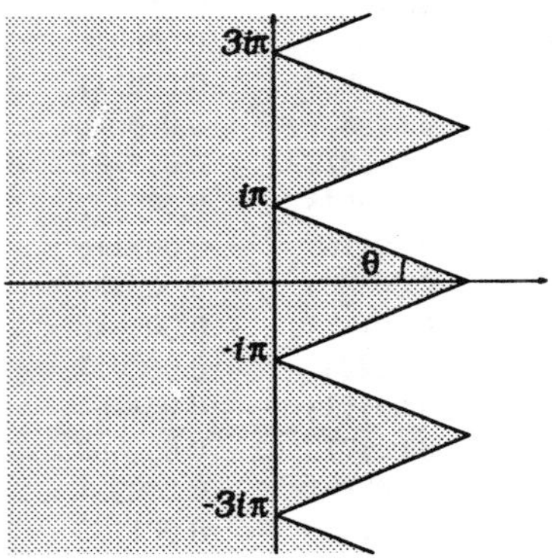

Figure 1: Domain Ω - the harmonic situation

Ω be the domain $\Omega = \{(u,v);\ -\infty < u < u_0(v);\ v \in \Re\}$ where $u_0(v) = (\pi - |v|)\cot\theta$ if $|v| \le \pi$ and where $u_0(v + 2k\pi) = u_0(v)$, $k \in \mathbf{Z}$. The angle θ is half the inside angle of each peak of $\partial\Omega$ at the points $w = 2k\pi$, $k \in \Re$. We want to approximate the univalent harmonic mapping F from $D = \{z;\ Re\ z < 0\}$ onto Ω, which is a solution of $\overline{F_{\bar{z}}} = AF_z$ and which satisfies $Re\ F(-\infty) = -\infty$, $G(-\infty) = 1$, and $H(-\infty) \in \Re$.

Let us take $a(z) = \frac{z}{2}$. Using the fact that Ω is symetric with respect to the real axis, we get, by Lemma 4, the linear program

$$\begin{aligned} &\min\ M \\ &-\pi \le v(t) \le \pi && \forall t \in [-\pi, \pi] \\ &(\pi - v(t))\cot\theta - u(t) \le M && \forall t \in [0, \pi] \\ &(\pi + v(t))\cot\theta - u(t) \le M && \forall t \in [-\pi, 0] \end{aligned}$$

where

$$u(t) \quad = \quad \frac{1}{2}\cos t + \sum_{k=1}^{n} a_k \left(\cos(kt) + \frac{k}{2(k+1)}\cos((k+1)t)\right)$$
$$v(t) = \quad = \quad t - \frac{1}{2}\sin t + \sum_{k=1}^{n} a_k \left(\sin(kt) - \frac{k}{2(k+1)}\sin((k+1)t)\right)$$

and where the coefficients a_k are real variables. By taking an equidistributed partition of $[-\pi, \pi]$, the program has now $n+1$ variales and $2k$ constraints (in fact, only k are possibly different). We use the simplex method and a C program to solve the problem for different cases:

a) $\theta = \frac{\pi}{3}$, $k = 100$

n	10	20	30	40
M	0.940541	0.938867	0.938480	0.938362
a_1	0.375335	0.372891	0.372382	0.372146
a_2	-0.176078	-0.179471	-0.180157	-0.180466
a_3	0.070791	0.068498	0.068016	0.067742
a_4	0.011874	0.010246	0.009911	0.009657
a_5	0.001796	0.000337	0.000078	-0.000213
a_6	0.005926	0.005372	0.005187	0.004946
a_7	0.006064	0.005648	0.005402	0.005106
a_8	0.000544	0.001168	0.001118	0.000908
a_9	0.002811	0.003506	0.003333	0.003106
a_{10}	0.000211	0.001513	0.001440	0.001285
a_{11}		0.002230	0.002087	0.001928
a_{12}		0.001133	0.001164	0.001023
a_{13}		0.001774	0.001583	0.001495
a_{14}		0.000801	0.000931	0.000823
a_{15}		0.001300	0.001222	0.001181

b) $\theta = \frac{\pi}{4}$, $k = 100$

n	10	20	30	40
M	1.255831	1.253171	1.252636	1.252457
a_1	1.009733	1.002828	1.001224	1.000634
a_2	-0.022191	-0.030949	-0.033255	-0.034065
a_3	-0.012290	-0.019191	-0.021544	-0.022343
a_4	0.041984	0.036927	0.034698	0.033950
a_5	0.033859	0.030221	0.028097	0.027357
a_6	0.014421	0.012301	0.010434	0.009717
a_7	0.011748	0.010875	0.009195	0.008456
a_8	0.009542	0.010016	0.008673	0.007905
a_9	0.006494	0.008871	0.007865	0.007103
a_{10}	0.001822	0.006605	0.006026	0.005282
a_{11}		0.005893	0.005501	0.004847
a_{12}		0.004664	0.004615	0.004077
a_{13}		0.003994	0.004255	0.003794
a_{14}		0.003052	0.003638	0.003223
a_{15}		0.002640	0.003420	0.003040

c) $\theta = \frac{\pi}{6}$, $k = 100$

n	10	20	30	40
M	1.487341	1.481190	1.480068	1.479649
a_1	1.658302	1.635696	1.631487	1.629773
a_2	0.495358	0.460115	0.453187	0.450210
a_3	0.164313	0.125928	0.117560	0.113865
a_4	0.091962	0.058169	0.049400	0.045463
a_5	0.082818	0.057735	0.049214	0.045292
a_6	0.069326	0.055333	0.047303	0.043556
a_7	0.051115	0.048046	0.040565	0.037093
a_8	0.031871	0.039301	0.032549	0.029355
a_9	0.016365	0.033107	0.027073	0.024228
a_{10}	0.004413	0.028238	0.023104	0.020607
a_{11}		0.024636	0.020500	0.018275
a_{12}		0.021148	0.018479	0.016213
a_{13}		0.018128	0.016937	0.014628
a_{14}		0.015114	0.015493	0.013170
a_{15}		0.012417	0.014350	0.012050

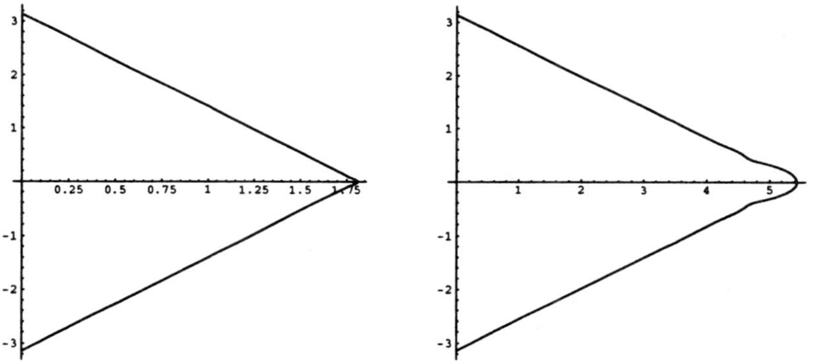

Figure 2: $\theta = \frac{\pi}{3}$ and $\theta = \frac{\pi}{6}$ for $n = 40$ and $k = 100$

Figure 2, shows the results (in the harmonic situation) for the cases $\theta = \frac{\pi}{6}$ and $\theta = \frac{\pi}{3}$ (both having $n = 40$ and $k = 100$). The next figure, illustrates the mapping (in the logharmonic situation) in a polar grid of the unit disk. Finally, we visualize the precision of the numerically calculated function by looking at the difference of the real image of $|z| = 1$ and the expected one (dashed line) in figure 4.

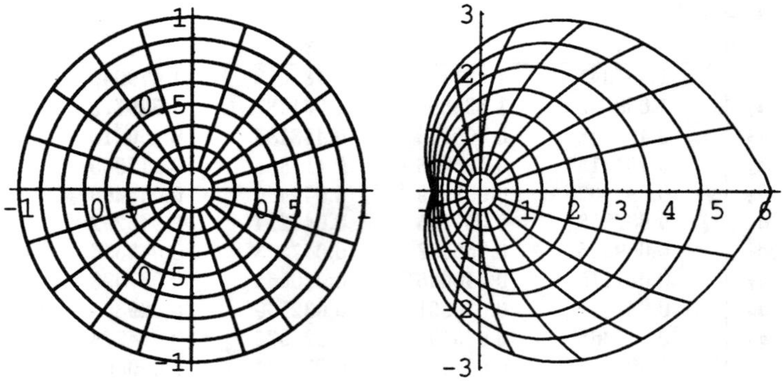

Figure 3: $\theta = \frac{\pi}{3}$, $k = 100$ and $n = 40$

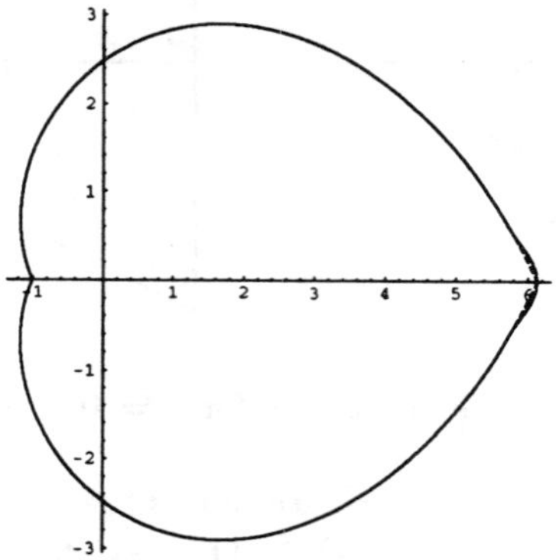

Figure 4: $\theta = \frac{\pi}{3}$, $k = 100$ and $n = 40$, the dashed line is the expected solution

References

[1] **Abdulhadi, Z. and D. Bshouty**, *Univalent mappings in $H \cdot \overline{H}$*, Trans. of Amer. Math. Soc. 305, 841-849 (1988).

[2] **Abdulhadi, Z. and W. Hengartner**, *Spirallike logharmonic mappings*, Complex Variables 9, 121-130 (1987).

[3] **Abdulhadi, Z. and W. Hengartner**, *Univalent harmonic mappings on the left half-plane with periodic dilatation, univalent functions, fractional calculus, and their applications*, Ellis Horwood Lim, 13-28 (1989).

[4] **Abdulhadi, Z. and W. Hengartner**, *One pointed univalent logharmonic mappings*, preprint.

[5] **Bojarski, B.V.**, *On solutions of a linear elliptic system of differential equations in the plane*, Dokl. Akad. Nauk. SSSR 102, 871-874 (Russian) (1955).

[6] **Bshouthy, D., N. Hengartner and W. Hengartner**, *A constructive method for starlike harmonic mappings*, Numer. Math. 54, 167-178 (1988).

[7] **Bshouthy, D., W. Hengartner**, *Univalent solutions of the Dirichlet problem for ring domains*, Complex Variables 21, 159-169 (1993).

[8] **Hengartner, W. and L. Nadeau**, *Univalent harmonic exterior mappings as solutions of an optimization problem*, Matem, Vesnik 40, 233-240 (1988).

[9] **Hengartner, W. and G. Schober**, *Univalent harmonic exterior and ring mappings*, J. Math. Anal. and Appl. 156 Nr.1, 154-171 (1991).

[10] **Nitsche, J.C.C.**, *Vorlesungen über Minimalflächen*, Springer Verlag, New-York (1975).

[11] **Opfer, G.**, *New extremal properties for constructing conformal mappings*, Numer. Math. 32, 423-429 (1979).

[12] **Opfer, G.**, *Conformal mappings onto prescribed regions via optimization techniques*, Numer. Math. 35, 189-200 (1980).

[13] **Opfer, G.**, *Solving complex approximation problems by semiinfinite-finite optimization techniques: A study on convergence*, Numer. Math. 39, 411-420 (1980).

[14] **Osserman, R.**, *A survey on minimal surfaces*, Dover (1986).

CHOQUET-TYPE INTEGRAL REPRESENTATION OF POLYEXCESSIVE FUNCTIONS

K. JANSSEN, H.-H. MÜLLER

ABSTRACT
We develop some aspects of multiparameter potential theory: We give a Choquet-type integral representation for functions which are excessive for a countable family of submarkovian resolvents of commuting kernels on a Lusin measurable space. For the subclass of polyexcessive functions we show that there is a 'unique' representation by minimal polyexcessive functions. As in the classical one-parameter setting we need for these results the existence of a reference measure.

0. INTRODUCTION

It is a well known fact in one-parameter potential theory that measures which are supermedian for a resolvent of kernels admit a 'unique' Choquet-type integral representation by minimal elements under very general assumptions. The corresponding dual result for excessive functions holds only if there exists a reference measure permitting to identify excessive functions with measures which are excessive for some dual resolvent (cf. [6, XII.11, 13, 14, 22, 24, 25, 26, 30] for such results under various regularity assumptions).

In this paper we consider a family $\mathbf{W} = \left(\mathbf{W}^{(i)} \right)^{i \in I}$ of at most countably many substochastic resolvents on a Lusin measurable space, and we assume that these resolvents commute. It seems that many approaches to multi-parameter potential theory are covered by this setting (cf. [7, 8, 9, 19, 23, 31]).
We call a positive function u excessive for $\mathbf{W}$, if u is excessive for each of the resolvents $\mathbf{W}^{(i)}$ for $i \in I$. It appears that for some potential theoretical

K. GowriSankaran et al. (eds.), Classical and Modern Potential Theory and Applications, 293–314.

questions the subclass of polyexcessive functions plays an important role (cf. (1.1) for the precise definition of polyexcessive functions).
The central result of this paper is theorem (4.2), where we prove existence of a Choquet-type integral representation of excessive functions by minimal excessive functions (i.g. there is no unicity), and where we prove that polyexcessive functions admit a 'unique' Choquet-type integral representation by minimal polyexcessive functions.
Moreover, in chapter 5 we study in some detail a concrete form of this abstract integral representation, namely we show that in the product setting a minimal polyexcessive function on the product space is the product of minimal polyexcessive functions on the factor spaces. This extends classical results obtained in [1, 10, 12, 16, 17, 18].

The proof of the central result (4.2) is based on the ideas which led to the corresponding results in the one-parameter setting. Namely, we show in chapter 2 - without any assumption concerning the existence of a reference measure - that for the given family **W** of resolvents and any supermedian measure μ there exists a family **V** of commuting substochastic resolvents such that **W** is in duality to **V** with respect to μ. In chapter 3 we show that under the assumption of the existence of a reference measure (cf. the beginning of chapter 3 for the precise formulation of this hypothesis) we can identify a.e. finite excessive or polyexcessive functions for **W** and excessive or polyexcessive measures for the dual family **V**. This permits to apply the results of [21], where a Choquet-type integral representation for supermedian and polysupermedian measures was developed.

Apart from chapter 5 there are many other general classes of examples where concrete versions of the abstract integral representations of polyexcessive functions or polyexcessive measures can be derived. In particular, in the special case where a resolvent $\mathbf{W}^{(i)}$ appears infinitely often in the family **W** it is easily seen that minimal polyexcessive functions are eigenfunctions for

$\mathbf{W}^{(i)}$, hence some of the results of [2, 3, 20] generalize considerably to completely polyexcessive functions. For details of some such results the reader should consult [21, 27]; also, the reader can find more elaborated proofs of some of the results of this paper in [27].

The notions of this paper are mostly standard, as basic references for our notations we use [4, 11, 15].
In particular, if $(E,\mathfrak{E})$ is a measurable space, then $p\mathfrak{E}$ denotes the class of $\mathfrak{E}$-measurable positive numerical functions on E; moreover, if μ is a measure on $(E,\mathfrak{E})$, then we also use the notation $\mu(f) = \int f \, d\mu$.

1. POLYEXCESSIVE FUNCTIONS: GENERAL NOTATIONS, EXAMPLES

In this chapter let $(E,\mathfrak{E})$ be a measurable space, and let I be some non-empty index set.
Moreover, let $\mathbf{W} = \left(\mathbf{W}^{(i)} \right)^{i \in I}$ be a family of commuting submarkovian resolvents on $(E,\mathfrak{E})$, i.e. $\mathbf{W} = \left(W_\alpha^{(i)} \right)_{\alpha>0}^{i \in I}$ is a family of kernels on $(E,\mathfrak{E})$ such that

$$\alpha W_\alpha^{(i)} 1 \leq 1, \quad W_\alpha^{(i)} = W_\beta^{(i)} + (\beta-\alpha) W_\alpha^{(i)} W_\beta^{(i)},$$
$$W_\alpha^{(i)} W_\beta^{(j)} = W_\beta^{(j)} W_\alpha^{(i)} \qquad \text{for } 0 < \alpha \leq \beta, \ \ i, j \in I.$$

We denote the associated potential kernels by $W^{(i)} := W_0^{(i)} := \uparrow\lim_{\alpha \downarrow 0} W_\alpha^{(i)}$.

The following abbreviations contain some redundance to facilitate the identification of objects: For any finite subset $H \subset I$ we denote by $W^{(H)}$ the composition of the kernels $W^{(i)}$ for $i \in H$; for $\alpha^{(H)} := \left(\alpha_i^{(H)} \right)_{i \in H} \in [0,\infty[^H$ we denote by $W_{\alpha^{(H)}}^{(H)}$ the kernel obtained by composition of $W_{\alpha_i^{(H)}}^{(i)}$ for $i \in H$; we also denote by $|\alpha^{(H)}|$ the product of the reals $\alpha_i^{(H)}$ for $i \in H$; for $K \subset H$ we denote $\alpha_K^{(H)} := \left(\alpha_i^{(H)} \right)_{i \in K} \in [0,\infty[^K$. For any constant $c \in [0,\infty]$ we also interpret c as the constant function $c = (c)_{i \in H} \in [0,\infty]^H$. Finally $|H|$ denotes cardinality of H.

1.1. DEFINITIONS

a) Let μ be a σ-finite measure on $(E,\mathfrak{E})$. We call μ *supermedian* for **W**, if $\mu\, \alpha\, W_\alpha^{(i)} \le \mu$ for all $\alpha > 0$ and $i \in I$; if $\mu\, \alpha\, W_\alpha^{(i)} \uparrow \mu$ for $\alpha \uparrow \infty$ for every $i \in I$, then μ is called *excessive* for **W**.
We call μ *polysupermedian* for **W**, if for every finite subset $H \subset I$ and any $\alpha^{(H)} \in\,]0,\infty[^H$

$$\mu \sum_{\substack{K \subset H \\ |K| \text{ odd}}} |\alpha_K^{(H)}|\; W^{(K)}_{\alpha_K^{(H)}} \le \mu \sum_{\substack{K \subset H \\ |K| \text{ even}}} |\alpha_K^{(H)}|\; W^{(K)}_{\alpha_K^{(H)}} ;$$

we call μ *polyexcessive* for **W**, if μ is excessive and polysupermedian for **W**.

b) Let $u \in p\mathfrak{E}$. We call u *supermedian* for **W**, if $\alpha\, W_\alpha^{(i)}\, u \le u$ for all $\alpha > 0$ and $i \in I$; if $\alpha\, W_\alpha^{(i)}\, u \uparrow u$ for $\alpha \uparrow \infty$ for every $i \in I$, then u is called *excessive* for **W**.
We call u *polysupermedian* for **W**, if for every finite subset $H \subset I$ and any $\alpha^{(H)} \in\,]0,\infty[^H$

$$\sum_{\substack{K \subset H \\ |K| \text{ odd}}} |\alpha_K^{(H)}|\; W^{(K)}_{\alpha_K^{(H)}}\, u \le \sum_{\substack{K \subset H \\ |K| \text{ even}}} |\alpha_K^{(H)}|\; W^{(K)}_{\alpha_K^{(H)}}\, u ;$$

we call u *polyexcessive* for **W**, if u is excessive and polysupermedian for **W**.

The calculations we do with (poly-)supermedian measures and functions are based on corresponding results for the one-parameter setting as in [XII11,15].

1.2. REMARKS

a) In examples the resolvents are typically the Laplace transforms of commuting substochastic measurable semigroups, i.e. $W_\alpha^{(i)} = \int e^{-\alpha t} P_t^{(i)} dt$. As in the 1-parameter setting $|I| = 1$ it is easily seen, that then all the above notions with the exceptions of 'supermedian function' and 'polysupermedian function' can also be characterized by the relations obtained when replacing $\alpha W_\alpha^{(i)}$ by $P_t^{(i)}$ in (1.1).

b) The setting of a commuting family of semigroups $\left(P_t^{(i)} \right)_{t>0}^{i \in I}$ is the same as the setting of one multiparameter semigroup $\left(Q_{\underline{t}} \right)_{\underline{t} \in T}$ for the semigroup $T := \{ (t_i) \in [0,\infty[^I : \{i: t_i > 0\}$ is finite$\}$: the identification is determined by $P_t^{(j)} = Q_{\underline{t}^{(j)}}$ where $\underline{t}_j^{(j)} = t$ and $\underline{t}_i^{(j)} = 0$ for $i \neq j$ for all $j \in I$.

It follows easily from Kolmogorov's theorem that a finite function $u \in p\mathfrak{E}$ is polyexcessive iff for all $x \in E$ the mapping $\underline{t} \mapsto Q_{\underline{t}}u(x)$ $(\underline{t} \in T)$ is the distribution function of a measure on $]0,\infty]^I$. A similar characterization holds in terms of the resolvents (cf. [21] for details for a version for polysupermedian measures).

Next we give some simple examples to illustrate the scope of the theory.

1.3. EXAMPLES

a) *The identical setting:* The most simple example of a family **W** of commuting resolvents is given by $\mathbf{W} = \left(W_\alpha^{(i)} \right)_{\alpha>0}^{i \in I}$, where for each $i \in I$ we have $W_\alpha^{(i)} = W_\alpha$ for all $\alpha > 0$ for one fixed resolvent $\left(W_\alpha \right)_{\alpha>0}$. Polyexcessive functions in this setting are examples for functions which are called 'poly-surharmoniques' in [7] (cf. also [19, 28] for special classes of polyharmonic functions for the Laplace equation for finite I).

In the case $I = \mathbb{N}$ polyexcessive measures and functions have been studied as '*completely excessive*' measures and functions (cf. [2, 3, 20]). In the special case $E =]0,\infty[$ and the semigroup of uniform right motion on E completely excessive functions are exactly the Laplace transforms of measures on $[0,\infty[$. Quite generally, completely excessive measures and functions can be written as a mixture of eigenelements (cf. also [21, 27]).

b) The group setting: If $(E,\mathfrak{E})$ is a measurable abelian group, then any family of measurable convolution semigroups $\left(\mu_{t,i}\right)_{t>0}$ of subprobability measures $(i \in I)$ gives rise to a family of semigroups of substochastic convolution kernels which obviously commute.

c) The subordination setting: Let $\left(P_t\right)_{t>0}$ be a measurable substochastic semigroup of kernels on the measurable space $(E,\mathfrak{E})$. For $i \in I$ let $\left(\mu_{t,i}\right)_{t>0}$ be a measurable convolution semigroup of subprobability measures on $\mathbb{R}$ carried by $\mathbb{R}_+$. Then $P_t^{(i)} := \int P_s\, \mu_{t,i}(ds)$ $(t > 0,\ i \in I)$ defines a commuting family of semigroups. In some cases the corresponding potential kernels and also the composition of such kernels can be computed explicitly, as f.e. in the setting of Riesz kernels, and then polyexcessive measures or functions can be determined explicitly (cf. [21, 27]).

d) The product setting: Let $(E,\mathfrak{E})$ be the product space of measurable spaces $(E_i,\mathfrak{E}_i)$ for $i \in I$. For each $i \in I$ let $W^{(i,0)} = \left(W_\alpha^{(i,0)}\right)_{\alpha>0}$ be a submarkovian resolvent on $(E_i,\mathfrak{E}_i)$. By tensoring the identity we obtain for each $i \in I$ a submarkovian resolvent $\left(W_\alpha^{(i)}\right)_{\alpha>0}$ on $(E,\mathfrak{E})$, and these resolvents obviously commute since they act on different components. For finite index sets I Choquet-type integral representations of excessive and polyexcessive functions in this setting have been studied by many authors (cf. [1, 10, 12, 16, 17, 18]).

Special case: For $|I| = 2$, for $E_1 = E_2 =]0,\infty[$ and for the semigroups of uniform right motion on E_i the integral representation of finite polyexcessive functions on $]0,\infty[^2$ is exactly the classical result that finite measures on $]0,\infty]^2$ can be described by their 2-dimensional distribution functions on $]0,\infty[^2$.

Generalization: Instead of a single resolvent $W^{(i,0)}$ there may be given a commuting family $\left(W^{(i,k)} \right)^{k \in K_i}$ of resolvents on each factor space $(E_i, \mathfrak{E}_i)$. In chapter 5 we shall see that the classical results carry over to this setting for finite I. As a particular example for this extension one may look at two-dimensional Laplace transforms.

1.4. REMARK

Our motivation to coin the name 'polyexcessive' for the above functions originated in the study of [7], where the classical notions of polyharmonic and polysuperharmonic functions for the Laplace equation of [28] were generalized to an axiomatic framework. It is easily seen and well known, that polyexcessive functions in the identical setting can be interpreted as ordinary excessive functions for a suitable resolvent on a bigger state space; in particular, for the identical setting there is a one-parameter process on an enlarged state space, which allows to do probabilistic potential theory.
Also, in the product setting one may consider associated independent processes on the factor spaces to develop probabilistic multi-parameter potential theory. But we see no general way, to associate nice processes with general nice commuting semigroups.

2. CONSTRUCTION OF DUAL RESOLVENTS

In the sequel we assume given a family $W = \left(W^{(i)} \right)^{i \in I}$ of commuting submarkovian resolvents on $(E, \mathfrak{E})$ as in chapter 1.
We assume that I is at most countable, and we also assume that $(E, \mathfrak{E})$ is a Lusin measurable space (i.e. E may be assumed to be a borel subset of some compact metric space equipped with it's borel field $\mathfrak{E}$; cf. the appendix of [29] for basic properties of such spaces).

2.1. PROPOSITION

Let the measure μ be supermedian for $\left(W^{(i)} \right)^{i\in I}$. Then there exists a family $\left(V^{(i)} \right)^{i\in I}$ of commuting submarkovian resolvents on $(E,\mathfrak{E})$ such that $\left(V^{(i)} \right)^{i\in I}$ is in duality to $\left(W^{(i)} \right)^{i\in I}$ with respect to μ, i.e.

$$\int f\, W_\alpha^{(i)} g\, d\mu = \int g\, V_\alpha^{(i)} f\, d\mu$$

for all $i \in I$, $\alpha > 0$, $f, g \in p\mathfrak{E}$.
Moreover, if $W^{(H)} \ll \mu$ for some finite subset $H \subset I$, then the above resolvents can be chosen as to satisfy $V^{(H)} \ll \mu$.

Proof: i) Let D denote the rationals in $]0,\infty[$. According to Th.IX.14 in [11] there exists for every $i \in I$ and $\alpha > 0$ a kernel $\hat{W}_\alpha^{(i)}$ which is in duality to $W_\alpha^{(i)}$ with respect to μ, and any two such dual kernels differ only on some μ-zero set.
Since I and D are at most countable, the defining duality equalities imply the appropriate relations of commutativity, resolvent equations and submarkovianity for $\left(\hat{W}_\alpha^{(i)} \right)_{\alpha\in D}^{i\in I}$ except on some μ-zero set. Replacing the measures $\varepsilon_x \hat{W}_\alpha^{(i)}$ for such exceptional $x \in E$ by the zero-measure, we obtain a new family of smaller kernels, which again are in duality with respect to μ to the kernels $\left(W_\alpha^{(i)} \right)_{\alpha\in D}^{i\in I}$, and which again are commuting and satisfy the resolvent equations μ-a.e..
Iterating this procedure, we obtain for every $i \in I$ and $\alpha \in D$ a decreasing sequence of kernels; the limit kernels $\left(V_\alpha^{(i)} \right)_{\alpha\in D}^{i\in I}$ are then in duality to $\left(W_\alpha^{(i)} \right)_{\alpha\in D}^{i\in I}$ with respect to μ and they satisfy the commutativity relations and the resolvent equations everywhere. Completing this family of kernels by $V_\beta^{(i)} := \uparrow\lim_{\substack{\alpha\downarrow\beta \\ \alpha\in D}} V_\alpha^{(i)}$ for $\beta \geq 0$ and $i \in I$ then gives the wanted dual resolvents.
ii) If moreover $W^{(H)} \ll \mu$ for some finite subset $H \subset I$, then Th.V.58 in [11] implies that the kernel $W_1^{(H)}$ is given by some jointly measurable density

$w_H \in p(\mathfrak{E} \otimes \mathfrak{E})$ with respect to μ. Consequently, by switching the variables we obtain a kernel $U \ll \mu$, which is in duality to $W_1^{(H)}$. Obviously, we have $1_A U = 1_A \hat{W}_1^{(H)}$ for some $A \in \mathfrak{E}$ such that $\mu(E\backslash A) = 0$; repeating the construction of i) in such a way that $\varepsilon_x \hat{W}_\alpha^{(i)} = 0$ for $x \in E\backslash A$, $i \in I$ and $\alpha \in D$ the stated result obtains.

2.2. REMARK

Since the function 1 is supermedian for $\mathbf{W}$, the measure μ is co-supermedian for $\mathbf{W}$ (i.e. μ is supermedian for any resolvent $\mathbf{V}^{(i)}$ which is in duality to $\mathbf{W}^{(i)}$ with respect to μ). If μ is excessive for $\mathbf{W}$, then μ is also co-excessive.

3. IDENTIFICATION OF (POLY-)EXCESSIVE FUNCTIONS AND (POLY-)EXCESSIVE MEASURES

Additionally to the assumptions of chapter 2 we assume in this chapter the existence of a reference measure μ for $\mathbf{W}$, i.e. we assume that there exists a finite subset $H \subset I$ and a supermedian measure μ such that $W^{(H)} \ll \mu$.

Obviously, we may then choose μ as a measure which is excessive for $W^{(i)}$, $i \in H$; applying (2.1) we can find a family $\mathbf{V}$ of commuting submarkovian resolvents which are in duality to $\mathbf{W}$ with respect to μ such that $V^{(H)} \ll \mu$.

In the sequel we denote by $\mathcal{S}$, $\mathcal{E}$, $\mathcal{PS}$, $\mathcal{PE}$ the convex cones of functions which are supermedian, excessive, polysupermedian or polyexcessive, respectively, with respect to $\mathbf{W}$ and which are finite μ-a.e.; we denote by $\mathcal{M}$ the convex cone of σ-finite measures on $(E,\mathfrak{E})$, and we denote by $\mathcal{MS}$, $\mathcal{ME}$, $\mathcal{MPS}$, $\mathcal{MPE}$ the convex cones of measures which are supermedian, excessive, polysupermedian or polyexcessive for $\mathbf{V}$. Moreover, for every measure ρ on $(E,\mathfrak{E})$ and any subset $\mathcal{G} \subset p\mathfrak{E}$ we denote $\mathcal{G}^\rho := \{ f \in \mathcal{G} : \rho(f) < \infty \}$; in duality, for every function $g \in p\mathfrak{E}$ and any subset $\mathcal{H} \subset \mathcal{M}$ we denote $\mathcal{H}^g := \{ \eta \in \mathcal{H} : \eta(g) < \infty \}$.

3.1. THEOREM

The mapping B: u $\mapsto$ uμ is an additive, positively homogeneous bijection from $\mathcal{E}$ onto $\mathcal{ME}$ and from $\mathcal{PE}$ onto $\mathcal{MPE}$, respectively. Moreover, B respects the natural and the specific order of these cones.

Proof: i) For u $\in \mathcal{E}$ we have u$\mu \in \mathcal{ME}$, since uμ is a σ-finite measure, and since for i $\in$ I we have for f $\in$ p$\mathfrak{E}$ and $\alpha \uparrow \infty$

$$(u\mu)\,(\alpha V_\alpha^{(i)} f) \;=\; \int f\,\alpha W_\alpha^{(i)} u\; d\mu \;\uparrow\; \int f\, u\; d\mu \;=\; (u\mu)\,(f);$$

if moreover u is polysupermedian, then we obtain in the same way by duality, that uμ is polysupermedian, i.e. B maps $\mathcal{PE}$ into $\mathcal{MPE}$.

ii) B is injective: If u, v are μ-a.e. finite excessive functions satisfying uμ = vμ, then $W^{(H)} \ll \mu$ implies

$$u = \uparrow \lim_{\alpha^{(H)}\uparrow\infty} |\alpha^{(H)}|\; W^{(H)}_{\alpha^{(H)}}\, u \;=\; \uparrow \lim_{\alpha^{(H)}\uparrow\infty} |\alpha^{(H)}|\; W^{(H)}_{\alpha^{(H)}}\, v = v\,.$$

iii) B is surjective: Whenever η is an excessive measure for **V**, then we conclude that $\eta \ll \mu$, hence η = sμ for some μ-a.e. finite function s; from the duality equalities we conclude for $0 < \alpha \le \beta$ and $n \uparrow \infty$

$$\alpha\, W_\alpha^{(i)}\, s \le \beta\, W_\beta^{(i)}\, s \le s \quad \mu\text{-a.e.} \quad \text{and} \quad n\, W_n^{(i)}\, s \uparrow s \quad \mu\text{-a.e.};$$

from the resolvent equations we see that $|\alpha^{(H)}|\; W^{(H)}_{\alpha(H)}$ s is monotone increasing in $\alpha^{(H)} \in\,]0,\infty[^H$ in each component separately, and it is easily verified that for $\alpha^{(H)}\uparrow\infty$ the limit function u is an excessive function such that u=s μ-a.e..
If moreover η is polysupermedian, then the above density s is μ-a.e. polysupermedian, and the above regularization u is in fact polyexcessive.

iv) Finally, B is obviously additive and positively homogeneous, hence B respects the specific order; as in ii) it is easily seen that excessive functions u, v satisfy u $\le$ v iff u $\le$ v μ-a.e. iff u$\mu \le$ vμ.

3.2. NOTATIONS

a) We endow the cone $\mathcal{M}$ of σ-finite measures on $(E,\mathcal{E})$ with the smallest σ-field $\mathfrak{M}$ such that the mappings $\eta \mapsto \eta(f)$ are measurable for every f in $p\mathfrak{E}$;
b) we endow $p\mathfrak{E}$ with the smallest σ-field $\mathfrak{F}$ such that $f \mapsto \eta(f)$ is measurable for every $\eta \in \mathcal{M}$ which is absolutely continuous with respect to μ;
c) we endow all subsets of $\mathcal{M}$ or $p\mathfrak{E}$ with the corresponding trace fields.

3.3. CONSEQUENCE

The bijection B from (3.1) is a measurable isomorphism.

Proof: Simply by using Radon-Nikodym's theorem we obtain the isomorphism of the generating systems for the σ-fields.

3.4. PROPOSITION

Let $g \in p\mathfrak{E}$ satisfy $0 < g < \infty$, and let $\nu := g\mu$. Then the evaluation functional $\delta : (x,u) \mapsto u(x)$ is a measurable function on $E \times \mathcal{E}^{\nu}$ for the product σ-field. This holds also for the restriction of δ to $E \times \mathcal{P}\mathcal{E}^{\nu}$.

Proof: Defining for $\eta \in \mathcal{M}\mathcal{E}$ the measure $\varepsilon_{\eta}R := \eta$ we obtain an $(\mathcal{M}\mathcal{E}, \mathfrak{M} \cap \mathcal{M}\mathcal{E})$-$(E,\mathfrak{E})$-kernel R which is absolutely continuous with respect to μ. Since $\mathfrak{E}$ is countably generated we conclude from $0 < g < \infty$ and the version Th.V.58 in [11] of Radon-Nikodym's theorem that there exists some function $s \in p(\mathfrak{E} \otimes \mathfrak{M} \cap \mathcal{M}\mathcal{E}^{g})$ such that $\eta = s(\cdot,\eta)\,\mu$ for all $\eta \in \mathcal{M}\mathcal{E}^{g}$; applying the same regularization procedure as in part iii) of the proof of (3.1), we may assume $s(\cdot,\eta) \in \mathcal{E}^{\nu}$, and then (3.3) gives the result.

4. INTEGRAL REPRESENTATION OF EXCESSIVE AND POLYEXCESSIVE FUNCTIONS

We make the same assumptions as in the last chapter; in particular, μ will be a supermedian reference measure for **W**.

4.1. REMARKS AND NOTATIONS

a) Let $w \not\equiv 0$ be an excessive function which is μ-a.e. finite. Then there exists a bounded strictly positive function $g \in p\mathfrak{E}$ such that $\mu(gw) = 1$. Any such function g is called a reference function for w, and any such measure $\nu := g\mu$ is called a reference measure for w. Moreover, we denote $\mathcal{MS}_1^g := \{ \eta \in \mathcal{MS} : \eta(g) = 1 \}$ and $\mathcal{E}_1^\nu := \{ u \in \mathcal{E} : \nu(u) = 1 \}$. Similarly, $\mathcal{ME}_1^g$, $\mathcal{MPE}_1^g$, and $\mathcal{PE}_1^\nu$ denote the corresponding suitably normalized convex sets.

   Finally, for any convex set $\mathcal{C}$ we denote by ${}_e\mathcal{C}$ the set of extreme points in $\mathcal{C}$ (i.e. $x \in {}_e\mathcal{C}$ iff $x = (y+z)/2$ for some $y, z \in \mathcal{C}$ can only hold for $x = y$).

b) From (3.1) and (3.3) we conclude that $\mathcal{ME}_1^\nu$ and $\mathcal{E}_1^g$ as well as ${}_e\mathcal{ME}_1^\nu$ and ${}_e\mathcal{E}_1^g$ are isomorphic via the mapping B; the same is true for $\mathcal{MPE}_1^\nu$ and $\mathcal{PE}_1^g$ as well as for ${}_e\mathcal{MPE}_1^\nu$ and ${}_e\mathcal{PE}_1^g$.

c) As a consequence we conclude from [21] that $\mathcal{E}_1^g$ and $\mathcal{PE}_1^g$ are Lusin measurable spaces; this also holds for their subsets of extreme points ${}_e\mathcal{E}_1^g$ and ${}_e\mathcal{PE}_1^g$.

4.2. THEOREM

Let $w \not\equiv 0$ be an excessive function which is finite μ-a.e, let g be a strictly positive bounded reference function for w and let $\nu := g\mu$ the corresponding reference measure such that $\nu(w)=1$.

a) There exists a representing probability measure m for w on ${}_e\mathcal{E}_1^\nu$, i.e. there exists a probability measure m on $({}_e\mathcal{E}_1^\nu, \mathfrak{F} \cap {}_e\mathcal{E}_1^\nu)$ such that

$$w(x) = \int u(x)\ dm(u) \quad \text{for all } x \in E.$$

b) If moreover w is polyexcessive, then there exists a unique representing probability measure m for w on ${}_e\mathcal{PE}^{\nu}_1$, i.e. there exists a unique probability measure m on $({}_e\mathcal{PE}^{\nu}_1, \mathfrak{F} \cap {}_e\mathcal{PE}^{\nu}_1)$ such that

$$w(x) = \int u(x)\, dm(u) \quad \text{for all } x \in E.$$

Proof: Using the bijection $B : u \mapsto u\mu$ from (3.1) the result is an easy consequence of the integral representation of (poly-)supermedian measures given in [21]: If w is excessive such that $\nu(w)=1$, then $w\mu \in \mathcal{ME}^g_1$, hence we know

$$w\mu = \int \eta \, d\tau(\eta)$$

for some probability measure τ on $({}_e\mathcal{MS}^g_1, \mathfrak{M} \cap {}_e\mathcal{MS}^g_1)$; if moreover w is polyexcessive, then $w\mu \in \mathcal{MPS}^g_1$, and the above formula holds for a unique probability measure τ on $({}_e\mathcal{MPS}^g_1, \mathfrak{M} \cap {}_e\mathcal{MPS}^g_1)$. Since w is excessive we conclude in both cases

$$\eta\, (\bigcup_{i \in I} V^{(i)} 1 = 0\,) = 0$$

for τ-a.a. η, hence τ is concentrated on the measurable subsets ${}_e\mathcal{ME}^g_1$ and ${}_e\mathcal{MPE}^g_1$, respectively. We define m as the image measure of τ under the inverse mapping B^{-1} of the mapping $B : u \mapsto u\mu$ from (3.1). Then we obtain from Fubini's theorem and (3.1), (3.2) and (3.3)

$$\int w\, f\, d\mu = \iint u\, f\, d\mu\, dm(u) = \int (\int u\, dm(u))\, f\, d\mu$$

for all $f \in p\mathfrak{E}$, i.e. $w = \int u\, dm(u)$ μ-a.e..

Since both w and $\int u\, dm(u)$ are excessive functions, we conclude that they coincide on E. Obviously, m is a probability measure on ${}_e\mathcal{E}^{\nu}_1$.

If moreover w is polyexcessive, then the above construction yields a probability measure m on ${}_e\mathcal{PE}^{\nu}_1$, and unicity of the representing measure follows from unicity of τ in this case and from (3.1) again.

4.3. REMARK

Simple examples show that i.g. there is no unicity of the representing measure for an excessive function. In particular, there is no unicity in the product setting of example (1.3.d), when uniform motion or Brownian motion is considered on two factor spaces $]0,\infty[$.

Inspection of the proof of (4.2) and of the corresponding result of [21] gives by standard but tedious arguments the following

4.4. CONSEQUENCE

Let $g \in p\mathfrak{E}$ be strictly positive, let $\nu := g\mu$, let $(\Omega,\mathfrak{A})$ be a measurable space, and let $t : \Omega \mapsto \mathcal{P}\mathcal{E}_1^{\nu}$ be a measurable mapping. If for $\omega \in \Omega$ we denote by m_ω the uniquely determined representing measure for $t(\omega)$ on ${}_e\mathcal{P}\mathcal{E}_1^{\nu}$, then $(m_\omega)_{\omega\in\Omega}$ is a stochastic kernel.

5. THE PRODUCT SETTING

In order to turn the abstract integral representation of chapter 4 into a concrete one, the class of extremal polyexcessive functions has to be determined. This can be done to some extent in many situations. For some such results for polysupermedian measures or functions the reader may consult [21, 27].

In this chapter we study in some detail a general product setting for two factor spaces. By induction the results carry over to finitely many factors. The special case $|I_1| = 1 = |I_2|$ of the following assumptions was studied by Cairoli [10] and Gowrisankaran [17] a.o. under additional strong regularity assumptions.

5.1. ASSUMPTIONS

Let I_1 and I_2 be disjoint at most countable non-empty sets.
For $j := 1, 2$ we assume: $(E_j, \mathfrak{E}_j)$ is a Lusin measurable space, $\left(\mathbf{U}^{(i)} \right)^{i \in I_j}$ is a family of commuting submarkovian resolvents on $(E_j, \mathfrak{E}_j)$, and there exists a finite subset $H_j \subset I_j$ and a measure μ_j on $(E_j, \mathfrak{E}_j)$ such that μ_j is excessive for $\left(\mathbf{U}^{(i)} \right)^{i \in I_j}$ and $\mathrm{U}^{(H_j)} \ll \mu_j$.
We define $(E, \mathcal{E}) := (E_1 \times E_2, \mathfrak{E}_1 \otimes \mathfrak{E}_2)$, $I := I_1 \cup I_2$, $\mathbf{W}^{(i)} := \mathbf{U}^{(i)} \otimes \mathrm{Id}$ for $i \in I_1$ and $\mathbf{W}^{(i)} := \mathrm{Id} \otimes \mathbf{U}^{(i)}$ for $i \in I_2$.
Then $\left(\mathbf{W}^{(i)} \right)^{i \in I}$ is a family of commuting submarkovian resolvents on $(E, \mathcal{E})$, and we have $\mathrm{W}^{(H)} \ll \mu$ for $H := H_1 \cup H_2$ and the excessive measure $\mu := \mu_1 \otimes \mu_2$.

5.2. DEFINITION

We call a function w on E *separately excessive* (*separately polyexcessive*) if

i) $w(x, \cdot)$ is excessive (polyexcessive) on E_2 for every $x \in E_1$ and
ii) $w(\cdot, y)$ is excessive (polyexcessive) on E_1 for every $y \in E_2$.

5.3. REMARK

Every excessive (polyexcessive) function is a separately excessive (separately polyexcessive) function.
Conversely, every $\mathfrak{E}$-measurable separately excessive function is excessive.

For the next result remember that a non-zero element u of a convex cone $\mathcal{C}$ is said to be minimal in $\mathcal{C}$, if $u = v + w$ for $v, w \in \mathcal{C}$ can only hold if v is proportional to u.
Also remember the relations between 'extreme points' and 'minimal points'; in particular we have ${}_e\mathcal{P}\mathcal{E}_1^{\nu} = \{ u \in \mathcal{P}\mathcal{E} : u \text{ is minimal and } \nu(u) = 1 \}$.

5.4. PROPOSITION

Let u_1, u_2 be excessive (or polyexcessive) functions on E_1 and E_2, respectively, and denote by $u := u_1 \otimes u_2$ the product function on E. Then we have

a) u is excessive (or polyexcessive);

b) if $u_j < \infty$ μ_j-a.e. for j = 1, 2, then $u < \infty$ μ-a.e.;
   u = 0 holds iff u_1 or u_2 vanish identically;
   if $u \not\equiv 0$ and $u < \infty$ μ-a.e. then $u_j < \infty$ μ_j-a.e. for j = 1, 2.

c) if $u < \infty$ μ-a.e. and if u_1 and u_2 are minimal, then u is also minimal.

The proof follows straightforward from the definitions.

5.5. THEOREM

Let ρ_j be a σ-finite measure on $(E_j,\mathfrak{E}_j)$ for j = 1, 2 such that $\mu \ll \rho := \rho_1 \otimes \rho_2$. Let u be a minimal polyexcessive function in $\mathcal{PE}^\rho$. Then $u = u_1 \otimes u_2$ for minimal polyexcessive functions u_j on $(E_j,\mathfrak{E}_j)$ for j = 1, 2.

Proof: i) Choose bounded strictly positive functions $g_j \in p\mathfrak{E}_j$ for j = 1, 2 such that $\nu := \nu_1 \otimes \nu_2 := g_1\mu_1 \otimes g_2\mu_2$ satisfies $\nu(u) = 1$. Then

$$r_2(y) := \int u(x,y)\, d\nu_1(x) \qquad (y \in E_2)$$

defines a polyexcessive function on E_2 which is finite ν_2-a.e..
Let $y \in E_2$ satisfy $r_2(y) < \infty$. Then we know from (5.3) that $u(\cdot,y)$ is a ν_1-integrable polyexcessive function on E_1, hence there exists by (4.2) a unique representing measure $M_1(y,\cdot)$ with total mass $r_2(y)$ on ${}_e\mathcal{PE}_1^{\nu_1}$, i.e.

$$u(x,y) = \int u_1(x)\, M_1(y,du_1) \qquad (x \in E_1) .$$

If we define $M_1(y,\cdot) = 0$ for $y \in \{ r_2 = \infty \}$, then $(M_1(y,\cdot))_{y\in E_2}$ defines a kernel by (4.4).

ii) Next we show that M_1 can be regularized to give what might be called a 'polyexcessive kernel'.

First consider $y \in E_2$ such that $r_2(y) < \infty$; denote by $\delta_1 : E_1 \times {}_e\mathcal{PE}_1^{\nu_1} \mapsto \bar{\mathbb{R}}_+$ the evaluation functional $(x,u_1) \mapsto u_1(x)$ and let $\alpha^{(H_2)}, \beta^{(H_2)} \in]0,\infty[^{H_2}$, and $i_0 \in H_2$ such that $\alpha_{i_0}^{(H_2)} \geq \beta_{i_0}^{(H_2)}$ and $\alpha_i^{(H_2)} = \beta_i^{(H_2)}$ for $i \neq i_0$; then we have the inequalities

$$|\alpha^{(H_2)}|\ W^{(H_2)}_{\alpha^{(H_2)}}\, u(\cdot,y) \ \leq\ |\beta^{(H_2)}|\ W^{(H_2)}_{\beta^{(H_2)}}\, u(\cdot,y) \ \leq\ u(\cdot,y)\ ,$$

where each of these functions is polyexcessive on E_1; this chain of inequalities can be completed to equalities by adding suitable functions which are polyexcessive on E_1; since $\{ r_2 = \infty \}$ is a μ_2-zero set, existence and uniqueness of the corresponding representing measures implies

$$\varepsilon_y\ |\alpha^{(H_2)}|\ U^{(H_2)}_{\alpha^{(H_2)}}\, M_1 \ \leq\ \varepsilon_y\ |\beta^{(H_2)}|\ U^{(H_2)}_{\beta^{(H_2)}}\, M_1 \ \leq\ \varepsilon_y\ M_1\ ;$$

since u is excessive, the polyexcessive function $u(\cdot,y)$ is also represented by the measure

$$\uparrow \lim_{\alpha^{(H_2)}\uparrow\infty} \varepsilon_y\ |\alpha^{(H_2)}|\ U^{(H_2)}_{\alpha^{(H_2)}}\, M_1\ ,$$

which consequently has to agree with $M_1(y,\cdot)$. Again from $\mu_2(\{ r_2 = \infty \}) = 0$ we conclude that $M := \uparrow \lim_{\alpha^{(H_2)}\uparrow\infty} |\alpha^{(H_2)}|\ U^{(H_2)}_{\alpha^{(H_2)}}\, M_1$ defines a kernel such that

$$\uparrow \lim_{\alpha^{(H_2)}\uparrow\infty} |\alpha^{(H_2)}|\ U^{(H_2)}_{\alpha^{(H_2)}}\, M \ =\ M\ , \text{ and } M1 = r_2.$$

Moreover we conclude for $f \in p(\mathfrak{F}_1 \cap {}_e\mathcal{PE}_1^{\nu_1})$ that Mf is polyexcessive on E_2: for $f \leq 1$ this follows from $1 = f + (1 - f)$ by (4.2), and for general f this follows from the fact that a countable sum of polyexcessive functions is again a polyexcessive function.

iii) Fubini's theorem implies that for $h \in p(\mathfrak{E}_1 \otimes \mathfrak{F}_1 \cap {}_e\mathcal{PE}_1^{\nu_1})$ the function $Mh : E_1 \times E_2 \mapsto \overline{\mathbb{R}}_+$, $Mh : (x,y) \mapsto \varepsilon_y Mh(x,\cdot)$ is $\mathfrak{E}_1 \otimes \mathfrak{E}_2$-measurable; if moreover $h(\cdot,u_1)$ is polyexcessive on E_1 for all $u_1 \in {}_e\mathcal{PE}_1^{\nu_1}$, then Mh is polyexcessive on E since M and $W_\alpha^{(i)}$ commute ($\alpha > 0$, $i \in I_1$). In particular, $M\delta_1$ is polyexcessive on E. From $u = M\delta_1$ except on the μ-zero set $E_1 \times \{ r_2 = \infty \}$ we conclude $u = M\delta_1$ on E.

iv) Next we prove: There exists $u_1 \in {}_e\mathcal{PE}_1^{\nu_1}$ such that $M(y,\cdot) = r_2(y)\varepsilon_{u_1}$ for all $y \in \{ r_2 < \infty \}$. If $f \in p(\mathfrak{F}_1 \cap {}_e\mathcal{PE}_1^{\nu_1})$ satisfies $f \le 1$, then u decomposes into polyexcessive functions by

$$u = M\delta_1 = M(\delta_1 f) + M(\delta_1(1-f)),$$

hence minimality of u gives us some constant $\gamma_f \in [0,1]$ such that $M(\delta_1 f) = \gamma_f u$. Unicity of representing measures implies for $y \in \{ r_2 < \infty \}$ the equality $f = \gamma_f$ $M(y,\cdot)$-a.e. for all such f, hence $M(y,\cdot)$ is proportional to some point mass $\varepsilon_{u_{1,y}}$ for some point $u_{1,y} \in {}_e\mathcal{PE}_1^{\nu_1}$, and the proportionality factor equals then $r_2(y)$. For any two points y, z in $\{ 0 < r_2 < \infty \}$ we conclude for $f := 1_{\{u_y\}}$ by the above reasoning

$$f = \gamma_f \quad M(y,\cdot) - \text{a.e. and } M(z,\cdot) - \text{a.e.},$$

hence $\gamma_f = 1$ and $u_{1,y} = u_{1,z}$. As a consequence there exists $u_1 \in {}_e\mathcal{PE}_1^{\nu_1}$ such that $M(y,\cdot) = r_2(y)\varepsilon_{u_1}$ for all $y \in \{ r_2 < \infty \}$.

v) We have seen $u = u_1 \otimes r_2$ for some $u_1 \in {}_e\mathcal{PE}_1^{\nu_1}$ except on the μ-zero set $E_1 \times \{ r_2 = \infty \}$. Since r_2 is polyexcessive on E_2, the function $u_1 \otimes r_2$ is polyexcessive on E, hence $u = u_1 \otimes r_2$ on E. By symmetry we also have $u = r_1 \otimes u_2$ for suitable $u_2 \in {}_e\mathcal{PE}_1^{\nu_2}$ and $r_1 \in \mathcal{PE}^{\nu_1}$; unicity of the representing measures implies $u_1 = c_1 r_1$, $u_2 = c_2 r_2$ for constants c_1, c_2, hence the result.

5.6. THEOREM

Let $w \not\equiv 0$ be a polyexcessive function. We assume that there exists a product of σ-finite measures $\rho = \rho_1 \otimes \rho_2$ on $(E, \mathfrak{E})$ such that $\mu \ll \rho$ and $\rho(w) < \infty$. Then there exists a product measure $\nu = \nu_1 \otimes \nu_2$ such that $\nu(w) = 1$ and $\nu_j = g_j \mu_j$ with strictly positive and bounded functions $g_j \in p(\mathfrak{E}_j)$ for $j = 1, 2$ and there exists a unique probability measure m on ${}_e\mathcal{P}\mathcal{E}_1^{\nu_1} \times {}_e\mathcal{P}\mathcal{E}_1^{\nu_2}$ equipped with the canonical product σ-field such that

$$w(x,y) = \int u_1(x)\, u_2(y)\, dm(u_1,u_2) \qquad \text{for } (x,y) \in E.$$

Proof: The construction of g_1 and g_2 follows from Radon Nikodym's theorem. According to (4.2) we find a unique probability measure $\widetilde{m}$ on ${}_e\mathcal{P}\mathcal{E}_1^{\nu}$ such that

$$w(x,y) = \int u(x,y)\, d\widetilde{m}(u) \qquad \text{for } (x,y) \in E.$$

From (5.4) and (5.5) we conclude that ${}_e\mathcal{P}\mathcal{E}_1^{\nu}$ is measurably isomorphic to ${}_e\mathcal{P}\mathcal{E}_1^{\nu_1} \times {}_e\mathcal{P}\mathcal{E}_1^{\nu_2}$, hence the result.

5.7. REMARK

As a consequence we conclude that for $|I_1| = 1 = |I_2|$ the class of polyexcessive functions on E does not depend on the particular time scale given by the two resolvents, i.e. if we change the original resolvents such that the class of excessive functions is not changed, then the class of polyexcessive functions does not change.

In particular, if $(E_j, \mathcal{W}_j)$ is a balayage space in the sense of [4] for $j = 1, 2$, then there is a well defined class of polyexcessive functions on $E_1 \times E_2$ which can be defined by inequalities involving the harmonic measures similarly as in [17].

Moreover, if $W := U^{(1)} \otimes U^{(2)}$ denotes the product of two associated potential kernels, then it can be shown by construction of a suitable reference measure: Every W-a.e. finite polyexcessive function has a 'unique' Choquet-type integral representation by minimal polyexcessive functions.

More generally, this last result is also true if we start with two standard H-cones of functions in the sense of [6] on Lusin measurable spaces (cf. [27] for details).

REFERENCES

[1] T. Barth: *Integraldarstellung mehrfach superharmonischer Funktionen,* Dissertation, Universität Erlangen (1977).

[2] H. Ben Saad, K. Janßen: *Bernsteins theorem for completely excessive measures,* Nagoya Math. J. 119 (1990), 133-141.

[3] L. Beznea: *Ultrapotentials and positive eigenfunctions for an absolutely continuous resolvent of kernels,* Nagoya Math. J. 112 (1988), 125-142.

[4] J. Bliedtner, W. Hansen: *Potential Theory: An Analytic and Probabilistic Approach to Balayage,* Springer, Berlin - Heidelberg - New York - Tokyo (1986).

[5] R. M. Blumenthal, R. K. Getoor: *Markov Processes and Potential Theory,* Academic Press, New York - London (1968).

[6] N. Boboc, Gh. Bucur, A. Cornea: *Order and Convexity in Potential Theory: H-Cones,* Lecture Notes in Math. 853, Springer, Berlin - Heidelberg - New York (1981).

[7] N. Boboc, P. Mustaţă: *Considérations axiomatiques sur les fonctions poly-surharmoniques,* Rev. Roum. Math. Pures et Appl. XVI (1971), 1167-1184.

[8] R. Cairoli: *Produits de semi-groupes de transition et produits de processus,* Publ. Inst. Stat. Univ. Paris 15 (1966), 311-384.

[9] R. Cairoli: *Semi-groupes de transition et fonctions excessives,* in "Séminaire de Probabilités I Université de Strasbourg", Lecture Notes in Math. 39, Springer, Berlin - Heidelberg - New York (1967), 18-33.

[10] R. Cairoli: *Une représentation intégrale pour fonctions séparément excessives,* Ann. Inst. Fourier Grenoble 18,1 (1968), 317-338.

[11] C. Dellacherie, P.-A. Meyer: *Probabilités et potentiel,* Chapitres I-IV (1975), Chapitres V-VIII (1980), Chapitres IX-XI (1983), Chapitres XII-XVI (1987), Hermann, Paris.

[12] A. Drinkwater Roberts: *Integral Representation for Multiply Superharmonic Functions,* Math. Ann. 215 (1975), 69-78.

[13] E. B. Dynkin: *Integral Representation of Excessive Measures and Excessive Functions,* Russian Math. Surveys 27,1 (1972), 43-84.

[14] E. B. Dynkin: *Minimal excessive measures and functions,* Trans. Amer. Math. Soc. 258 (1980), 217-244.

[15] R. K. Getoor: *Excessive Measures,* Birkhäuser, Boston - Basel - Berlin (1990).

[16] K. Gowrisankaran: *Multiply harmonic functions,* Nagoya Math. J. 28 (1966), 27-48.

[17] K. Gowrisankaran: *Integral representation for a class of multiply superharmonic functions,* Ann. Inst. Fourier 23,4 (1973), 105-143.

[18] K. Gowrisankaran: *Multiply superharmonic functions,* Ann. Inst. Fourier 25, 3 et 4 (1975), 235-244.

[19] M. Itô: *Sur les Fonctions Polyharmoniques et le Problème de Riquier,* Nagoya Math. J. 37 (1970), 81-90.

[20] M. Itô: *Positive eigen elements for an infinitesimal generator of a diffusion semi-group and their integral representations,* in "Potential Theory Copenhagen 1979", Lecture Notes in Math. 787, Springer, Berlin - Heidelberg - New York (1980), 163-184.

[21] K. Janßen: *Choquet-type integral representation of polysupermedian measures,* to appear in Potential Analysis.

[22] H. Kunita, T. Watanabe: *Markov processes and Martin boundaries I,* Ill. J. Math. 9 (1965), 485-526.

[23] G. Mazziotto: *Two-parameter Hunt processes and a potential theory,* Ann. Prob. 16 (1988), 600-619.

[24] P.-A. Meyer: *Processus de Markov: la frontière de Martin,* Lecture Notes in Math. 77, Springer, Berlin - Heidelberg - New York (1968).

[25] P.-A. Meyer: *Représentation intégrale des fonctions excessives, résultats de Mokobodzki,* in "Séminaire de Probabilités V Université de Strasbourg", Lecture Notes in Math. 191, Springer, Berlin - Heidelberg - New York (1971), 196-208.

[26] G. Mokobodzki: *Dualité formelle et représentation intégrale des fonctions excessives,* Actes, Congrès intern. Math. 2 D5 (1970), Gauthier-Villars, Paris (1971), 531-535.

[27] H.-H. Müller: *Integraldarstellung polyexzessiver Funktionen,* Dissertation, Universität Düsseldorf (1992).

[28] M. Nicolesco: *Recherche sur les fonctions polyharmoniques,* Ann. Sci. École Norm. Sup. (3) 52 (1935), 183-220.

[29] M. Sharpe: *General Theory of Markov Processes,* Academic Press, San Diego - London (1988).

[30] J. Steffens: *Duality and integral representation for excessive measures,* Math. Z. 210 (1992), 495-512.

[31] J. B. Walsh: *Probability and a Dirichlet Problem for multiply superharmonic functions,* Ann. Inst. Fourier, 18,2 (1968), 221-279.

Klaus Janßen, Hans-Helge Müller
Mathematisches Institut
Universität Düsseldorf
Universitätsstr.1
40225 Düsseldorf
Germany

REFINING THE LOCAL UNIFORM CONVERGENCE TOPOLOGY

PETER A. LOEB
Department of Mathematics
University of Illinois
1409 West Green Street
Urbana, Ill. 61801, USA

HORST OSSWALD
Mathematisches Institut
der Universität München
Theresienstr. 39
80333 München, Germany

A prominent example for the authors' work in [5] is provided by the space $\mathcal{H}$ consisting of differences of positive harmonic functions where the topology is that of uniform convergence on compact sets. With this topology, often called the u.c.c. topology, $\mathcal{H}$ is not locally solid. That is, neighborhoods U of 0 in $\mathcal{H}$ lack the following property stated in terms of the lattice absolute value $|\cdot|_{\mathcal{H}}$ and elements $h \in U$ and $g \in \mathcal{H}$: If $|g|_{\mathcal{H}} \leq |h|_{\mathcal{H}}$, then $g \in U$. One advantage of locally solid topologies is that lattice operations are always continuous. The general theory in [5] provides a method for constructing the coarsest possible, locally solid refinement of the u.c.c. topology on $\mathcal{H}$. We show in this note that this refinement can be obtained from an integral representation using the total variation norm on the corresponding finite signed measures. We also give a description of the resulting topology without reference to an integral representation.

If no further generality is desired, the reader may assume that we are speaking about ordinary harmonic functions on a domain W in Euclidean space. Our discussion, however, is valid for quite general potential-theoretic settings. For convenience, we will speak in terms of a Brelot [2] harmonic space on a σ-compact domain W and assume that 1 is superharmonic. With some modifications, as in [4], the discussion is also valid for a Bauer [1] harmonic space.

For a simple case, let W be the open unit disk D in the complex plane with ordinary harmonic functions. To demonstrate that neighborhoods of 0 are not solid, we partition the unit circle C into intervals of length $\pi/2^n$. Starting from the point 1 on C, set $f_n = 1$ on the even numbered intervals and set $f_n = -1$ on the odd numbered intervals. Let h_n denote the harmonic extension of f_n. The sequence $< h_n >_{n \in \mathbf{N}}$ converges to 0 in the u.c.c. topology. On the other hand, for each $n \in \mathbf{N}$, the lattice absolute value $|h_n|_{\mathcal{H}}$ is the harmonic extension of $|f_n|$; i.e., $|h_n|_{\mathcal{H}} \equiv 1$ on the disk D. Thus we have an example of a sequence converging to 0 for which the lattice absolute value does not converge to 0. A discussion of the measurability of the lattice operations can be found in Gowrisankaran's article [3].

For the general case, we fix a point w_0 in W and let Φ consist of the positive harmonic functions on W which take the value 1 at w_0. Since Φ is a Choquet simplex, there is a one-to-one correspondence with the space of Borel probability measures on the extreme elements Φ_e of Φ. Recall that an arbitrary harmonic function h has a lattice absolute

K. GowriSankaran et al. (eds.), Classical and Modern Potential Theory and Applications, 315–316.

value if and only if it is in $\mathcal{H}$. Moreover, $\mathcal{H}$ corresponds to the space M of finite signed Borel measures on Φ_e. That is, given $h \in \mathcal{H}$, let h^+ be the least harmonic majorant of $h \vee 0$, and h^- be the least harmonic majorant of $-h \vee 0$, so $h = h^+ - h^-$. Let ν^+ and ν^- be the nonnegative elements of M representing h^+ and h^- respectively. Then $\nu^+ - \nu^-$ represents h, $\nu^+ + \nu^-$ represents $|h|_{\mathcal{H}}$, and $(\nu^+ + \nu^-)(\Phi_e)$ is the total variation norm of $\nu^+ + \nu^-$. With the topology given by the total variation norm, the space M is a locally solid topological vector lattice. Here is a way to construct the corresponding topology on $\mathcal{H}$ without reference to an integral representation.

Theorem. *For each h in $\mathcal{H}$ set $\|h\| = |h|_{\mathcal{H}}(w_0)$. This norm on $\mathcal{H}$ generates the coarsest possible, locally solid refinement of the u.c.c. topology. Moreover, $\mathcal{H}$ with this norm is isometric to M supplied with the total variation norm.*

Proof. Let $\mathcal{U}$ be the usual base for the u.c.c. neighborhoods of 0 in $\mathcal{H}$. For each $U \in \mathcal{U}$, let $\widetilde{U} := \{h \in \mathcal{H} : |h|_{\mathcal{H}} \in U\} \subseteq U$. The topology generated by the family $\widetilde{\mathcal{U}} := \{\widetilde{U} : U \in \mathcal{U}\}$ is finer than the u.c.c. topology and is locally solid. Any locally solid refinement of the u.c.c. topology produces for each $U \in \mathcal{U}$ a smaller neighborhood V of 0 such that if $g \in V \subseteq U$, then $|g|_{\mathcal{H}} \in V \subseteq U$, whence, $V \subseteq \widetilde{U}$. By Harnack's inequality, the topology generated by $\widetilde{\mathcal{U}}$ is metrizable with respect to the norm $h \mapsto \|h\|$. Given $h \in \mathcal{H}$, let ν^+ and ν^- be the nonnegative elements of M representing h^+ and h^- respectively. Then

$$\|h\| = (h^+ + h^-)(w_0) = (\nu^+ + \nu^-)(\Phi_e). \quad \square$$

The first author's work was supported in part by a grant from the U.S. National Science Foundation.

References

[1] H. Bauer, *Harmonische Räume und ihre Potential Theorie*, Springer-Verlag, Berlin, 1966.

[2] M. Brelot, *Lectures on Potential Theory*, Tata Institute, Bombay, 1961.

[3] K. Gowrisankaran, Measurability of lattice operations in a cone, *Proc. Amer. Math. Soc.* 41(1973), 237-240.

[4] P. A. Loeb, A regular metrizable boundary for solutions of elliptic and parabolic differential equations, *Math. Annalen* 251(1980), 43-50.

[5] P. A. Loeb and H. Osswald, Nonstandard integration theory in topological vector lattices, to appear.

Daily rheological phenomena

TAKAFUMI MURAI

Rheology Laboratory, Nagoya University, 464-01, Japan

ABSTRACT. We are interested in rheological phenomena existing in our daily lives. Such phenomena include many problems which must be discussed prior to theoretical formulation: eg. motivations, meanings of settings. This paper is written as an attempt to explain the process from a real motivation till a rough theory.

1. Daily lives

There are much deformation and flow around us. Interpreting such phenomena in our daily lives is our rheology. "Rhe" does not mean only fluid motion such as flow of water and vortex motion of oil, but means every change; flow of people, transit of society etc.. The world is filled with mysteries and riddles, when we look around with curiosity. We are happy and satisfied, when invariant is found in a rheological phenomenon. I have a camera in my walk. "After a long study, now comes daily rheology." (The author would like to express his thanks to the organizing committee of NATO ARW for giving him a chance at a mathematical conference to show his attempt at daily rheology.)

2. Stability of the stratospheric ozone layer

Our family went out to Jumbo Swimming Pool in Nagashima Spa (Jul. 1992) after a long time. Since we were sunburnt very much, we brooded over the world environment, considering that our sunburn was caused by the decrease of the ozone layer. After our vacation, I read [S] (which is an introduction to the ozone layer for beginners), visited Prof. Iwasaka (Solar-Terrestrial Env.) to ask some questions, and next went to the Japan Meteorological Agency to get some data. One of my worries is as follows: When CFCs in the atmosphere will disappear (according to a plan in Montreal Protocol), will the ozone layer return to the previous situation ? This question is called the stability problem and is not nonsense, because in dynamics, the following so-called phase transition frequently appears: Once a balance is destroyed, then a situation arrives at a different phase from the previous one, even if the dynamical condition returns to the previous one. The ozone layer protects us from ultra-violet radiations absorbing these radiations, and a theory asserts that megalosaurs died by a phase transition of the ozone layer. Thus, if a phase transition of the ozone layer appears again, our effort will be a past cure. Answering this stability problem and estimating the phase transition point (i.e. the critical point) are big themes which require an international project team. Since our laboratory is tiny, we try only to do our best. Our opinion (based on the discussion in 2.1–2.6) is as follows: The stratospheric ozone layer is stable as long as damage is small, i.e. the ozone layer will return to the previous situation when CFCs will disappear, if the ozone depletion by CFCs is sufficiently small. The stability is reduced to an inequality $k_1[O][O_2][M] \gg k_2[O][O_3]$,

K. GowriSankaran et al. (eds.), Classical and Modern Potential Theory and Applications, 317–352.

where k_1, k_2 denote the rate constants of two reactions $O + O_2 + M \to O_3 + M$ and $O + O_3 \to O_2 + O_2$, respectively. "Sunburn causes anxiety about the ozone layer:

$$\frac{\partial}{\partial t}\begin{pmatrix} [O] \\ [O(^1D)] \\ [O_3] \\ [NO_x] \\ [HO_x] \\ [ClO_x] \end{pmatrix} = \mathbf{F}([O],[O(^1D)],[O_3],[NO_x],[HO_x],[ClO_x])".$$

2.1. Basic information

We begin by noting roughly basic information concerning the ozone layer. The atmosphere consists of many gases (N_2, O_2, Ar, CO_2, Ne, He, CH_4, Kr, H_2, N_2O, CO, Xe, O_3, NH_3, H_2O, $\cdots$) and the ozone O_3 is one of principal constituents. The atmosphere is also classified into the troposphere (~ 10 km), the stratosphere (10 km ~ 60 km), the mesosphere, the thermosphere, $\cdots$ and the exosphere (600 km ~). The vertical distribution of the concentration of O_3 is similar to a Gaussian distribution, and its concentration is maximal at the altitude about 25 km. It may be reasonable to call the part (10 km ~ 40 km) the (stratospheric) ozone layer. After the detection of a halocarbon CCl_3F in the North Atlantic Ocean in the beginning of 70th, the destruction of the ozone layer was predicted, and this prediction came true by the outbreak of the ozone hole at South Pole (spring, late 70th). From this fact, the Montreal Protocol (1987) etc. oblige each country to observe the ozone layer and to superintend the production of chlorofluorocarbons(CFCs). At present, the ozone layer is analyzed mainly by using TOMS(Total Ozone Mapping Spectrometer)-data which is sent from Nimbus 7 once in a day. The ozone data are gathered at WODC(World Ozone Data Center) in WMO(World Meteorology Organization), Canada, and these data are published under the title "Ozone data for the world". In this chapter, we refer to [JEA1, 2, JMA1, 2] which are published by the Japan Environmental Agency and the Japan Meteorological Agency.

Dobson unit is used to show the width of the ozone layer. Take a pillar of the air with section area 1 cm^2, and measure the volume of O_3 in this pillar under the standard condition (0° C, 10^3 hPa). If the volume is $x \cdot 10^{-3}$ cm^3 (i.e. the weight is $x \times \frac{48.0}{22.4} \times 10^{-6}$ g), we say that the ozone layer has x Dobson units (at the point where the pillar stands) or Dobson number is x. The profile of actual Dobson number is very complicated. (cf. Data 1 in DATA page; this was copied from [JEA1, p. 13] with a suitable modification.) Thus a suitably mollified mean $F(t, P)$ of Dobson number is studied usually, where t denotes the last two figures of "year" and P is the point where the pillar stands; the smooth curves in Data 1 correspond to F. The outbreak year T_0 of the ozone depletion is about 70. Data 2 shows the mean profile of F over the world from Jan. 79 to Dec. 90 by TOMS-data. (This was traced from [JEA1, p. 7] with a suitable modification. Since

Mt. Pinatubo, Luzon exploded in Jun. 1991, it is better not to use recent data (91$\sim$93).) Data 3 shows an ozone hole at the South Pole (Oct. 90); this is a rough sketch (cf. [JEA1, p. 32]).

The profile of Dobson number is related to various factors which are classified into the following two parts: (A) Factors related to chemical reactions, (B) Factors related to atmospheric flow. Factors in (A) are also classified into the following three parts (cf. Section 2.2): (A1) Chapman mechanism, (A2) Dissociation of nitrogen oxides NO_x and hydrogen oxides HO_x, (A3) Dissociation of ClO_x, BrO_x etc.. There are many artificial trace constituents which are decomposed in the stratosphere and whose atmospheric lifetimes are very long: eg. CFC-11(= CCl_3F), CFC-12(=CCl_2F_2), methylchloroform(=CH_3CCl_3), carbon tetrachloride(=CCl_4), CFC-113(=CCl_2FCClF_2), $\cdots$. At present, it is considered that such constituents cause the ozone depletion. Data 4 shows CFC-concentrations (pptv/v $= 10^{-12}$ v/v) at Sapporo (43° N) and Showa Base (70° S). This was written by referring to [JEA2, p. 37] (fat lines) and WRI's data "World Resources 90–91" (dots). Thin lines are linear extensions to see the first year T_{CFC} when CFCs began to accumulate in the troposphere. There is no manufacturing zone around these places, and hence these trace gases are considered as background contaminants. Data 4 shows that CFCs began to accumulate in the troposphere at about 64. It takes $T_0 - T_{CFC}$ years for CFCs to arrive at the stratosphere according to the vertical eddy diffusion; $T_0 - T_{CFC}$ is called the latent period. Factors in (B) (eg. QBO(quasi-biennial oscillation), atmospheric circulations) cause the long-range transportation of O_3. (See DATA page.)

2.2. Ozone Equation

According to [AH, CWL, S], more than 100 chemical reactions and photolyses are related to the generation of O_3. Referring to [JEA1, 2], we pick up the following 26 reactions consisting of O, N, H, Cl:

Data 1 (Ozone Deviation:%)

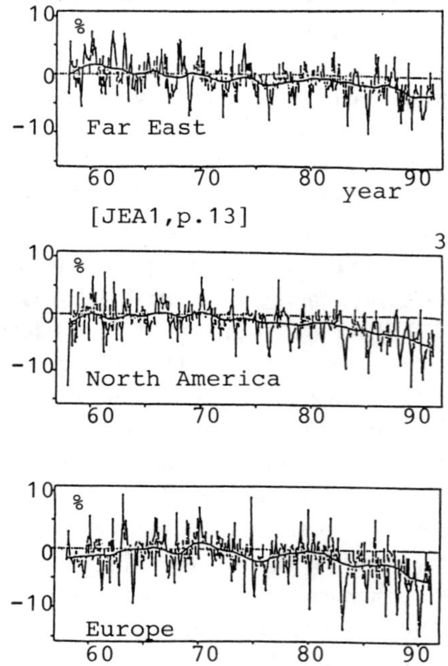

Data 2 (Ozone Map: Dobson unit)

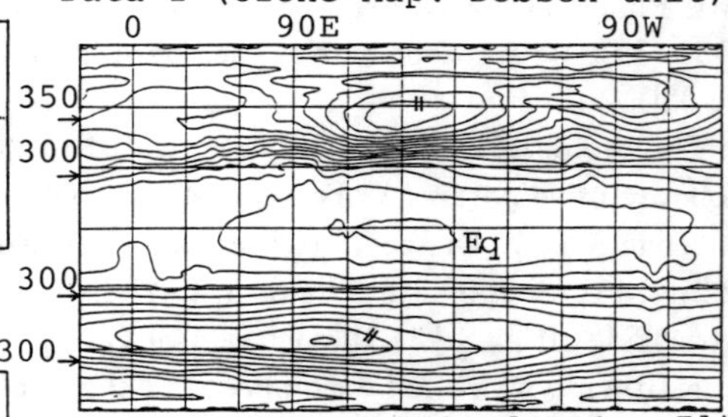

(//350, ‖400 [JEA1,p.7])

Data 3

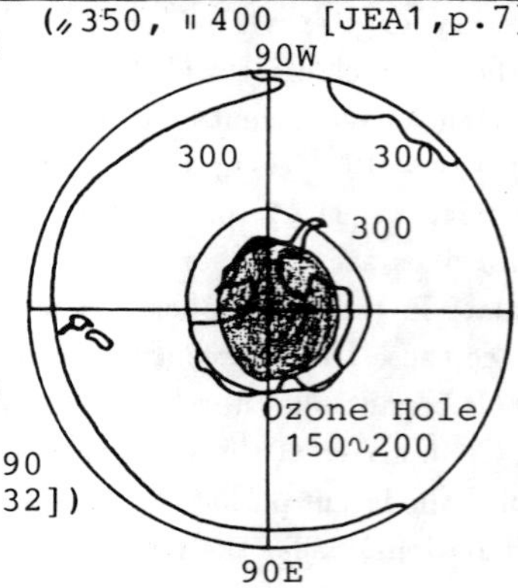

(Ozone Hole Oct. 90 [JEA1,p.32])

Data 4 (CFC-concentrations: pptv/v)

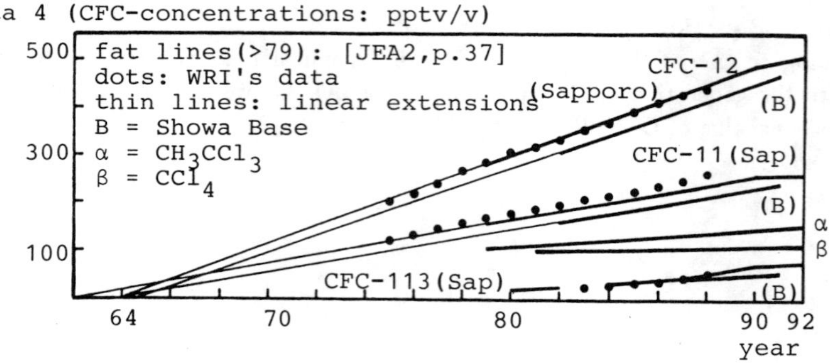

(WRI's data:pptv/v)

year	75	76	77	78	79	80	81	82	83	84	85	86	87	88
CFC-12	200	217	239	266	283	307	315	330	350	366	384	404	421	439
CFC-11	120	133	148	159	167	179	185	193	205	213	223	232	247	263
CFC-113	-	-	-	-	-	-	-	-	24	27	31	35	41	52

We would like to express our thanks to Prof. Iwasaka for calling our attention to various important data.

1) $O_2 + h_\nu(\text{205–242nm}) \xrightarrow{J_1} O + O$

2) $O + O_2 + M \xrightarrow{k_1} O_3 + M$

3) $O_3 + h_\nu(\text{405–850nm}) \xrightarrow{J_2} O + O_2^*$

4) $O + O_3 \xrightarrow{k_2} O_2^* + O_2^*$

5) $O_3 + h_\nu(< \text{310nm}) \xrightarrow{J_3} O(^1D) + O_2(^1\Delta g)^*$

6) $O(^1D) + M \xrightarrow{k_3} O + M$

7) $O(^1D) + N_2O \xrightarrow{k_4} NO + NO$

8) $NO + O_3 \xrightarrow{k_5} NO_2 + O_2^*$

9) $NO_2 + h_\nu(\text{190–400nm}) \xrightarrow{J_4} NO + O$

10) $O(^1D) + H_2O \xrightarrow{k_6} OH + OH$

11) $OH + O_3 \xrightarrow{k_7} HO_2 + O_2^*$

12) $HO_2 + O \xrightarrow{k_8} OH + O_2^*$

13) $HO_2 + OH \xrightarrow{k_9} H_2O^* + O_2^*$

14) $NO_2 + OH + M \xrightarrow{k_{10}} HNO_3^* + M$

15) $Cl + O_3 \xrightarrow{k_{11}} ClO + O_2^*$

16) $ClO + O \xrightarrow{k_{12}} Cl + O_2^*$

17) $ClO + NO \xrightarrow{k_{13}} Cl + NO_2$

18) $ClO + NO_2 + M \xrightarrow{k_{14}} ClONO_2^* + M$

19) $Cl + HO_2 \xrightarrow{k_{15}} HCl^* + O_2^*$

20) $ClONO_2 + HCl \xrightarrow{\ell_1} HNO_3^* + Cl_2$

21) $Cl_2 + h_\nu(\text{240–450nm}) \xrightarrow{\ell_2} Cl + Cl$

22) $ClONO_2 + H_2O \xrightarrow{\ell_3} HNO_3^* + HOCl$

23) $HOCl + h_\nu(\text{200–410nm}) \xrightarrow{\ell_4} OH + Cl$

24) $ClO + ClO + M \xrightarrow{\ell_5} Cl_2O_2 + M$

25) $Cl_2O_2 + h_\nu(\text{200–360nm}) \xrightarrow{\ell_6} Cl + ClOO$

26) $ClOO + M \xrightarrow{\ell_7} Cl + O_2^* + M.$

Here J_j, k_j, ℓ_j show the rates of reactions per a second; these values depend on (t, Q) (Q denotes a point in the stratosphere). The system 1)–26) is not closed, and the notation X^* ($X = O_2$, $O_2(^1\Delta g)$, H_2O, HNO_3, $ClONO_2$, HCl) shows substances going out this system. There exist the following 24 substances in this system: O, $O(^1D)$, O_2, O_2^*, $O_2(^1\Delta g)^*$, O_3, NO, NO_2, N_2O, OH, HO_2, H_2O, H_2O^*, Cl, Cl_2, ClO, $ClOO$, Cl_2O_2, NHO_3^*, $ClONO_2$, $ClONO_2^*$, HCl, HCl^*, M = (atmosphere). We consider that $[O_2]$, $[N_2O]$, $[H_2O]$, $[M]$ are determined by a global balance, where $[\cdot] = [\cdot](t, Q)$ shows the number-concentration. Formulae 20)–26) are reactions existing only in ozone holes. We consider that $[ClONO_2] = \ell_0[ClONO_2^*]$, $[HCl] = \ell_0'[HCl^*]$, with suitable constants ℓ_0, ℓ_0' less than 1. The system 1)–4) is Chapman mechanism. Formulae 5)–14) show reactions related to NO_x, HO_x. Chlorine Cl dissociates from CFCs by photolyses such as

$$CCl_3F + h_\nu(\text{150–226nm}) \to Cl + Cl + CClF, \quad CCl_2F_2 + h_\nu(\text{150–226nm}) \to Cl + Cl + CF_2$$

and decomposes O_3 by a reaction 15). Adding 15) and 16), we see that Cl plays the role of a catalyst. Reactions of Cl in ozone holes are also catalytic. Reactions 14), 18), 19), 20) and 22) are purification operations (related to the acid rain). Unknown functions in this system are $[O]$, $[O(^1D)]$, $[O_3]$, $[NO]$, $[NO_2]$, $[OH]$, $[HO_2]$, $[Cl]$, $[Cl_2]$, $[ClO]$, $[ClOO]$, $[Cl_2O_2]$, $[HOCl]$, $[ClONO_2]$ $(= \ell_0[ClONO_2^*])$, $[HCl](= \ell_0'[HCl^*])$. Since there exist 15 unknown functions, 15 equations are necessary to study. The law of mass action shows that

$$\frac{1}{\rho_0}\frac{\partial}{\partial t}[\mathrm{O}] = 2J_1[\mathrm{O_2}] - k_1[\mathrm{O}][\mathrm{O_2}][\mathrm{M}] + J_2[\mathrm{O_3}] - k_2[\mathrm{O}][\mathrm{O_3}] + k_3[\mathrm{O}(^1\mathrm{D})][\mathrm{M}] + J_4[\mathrm{NO_2}] - k_8[\mathrm{HO_2}][\mathrm{O}] - k_{12}[\mathrm{ClO}][\mathrm{O}] + \lambda_{\mathrm{O}},$$

$$\frac{1}{\rho_0}\frac{\partial}{\partial t}[\mathrm{O}(^1\mathrm{D})] = J_3[\mathrm{O_3}] - [\mathrm{O}(^1\mathrm{D})]\{k_3[\mathrm{M}] + k_4[\mathrm{N_2O}] + k_6[\mathrm{H_2O}]\} + \lambda_{\mathrm{O}(^1\mathrm{D})},$$

$$\frac{1}{\rho_0}\frac{\partial}{\partial t}[\mathrm{O_3}] = k_1[\mathrm{O}][\mathrm{O_2}][\mathrm{M}] - J_2[\mathrm{O_3}] - k_2[\mathrm{O}][\mathrm{O_3}] - J_3[\mathrm{O_3}] - k_5[\mathrm{NO}][\mathrm{O_3}] - k_7[\mathrm{OH}][\mathrm{O_3}] - k_{11}[\mathrm{Cl}][\mathrm{O_3}] + \lambda_{\mathrm{O_3}},$$

$$\frac{1}{\rho_0}\frac{\partial}{\partial t}[\mathrm{NO}] = 2k_4[\mathrm{O}(^1\mathrm{D})][\mathrm{N_2O}] - k_5[\mathrm{NO}][\mathrm{O_3}] + J_4[\mathrm{NO_2}] - k_{13}[\mathrm{ClO}][\mathrm{NO}] + \lambda_{\mathrm{NO}},$$

$$\frac{1}{\rho_0}\frac{\partial}{\partial t}[\mathrm{NO_2}] = k_5[\mathrm{NO}][\mathrm{O_3}] - J_4[\mathrm{NO_2}] - k_{10}[\mathrm{OH}][\mathrm{NO_2}][\mathrm{M}] + k_{13}[\mathrm{ClO}][\mathrm{NO}] - k_{14}[\mathrm{ClO}][\mathrm{NO_2}][\mathrm{M}] + \lambda_{\mathrm{NO_2}},$$

$$\frac{1}{\rho_0}\frac{\partial}{\partial t}[\mathrm{OH}] = 2k_6[\mathrm{O}(^1\mathrm{D})][\mathrm{H_2O}] - k_7[\mathrm{OH}][\mathrm{O_3}] + k_8[\mathrm{HO_2}][\mathrm{O}] - k_9[\mathrm{HO_2}][\mathrm{OH}] - k_{10}[\mathrm{NO_2}][\mathrm{OH}][\mathrm{M}] + \ell_4[\mathrm{HOCl}] + \lambda_{\mathrm{OH}},$$

$$\frac{1}{\rho_0}\frac{\partial}{\partial t}[\mathrm{HO_2}] = k_7[\mathrm{OH}][\mathrm{O_3}] - k_8[\mathrm{HO_2}][\mathrm{O}] - k_9[\mathrm{HO_2}][\mathrm{OH}] - k_{15}[\mathrm{Cl}][\mathrm{HO_2}] + \lambda_{\mathrm{HO_2}},$$

$$\frac{1}{\rho_0}\frac{\partial}{\partial t}[\mathrm{Cl}] = k^0_{\mathrm{CFC}}\tau^6(\mathrm{CFC}) - k_{11}[\mathrm{Cl}][\mathrm{O_3}] + k_{12}[\mathrm{ClO}][\mathrm{O}] + k_{13}[\mathrm{ClO}][\mathrm{NO}] - k_{15}[\mathrm{Cl}][\mathrm{HO_2}] + 2\ell_2[\mathrm{Cl_2}] + \ell_4[\mathrm{HOCl}] + \ell_6[\mathrm{Cl_2O_2}] + \ell_7[\mathrm{ClOO}][\mathrm{M}] + \lambda_{\mathrm{Cl}} \quad (\tau = 10),$$

$$\frac{1}{\rho_0}\frac{\partial}{\partial t}[\mathrm{Cl_2}] = \ell_1[\mathrm{ClONO_2}] - \ell_2[\mathrm{Cl_2}] + \lambda_{\mathrm{Cl_2}},$$

$$\frac{1}{\rho_0}\frac{\partial}{\partial t}[\mathrm{ClO}] = k_{11}[\mathrm{Cl}][\mathrm{O_3}] - k_{12}[\mathrm{ClO}][\mathrm{O}] - k_{13}[\mathrm{ClO}][\mathrm{NO}] - k_{14}[\mathrm{ClO}][\mathrm{NO_2}][\mathrm{M}] - 2\ell_5[\mathrm{ClO}]^2[\mathrm{M}] + \lambda_{\mathrm{ClO}},$$

$$\frac{1}{\rho_0}\frac{\partial}{\partial t}[\mathrm{ClOO}] = \ell_6[\mathrm{Cl_2O_2}] - \ell_7[\mathrm{ClOO}][\mathrm{M}] + \lambda_{\mathrm{ClOO}},$$

$$\frac{1}{\rho_0}\frac{\partial}{\partial t}[\mathrm{Cl_2O_2}] = \ell_5[\mathrm{ClO}]^2[\mathrm{M}] - \ell_6[\mathrm{Cl_2O_2}] + \lambda_{\mathrm{Cl_2O_2}},$$

$$\frac{1}{\rho_0}\frac{\partial}{\partial t}[\mathrm{ClONO_2}] = \ell_0 k_{14}[\mathrm{ClO}][\mathrm{NO_2}][\mathrm{M}] - \ell_1[\mathrm{ClONO_2}][\mathrm{HCl}] - \ell_3[\mathrm{ClONO_2}][\mathrm{H_2O}] + \lambda_{\mathrm{ClONO_2}},$$

$$\frac{1}{\rho_0}\frac{\partial}{\partial t}[\mathrm{HCl}] = \ell_0' k_{15}[\mathrm{Cl}][\mathrm{HO_2}] - \ell_1[\mathrm{ClONO_2}][\mathrm{HCl}] + \lambda_{\mathrm{HCl}},$$

$$\frac{1}{\rho_0}\frac{\partial}{\partial t}[\mathrm{HOCl}] = \ell_3[\mathrm{ClONO_2}][\mathrm{H_2O}] - \ell_4[\mathrm{HOCl}] + \lambda_{\mathrm{HOCl}},$$

where λ_X shows the term concerning the diffusion, the transportation, Coriolis force and other factors in (B), (CFC) denotes a function related to CFC-concentrations in the stratosphere and $\rho_0 = 365\times 24\times 60^2 = \eta_0\tau^7$ ($\eta_0 \simeq 3$); this constant is necessary, since $\partial/\partial t$ corresponds to the rate per a year. Considering the mean situation of the stratosphere, we neglect the variable Q (i.e. $[\cdot](t,Q) = [\cdot](t)$) and put $\lambda_X \equiv 0$. The quantities J_j, k_j are read as functions of t, and each ℓ_j is replaced by a smaller function of t, which is also denoted by the same notation ℓ_j; this procedure is necessary, since 20)–26) exist only in ozone holes. Forming families $NO_x = NO + NO_2$, $HO_x = OH + HO_2$, $ClO_x = Cl + Cl_2 + ClO + ClOO + Cl_2O_2$, PSC = $ClONO_2$ + HCl + HOCl (PSC means substances existing in Polar Stratospheric Clouds), we read the above equations as:
(Ozone Equation)

$$(1)\qquad \frac{1}{\rho_0}\frac{d}{dt}[\mathrm{O}] = 2J_1[\mathrm{O_2}] + J_2[\mathrm{O_3}] + k_3[\mathrm{O}(^1\mathrm{D})][\mathrm{M}] + J_4'[\mathrm{NO}_x] - \{k_1[\mathrm{O_2}][\mathrm{M}] + k_2[\mathrm{O_3}] + k_8'[\mathrm{HO}_x] + k_{12}'[\mathrm{ClO}_x]\}[\mathrm{O}],$$

$$(2)\qquad \frac{1}{\rho_0}\frac{d}{dt}[\mathrm{O}(^1\mathrm{D})] = J_3[\mathrm{O_3}] - \{k_3[\mathrm{M}] + k_4[\mathrm{N_2O}] + k_6[\mathrm{H_2O}]\}[\mathrm{O}(^1\mathrm{D})],$$

$$(3)\qquad \frac{1}{\rho_0}\frac{d}{dt}[\mathrm{O_3}] = k_1[\mathrm{O_2}][\mathrm{M}][\mathrm{O}] - \{J_2 + J_3 + k_2[\mathrm{O}] + k_5'[\mathrm{NO}_x] + k_7'[\mathrm{HO}_x] + k_{11}'[\mathrm{ClO}_x]\}[\mathrm{O_3}],$$

$$(4)\qquad \frac{1}{\rho_0}\frac{d}{dt}[\mathrm{NO}_x] = 2k_4[\mathrm{N_2O}][\mathrm{O}(^1\mathrm{D})] - k_{10}'[\mathrm{M}][\mathrm{NO}_x][\mathrm{HO}_x] - k_{14}'[\mathrm{M}][\mathrm{NO}_x][\mathrm{ClO}_x],$$

$$(5)\qquad \frac{1}{\rho_0}\frac{d}{dt}[\mathrm{HO}_x] = 2k_6[\mathrm{H_2O}][\mathrm{O}(^1\mathrm{D})] - 2k_9'[\mathrm{HO}_x]^2 - k_{10}'[\mathrm{M}][\mathrm{NO}_x][\mathrm{HO}_x] - k_{15}'[\mathrm{HO}_x][\mathrm{ClO}_x] + \ell_4'[\mathrm{PSC}],$$

$$(6)\qquad \frac{1}{\rho_0}\frac{d}{dt}[\mathrm{ClO}_x] = k^0_{\mathrm{CFC}}\tau^6(\mathrm{CFC}) + (\ell_2' + \ell_6')[\mathrm{ClO}_x] + (\ell_1' + \ell_4')[\mathrm{PSC}] - k_{14}'[\mathrm{M}][\mathrm{NO}_x][\mathrm{ClO}_x] - k_{15}'[\mathrm{HO}_x][\mathrm{ClO}_x] - \ell_5'[\mathrm{M}][\mathrm{ClO}_x]^2,$$

$$(7)\qquad \frac{1}{\rho_0}\frac{d}{dt}[\mathrm{PSC}] = L_0[\mathrm{M}][\mathrm{NO}_x][\mathrm{ClO}_x] + L_0'[\mathrm{HO}_x][\mathrm{ClO}_x] - \ell_4'[\mathrm{PSC}] - 2\ell_1'[\mathrm{PSC}]^2,$$

where

$$J_4' = J_4[\mathrm{NO_2}]/[\mathrm{NO}_x],\qquad k_5' = k_5[\mathrm{NO}]/[\mathrm{NO}_x],\quad k_7' = k_7[\mathrm{OH}]/[\mathrm{HO}_x],$$
$$k_8' = k_8[\mathrm{HO_2}]/[\mathrm{HO}_x],\qquad k_9' = k_9[\mathrm{HO_2}][\mathrm{OH}]/[\mathrm{HO}_x]^2,$$
$$k_{10}' = k_{10}[\mathrm{NO_2}][\mathrm{OH}]/([\mathrm{NO}_x][\mathrm{HO}_x]),\quad k_{11}' = k_{11}[\mathrm{Cl}]/[\mathrm{ClO}_x],\quad k_{12}' = k_{12}[\mathrm{ClO}]/[\mathrm{ClO}_x],$$
$$k_{14}' = k_{14}[\mathrm{NO_2}][\mathrm{ClO}]/([\mathrm{NO}_x][\mathrm{ClO}_x]),\quad k_{15}' = k_{15}[\mathrm{HO_2}][\mathrm{Cl}]/([\mathrm{HO}_x][\mathrm{ClO}_x])$$

and ℓ'_j, L_0, L'_0 are analogously defined. Considering the mean rates of reactions, we regard J_j, J'_j, k_j, k'_j, ℓ'_j, L_0, L'_0 as constants. We read [M], $[O_2]$, $[N_2O]$ and $[H_2O]$ as constants also. Then Ozone Eq. (1)–(7) turns out to be a complete system of 7 unknown functions [O], $[O(^1D)]$, $[O_3]$, $[NO_x]$, $[HO_x]$, $[ClO_x]$, [PSC]. This completes the deduction of Ozone Eq..

2.3. Rate constants

According to [CWL, S], we have approximately

[M] $2\tau^{18}$	$[O_2]$ $3\tau^{17}$	$[N_2O]$ $5\tau^{11}$	$[H_2O]$ $7\tau^{12}$	$[O_3]_{T_0}$ $4\tau^{12}$	$[NO_2]_{T_0}$ $7\tau^8$	$[NO]_{T_0}$ $4\tau^8$	$[HO_2]_{T_0}$ $3\tau^7$	$[OH]_{T_0}$ $8\tau^5$	— —
J_1 $5\tau^{-13}$	J_2 $3\tau^{-4}$	J_3 $2\tau^{-5}$	J_4 $7\tau^{-3}$	k_1 τ^{-33}	k_2 $5\tau^{-16}$	k_3 $3\tau^{-11}$	k_4 $5\tau^{-11}$	k_5 $3\tau^{-15}$	k_6 $2\tau^{-10}$
k_7 $2\tau^{-14}$	k_8 $8\tau^{-11}$	k_9 $7\tau^{-11}$	k_{10} $4\tau^{-30}$	k_{11} $9\tau^{-12}$	k_{12} $4\tau^{-11}$	k_{13} $2\tau^{-11}$	k_{14} $4\tau^{-31}$	k_{15} $4\tau^{-11}$	— —

($\tau = 10$) at the altitude 20 km. We use these orders, since the number-concentration of O_3 is maximal at the altitude about 25 km. To estimate (CFC), it is better to study CFC-concentrations in the higher stratosphere, since the dissociation of Cl from CFCs is active there. (In this chapter, we regard (CFC) as a given function and do not study the relation with Data 4.) Since ozone holes appear only at South Pole in spring, we consider that ℓ'_j, L_0, L'_0 are very small and do not estimate. We rewrite the above table in the following form:

[M] $a_1\tau^{18}$	$[O_2]$ $a_2\tau^{17}$	$[N_2O]$ $a_3\tau^{12}$	$[H_2O]$ $a_4\tau^{13}$	$[O_3]_{T_0}$ $A\tau^{12}$	$[NO_x]_{T_0}$ $a_5\tau^9$	$[HO_x]_{T_0}$ $a_6\tau^7$	— —	— —
J_1 $b_1\tau^{-12}$	J_2 $b_2\tau^{-4}$	J_3 $b_3\tau^{-5}$	J'_4 $b_4\tau^{-3}$	k_1 $c_1\tau^{-33}$	k_2 $c_2\tau^{-15}$	k_3 $c_3\tau^{-11}$	k_4 $c_4\tau^{-10}$	k'_5 $c_5\tau^{-15}$
k_6 $c_6\tau^{-10}$	k'_7 $c_7\tau^{-15}$	k'_8 $c_8\tau^{-10}$	k'_9 $c_9\tau^{-12}$	k'_{10} $c_{10}\tau^{-31}$	k'_{11} $c_{11}\tau^{-12}$	k'_{12} $c_{12}\tau^{-11}$	k'_{14} $c_{14}\tau^{-31}$	k'_{15} $c_{15}\tau^{-11}$

(Here the order of each k'_j is determined from k_j and its definition, and $[Cl]/[ClO_x] = 1/2$ is used.) Then a_j, b_j, c_j, A take values in a range $1/10 \sim 10$. The initial condition is the chemical equilibrium at $t = T_0$ and $[ClO_x]_{T_0} = [PSC]_{T_0} = (CFC)_{T_0} = 0$, i.e.

$$\begin{aligned}(8)\qquad \frac{d}{dt}[O]_{T_0} &= \frac{d}{dt}[O(^1D)]_{T_0} = \frac{d}{dt}[O_3]_{T_0} = \frac{d}{dt}[NO_x]_{T_0} = \frac{d}{dt}[HO_x]_{T_0} \\ &= [ClO_x]_{T_0} = [PSC]_{T_0} = (CFC)_{T_0} = 0.\end{aligned}$$

We now study the relation among a_j, b_j, c_j, A which is required from the initial condition.

Condition (8) is equivalent to

$$\begin{aligned}
2J_1[\mathrm{O}_2] + J_2[\mathrm{O}_3]_{T_0} + k_3[\mathrm{O}(^1\mathrm{D})]_{T_0}[\mathrm{M}] + J_4'[\mathrm{NO}_x]_{T_0} \\
= \{k_1[\mathrm{O}_2][\mathrm{M}] + k_2[\mathrm{O}_3]_{T_0} + k_8'[\mathrm{HO}_x]_{T_0}\}[\mathrm{O}]_{T_0},\\
J_3[\mathrm{O}_3]_{T_0} &= \{k_3[\mathrm{M}] + k_4[\mathrm{N_2O}] + k_6[\mathrm{H_2O}]\}[\mathrm{O}(^1\mathrm{D})]_{T_0},\\
k_1[\mathrm{O}_2][\mathrm{M}][\mathrm{O}]_{T_0} &= \{J_2 + J_3 + k_2[\mathrm{O}]_{T_0} + k_5'[\mathrm{NO}_x]_{T_0} + k_7'[\mathrm{HO}_x]_{T_0}\}[\mathrm{O}_3]_{T_0},\\
2k_4[\mathrm{N_2O}][\mathrm{O}(^1\mathrm{D})]_{T_0} &= k_{10}'[\mathrm{M}][\mathrm{NO}_x]_{T_0}[\mathrm{HO}_x]_{T_0},\\
2k_6[\mathrm{H_2O}][\mathrm{O}(^1\mathrm{D})]_{T_0} &= 2k_9'[\mathrm{HO}_x]^2 + k_{10}'[\mathrm{M}][\mathrm{NO}_x]_{T_0}[\mathrm{HO}_x]_{T_0},\\
[\mathrm{ClO}_x]_{T_0} &= [\mathrm{PSC}]_{T_0} = (\mathrm{CFC})_{T_0} = 0.
\end{aligned}$$

These equations yield that

(9)
$$\begin{aligned}
[\mathrm{O}]_{T_0} &= (Ab_2 + a_1c_3[\mathrm{O}(^1\mathrm{D})]_{T_0}\tau^{-1} + a_5b_4\tau^{-2} + 2a_2b_1\tau^{-3})\\
&\quad\times\{a_1a_2c_1 + (Ac_2 + a_6c_8)\tau^{-5}\}^{-1}\tau^6\\
&= A(b_2 + b_3\tau^{-1} + a_5c_5\tau^{-2} + a_6c_7\tau^{-4})\\
&\quad\times(a_1a_2c_1 - Ac_2\tau^{-5})^{-1}\tau^6\ (= B\tau^6,\ \text{say}),
\end{aligned}$$

(10)
$$\begin{aligned}
[\mathrm{O}(^1\mathrm{D})]_{T_0} &= Ab_3(a_1c_3 + a_3c_4\tau^{-5} + a_4c_6\tau^{-4})^{-1} = 5a_1a_3^{-1}a_5a_6c_4^{-1}c_{10}\\
&= a_4^{-1}a_6c_6^{-1}(a_6c_9 + 5a_1a_5c_{10})\tau^{-1}\ (= C,\ \text{say}).
\end{aligned}$$

Replacing $[\mathrm{O}(^1\mathrm{D})]_{T_0}$ in (9) by $Ab_3(a_1c_3 + a_3c_4\tau^{-5} + a_4c_6\tau^{-4})^{-1}$ in (10), we have

$$\begin{aligned}
&\Big\{Ab_2 + Ab_3\tau^{-1}\frac{a_1c_3}{a_1c_3 + a_3c_4\tau^{-5} + a_4c_6\tau^{-4}} + a_5b_4\tau^{-2} + 2a_2b_1\tau^{-3}\Big\}(a_1a_2c_1 - Ac_2\tau^{-5})\\
&\quad= A(b_2 + b_3\tau^{-1} + a_5c_5\tau^{-2} + a_6c_7\tau^{-4})\{a_1a_2c_1 + (Ac_2 + a_6c_8)\tau^{-5}\},
\end{aligned}$$

which shows that $\alpha A^2 + \tau^3\beta A - \tau^3\gamma = 0$ with

$$\begin{aligned}
\alpha &= c_2\Big\{2b_2 + 2b_3\tau^{-1} + a_5c_5\tau^{-2} + a_6c_7\tau^{-4} - \frac{b_3(a_3c_4\tau^{-6} + a_4c_6\tau^{-5})}{a_1c_3 + a_3c_4\tau^{-5} + a_4c_6\tau^{-4}}\Big\},\\
\beta &= a_1a_2c_1\Big\{a_5c_5 + a_6c_7\tau^{-2} + \frac{b_3(a_3c_4\tau^{-4} + a_4c_6\tau^{-3})}{a_1c_3 + a_3c_4\tau^{-5} + a_4c_6\tau^{-4}}\Big\}\\
&\quad+ a_6c_8(b_2 + b_3\tau^{-1} + a_5c_5\tau^{-2} + a_6c_7\tau^{-4})\tau^{-3} + a_5b_4\tau^{-5} + 2a_2b_1c_2\tau^{-6},\\
\gamma &= a_1a_2c_1(a_5b_4 + 2a_2b_1\tau^{-1}).
\end{aligned}$$

Thus the condition required for a_j, b_j, c_j, A is expressed as

$$\begin{aligned}
A &= 5a_1a_3^{-1}a_5a_6b_3^{-1}c_4^{-1}c_{10}(a_1c_3 + a_3c_4\tau^{-5} + a_4c_6\tau^{-4})\\
&= a_4^{-1}a_6b_3^{-1}c_6^{-1}(a_1c_3 + a_3c_4\tau^{-5} + a_4c_6\tau^{-4})(a_6c_9 + 5a_1a_5c_{10})\tau^{-1}\\
&= \frac{1}{2\alpha}\Big\{\sqrt{\tau^6\beta^2 + 4\tau^3\alpha\gamma} - \tau^3\beta\Big\}\ \Big(\simeq \frac{\gamma}{\beta} \simeq a_5^{-1}c_5^{-1}(a_5b_4 + 2a_2b_1\tau^{-1})\Big).
\end{aligned}$$

2.4. Deduction of Mollified Ozone Eq.

It is not useful to study Ozone Eq. (1)–(7) in itself, because the orders of $[\cdot]$, J_j, J'_j, k_j, k'_j are mutually very different. The figures of solutions of Ozone Eq. are very irregular, and the actual profile of Dobson number is also (cf. Data 1). Our next problem is to define a modified equation from Ozone Eq. so that solutions of the resulting equation are sufficiently moderate. To do this, we take a differential equation $g'(t) = a(t) - bg(t)$ $(t \geq T_0)$ such that $g'(T_0) = 0$, where b is a positive constant. Then

$$g(t) - g(T_0) = \int_{T_0}^{t} \{a(s) - a(T_0)\} e^{-b(t-s)}\, ds.$$

If b is large, then the right-hand side is nearly equal to

$$\begin{aligned}\{a(t) - a(T_0)\} \int_{T_0}^{t} e^{-b(t-s)}\, ds &= \frac{1}{b}\{a(t) - a(T_0)\}\{1 - e^{-b(t-T_0)}\} \\ &\simeq \frac{1}{b}\{a(t) - a(T_0)\} \quad (t \geq T_0 + \frac{1}{\sqrt{b}}).\end{aligned}$$

Thus

$$g(t) \simeq g(T_0) + \frac{1}{b}\{a(t) - a(T_0)\} = \frac{1}{b} a(t) \quad (t \geq T_0 + \frac{1}{\sqrt{b}}). \tag{11}$$

To use this method, we write (1)–(5) in the form

$$\frac{d}{dt}[\mathrm{X}] = a_{\mathrm{X}}(t) - b_{\mathrm{X}}(t)[\mathrm{X}] \quad (\mathrm{X} = \mathrm{O},\ \mathrm{O}(^1\mathrm{D}),\ \mathrm{O}_3,\ \mathrm{NO}_x,\ \mathrm{HO}_x)$$

and study the orders of $b_{\mathrm{X}}(T_0)$ $(\mathrm{X} = \mathrm{O}, \cdots, \mathrm{HO}_x)$. Then

$$\begin{aligned}
b_{\mathrm{O}}(T_0) &= \rho_0\{k_1[\mathrm{O}_2][\mathrm{M}] + k_2[\mathrm{O}_3]_{T_0} + k'_8[\mathrm{HO}_x]_{T_0}\} \\
&= \eta_0\{a_1a_2c_1 + (Ac_2 + a_6c_8)\tau^{-5}\}10^9, \\
b_{\mathrm{O}(^1\mathrm{D})}(t) &\equiv \rho_0\{k_3[\mathrm{M}] + k_4[\mathrm{N}_2\mathrm{O}] + k_6[\mathrm{H}_2\mathrm{O}]\} = \eta_0(a_1c_3 + a_3c_4\tau^{-5} + a_4c_6\tau^{-4})10^{14}, \\
b_{\mathrm{O}_3}(T_0) &= \rho_0\{J_2 + J_3 + k_2[\mathrm{O}]_{T_0} + k'_5[\mathrm{NO}_x]_{T_0} + k'_7[\mathrm{HO}_x]_{T_0}\} \\
&= \eta_0(b_2 + b_3\tau^{-1} + Bc_2\tau^{-5} + a_5c_5\tau^{-2} + a_6c_7\tau^{-4})10^3, \\
b_{\mathrm{NO}_x}(T_0) &= \rho_0k'_{10}[\mathrm{M}][\mathrm{HO}_x]_{T_0} = \eta_0a_1a_6c_{10}10, \\
b_{\mathrm{HO}_x}(T_0) &= \rho_0\{2k'_9[\mathrm{HO}_x]_{T_0} + k'_{10}[\mathrm{M}][\mathrm{NO}_x]_{T_0}\} = \eta_0(2a_6c_9\tau^{-1} + a_1a_5c_{10})10^3.
\end{aligned}$$

Since the orders of $b_{\mathrm{O}}(T_0)$, $b_{\mathrm{O}(^1\mathrm{D})}(t)(\equiv b_{\mathrm{O}(^1\mathrm{D})}(T_0))$ are very large, we use (11) with "$t \geq T_0 + 1/\sqrt{b}$" replaced by "$t \geq T_0$". Taking account of the estimate of $b_{\mathrm{O}}(T_0)$, we neglect the term $\rho_0\{k_2[\mathrm{O}_3] + k'_8[\mathrm{HO}_x] + k'_{12}[\mathrm{ClO}_x]\}$ in $b_{\mathrm{O}}(t)$ and apply (11). Then

$$[\mathrm{O}] \simeq \rho_0\{2J_1[\mathrm{O}_2] + J_2[\mathrm{O}_3] + k_3[\mathrm{M}][\mathrm{O}(^1\mathrm{D})] + J'_4[\mathrm{NO}_x]\}\{\rho_0k_1[\mathrm{O}_2][\mathrm{M}]\}^{-1}.$$

Reviving $\rho_0\{k_2[\mathrm{O_3}] + k'_8[\mathrm{HO}_x] + k'_{12}[\mathrm{ClO}_x]\}$, we have

$$[\mathrm{O}] \simeq \{2J_1[\mathrm{O_2}] + J_2[\mathrm{O_3}] + k_3[\mathrm{M}][\mathrm{O(^1D)}] + J'_4[\mathrm{NO}_x]\} \\ \times \{k_1[\mathrm{O_2}][\mathrm{M}] + k_2[\mathrm{O_3}] + k'_8[\mathrm{HO}_x] + k'_{12}[\mathrm{ClO}_x]\}^{-1}.$$

Applying (11) to (2), we have

$$[\mathrm{O(^1D)}] \simeq \rho_0 J_3[\mathrm{O_3}]\rho_0^{-1}\{k_3[\mathrm{M}] + k_4[\mathrm{N_2O}] + k_6[\mathrm{H_2O}]\}^{-1}.$$

Since

$$k_4[\mathrm{N_2O}] + k_6[\mathrm{H_2O}] \le \{k_4[\mathrm{N_2O}] + k_6[\mathrm{H_2O}]\}[\mathrm{NO}_x]_{T_0}^{-1}[\mathrm{O_3}]_{T_0} \ll k_3[\mathrm{M}],$$

we have

$$[\mathrm{O(^1D)}] \simeq J_3[\mathrm{O_3}]\{k_3[\mathrm{M}] + (k_4[\mathrm{N_2O}] + k_6[\mathrm{H_2O}])[\mathrm{NO}_x]_{T_0}^{-1}[\mathrm{O_3}]_{T_0}\}^{-1}.$$

The estimates of $[\mathrm{ClO}_x]$ and [PSC] are more rough. If $d[\mathrm{PSC}]/dt = 0$, then (7) shows that

$$2\ell'_1[\mathrm{PSC}]^2 + \ell'_4[\mathrm{PSC}] - \{L_0[\mathrm{M}][\mathrm{NO}_x] + L'_0[\mathrm{HO}_x]\}[\mathrm{ClO}_x] = 0,$$

which yields that

$$[\mathrm{PSC}] = \frac{1}{4\ell'_1}\{\sqrt{\ell'^2_4 + 8\ell'_1(L_0[\mathrm{M}][\mathrm{NO}_x] + L'_0[\mathrm{HO}_x])[\mathrm{ClO}_x]} - \ell'_4\} \\ \simeq \frac{1}{\ell'_4}\{L_0[\mathrm{M}][\mathrm{NO}_x] + L'_0[\mathrm{HO}_x]\}[\mathrm{ClO}_x].$$

(In this discussion, we assume that L_0 and L'_0 are sufficiently small.) Considering that $d[\mathrm{PSC}]/dt$ is small, we use an estimate

$$[\mathrm{PSC}] \simeq \frac{1}{\ell'_4}\{L_0[\mathrm{M}][\mathrm{NO}_x] + L'_0[\mathrm{HO}_x]\}[\mathrm{ClO}_x].$$

Analogously, it is reasonable, from (6), to assume that

$$[\mathrm{ClO}_x] \simeq \frac{k^0_{\mathrm{CFC}}\tau^6(\mathrm{CFC}) + (\ell'_1 + \ell'_4)[\mathrm{PSC}]}{k'_{14}[\mathrm{M}][\mathrm{NO}_x] + k'_{15}[\mathrm{HO}_x] - (\ell'_2 + \ell'_6)}.$$

Considering that

$$(\ell'_1 + \ell'_4)[\mathrm{PSC}] \ll k^0_{\mathrm{CFC}}\tau^6(\mathrm{CFC}),\ \ \ell'_2 + \ell'_6 \ll k'_{14}[\mathrm{M}][\mathrm{NO}_x] + k'_{15}[\mathrm{HO}_x],$$

we take an estimate

$$[\mathrm{ClO}_x] \simeq \frac{k^0_{\mathrm{CFC}}\tau^6(\mathrm{CFC}) + \ell'_4[\mathrm{PSC}]}{k'_{14}[\mathrm{M}][\mathrm{NO}_x] + k'_{15}[\mathrm{HO}_x]}.$$

Taking account of these estimates and Ozone Eq., we define 7 equations of 7 unknown functions $[\mathrm{O}']$, $[\mathrm{O}(^1\mathrm{D})']$, $[\mathrm{O}'_3]$, $[\mathrm{NO}'_x]$, $[\mathrm{HO}'_x]$, $[\mathrm{ClO}'_x]$, $[\mathrm{PSC}']$ by

$$\begin{aligned}
[\mathrm{O}'] &= \{2J_1[\mathrm{O}_2] + J_2[\mathrm{O}'_3] + k_3[\mathrm{M}][\mathrm{O}(^1\mathrm{D})'] + J'_4[\mathrm{NO}'_x]\} \\
&\quad \times \{k_1[\mathrm{O}_2][\mathrm{M}] + k_2[\mathrm{O}'_3] + k'_8[\mathrm{HO}'_x] + k'_{12}[\mathrm{ClO}'_x]\}^{-1}, \\
[\mathrm{O}(^1\mathrm{D})'] &= J_3[\mathrm{O}'_3]\{k_3[\mathrm{M}] + (k_4[\mathrm{N}_2\mathrm{O}] + k_6[\mathrm{H}_2\mathrm{O}])[\mathrm{NO}_x]^{-1}_{T_0}[\mathrm{O}_3]_{T_0}\}^{-1}, \\
\frac{1}{\rho_0}\frac{d}{dt}[\mathrm{O}'_3] &= k_1[\mathrm{O}_2][\mathrm{M}][\mathrm{O}'] - \{J_2 + J_3 + k_2[\mathrm{O}'] \\
&\quad + k'_5[\mathrm{NO}'_x] + k'_7[\mathrm{HO}'_x] + k'_{11}[\mathrm{ClO}'_x]\}[\mathrm{O}'_3], \\
\frac{1}{\rho_0}\frac{d}{dt}[\mathrm{NO}'_x] &= 2k_4[\mathrm{N}_2\mathrm{O}][\mathrm{O}(^1\mathrm{D})'] - k'_{10}[\mathrm{M}][\mathrm{NO}'_x][\mathrm{HO}'_x] - k'_{14}[\mathrm{M}][\mathrm{NO}'_x][\mathrm{ClO}'_x], \\
\frac{1}{\rho_0}\frac{d}{dt}[\mathrm{HO}'_x] &= 2k_6[\mathrm{H}_2\mathrm{O}][\mathrm{O}(^1\mathrm{D})'] - 2k'_9[\mathrm{HO}'_x]^2 - k'_{10}[\mathrm{M}][\mathrm{NO}'_x][\mathrm{HO}'_x] \\
&\quad - k'_{15}[\mathrm{HO}'_x][\mathrm{ClO}'_x] + \ell'_4[\mathrm{PSC}'], \\
[\mathrm{ClO}'_x] &= \frac{k^0_{\mathrm{CFC}}\tau^6(\mathrm{CFC}) + \ell'_4[\mathrm{PSC}']}{k'_{14}[\mathrm{M}][\mathrm{NO}'_x] + k'_{15}[\mathrm{HO}'_x]}, \\
[\mathrm{PSC}'] &= \frac{1}{\ell'_4}\{L_0[\mathrm{M}][\mathrm{NO}'_x] + L'_0[\mathrm{HO}'_x]\}[\mathrm{ClO}'_x],
\end{aligned}$$

where each $[\mathrm{X}']$ is chosen so that $[\mathrm{X}']_{T_0} = [\mathrm{X}]_{T_0}$. Since this expression is complicated, we define the normalization of each $[\mathrm{X}']$ and rewrite this system. Normalization is defined as follows:

$$(\mathrm{O}_3) = [\mathrm{O}'_3]/[\mathrm{O}_3]_{T_0},\ \ (\mathrm{NO}_x) = [\mathrm{NO}'_x]/[\mathrm{NO}_x]_{T_0},\ \ (\mathrm{HO}_x) = [\mathrm{HO}'_x]/[\mathrm{NO}_x]_{T_0},$$
$$(\mathrm{O}(^1\mathrm{D})) = [\mathrm{O}(^1\mathrm{D})'],\ \ (\mathrm{O}) = \rho_0[\mathrm{O}']/[\mathrm{O}_3]_{T_0},\ \ (\mathrm{ClO}_x) = [\mathrm{ClO}'_x]\tau^{-6},\ \ (\mathrm{PSC}) = [\mathrm{PSC}']\tau^{-6}.$$

Then

$$\begin{aligned}
(\mathrm{O}_3)_{T_0} &= (\mathrm{NO}_x)_{T_0} = 1, & (\mathrm{HO}_x)_{T_0} &= (a_5^{-1}a_6)\tau^{-2}\,(=\xi,\ \text{say}) \\
(\mathrm{O}(^1\mathrm{D}))_{T_0} &= C, & (\mathrm{O})_{T_0} &= 10\eta_0 A^{-1}B.
\end{aligned}$$

The above system is rewritten as:

(Mollified Ozone Equation)

$$
\begin{aligned}
(\mathrm{O}) &= \{2J_{\mathrm{O}_2} + J_{\mathrm{O}}(\mathrm{O}_3) + k_{\mathrm{O}^1}(\mathrm{O}(^1\mathrm{D})) + J_{\mathrm{N}}(\mathrm{NO}_x)\} \\
&\quad \times \{k_{\mathrm{O}_3} + k_{\mathrm{O}_2}(\mathrm{O}_3) + k^*_{\mathrm{H}}(\mathrm{HO}_x) + k^*_{\mathrm{Cl}}(\mathrm{ClO}_x)\}^{-1}, \\
(\mathrm{O}(^1\mathrm{D})) &= J_{\mathrm{O}^1}(k_{\mathrm{O}^1} + m_{\mathrm{N}} + m_{\mathrm{H}})^{-1}(\mathrm{O}_3), \\
\frac{d}{dt}(\mathrm{O}_3) &= k_{\mathrm{O}_3}(\mathrm{O}) - \{J_{\mathrm{O}} + J_{\mathrm{O}^1} + k_{\mathrm{O}_2}(\mathrm{O}) \\
&\quad + k_{\mathrm{N}}(\mathrm{NO}_x) + k_{\mathrm{H}}(\mathrm{HO}_x) + k_{\mathrm{Cl}}(\mathrm{ClO}_x)\}(\mathrm{O}_3), \\
\frac{d}{dt}(\mathrm{NO}_x) &= 2m_{\mathrm{N}}(\mathrm{O}(^1\mathrm{D})) - k_{\mathrm{NH}}(\mathrm{NO}_x)(\mathrm{HO}_x) - k_{\mathrm{NCl}}(\mathrm{NO}_x)(\mathrm{ClO}_x), \\
\frac{d}{dt}(\mathrm{HO}_x) &= 2m_{\mathrm{H}}(\mathrm{O}(^1\mathrm{D})) - 2k_{\mathrm{H}^2}(\mathrm{HO}_x)^2 - k_{\mathrm{NH}}(\mathrm{NO}_x)(\mathrm{HO}_x) \\
&\quad - k_{\mathrm{HCl}}(\mathrm{HO}_x)(\mathrm{ClO}_x) + \delta_{\mathrm{PSC}}(\mathrm{PSC}), \\
(\mathrm{ClO}_x) &= \frac{k_{\mathrm{CFC}}(\mathrm{CFC}) + \delta_{\mathrm{PSC}}(\mathrm{PSC})}{k_{\mathrm{NCl}}(\mathrm{NO}_x) + k_{\mathrm{HCl}}(\mathrm{HO}_x)}, \\
(\mathrm{PSC}) &= \delta^{-1}_{\mathrm{PSC}}\{\varepsilon_{\mathrm{N}}(\mathrm{NO}_x) + \varepsilon_{\mathrm{H}}(\mathrm{HO}_x)\}(\mathrm{ClO}_x),
\end{aligned}
$$

where

$$
\begin{aligned}
J_{\mathrm{O}_2} &= \rho_0 J_1[\mathrm{O}_2][\mathrm{O}_3]^{-1}_{T_0} = \eta_0 A^{-1} a_2 b_1, & J_{\mathrm{O}} &= \rho_0 J_2 = \eta_0 b_2 10^3, \\
J_{\mathrm{O}^1} &= \rho_0 J_3 = \eta_0 b_3 10^2, & J_{\mathrm{N}} &= \rho_0 J'_4[\mathrm{NO}_x]_{T_0}[\mathrm{O}_3]^{-1}_{T_0} = \eta_0 A^{-1} a_5 b_4 10, \\
k_{\mathrm{O}_3} &= k_1[\mathrm{O}_2][\mathrm{M}] = a_1 a_2 c_1 10^2, & k_{\mathrm{O}_2} &= k_2[\mathrm{O}_3]_{T_0} = A c_2 10^{-3}, \\
k_{\mathrm{O}^1} &= \rho_0 k_3[\mathrm{M}][\mathrm{O}_3]^{-1}_{T_0} = \eta_0 A^{-1} a_1 c_3 10^2, & & \\
m_{\mathrm{N}} &= \rho_0 k_4[\mathrm{N}_2\mathrm{O}][\mathrm{NO}_x]^{-1}_{T_0} = \eta_0 a_3 a_5^{-1} c_4, & m_{\mathrm{H}} &= \rho_0 k_6[\mathrm{H}_2\mathrm{O}][\mathrm{NO}_x]^{-1}_{T_0} = \eta_0 a_4 a_5^{-1} c_6 10, \\
k_{\mathrm{N}} &= \rho_0 k'_5[\mathrm{NO}_x]_{T_0} = \eta_0 a_3 c_5 10, & k_{\mathrm{H}} &= \rho_0 k'_7[\mathrm{NO}_x]_{T_0} = \eta_0 a_5 c_7 10, \\
k^*_{\mathrm{H}} &= k'_8[\mathrm{NO}_x]_{T_0} = a_5 c_8 10^{-1}, & k_{\mathrm{H}^2} &= \rho_0 k'_9[\mathrm{NO}_x]_{T_0} = \eta_0 a_5 c_9 10^4, \\
k_{\mathrm{NH}} &= \rho_0 k'_{10}[\mathrm{M}][\mathrm{NO}_x]_{T_0} = \eta_0 a_1 a_5 c_{10} 10^3, & k_{\mathrm{Cl}} &= \rho_0 k'_{11}\tau^6 = \eta_0 c_{11} 10, \\
k_{\mathrm{NCl}} &= \rho_0 k'_{14}[\mathrm{M}]\tau^6 = \eta_0 a_1 c_{14}, & k_{\mathrm{HCl}} &= \rho_0 k'_{15}\tau^6 = \eta_0 c_{15} 10^2, \\
k^*_{\mathrm{Cl}} &= k'_{12}\tau^6 = c_{12} 10^{-5}, & \delta_{\mathrm{PSC}} &= \rho_0 \ell'_4[\mathrm{NO}_x]^{-1}_{T_0}\tau^6, \\
k_{\mathrm{CFC}} &= \rho_0 k^0_{\mathrm{CFC}}[\mathrm{NO}_x]^{-1}_{T_0}\tau^6, & \varepsilon_{\mathrm{N}} &= \rho_0 L_0[\mathrm{M}]\tau^6, \qquad \varepsilon_{\mathrm{H}} = \rho_0 L'_0\tau^6.
\end{aligned}
$$

Since a_j, b_j, c_j, a_5^{-1}, A, A^{-1} are not so large, we may consider that new rate constants $J_\cdot$, $k_\cdot$ are not so large. We regard that δ_{PSC}, ε_{N}, ε_{H} are small.

2.5. Restart from Mollified Ozone Eq.

Mollified Ozone Eq. is understood as a system deduced from the following formal reactions:

$$1)\ O_2 + h_\nu \xrightarrow{J_{O_2}} O + O \qquad 2)\ O + O_2 + M \xrightarrow{k_{O_3}} O_3 + M$$
$$3)\ O_3 + h_\nu \xrightarrow{J_O} O + O_2^* \qquad 4)\ O + O_3 \xrightarrow{k_{O_2}} O_2^* + O_2^*$$
$$5)\ O_3 + h_\nu \xrightarrow{J_{O^1}} O(^1D) + O_2(^1\Delta g)^* \qquad 6)\ O(^1D) + M \xrightarrow{k_{O^1}} O + M$$
$$7)\ O(^1D) + N_2O \xrightarrow{m_N} NO_x + NO_x \qquad 8)\ O(^1D) + H_2O \xrightarrow{m_H} HO_x + HO_x$$
$$9)\ NO_x + h_\nu \xrightarrow{J_N} NO_x + O \qquad 10)\ NO_x + O_3 \xrightarrow{k_N} NO_x$$
$$11)\ HO_x + O_3 \xrightarrow{k_H} HO_x \qquad 12)\ HO_x + O \xrightarrow{k_H^*} HO_x$$
$$13)\ HO_x + HO_x \xrightarrow{k_{H^2}} H_2O^* + O_2^* \qquad 14)\ NO_x + HO_x \xrightarrow{k_{NH}} HNO_x^*$$
$$15)\ CFC + h_\nu \xrightarrow{k_{CFC}} ClO_x \qquad 16)\ ClO_x + O_3 \xrightarrow{k_{Cl}} ClO_x$$
$$17)\ ClO_x + O \xrightarrow{k_{Cl}^*} ClO_x \qquad 18)\ NO_x + ClO_x \xrightarrow{k_{NCl}} ClONO_x^*$$
$$19)\ HO_x + ClO_x \xrightarrow{k_{HCl}} HClO_x^* \qquad 20)\ NO_x + ClO_x \xrightarrow{\varepsilon_N} PSC + (NO_x + ClO_x)$$
$$21)\ HO_x + ClO_x \xrightarrow{\varepsilon_H} PSC + (HO_x + ClO_x) \quad 22)\ PSC + h_\nu \xrightarrow{\delta_{PSC}} HO_x + ClO_x.$$

Here $(\cdot)$ is read as $[\cdot]$ with $(M) = (O_2) = (N_2O) = (H_2O) = 1$, and $J_\cdot, k_\cdot, \cdots, \varepsilon_\cdot$ are regarded as rate constants per a year. (The notation X^* shows a constituent X going out this system.) Considering that k_{O_3} and k_{O^1} are somewhat large, we apply (11) to equalities for (O) and $(O(^1D))$. Two equalities for (ClO_x) and (PSC) are rough.

We now restart from Mollified Ozone Eq., regarding $J_\cdot, k_\cdot, \cdots, \varepsilon_\cdot$ as given constants. The initial condition is as follows:

$$\frac{d}{dt}(O_3)_{T_0} = \frac{d}{dt}(NO_x)_{T_0} = \frac{d}{dt}(HO_x)_{T_0} = (ClO_x)_{T_0} = (CFC)_{T_0} = 0$$
$$(O_3)_{T_0} = (NO_x)_{T_0} = 1, \quad (HO_x)_{T_0} = \xi.$$

Combined with Mollified Ozone Eq., this condition yields some relations among rates constants. These are expressed as

$$\begin{aligned} B_0 &= (2J_{O_2} + J_O + k_{O^1}C_0 + J_N)(k_{O_3} + k_{O_2} + k_H^*\xi)^{-1} \\ &= k_{O_3}^{-1}(J_O + J_{O^1} + k_{O_2}B_0 + k_N + k_H\xi), \\ C_0 &= J_{O^1}(k_{O^1} + m_N + m_H)^{-1} = (2m_N)^{-1}k_{NH}\xi \\ &= (2m_H)^{-1}(2k_{H^2}\xi^2 + k_{NH}\xi), \end{aligned}$$

where $B_0 = (O)_{T_0}$ and $C_0 = (O(^1D))_{T_0}$ (cf. (9) and (10)). Mollified Ozone Eq. is considered as a differential equation concerning (O_3), (NO_x), (HO_x). The concrete form is as follows:

$$\frac{d}{dt}(\mathrm{O}_3) = \frac{k_{\mathrm{O}_3}\{2J_{\mathrm{O}_2} + J_{\mathrm{O}}^*(\mathrm{O}_3) + J_{\mathrm{N}}(\mathrm{NO}_x)\}}{k_{\mathrm{O}_3} + k_{\mathrm{O}_2}(\mathrm{O}_3) + k_{\mathrm{H}}^*(\mathrm{HO}_x) + k_{\mathrm{Cl}}^*(\mathrm{ClO}_x)}$$
$$- \Big[J_{\mathrm{O}} + J_{\mathrm{O}^1} + \frac{k_{\mathrm{O}_2}\{2J_{\mathrm{O}_2} + J_{\mathrm{O}}^*(\mathrm{O}_3) + J_{\mathrm{N}}(\mathrm{NO}_x)\}}{k_{\mathrm{O}_3} + k_{\mathrm{O}_2}(\mathrm{O}_3) + k_{\mathrm{H}}^*(\mathrm{HO}_x) + k_{\mathrm{Cl}}^*(\mathrm{ClO}_x)}$$
$$+ k_{\mathrm{N}}(\mathrm{NO}_x) + k_{\mathrm{H}}(\mathrm{HO}_x) + k_{\mathrm{Cl}}(\mathrm{ClO}_x)\Big](\mathrm{O}_3),$$

(12)
$$\frac{d}{dt}(\mathrm{NO}_x) = m_{\mathrm{N}}^*(\mathrm{O}_3) - k_{\mathrm{NH}}(\mathrm{NO}_x)(\mathrm{HO}_x) - k_{\mathrm{NCl}}(\mathrm{NO}_x)(\mathrm{ClO}_x),$$
$$\frac{d}{dt}(\mathrm{HO}_x) = m_{\mathrm{H}}^*(\mathrm{O}_3) - 2k_{\mathrm{H}^2}(\mathrm{HO}_x)^2 - k_{\mathrm{NH}}(\mathrm{NO}_x)(\mathrm{HO}_x)$$
$$- k_{\mathrm{HCl}}(\mathrm{HO}_x)(\mathrm{ClO}_x) + \{\varepsilon_{\mathrm{N}}(\mathrm{NO}_x) + \varepsilon_{\mathrm{H}}(\mathrm{HO}_x)\}(\mathrm{ClO}_x),$$
$$(\mathrm{ClO}_x) = \frac{k_{\mathrm{CFC}}(\mathrm{CFC})}{(k_{\mathrm{NCl}} - \varepsilon_{\mathrm{N}})(\mathrm{NO}_x) + (k_{\mathrm{HCl}} - \varepsilon_{\mathrm{H}})(\mathrm{HO}_x)},$$

where
$$J_{\mathrm{O}}^* = J_{\mathrm{O}} + J_{\mathrm{O}^1}k_{\mathrm{O}^1}(k_{\mathrm{O}^1} + m_{\mathrm{N}} + m_{\mathrm{H}})^{-1},$$
$$m_{\mathrm{X}}^* = 2J_{\mathrm{O}^1}m_{\mathrm{X}}(k_{\mathrm{O}^1} + m_{\mathrm{N}} + m_{\mathrm{H}})^{-1} \quad (\mathrm{X} = \mathrm{N},\ \mathrm{H})$$
and (CFC) is a given function. It is interesting to observe that (ClO_x) is decreasing in a range where (NO_x) and (HO_x) are increasing. In the case where $(\mathrm{CFC})_t = 0$ $(t \geq T)$ for some $T(\geq 100)$, (12) is simplified in $[T, \infty)$.

2.6. Stability of Mollified Ozone Eq.

We study the stability of the ozone layer supposing that CFCs in the atmosphere will disappear some day. Our problem is to investigate whether the ozone layer returns to the previous situation. Recall a mollified mean $F(t, P)$ of Dobson number in 2.1. Assuming that
$$(\mathrm{O}_3) = f(t)/f(T_0), \quad f(t) = \int F(t, P)\, d\sigma(P)$$
for a suitable measure $d\sigma(P)$, we read this problem as the stability problem of Mollified Ozone Eq., and apply the linear stability theory. Suppose that $(\mathrm{CFC})_t = 0$ $(t \geq T)$ for some $T(\geq 100)$. Then (12) is simplified as:

$$\frac{d}{dt}(\mathrm{O}_3) = \frac{\gamma_{\mathrm{O}} + J_{\mathrm{O}}^*\{(\mathrm{O}_3) - 1\} + J_{\mathrm{N}}\{(\mathrm{NO}_x) - 1\}}{1 + k'_{\mathrm{O}_2}(\mathrm{O}_3) + k'_{\mathrm{H}}(\mathrm{HO}_x)}$$
$$- \Big[J_{\mathrm{O}} + J_{\mathrm{O}^1} + \frac{k'_{\mathrm{O}_2}\{\gamma_{\mathrm{O}} + J_{\mathrm{O}}^*((\mathrm{O}_3) - 1) + J_{\mathrm{N}}((\mathrm{NO}_x) - 1)\}}{1 + k'_{\mathrm{O}_2}(\mathrm{O}_3) + k'_{\mathrm{H}}(\mathrm{HO}_x)}$$

(13)
$$+ k_{\mathrm{N}}(\mathrm{NO}_x) + k_{\mathrm{H}}(\mathrm{HO}_x)\Big](\mathrm{O}_3),$$
$$\frac{d}{dt}(\mathrm{NO}_x) = m_{\mathrm{N}}^*(\mathrm{O}_3) - k_{\mathrm{NH}}(\mathrm{NO}_x)(\mathrm{HO}_x),$$
$$\frac{d}{dt}(\mathrm{HO}_x) = m_{\mathrm{H}}^*(\mathrm{O}_3) - 2k_{\mathrm{H}^2}(\mathrm{HO}_x)^2 - k_{\mathrm{NH}}(\mathrm{NO}_x)(\mathrm{HO}_x) \quad (t \geq T),$$

where

$$\gamma_{\rm O} = 2J_{\rm O_2} + J_{\rm O}^* + J_{\rm N}, \quad k'_{\rm O_2} = k_{\rm O_2} k_{\rm O_3}^{-1}, \quad k'_{\rm H} = k_{\rm H}^* k_{\rm O_3}^{-1}.$$

The initial condition is equivalent to:

$$(14) \qquad \begin{gathered} (1 - k'_{\rm O_2})\gamma_{\rm O} = (1 + k'_{\rm O_2} + k'_{\rm H}\xi)(J_{\rm O} + J_{\rm O^1} + k_{\rm N} + k_{\rm H}\xi) \\ m_{\rm N}^* = k_{\rm NH}\xi, \quad m_{\rm H}^* = 2k_{\rm H^2}\xi^2 + k_{\rm NH}\xi. \end{gathered}$$

By the definition of $J_{\rm O}^*$, it follows that $J_{\rm O}^* < J_{\rm O} + J_{\rm O^1}$. Condition (14) shows that $k'_{\rm O_2} < 1$. We consider that a 12-tuple

$$(J_{\rm O}, J_{\rm O^1}, J_{\rm O}^*, J_{\rm O_2}, J_{\rm N}, k'_{\rm O_2}, k_{\rm N}, k_{\rm H}, k'_{\rm H}, k_{\rm NH}, k_{\rm H^2}, \xi)$$

of positive numbers is given so that $J_{\rm O}^* < J_{\rm O} + J_{\rm O^1}$, $k'_{\rm O_2} < 1$,

$$(1 - k'_{\rm O_2})(2J_{\rm O_2} + J_{\rm O}^* + J_{\rm N}) = (1 + k'_{\rm O_2} + k'_{\rm H}\xi)(J_{\rm O} + J_{\rm O^1} + k_{\rm N} + k_{\rm H}\xi)$$

and regard that a triple $(\gamma_{\rm O}, m_{\rm N}^*, m_{\rm H}^*)$ is defined by

$$\gamma_{\rm O} = 2J_{\rm O_2} + J_{\rm O}^* + J_{\rm N}, \quad m_{\rm N}^* = k_{\rm NH}\xi, \quad m_{\rm H}^* = 2k_{\rm H^2}\xi^2 + k_{\rm NH}\xi.$$

Constant $(1,1,\xi)^{tr}$ is a solution of (13), i.e. an equilibrium point of (13). ($(\cdot,\cdot,\cdot)^{tr}$ denotes a row vector.) Conversely, an equilibrium point is expressed as $(\omega^2, \omega, \omega\xi)^{tr}$ for some $\omega > 0$. From the right-hand side of the first equation in (13), we obtain

$$\begin{aligned} &(1 - \omega^2 k'_{\rm O_2})(2J_{\rm O_2} + J_{\rm O}^*\omega^2 + J_{\rm N}\omega) \\ &= \omega^2(1 + k'_{\rm O_2}\omega^2 + k'_{\rm H}\omega\xi)(J_{\rm O} + J_{\rm O^1} + k_{\rm N}\omega + k_{\rm H}\omega\xi). \end{aligned}$$

Since this is a polynomial of degree 5, the other equilibrium points exist possibly.

Assuming that $((\mathrm{O_3})_T, (\mathrm{NO}_x)_T, (\mathrm{HO}_x)_T)^{tr}$ is sufficiently near to $(1,1,\xi)^{tr}$, we discuss whether $((\mathrm{O_3})_t, (\mathrm{NO}_x)_t, (\mathrm{HO}_x)_t)^{tr}$ $(t \geq T)$ converges to $(1,1,\xi)^{tr}$ or not. Putting $\langle \mathrm{X} \rangle = (\mathrm{X}) - (\mathrm{X})_{T_0}$ ($\mathrm{X} = \mathrm{O_3}$, NO_x, HO_x), we rewrite (13) in terms of $\langle \cdot \rangle$. Neglecting terms of degree larger than 1, we obtain a linear differential equation:

$$\begin{aligned} \frac{d}{dt}\begin{pmatrix} \langle \mathrm{O_3} \rangle \\ \langle \mathrm{NO}_x \rangle \\ \langle \mathrm{HO}_x \rangle \end{pmatrix} &= Q \begin{pmatrix} \langle \mathrm{O_3} \rangle \\ \langle \mathrm{NO}_x \rangle \\ \langle \mathrm{HO}_x \rangle \end{pmatrix} \\ &= \begin{pmatrix} -p, & q, & -r \\ k_{\rm NH}\xi, & -k_{\rm NH}\xi, & -k_{\rm NH} \\ 2k_{\rm H^2}\xi^2 + k_{\rm NH}\xi, & -k_{\rm NH}\xi, & -4k_{\rm H^2}\xi - k_{\rm NH} \end{pmatrix} \begin{pmatrix} \langle \mathrm{O_3} \rangle \\ \langle \mathrm{NO}_x \rangle \\ \langle \mathrm{HO}_x \rangle \end{pmatrix}, \end{aligned}$$

where

$$p = J_{\mathrm{O}} + J_{\mathrm{O}^1} + k_{\mathrm{N}} + k_{\mathrm{H}}\xi - (1 - k'_{\mathrm{O}_2})J^*_{\mathrm{O}}\theta + k'_{\mathrm{O}_2}\gamma_{\mathrm{O}}\theta + (1 - k'_{\mathrm{O}_2})k'_{\mathrm{O}_2}\gamma_{\mathrm{O}}\theta^2,$$
$$q = (1 - k'_{\mathrm{O}_2})J_{\mathrm{N}}\theta - k_{\mathrm{N}}, \quad r = (1 - k'_{\mathrm{O}_2})\gamma_{\mathrm{O}}k'_{\mathrm{H}}\theta^2 + k_{\mathrm{H}}, \quad \theta = (1 + k'_{\mathrm{O}_2} + k'_{\mathrm{H}}\xi)^{-1}.$$

According to the linear stability theory, a solution $((\mathrm{O}_3)_t, (\mathrm{NO}_x)_t, (\mathrm{HO}_x)_t)^{tr}$ of (13) converges to $(1, 1, \xi)^{tr}$, if the real parts of three eigenvalues of Q are negative. Let

$$\det(Q - \lambda I) = -(\lambda^3 + \alpha'\lambda^2 + \beta'\lambda + \gamma'),$$

where I denotes the unit matrix. Then

$$\alpha' = p + k_{\mathrm{NH}}\xi + (4k_{\mathrm{H}^2}\xi + k_{\mathrm{NH}}),$$
$$\beta' = k_{\mathrm{NH}}\xi(p - q) + (4k_{\mathrm{H}^2}\xi + k_{\mathrm{NH}})p + (2k_{\mathrm{H}^2}\xi^2 + k_{\mathrm{NH}}\xi)r + 4k_{\mathrm{H}^2}k_{\mathrm{NH}}\xi^2,$$
$$\gamma' = 2k_{\mathrm{H}^2}k_{\mathrm{NH}}\xi^2(2p - q) + 2k_{\mathrm{H}^2}k_{\mathrm{NH}}\xi^3 r.$$

The numbers p, r are positive, since $J^*_{\mathrm{O}} < J_{\mathrm{O}} + J_{\mathrm{O}^1}$, $k'_{\mathrm{O}_2} < 1$ and

$$(1 - k'_{\mathrm{O}_2})J^*_{\mathrm{O}}\theta < (1 - k'_{\mathrm{O}_2})\gamma_{\mathrm{O}}\theta = J_{\mathrm{O}} + J_{\mathrm{O}^1} + k_{\mathrm{N}} + k_{\mathrm{H}}\xi$$

(by (14)). Condition (14) shows also that

$$2p - q > p - q > J_{\mathrm{O}} + J_{\mathrm{O}^1} + k_{\mathrm{N}} + k_{\mathrm{H}}\xi$$
$$- (1 - k'_{\mathrm{O}_2})(J^*_{\mathrm{O}} + J_{\mathrm{N}})\theta + k_{\mathrm{N}} = (1 - k'_{\mathrm{O}_2})2J_{\mathrm{O}_2}\theta + k_{\mathrm{N}} > 0.$$

Thus α', β', γ' are positive. In this case, we see that the real parts of all eigenvalues of Q are negative if and only if $\alpha'\beta' > \gamma'$. In fact, there exists at least one negative eigenvalue μ in this case. Suppose that there exist mutually conjugate two eigenvalues ν, $\overline{\nu}$. Since

$$\mu + \nu + \overline{\nu} = -\alpha', \quad \mu(\nu + \overline{\nu}) + |\nu|^2 = \beta', \quad \mu|\nu|^2 = -\gamma',$$

we have

$$-(\nu + \overline{\nu}) = \alpha' - (-\mu) = \frac{\beta' - |\nu|^2}{-\mu}.$$

Since $-\mu$ is positive, this relation shows that the following three inequalities are equivalent: (i) $\nu + \overline{\nu} < 0$, (ii) $\alpha' > -\mu$, (iii) $\beta' > |\nu|^2$. Thus $\nu + \overline{\nu} < 0$ if and only if

$$\alpha'\beta' > -\mu|\nu|^2 \, (= \gamma').$$

Suppose that the other two eigenvalues ν, ν' are real. Then $\nu\nu' > 0$ by $\gamma' > 0$. In the same manner as above, we see that $\nu + \nu' < 0$ if and only if $\alpha'\beta' > \gamma'$. Consequently, the above assertion yields.

A simple calculation shows that

$$\begin{aligned}\alpha'\beta' &= \{p + k_{\mathrm{NH}}\xi + (4k_{\mathrm{H}^2}\xi + k_{\mathrm{NH}})\}\\ &\quad\times \{k_{\mathrm{NH}}\xi(p-q) + (4k_{\mathrm{H}^2}\xi + k_{\mathrm{NH}})p + (2k_{\mathrm{H}^2}\xi^2 + k_{\mathrm{NH}}\xi)r + 4k_{\mathrm{H}^2}k_{\mathrm{NH}}\xi^2\}\\ &> (2k_{\mathrm{H}^2}\xi + k_{\mathrm{NH}}\xi)\{k_{\mathrm{NH}}\xi(p-q) + 2k_{\mathrm{H}^2}\xi p + 2k_{\mathrm{H}^2}\xi^2 r\}\\ &> 2k_{\mathrm{H}^2}k_{\mathrm{NH}}\xi^2(p-q) + 2k_{\mathrm{H}^2}k_{\mathrm{NH}}\xi^2 p + 2k_{\mathrm{H}^2}k_{\mathrm{NH}}\xi^3 r = \gamma'.\end{aligned}$$

Thus a solution $((O_3)_t, (NO_x)_t, (HO_x)_t)^{tr}$ of (13) converges to $(1, 1, \xi)^{tr}$ as long as the initial value $((O_3)_T, (NO_x)_T, (HO_x)_T)^{tr}$ is sufficiently near to $(1, 1, \xi)^{tr}$. In the proof, we use only $k'_{O_2}(= k_{O_2}k_{O_3}^{-1}) < 1$ as a condition for rate constants. This means that the stability is based on the inequality $k_1[O][O_2][M] > k_2[O][O_3]$. It is very important to study whether the ozone depletion by CFCs is sufficiently small or not.

References

[**AH**] H. Akimoto and S. Hayashida, *Chemical reactions in the stratosphere*, in "Chlorofluorocarbons: Global environmental impact and countermeasures," Chemical Compendium 11, 1990, pp. 46–61.

[**CWL**] R. J. Cicerone, S. Walter and S. C. Liu, *Nonlinear response of stratospheric ozone column to chlorine injection*, J. of Geophy. Res. **88–6** (1983), 3647–3661.

[**JEA1**] JEA, "Annual report concerning the superintendence of the ozone layer 91," Japan Environmental Agency, Tokyo, 1992.

[**JEA2**] —, "Annual report concerning the superintendence of the ozone layer 92," Japan Environmental Agency, Tokyo, 1993.

[**JMA1**] JMA, "Figures of global ozone distributions," Technical Data Series No. 56, Japan Meteorological Agency, Tokyo, 1991.

[**JMA2**] —, "Observation and analysis of the ozone layer," Technical Report No. 112, Japan Meteorological Agency, Tokyo, 1992.

[**S**] T. Shimazaki, "The stratospheric ozone layer, Second Edition," Tokyo Univ. Press, Tokyo, 1989.

[**YW**] N. Yoshida and E. Wada, *Major atmospheric constituents*, in "Chemistry of Atmosphere," Chemical Compendium 10, 1990, pp. 46–61.

3. Hydrodynamical phase transition and life-style of marine organisms

At a summer seminar of rheology (Aug. 1992), Prof. Mihashi (Biology) gave a lecture concerning a biological similarity law: The speed (of motion) of an animal is comparable with his size. Prof. Shikata (Mathematics) commented that the distribution of sizes of marine organisms seems too large to apply the similarity law. After the summer seminar, I read [Mo] (following Prof. Mihashi's advice) where it is pointed out that an organism has his proper time on which the similarity law is based. In fact, tortoise(testudo elephantopus, carapace 1.2 m, weight 180 kg, Islands Galápagos) moves very slowly and lives more than 300 years. There are many articles concerning this similarity law. In [R], the similarity law is discussed from the viewpoint of dimensional analysis (i.e. scaling). In [O], the similarity law for marine organisms is studied from the viewpoint of viscosity. Life-style of microorganisms is investigated in [P]. Alexander's book [Ax] is a classical compendium concerning animal mechanics.

The methods of motion are generally classified into 'run', 'fly', and 'swim', and the rate constants (of the similarity laws) are mutually very different. In fact, the flying speed of a sparrow is about ten times of a saurel with same size. Thus it seems necessary to discuss each method very carefully. In this chapter, I focus mainly on 'swim' i.e. the similarity law for marine organisms. Reynolds number Re of a marine organism is defined by $Re = u\ell/\nu$, where u (cm/s) denotes his speed, ℓ (cm) is his size and $\nu(\simeq 10^{-2})$ (cm^2/s) is the kinetic viscosity of the sea water. The range of Reynolds numbers of marine organisms is very wide: Reynolds number of a bacterium is about 10^{-4} and Reynolds number of a whale(eg. physeter, size 18 m, speed 29.5 km/h) is about 10^8. Thus it is quite natural to consider that a hydrodynamical phase transition may exist in a range $[10^{-4}, 10^8]$. In fact, in the case of thin boards, Reynolds transition exists and the transition point (i.e. the critical point) is about $1 \sim 10^4$. In this chapter, the similarity law (for all marine organisms) will be divided into two similarity laws at a transition point and, after the discussion of each similarity law, the global similarity law will be studied. A relation with electromagnetic theory will be also suggested. "Equality. It is lifetime: $\sqrt{\ell/(\nu u)}$ (s/cm)".

3.1. Stress transition and life-style transition

The similarity laws are closely related to life-styles, and their bases depend on methods: 'run', 'fly', 'swim' (cf. [Ax]). Take, for example, a runner. Setting a coordinate frame on his head, we observe the motion of his two feet, which are very accelerative. Then it is clear that the accelerations of two feet are mutually inverse. Since two feet go and return, they do not separate from his body. His weight is always on his one foot whose direction is same as the direction of his face. (This method is same as the method to row a boat.) In the case of 'run', the similarity law is reduced to a relation between internal force and weight. Take, for example, a kite flying with all might. Setting a coordinate frame on his head, we observe the motion of his two wings, which are very accelerative.

Different from 'run', the direction of their accelerations is almost perpendicular to the ground, and he controls the direction of his wings so that the lift is balanced with his weight. According to the lifting line theory [M, Chap. XI], the aero-drag is induced from wakes appearing behind him. In this case, the similarity law is reduced to a relation between internal force and momentum of wakes, and this law has almost no relation with weight and aero-viscosity.

Now we investigate the similarity law in the case of 'swim' (with all might). Since the motion of the sea water obeys Navier-Stokes' equation [Sch, Chap. IV], we must discuss viscosity drag. Let μ (g/cm·s), ν (cm^2/s), ρ (g/cm^3) denote the static viscosity, the kinetic viscosity and the density of the sea water, respectively. (CGS system of units is mainly used from the feeling of daily lives.) Then $\nu = \mu/\rho$, $\mu \simeq 10^{-2}, \rho \simeq 1$. Suppose that a fish (a marine organism) with size ℓ (cm) is swimming with speed u (cm/s) in the sea. His surface area and his volume are regarded as ℓ^2 (cm^2) and ℓ^3 (cm^3), respectively. Reynolds number $Re = u\ell/\nu$ plays an important role in the motion obeying Navier-Stokes' equation and this number is dimensionless. In the case of moving thin boards, there exists a transition point around $Re = 1 \sim 10^4$, and the flowing pattern changes drastically from laminar flow to turbulent flow at this point. Considering that a phase transition exists at the analogous position in the case of marine organisms also, we discuss the similarity law, distinguishing the case of $Re \geq 10^4$ and the case of $Re \leq 1$.

We begin with the case of $Re \geq 10^4$. In this case, flow should be turbulent, and Reynolds stress [Sch, Chap. XXIII] should exist. But, fish swims faster than we expect. This fact is given attention for a long time, and at present, this is considered as control of turbulence (Toms effect) by his visco-elastic body and a secretion from there (cf. [N, p. 285]). From this fact, we consider that flow is still laminar with a fat boundary-layer and apply Prandtl's boundary-layer theory [Sch, Chap. VII]. The boundary-layer width δ (cm) is introduced, in Navier-Stokes' theory, as a quantity showing the distance from an obstacle to uniform flow. According to Prandtl's theory, $\delta \propto (\nu\ell u^{-1})^{1/2}$. In our case, we regard δ as the distance from our fish to the static water, and use this comparison theorem. By viscosity, the speed of water attaching to our fish is u (cm/s). Thus, the tangential stress per unit area (cm^2) (which is caused by the motion of water with the tangential direction) is comparable with $\mu u(\nu\ell u^{-1})^{-1/2}$ (g/cm·s^2). Integrating this stress over his surface, we see that the total tangential stress from water is comparable with $\rho\nu^{1/2}(u\ell)^{3/2}$ (dyn).

Next we try to find the internal force balanced with this stress. Setting a coordinate frame on his head, we observe the motion of his tail, which is very accelerative. Different from the case of 'run', the direction of its acceleration is perpendicular to that of his face, and different from the case of 'fly', the motion of his tail works only to move ahead. The buoyancy is controlled by a bladder inside his body. In order to study his internal force, we slice him perpendicularly to his backbone into ℓ pieces. (Since Re is large, we assume $\ell \gg 1$.) Then the surface area of each thin piece is nearly comparable with ℓ^2. Each piece would contain a bone (like a knitting needle with two sharp edges and with a hole to thread it in the middle) of length $\propto \ell$ and of section diameter $\propto \ell$. The elastic force of this bone is comparable with (section diameter)$\times$(length) ($\propto \ell^2$), and hence the elastic force of this thin piece is also comparable with ℓ^2. Adding ℓ thin pieces, we see that the total elastic force is comparable with ℓ^3.

From these observations, we put $\rho\nu^{1/2}(u\ell)^{3/2} = f\ell^3$. Then $u = (f^{2/3}\rho^{-2/3}\nu^{-1/3})\ell$. Since the dimension of f is dyn/cm^3, we regard f as the elastic force per unit volume (cm^3) and call this quantity the proper elastic force. This is expressed as $f = (\rho^2\nu)^{1/2}(u\ell^{-1})^{3/2}$. The similarity law ($Re \geq 10^4$) is equivalent to the invariance of the proper elastic force for marine organisms ($Re \geq 10^4$).

Next we discuss the case of $Re \leq 1$. In this case, flow is laminar, highly viscous, and Stokes' approximation fits very well. This is an equation obtained from Navier-Stokes' equation by removing the inertia term, and is nothing less than Poisson's equation in the case of steady flow. Since Reynolds number is small, it is better to image a marine microorganism. In Stokes' approximation, the boundary-layer width δ is comparable with ℓ, and hence the tangential stress per unit area is comparable with $\mu u\ell^{-1}$. Thus, integrating this stress over his surface, we see that the total tangential stress is comparable with $\mu u\ell$ (dyn). (In the case of a ball, this is Newton-Stokes' drag theorem.) Now we try to find the internal force balanced with this stress. If he adopts $f\ell^3$ in the same manner as above, his speed turns out comparable with ℓ^2. Since he is very small, this method is quite inefficient. Probably, he will try to use his surface force. To assert that $f^*\ell^2$ is a correct quantity, we take a planula (a larva of medusa), whose surface is covered by many cilia. Let ε (cm^2) and n be the surface area of a cilium and the number of cilia, respectively. Assume also that εn is comparable with his surface area. Since a cilium corresponds to a thin piece in the above discussion, its strain should be comparable with ε. Considering that only cilia have the strain, we obtain that the total strain over his surface is comparable with $\varepsilon n(\propto \ell^2)$. (The difference with the first case is the fact that $n \neq \ell$.)

Taking account of this discussion, we choose $f^*\ell^2$ as his internal force, and put $\mu u\ell = f^*\ell^2$. Then $u = (f^*\mu^{-1})\ell$. In the case of flagellata, a flagellum corresponds to cilia of ciliate. A marine bacterium without cilium and flagellum is covered by a thin membrane which corresponds to a flagellum. Thus, in any case, $f^*\ell^2$ is a required quantity, and the equality $u = (f^*\mu^{-1})\ell$ holds. Since the dimension of f^* is dyn/cm^2, we regard f^* as the strain per unit area and call this quantity the proper strain. This quantity is given by $f^* = \mu u\ell^{-1}$. The similarity law ($Re \leq 1$) is equivalent to the invariance of the proper strain for marine microorganisms ($Re \leq 1$).

3.2. Law of the power 3/2 and Coulomb capacity

In 3.1, we suggested that two similarity laws exist. Now we try to verify these two similarity laws by using real data, and try to show that two rate constants are equal, i.e. the global similarity law (= the similarity law for all marine organisms) holds. The following table is mainly made by reading the values of dots of a figure in [O] and comparing with figures and data in [Mo, Ok]. Here

$$a = \log_{10}\{\nu^{-1}u\ell\},\ b = \log_{10}\{\nu^{-1}(u^{-1}\ell)\},\ U = \log_{10} u,\ L = \log_{10} \ell.$$

(The quantity $\sqrt{\nu^{-1}(u^{-1}\ell)} = e^{b/2}$ corresponds to the proper lifetime.)

No	a	b	U	L	No	a	b	U	L	No	a	b	U	L
1	-5.0	0.8	-2.9	-4.1	22	0.0	1.0	-0.5	-1.5	43	5.1	1.5	1.8	1.3
2	-4.7	0.5	-2.6	-4.1	23	0.1	1.7	-0.8	-1.1	44	5.2	1.8	1.7	1.5
3	-4.3	0.5	-2.4	-3.9	24	0.5	2.3	-0.9	-0.6	45	5.3	1.7	1.8	1.5
4	-3.5	1.5	-2.5	-3.0	25	0.5	1.7	-0.6	-0.9	46	5.3	1.3	2.0	1.3
5	-3.0	1.8	-2.4	-2.6	26	1.4	1.8	-0.2	-0.4	47	5.4	1.4	2.0	1.4
6	-2.9	0.5	-1.7	-3.2	27	1.8	0.2	0.8	-1.0	48	5.5	1.7	1.9	1.6
7	-2.7	2.1	-2.4	-2.3	28	2.2	1.4	0.4	-0.2	49	5.6	1.2	2.2	1.4
8	-2.7	1.5	-2.1	-2.6	29	2.5	1.9	0.3	0.2	50	5.7	1.5	2.1	1.6
9	-1.8	2.4	-2.1	-1.7	30	2.6	1.0	0.8	-0.2	51	6.0	1.2	2.4	1.6
10	-1.6	1.8	-1.7	-1.9	31	3.4	1.4	1.0	0.4	52	6.2	1.6	2.3	1.9
11	-1.3	1.1	-1.2	-2.1	32	3.7	1.9	0.9	0.8	53	6.4	2.0	2.2	2.2
12	-1.1	1.5	-1.3	-1.8	33	3.7	1.5	1.1	0.6	54	6.6	1.8	2.4	2.2
13	-0.9	1.1	-1.0	-1.9	34	3.9	1.3	1.3	0.6	55	6.6	1.4	2.6	2.0
14	-0.8	1.6	-1.2	-1.6	35	4.1	1.5	1.1	0.6	56	6.6	1.0	2.8	1.8
15	-0.7	0.7	-0.7	-2.0	36	4.3	1.9	1.2	1.1	57	6.6	0.6	3.0	1.6
16	-0.6	1.6	-1.1	-1.5	37	4.7	1.7	1.5	1.2	58	6.8	1.6	2.6	2.2
17	-0.6	1.0	-0.8	-1.8	38	4.7	1.1	1.8	0.9	59	7.1	1.1	3.0	2.1
18	-0.4	1.6	-1.0	-1.4	39	4.8	1.8	1.5	1.3	60	7.3	1.3	3.0	2.3
19	-0.4	1.2	-0.8	-1.6	40	4.8	1.4	1.7	1.1	61	7.4	0.8	3.3	2.1
20	-0.2	1.0	-0.6	-1.6	41	5.0	1.6	1.7	1.3	62	7.8	2.6	2.6	3.2
21	-0.2	0.8	-0.5	-1.7	42	5.0	1.2	1.9	1.1	63	8.2	2.4	2.9	3.3

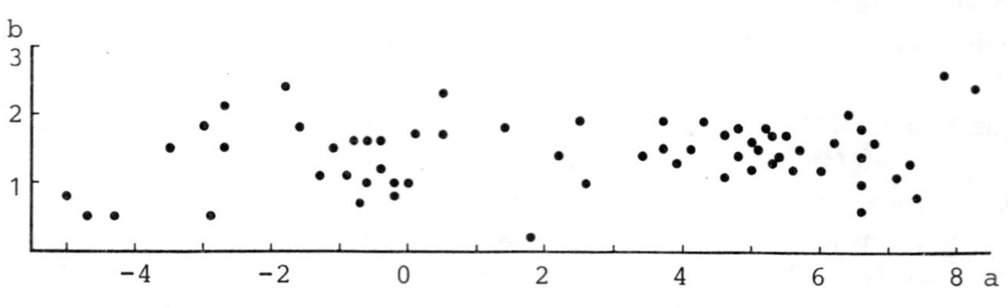

The identity numbers 1 ~ 3 are bacteria, Nos. 62, 63 are whale(eubalaena, size 15 m, speed 15 km/h; $a = 7.8$, $b = 2.6$) and whale(physeter, size 18 m, speed 29.5 km/h; $a = 8.2$, $b = 2.4$) (cf. [Ok]). The investigation of data is very complicated. (According to [Ok], dolphin(delphinus, size 1.3~2 m) has a reserve to swim zigzag in front of a steamship with speed 15~30 knots/h. Calculating with 1.3 m, 15 knots/h, we have $a = 7.0$, $b = 1.2$. Calculating with 2 m, 30 knots/h, we have $a = 7.5$, $b = 1.1$. The work consists of determining figures from rough data of this kind.) Assuming that this table is true, we work in this paragraph. If dots are scattered constantly, we can assert a similarity law. In order to judge whether dots are scattered constantly or not, some information criterion is necessary. Here we apply the idea of AIC (cf. [A]), which is a notion that a model is better if (the number of unknown parameters)+(entropy) is smaller. Taking two statistic data $A = \{a_i;\ i = 1, \cdots, N\}, B = \{b_i;\ i = 1, \cdots, N\}$ with N samples, we pose a problem whether B is scattered constantly or scattered comparably with A. First, for a pair (α', β') of parameters, we define γ-entropy ($\gamma > 1$) by

$$I(\alpha', \beta') = \log_\gamma \{\frac{1}{N} \sum_{i=1}^{N} (b_i - \alpha' a_i - \beta')^2\}$$

and choose the pair (α, β) which minimizes γ-entropy in all pairs. A simple calculation shows that

$$\beta = \frac{1}{N} \sum_{i=1}^{N} (b_i - \alpha a_i), \quad \alpha = \{\sum_{i=1}^{N} a_i b_i - \frac{1}{N} \sum_{i=1}^{N} a_i \sum_{i=1}^{N} b_i\}\{\sum_{i=1}^{N} b_i^2 - \frac{1}{N} (\sum_{i=1}^{N} b_i)^2\}^{-1}.$$

Information criterion for 'comparably' is defined by $IC(2) = 2 + I(\alpha, \beta)$. The value 2 means the number of parameters. Next we choose the number m whose combination with 0 minimizes γ-entropy in all pairs of form $(0, \cdot)$. Then $m = (1/N) \sum_{i=1}^{N} b_i$ (the mean of b_i) and $I(0, m) = \log_\gamma \sigma^2$, where σ is the deviation of B. Information criterion for 'constantly' is defined by $IC(1) = 1 + I(0, m)$. Now the value $\gamma > 1$ is chosen so that $IC(1) = IC(2)$, and $\tau = 1/\gamma$ is called the probability of 'constantly'. Our notion is as follows: One can assert, with probability τ, that B is scattered constantly. This information criterion is slightly different from AIC, however, the basic idea is same, and the computation is easier. The probability τ is given by

$$\tau = 1 - \{\sum_{i=1}^{N} (a_i - n)(b_i - m)\}^2 \{\sum_{i=1}^{N} (a_i - n)^2 \sum_{i=1}^{N} (b_i - m)^2\}^{-1} \quad (n = \frac{1}{N} \sum_{i=1}^{N} a_i).$$

We apply this information criterion to the case of $Re \leq 1$. Observing the distribution of figures in the table, we take Nos. 1~23 as samples ($N = 23$). Then the probability is larger than 0.9. In order to apply this information criterion to the case of $Re \geq 10^4$, we take Nos. 31~63 ($N = 33$). Then the probability is larger than 0.9. Thus, in any case, we can assert, with probability > 0.9, that a similarity law holds. How is the probability, if we apply this information criterion to all marine organisms (i.e. Nos. 1~63)? A calculation shows that the probability is still larger than 0.9. Thus we can assert, with probability > 0.9, that the global similarity law holds. (The methods 'run' and 'fly' are also adaptation of organisms to phase transitions from 'water' to 'ground' and 'air'. From this viewpoint, the universal similarity law for all organisms may hold (with suitable normalization).)

Now we show a corollary of the global similarity law. This law implies that marine organisms adapt themselves to a hydrodynamical phase transition. Recall the following two equalities:

$$u(i) = (f^*(i)\mu^{-1})\ell(i) \quad (i = 1, \cdots, 23),$$
$$u(j) = (f(j)^{2/3}\rho^{-2/3}\nu^{-1/3})\ell(j) \quad (j = 31, \cdots, 63),$$

where i, j denote the identity numbers. The global similarity law is equivalent to the equality

$$f^*(i)\mu^{-1} = f(j)^{2/3}\rho^{-2/3}\nu^{-1/3},$$

which is rewritten as

$$f(j) = (\rho\nu^2)^{-1/2} f^{*3/2}.$$

The dimension of a constant $\rho\nu^2$ is same as that of force (i.e. dyn). The proper elastic force $f(j)$ shows the elastic force of an organism j per unit volume, and the proper strain $f^*(i)$ shows the strain of a microorganism i per unit surface area. Thus the global similarity law is equivalent to the law of the power 3/2 between the proper elastic force and the proper strain. There are many laws of the power 3/2 which yield from balances between 2-dimensional quantities and 3-dimensional quantities: eg. (diameter of backbone) $\propto$ (size)$^{3/2}$. Our law is also in this category.

Finally, we show a model of swimming force of marine organisms. Regarding a set E in the 3-dimensional Euclidean space as a marine organism, we denote by $C(E)$ (dyn) his swimming force. ($C(\cdot)$ is a scalar.) If his size is large, then his swimming force is equal to $f \times$(volume). If his size is small, then his swimming force is equal to $f^* \times$(surface area). Thus $C(E)$ is 3-dimensional if the diameter of E is large, and $C(E)$ is 2-dimensional if that of E is small. This shows that $C(\cdot)$ is not a measure. Assuming that $C(\cdot)$ is a potential-theoretic capacity [L, Chap. II], we try to find a suitable capacity.

Modified Coulomb capacity is defined by

$$C(E) = \sup\{\int d\mu; U\mu \leq 1 \text{ on } E, \ (\text{support of } \mu) \subseteq E\},$$

where $U\mu$ denotes the potential of a measure μ with a convolution kernel

$$k(x) = |x|^{-2} \ \ (|x| \leq 1), \quad k(x) = 0 \ \ (|x| > 1).$$

This capacity is one of the simplest capacities satisfying the above condition. If we take this capacity as a model of swimming force, then we see that swimming force is related to electromagnetic force. The measure μ_E attaining the supreme (i.e. the equilibrium measure of E) shows the distribution of the power. Swimming force does not, in general, preserve the orders of volume and surface area. For a contraction mapping T (i.e. a mapping satisfying $|T(x) - T(y)| \leq |x - y|$), the inequality $C(T(E)) \leq C(E)$ holds. This means that swimming force is contractible for not only a similar contraction but also a general contraction. Introducing the modified Coulomb capacity, we can study swimming force as application of electromagnetic dynamics.

REFERENCES

[**A**] H. Akaike, *On entropy maximization principle*, in "Application of Statistics," North-Holland, Amsterdam, 1977, pp. 27–41.

[**Ax**] R. McN. Alexander, "Animal mechanics, Second Edition," Blackwell Sci. Publ., Oxford, 1983.

[**L**] N. S. Landkov, "Foundations of modern potential theory," Springer-Verlag, Berlin, 1972.

[**M**] L. M. Milne-Thomson, "Theoretical aerodynamics, Fourth Edition," Dover, New York, 1966.

[**Mo**] T. Motokawa, "(Japanese) Elephant's time and mouse's time," Chuo New Book Ser. 1807, Chuo, Tokyo, 1992.

[**N**] T. Nakagawa, "(Japanese) Rheology, Second edition," Iwanami Zensho 249, Iwanami, Tokyo, 1992.

[**O**] A. Ohkubo, *Marine organisms and fluid dynamics*, in "(Japanese) Mathematical fluid dynamics," Mathematical Science, Science, Tokyo, 1987, pp. 104–111.

[**Ok**] K. Okamoto, "New illustrated encyclopedia of the fauna of Japan," Hokuryu, Tokyo, 1960.

[**P**] E. M. Purcell, *Life at low Reynolds number*, Amer. J. of Phy. **45-1** (1977), 3–11.

[**R**] J. Rallison, *Scaling methods in physics and biology*, in "New applications of mathematics," Penguin Books Ltd., London, 1991, pp. 124–140.

[**Sch**] H. Schlichting, "Boundary-layer theory," McGraw-Hill, New York, 1979.

4. Layers and phase transitions by proper motion of powder

One day, I read a book [M], where various interesting properties of powders are written from the viewpoint of social science, and then studied a compendium [D] of powder technology. One of the most popular powders is sand. Sand exists everywhere and troubles us very much. Its physical properties are interesting and sand flow is complicated as follows: (i) Sand in a desert does not cause silicosis to camels, but sand with same particle-size in a site of construction does. (ii) Physical properties change by a shock and a pressure. Liquefaction by an earthquake sweeps away our houses frequently (without water), and quick-sand drowns good swimmers occasionally when they stand up. (iii) The speed of sand flowing out from a small orifice of a board is independent of the amount of sand accumulating on the board. This is the sand-glass principle and different from Torricelli's theorem on fluids.

To learn a detail, I visited the author of [M] Prof. Miwa (Powder Tech.) (Nov. 1992). He, with Prof. Hidaka (Powder Tech.), kindly advised to study mechanisms of sand-flow and singing-sand. This is a starting point of a topic in this chapter. "Sand-glass. Flow is white bacteria:

$$\rho\frac{D\mathbf{q}}{Dt} = \nabla_j\tau_{ij} + \mathbf{f},\ \rho^*\varepsilon\frac{D\mathbf{p}}{Dt} = \nabla_j\tau^*_{ij}."$$

4.1. A preliminary experimentation

We begin by observing the motion of sand. One of the simplest experimentations is an experimentation of 2-dimensional gravitational flow (i.e. an observation of a thin sand-glass). Preparations: sand (a proper quantity), acrylic panels (3mm×50cm×50cm, 2 pieces; 5mm×50cm×50cm, a piece), two stands (1 m), four holders, ten clips (to pinch an acrylic panel of width 13 mm), ink (red, blue), beakers and daily carpenter's tools etc.. Any sand works, however, Flattery Sand is best; this is natural beach in Australia. Cut three pieces of acrylic sticks (5mm×2cm×50cm) off from an acrylic panel of width 5 mm, and cut a piece into two parts (5mm×2cm×25cm). Put these four acrylic sticks between two panels (3mm×50cm×50cm), and make a vessel with a crack of width 5 mm in the bottom. Fix this vessel by ten clips, and finally put it between two stands by four holders. This is our 2-dimensional sand-glass. Since we must wipe the inside, it is better not to use a paste. Next we prepare red/blue sand which plays the role of tracers. The method is simple. Put a pan on the fire, boil sand with red ink and keep stirring it all the while. Stop the fire before sand becomes well-done, and dry red sand on a board. Since it takes several days for us to dry tracers, it is better to practice an experimentation only by white sand. Close the orifice by a vinyl tape, and pour white sand into our sand-glass by a funnel. Flatten sand by a thin broom and put a beaker under the closed orifice. The height of sand is about 30 cm. Peel the vinyl tape and observe the motion of sand. Observe sand-particles very carefully and repeatedly. Some sand-particles flow, some vibrate, some rotate, some slide and some do not move. The panorama is like a world of bacteria in a microscope. When tracers are dried up, we can experiment more in detail.

It takes several months for us to improve our sand-glass and to master an experimental technique. In parallel with experimentations, it seems better to do a trial and error of forming the equation of powder-motion.

4.2. Proper motion of powder

We call a set of sand-particles a powder, distinguishing with a fluid. A static fluid is still moving in the sense of Brownian motion, but a static power does not. Considering a particle $\mathbf{a}$ as a singleton in the 3-dimensional Euclidean space, we denote by $\mathbf{r} = \mathbf{r}(\mathbf{a}, t)$ its position at instant t. There are some scalars (eg. density (g/cm^3), volume (cm^3), temperature (° K)) and some vectors (eg. velocity $\partial \mathbf{r}(\mathbf{a}, t)/\partial t$ (cm/s), acceleration $\partial^2 \mathbf{r}(\mathbf{a}, t)/\partial t^2$ (cm/s^2)), which show physical states of $\mathbf{a}$ at instant t. We assume that the density ρ, the volume and the temperature of our powder are independent of $(\mathbf{a}, t)$. The velocity and the acceleration of a particle existing at a point $\mathbf{x}$ at instant t are denoted by $\mathbf{q}(\mathbf{x}, t)$ (cm/s), $D\mathbf{q}(\mathbf{x}, t)/Dt$ (cm/s^2), respectively. These are vector fields with respect to the variable $\mathbf{x}$. From an astronomical observation, Cosserat brothers [C] discussed the angular velocity of a particle, by introducing a proper coordinate axis associated with the particle. This idea corresponds to introducing an internal structure in a quantum. For a long time, their idea was buried, but now it is pointed out that their idea is applicable to semi-microdynamics. Since the directions of rotation of two adjacent particles are mutually inverse, a modification is necessary to combine their idea with continuum dynamics. In this chapter, we regard their angular velocity as an infinitesimal motion of a particle. There are many infinitesimal motions of particles. A powder is suspension in a medium/fluid (eg. water, atmosphere), and the powder-motion is closely related to flow of the medium. (A sand-glass does not work on the moon, because there is no atmosphere.) When particles are sufficiently dense in a medium, each particle vibrates by the influence of turbulence of the medium. This vibration is also regarded as an infinitesimal motion. Other examples are slide and recoil. Infinitesimal motions are caused by turbulence of a medium, particle-shapes, surface activity, static electricity, trace elements in a medium etc., and these are observed as vibration, rotation, slide, recoil etc.. Since there are many infinitesimal motions, we image their mean, and call it the proper motion. The proper motion is supposed to be a vector field expressed in the form $\varepsilon \mathbf{p}(\mathbf{x}, t)$, where $\varepsilon > 0$ is infinitesimal with dimension cm and the dimension of $\mathbf{p}$ is 1/s. The density ρ shows the mean of the densities of a powder and a medium. Thus the proper density ρ^* (g/cm^3) of a powder is also defined. The scalars ε, ρ^* are assumed to be constants. The proper acceleration is defined by

$$\varepsilon \frac{D\mathbf{p}}{Dt}(\mathbf{x}, t) = \varepsilon \Big\{ \frac{\partial \mathbf{p}}{\partial t} + (\mathbf{q}\nabla)\mathbf{p} \Big\}.$$

The proper position (field) $\varepsilon\ell$ is defined by

$$\varepsilon\ell(\mathbf{x},t) = \varepsilon\int_0^t \mathbf{p}(\mathbf{r}(\mathbf{a},s),s)\,ds,$$

where $\mathbf{a}$ denotes the particle whose position is $\mathbf{x}$ at instant t. Equality

$$\frac{D\ell}{Dt}\Big(\equiv \frac{\partial\ell}{\partial t} + (\mathbf{q}\nabla)\ell\Big) = \mathbf{p}$$

holds. The powder-motion is expressed by a pair $(\rho D\mathbf{q}/Dt, \rho^*\varepsilon D\mathbf{p}/Dt)$ of two vector fields; $\rho D\mathbf{q}/Dt$ shows the force of a powder domain per unit volume, and $\rho^*\varepsilon D\mathbf{p}/Dt$ shows the proper force inducing the proper motion per unit volume. According to Hooke's experimental law, force is equal to a summation of surface stress and external (body) force. Taking account of this law, we put

(1) $$\rho\frac{D\mathbf{q}}{Dt} = \nabla_j\tau_{ij} + \mathbf{f}, \quad \rho^*\varepsilon\frac{D\mathbf{p}}{Dt} = \nabla_j\tau^*_{ij}.$$

Here $\mathbf{f}$ denotes external force and it is assumed that the proper motion is independent of the external force. Two (3,3)-matrices τ_{ij}, τ^*_{ij} denote surface stress tensor fields of order 2 (dyn/cm^2). In (1), summations are taken for indices appearing twice.

4.3. Equation of powder-motion

Formula (1) is Hooke's experimental law and does not work by itself. It is necessary to find suitable surface stress tensors τ_{ij}, τ^*_{ij} expressed in terms of $\mathbf{q}$, $\mathbf{p}$. Not only even tensor fields, but also odd tensor fields are discussed from the viewpoint of semi-microdynamics. Natural tensor fields (of order 2) generated from $\mathbf{p} = (p_i)$ are only

(2) $$\begin{pmatrix} 0 & -p_3 & p_2 \\ p_3 & 0 & -p_1 \\ -p_2 & p_1 & 0 \end{pmatrix} \quad \text{(and its constant multiples).}$$

More precisely, a tensor field $c_{ijk}p_k$ with isotropic coefficients c_{ijk} with respect to the 3-dimensional rotation group is expressed as (2), because a tensor c_{ijk} is reduced to ε-tensor or its constant multiples. If we take (2) as τ_{ij}, then $\nabla_j\tau_{ij} = \nabla\wedge\mathbf{p}$ (or its constant multiples). Natural tensor fields generated from $\nabla = (\nabla_i)$, $\mathbf{p} = (p_i)$ are expressed in the form

(3) $$\lambda(\nabla\mathbf{p})I + \lambda'\nabla_i\mathbf{p}_j + \lambda''\nabla_j\mathbf{p}_i \quad (\lambda,\ \lambda',\ \lambda'' \text{ are constants, } I \text{ is the unit tensor}).$$

More precisely, a tensor field $c_{ijk\ell}\nabla_k p_\ell$ with isotropic coefficients $c_{ijk\ell}$ with respect to the rotation group is expressed in the form (3), because a tensor $c_{ijk\ell}$ is reduced to

$\lambda\delta_{ij}\delta_{k\ell}+\lambda'\delta_{ik}\delta_{j\ell}+\lambda''\delta_{i\ell}\delta_{jk}$, where δ denotes Dirac's notation. If we take (3) as τ_{ij}, then

$$\nabla_j\tau_{ij}=(\lambda+\lambda')\nabla(\nabla\mathbf{p})+\lambda''\Delta\mathbf{p}.$$

The same argument works for $\mathbf{q}$, ℓ. Since particles throng, the pressure tensor field σ_{ij} would exist. Considering that τ_{ij} in (1) is expressed as a summation of (2), (3) and the pressure tensor field, we put

$$\rho\frac{D\mathbf{q}}{Dt}=\nabla_j\sigma_{ij}+\sum_{k=1}^{3}\{\mu_k^1\Delta\mathbf{q}^k+\mu_k^2\nabla(\nabla\mathbf{q}^k)+\mu_k^3\nabla\wedge\mathbf{q}^k\}+\mathbf{f} \tag{4}$$
$$(\mathbf{q}^1=\mathbf{q},\ \mathbf{q}^2=\mathbf{p},\ \mathbf{q}^3=\ell).$$

Since a powder is suspension in a medium, strain by the position field corresponding to $\mathbf{q}$ is neglected. Strain by the proper position $\varepsilon\ell$ is not negligible, because some particles vibrate and recoil. Since (4) has too many terms and seems not realistic, we remove some terms looking unimportant. The operator $\nabla\wedge$ is the first space-derivative for vector fields and $-\Delta(=\nabla\wedge(\nabla\wedge\cdot)-\nabla(\nabla\cdot))$ corresponds to the second space-derivative. Since the acceleration $D\mathbf{q}/Dt$ is related to the second space-derivative in hydrodynamics, we remove $\nabla\wedge\mathbf{q}$ (and leave $\Delta\mathbf{q}$, $\nabla(\nabla\mathbf{q})$). If we regard $\varepsilon\mathbf{p}$ as turbulence of $\mathbf{q}$ and adopt the above procedure, the resulting equation is similar to an equation for emulsion (eg. milk+water). From the observation in 4.1, the powder-motion is quite different from the motion of emulsion. Thus, regarding $\varepsilon\mathbf{p}$ as a space-derivative of turbulence of $\mathbf{q}$, we remove $\Delta\mathbf{p}$, $\nabla(\nabla\mathbf{p})$ (and leave $\nabla\wedge\mathbf{p}$). Since the relation between $\varepsilon\nabla\wedge\mathbf{p}$ and $\varepsilon\nabla\wedge\ell$ corresponds to the relation between the viscosity term $\Delta\mathbf{q}$ and the elasticity term Δ(the position field), we remove $\Delta\ell$, $\nabla(\nabla\ell)$ (and leave $\nabla\wedge\ell$). Consequently, (4) is simplified in the form

$$\rho\frac{D\mathbf{q}}{Dt}=\nabla_j\sigma_{ij}+\lambda\Delta\mathbf{q}+\tilde{\lambda}\nabla(\nabla\mathbf{q})+\mu\nabla\wedge\mathbf{p}+\nu\nabla\wedge\ell+\mathbf{f}. \tag{5}$$

The analogous discussion is necessary to form an equation from the second formula in (1). Forming an equation corresponding to (4), we remove some terms looking unimportant. Since $\varepsilon\mathbf{p}$ is a vector field showing a relatively fine scale, we remove $\Delta\mathbf{q}$, $\nabla(\nabla\mathbf{q})$, $\nabla\wedge\mathbf{q}$. Since $\rho^*\varepsilon D\mathbf{p}/Dt$ shows the proper acceleration, we remove $\nabla\wedge\mathbf{p}$, $\nabla\wedge\ell$ (and leave $\Delta\mathbf{p}$, $\nabla(\nabla\mathbf{p})$, $\Delta\ell$, $\nabla(\nabla\ell)$) in the same manner as visco-elasticity theory (cf. [L]). Consequently, we obtain

$$\rho^*\varepsilon\frac{D\mathbf{p}}{Dt}=\nabla_j\sigma^*_{ij}+\alpha\Delta\mathbf{p}+\tilde{\alpha}\nabla(\nabla\mathbf{p})+\beta\Delta\ell+\tilde{\beta}\nabla(\nabla\ell), \tag{6}$$

where σ^*_{ij} corresponds to σ_{ij}. The system (5), (6) has just been obtained, but this is not complete. In fact, (5), (6) yield 6 scalar equations, however, unknown functions are

$(q_i, p_i, \sigma_{ij}, \sigma^*_{ij})$ (24 functions). (The external force $\mathbf{f}$ is a given vector field.) Thus it is necessary to introduce a different method to form additional equations.

Prior to showing a different method, we define here Galilei transform. Galilei transform $(\mathbf{q}, \varepsilon\mathbf{p}) \to (\mathbf{q}', \varepsilon\mathbf{p}')$ with respect to a constant $\mathbf{c}$ is defined by

$$\mathbf{q}'(\mathbf{x}', t) = \mathbf{q}(\mathbf{x}' + \mathbf{c}t, t) - \mathbf{c}, \quad \mathbf{p}'(\mathbf{x}', t) = \mathbf{p}(\mathbf{x}' + \mathbf{c}t, t).$$

The pair $(\mathbf{q}', \varepsilon\mathbf{p}')$ shows the powder-motion observed from an inertia coordinate system with the relative velocity $\mathbf{c}$ (from our frame). Since the proper motion is observed from a coordinate system taken on each particle, the vector $-\mathbf{c}$ is not necessary in the definition of $\mathbf{p}'$. It is easily seen that $(\rho D\mathbf{q}/Dt, \rho^*\varepsilon D\mathbf{p}/Dt)$ is Galilei (transform) invariant, i.e.

$$\rho\frac{D\mathbf{q}'}{Dt}(\mathbf{x}', t) = \rho\frac{D\mathbf{q}}{Dt}(\mathbf{x}, t)|_{\mathbf{x}=\mathbf{x}'+\mathbf{c}t}, \quad \rho^*\varepsilon\frac{D\mathbf{p}'}{Dt}(\mathbf{x}', t) = \rho^*\varepsilon\frac{D\mathbf{p}}{Dt}(\mathbf{x}, t)|_{\mathbf{x}=\mathbf{x}'+\mathbf{c}t}.$$

Now we show a different method. In the following argument, we use only ∇, $\mathbf{q}$, $\mathbf{p}$. Natural scalar fields generated from $\nabla = (\nabla_i)$, $\mathbf{q} = (q_i)$ are only $\nabla\mathbf{q}$ and its constant multiples; the discussion is same as above. Analogously, we obtain natural scalar fields $\nabla\mathbf{p}$, $\mathbf{qp}$. Since $\mathbf{qp}$ is not Galilei invariant, we do not use this scalar. Since $\nabla\mathbf{q}$, $\nabla\mathbf{p}$ are Galilei invariant, the condition

$$\nabla\mathbf{q} = \nabla\mathbf{p} = 0 \tag{7}$$

should have a physical meaning. In fact, this condition means the incompressibility of $\mathbf{q}$ and $\varepsilon\mathbf{p}$. Natural vector fields obtained from ∇, $\mathbf{q}$, $\mathbf{p}$ in the same manner are only $\nabla \wedge \mathbf{q}$, $\nabla \wedge \mathbf{p}$, $\mathbf{q} \wedge \mathbf{p}$ (and their constant multiples). Since $\mathbf{q} \wedge \mathbf{p}$ is not Galilei invariant, we do not use this vector field. Since $\nabla \wedge \mathbf{q}$, $\nabla \wedge \mathbf{p}$ are Galilei invariant, the conditions

$$(8) \quad \nabla \wedge \mathbf{q} = 0, \qquad (9) \quad \nabla \wedge \mathbf{p} = 0$$

should have physical meanings. In fact, these conditions mean the irrotationality of $\mathbf{q}$ and $\varepsilon\mathbf{p}$. These conditions (7)–(9) yield 8 scalar equations. Consequently, adding 6 scalar equations from (5), (6), we have now 14 scalar equations. If we use the proper position $\varepsilon\ell$, we obtain two more vector equations $\nabla\ell = 0$ and $\nabla \wedge \ell = 0$. Since the incompressibility and the irrotationality of the proper position are too strong, we do not adopt these equations as additional conditions (cf. the discussion for the static layer in 4.4).

Next we reduce the number of unknown functions in tensors σ_{ij}, σ^*_{ij}. In the case of a fluid, the pressure tensor field is expressed in the form σI with an unknown scalar σ. This is based on the isotropy of a static fluid (i.e. Brownian motion of a static fluid). A static powder is not isotropic (since it has no Brownian motion) and its static dynamics is closely related to the gravity. Taking account of these observations, we put $\sigma_{ij} = \sigma B$ with an unknown scalar σ and a given matrix $B = (b_{ij})$ (determined by a powder). Analogously, we put $\sigma^*_{ij} = \sigma^* B^*$. ($B^* = (b^*_{ij})$ is given. Since (6) shows the behaviour of a fine scale, the equality $B^* = I$ may fit.)

Finally, we determine the equation of powder-motion. We have 14 scalar equations from (5)–(9), and unknown functions are $(q_i, p_i, \sigma, \sigma^*)$ (8 functions). If we adopt (5) and (6), then there is no choice except the system (5)–(7), which means just 8 scalar equations. Thus we take the system (5)–(7) as Equation of powder-motion. It is considered that (8), (9) are conditions showing states of a powder. Equation of powder-motion (5)–(7) is a model of Cosserat-Kelvin-Voigt type and used in a suitable powder-domain; constants in (5) and (6) depend on the powder-domain (and hence these values are globally discontinuous and locally constant).

4.4. Invariants

A preliminary experimentation shows that the powder-motion is divided into a moving layer and a static layer/stratum. A moving layer may be divided into several layers. According to the change of the amount of powder, phase transitions may appear. We now discuss layers and phase transitions from Equation of powder-motion. By (7), two equations (5) and (6) are simplified as

$$\rho \frac{D\mathbf{q}}{Dt} = b_{ij} \nabla_j \sigma + \lambda \Delta \mathbf{q} + \mu \nabla \wedge \mathbf{p} + \nu \nabla \wedge \ell + \mathbf{f},$$
$$\rho^* \varepsilon \frac{D\mathbf{p}}{Dt} = b^*_{ij} \nabla_j \sigma^* + \alpha \Delta \mathbf{p} + \beta \Delta \ell + \tilde{\beta} \nabla(\nabla \ell).$$

The dimension of $(\lambda, \mu, \nu, \alpha, \beta, \tilde{\beta})$ is ((g/cm·s),(g/cm·s^2),(g/cm·s^2),(g/s),(g/s^2),(g/s^2)). The transform to non-dimension

$$(\mathbf{x}, t, \mathbf{q}, \mathbf{p}, \ell, \sigma, \sigma^*, \mathbf{f}) \longrightarrow (\mathbf{x}', t', \mathbf{q}', \mathbf{p}', \ell', \sigma', \sigma'^*, \mathbf{f}')$$

is defined by

$$\mathbf{x}' = \mathbf{x}/L, \qquad t' = Ut/L, \qquad \mathbf{q}' = \mathbf{q}/U, \qquad \mathbf{p}' = \mathbf{p}/W,$$
$$\ell' = U\ell/(LW), \quad \sigma' = \sigma/(\rho U^2), \quad \sigma'^* = \sigma^*/(\rho^* \varepsilon U W), \quad \mathbf{f}' = L\mathbf{f}/(\rho U^2).$$

Here (L, U, W) is a triple of constants with dimension (cm,cm/s,1/s) and shows a rough size of $(\|d\mathbf{x}\|, \|\mathbf{q}\|, \|\mathbf{p}\|)$ ($\|\cdot\|$ denotes the norm and $\|d\mathbf{x}\|$ corresponds to the particle-size) in a neighborhood which we observe. A triple $(d\mathbf{x}', \mathbf{q}', \mathbf{p}')$ is regarded as the normalization of $(d\mathbf{x}, \mathbf{q}, \mathbf{p})$. By the transform to non-dimension, Equation of powder-motion is transformed into the following dimensionless equation:

$$(10)\qquad \begin{aligned} &\frac{D\mathbf{q}'}{Dt'} = b_{ij}\nabla'_j\sigma' + R_\lambda^{-1}\Delta'\mathbf{q}' + R_\mu^{-1}\nabla'\wedge\mathbf{p}' + R_\nu^{-1}\nabla'\wedge\ell' + \mathbf{f}',\\ &\frac{D\mathbf{p}'}{Dt'} = b^*_{ij}\nabla'_j\sigma'^* + R_\alpha^{-1}\Delta'\mathbf{p}' + R_\beta^{-1}\Delta'\ell' + R_{\tilde\beta}^{-1}\nabla'(\nabla'\ell'),\\ &\nabla'\mathbf{q}' = \nabla'\mathbf{p}' = 0 \quad (\nabla' = (\nabla'_j) = (\partial/\partial x'^j),\ \Delta' = \nabla'\nabla'), \end{aligned}$$

where

$$R_\lambda = (\frac{\rho}{\lambda})UL, \qquad R_\mu = (\frac{\rho}{\mu})\frac{U^2}{W}, \qquad R_\nu = (\frac{\rho}{\nu})\frac{U^3}{WL},$$
$$R_\alpha = (\frac{\rho^*\varepsilon}{\alpha})UL, \quad R_\beta = (\frac{\rho^*\varepsilon}{\beta})U^2, \quad R_{\tilde\beta} = (\frac{\rho^*\varepsilon}{\tilde\beta})U^2.$$

(Number (10) shows only the first equation.) Thus the powder-motion is determined by 6 dimensionless constants (6 invariants) $(R_\lambda, R_\mu, R_\nu; R_\alpha, R_\beta, R_{\tilde\beta})$. The number of invariants would be related to the number of layers and that of phase transitions. Equation (10) shows mainly the behaviour of the standard scale, and the second equation shows that of the fine scale. Thus 3 invariants $(R_\lambda, R_\mu, R_\nu)$ in (10) would be more important than the other 3 invariants. Since $(R_\lambda, R_\mu, R_\nu)$ are defined by the inverses of coefficients, the term with the smallest invariant characterizes mainly (10). Thus R_λ (i.e. the R_λ^{-1} term) is dominant in a neighborhood where flow is rapid ($U \gg 1$), R_ν is dominant in a neighborhood where flow is slow ($U \ll 1$) and R_μ is main in a neighborhood where the proper motion is active ($W \gg 1$). A typical equation corresponding to R_λ is $\Delta'\mathbf{q}' = 0$, $\nabla'\mathbf{q}' = 0$. With R_λ, we associate a layer where particles behave like viscous flow and call this layer the viscous layer. A typical equation corresponding to R_μ is $\nabla'\wedge\mathbf{p}' = 0$, $\nabla'\mathbf{p}' = 0$, which implies that the proper motion of particles is ideal (i.e. irrotational and incompressible). With R_μ, we associate a layer where the proper motion is active and call this layer the proper motional layer. A typical equation corresponding to R_ν is $\nabla'\wedge\ell' = 0$, $\nabla'\mathbf{p}' = 0$ ($\mathbf{q}' = \ell'(\cdot,0) = 0$). Since this equation yields $\ell' = \mathbf{q}' = 0$ under a suitable boundary condition, we associate the static layer with R_ν. In the case where 3 invariants are large (i.e. $R_\cdot^{-1} \simeq 0$), (10) is nearly equivalent to $D\mathbf{q}'/Dt' = b_{ij}\nabla'_j\sigma' + \mathbf{f}'$, which is similar to Euler's equation of motion. Thus we associate a layer moving like inviscid flow with this case; a number c is used to express this case. Consequently, we have 4 layers (the inviscid l., the viscous l., the proper motional l., the static l.) associated with a 4-tuple $(c, R_\lambda, R_\mu, R_\nu)$ of invariants. (R_λ is called Reynolds number.)

At the boundary between the inviscid layer and the viscous layer, the inertia term $(\mathbf{q}'\nabla')\mathbf{q}'$ is balanced with the viscosity term $R_\lambda^{-1}\Delta'\mathbf{q}'$, which means that $R_\lambda \simeq (\Delta'\mathbf{q}')/(\mathbf{q}'\nabla')\mathbf{q}'$. Since $\mathbf{q}'$ shows the normalization, the right-hand side is nearly equal to some constant c depending only on states of a powder. Thus we have $R_\lambda \simeq c$. At the boundary between the viscous layer and the proper motional layer, it should hold that $R_\lambda \simeq R_\mu$, and hence

(11) $$W \simeq c'UL^{-1} \quad (c' = \lambda\mu^{-1}).$$

At the boundary between the proper motional layer and the static layer, it should hold that $R_\mu \simeq R_\nu$, and hence $U \simeq c''L$ $(c'' = \mu^{-1}\nu)$. There exist logically ${}_4C_2(= 6)$ boundaries, however, it is difficult to find 6 boundaries by experimentations.

The discussion for phase transitions is analogous. According to the change of (L, U, W), the main invariant in (1ρ) transfers. When the main invariant transfers from c to R_λ, a phase transition (Reynolds transition) appears. The critical point of Reynolds transition is given by $R_\lambda \simeq c$. Analogously, $R_\lambda R_\mu$-transition (the transition from R_λ to R_μ; $W \simeq c'UL^{-1}$), $R_\mu R_\nu$-transition ($U \simeq c''L$) etc. exist. Logically, 6 transitions exist. It is difficult to check these transitions by experimentations.

4.5. An experimental proof of (11)

In this section, we try to prove (11) experimentally. The quantities U and $LW(= V)$ show the speed of particles and the speed of proper motion in a neighborhood which we observe, because L corresponds to the particle-size. Thus considering that the ratio V/U is a scalar field, we read (11) as follows:

(12) The boundary between the viscous layer and the proper motional layer ($R_\lambda R_\mu$-boundary, say) consists of points whose ratio of two speeds are nearly equal to a constant.

(13) The $R_\lambda R_\mu$-transition appears in a neighborhood when the ratio V/U arrives at a constant.

The machine which we developed is drawn in PHOTO page. (Figure II shows the 2-dimensional projection.) This machine is made of acrylic panels. The part A shows a thin vessel of form $\simeq$ 5mm×30cm×45cm to observe the powder-motion. The parts B, C, D are cubic vessels. Six rectangles containing o show movable bars having knobs o to control the sizes α, β, γ, δ of orifices. (Notation • shows an acrylic bolt.) This machine is an application of the sand-glass principle; we can observe steadily various phenomena in the vessel A as long as a powder and tracers in the vessels B, C, D keep non-empty.

Photograph I and V show an experimentation to prove that each moving layer consists of smooth stream tubes as long as there is enough sand in A. Blue sand flows out of C, red sand flows out of D and white sand falls from B. In Photograph V, β is very small and white sand plays the role of a tracer. From these photographs, $R_\lambda R_\mu$-boundary would be thin.

Photograph III shows an experimentation to prove that θ in a handwritten figure IV is independent of U and $L(=$ (particle-size)$)$. Preparing two kinds of glass beads (particle-size: 1 mm and 2 mm), we experimented repeatedly varying α and $L(=1,2\,\text{mm})$ under the condition $\alpha=\beta>0$, $\delta=\gamma=0$. (Varying α means to change U.) The line $M\cup N$ shows the boundary between the part where red beads are dense and the part where red beads are not so dense. Particles moving along this line roll on M clockwise and roll on N anti-clockwise. In any case, this line is similar and $\theta\simeq 110°$. This phenomenon shows/suggests that (12) holds. In fact, the line $M\cup N$ would be $R_\lambda R_\mu$-boundary. The invariant of θ shows that $V/U\simeq c'$ at the corner for a constant c' independent of U and L.

Photographs V and VI show that there exists a number τ'' such that the powder-motion changes drastically when the amount of moving sand in A arrives at the value τ''. This is a phase transition.

In the experimentation of Photograph V, $\delta=\gamma>\alpha>0$ and β is very small. Even if we close the orifice of B, the situation does not change so much. Let U and U_0 denote the resulting speeds of moving sand near the orifice O_A of A and the orifice O_D of D respectively, and let τ be the volume of moving sand in A at an instant; this situation is called τ-situation. Now we show how to get the situation of Photograph VI. First we enlarge O_A. Then the volume of sand in A decreases. Next we return the size of O_A to α. Then we have again a steady flow such that the speed of sand near O_A is U (by the sand-glass principle) and the amount τ' of moving sand in A is less than τ; this is τ'-situation. Repeating this procedure, we arrive at the situation of Photograph VI (i.e. τ''-situation). The wave in this photograph appears periodically, and U, U_0 are invariant for all situations. (Since red sand has consumed, white sand is used in the vessel D in Photograph VI.)

Each particle flows out of O_D (or O_C) with speed U_0, gets potential energy at each instant and arrives at O_A with speed U. The value U is less than the speed calculated from potential energy. Thus energy is consumed by friction at each instant. Let E (dyn·cm) denote the total potential energy consumed by friction in the vessel A at an instant. The value of E would be invariant in all situations, because U and U_0 do not change. Since $\rho\tau'$ shows the total weight of moving sand at τ'-situation, $E/(\rho\tau')$ shows the mean energy consumed by a moving particle. Thus we may consider that the speed of proper motion is expressed as $V(\tau')=k\sqrt{E/(\rho\tau')}$ with a dimensionless constant k. Since U is invariant, our procedure turns out to be a method to increase the ratio V/U. Consequently, we see that Photographs V and VI are an experimental proof of (13).

We would like to express our thanks to Mme. Shikata and Marie Ito who took good shots (Photographs I, V and VI) by a slowmotion VTR. Taking Photograph III is very easy. If an interesting phenomenon has appeared in A under the condition $\alpha=\beta>0$, we close only the orifice O_A very quickly. Then everything stops and we can take a photograph at any time. We hope to take a photograph of a frozen Kárán sand-vortex sheet by this method. A student Mr. Mase is now studying the boundary-layer of powder-motion.

REFERENCES

[C] E. and F. Cosserat, "Théorie des corps déformables," Hermann, Paris, 1909.
[D] J. M. Dallavalle, "Micromeritics, the technology of fine particles, Second Edition," Pitman, New York, 1943.
[L] A. S. Lodge, "Elastic liquids," Academic Press, London, 1964.
[M] S. Miwa, "(Japanese) Mysteries of powder, riddles of sand," Heibonsha, Tokyo, 1977.

I

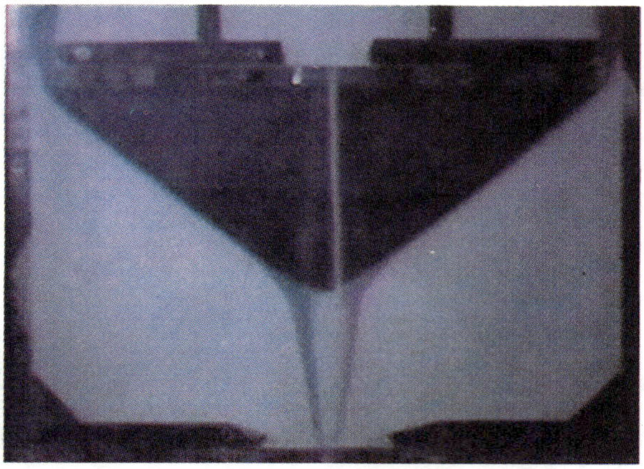

II

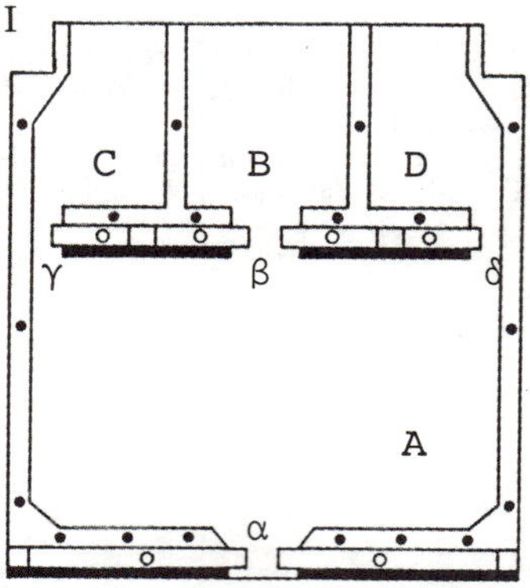

III

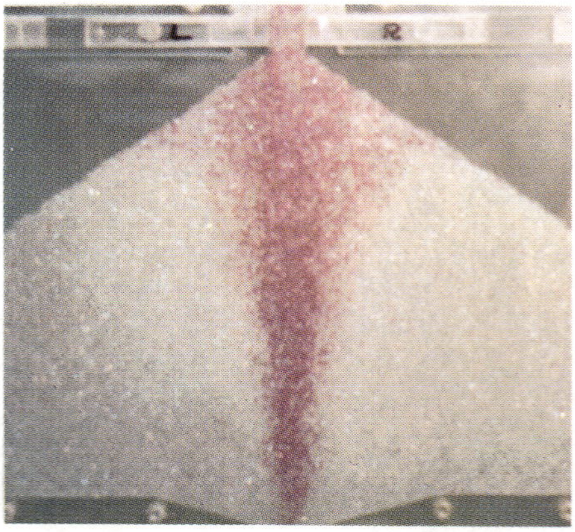

IV

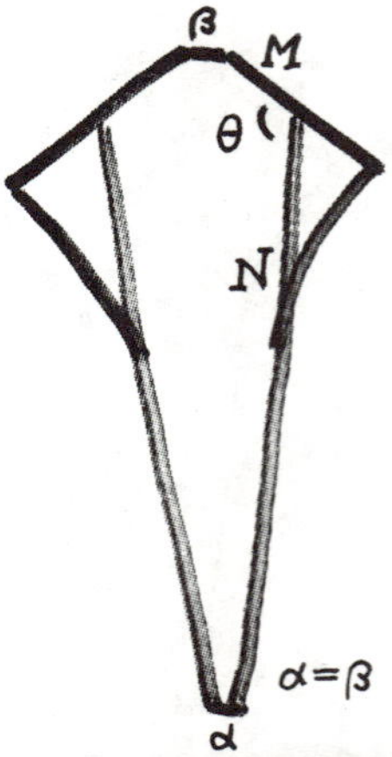

V

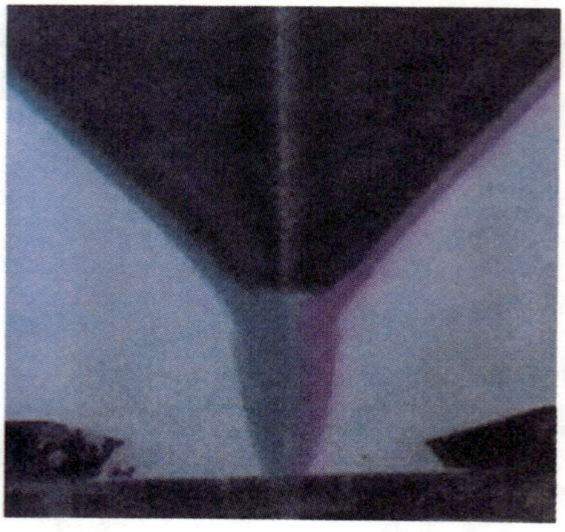

VI

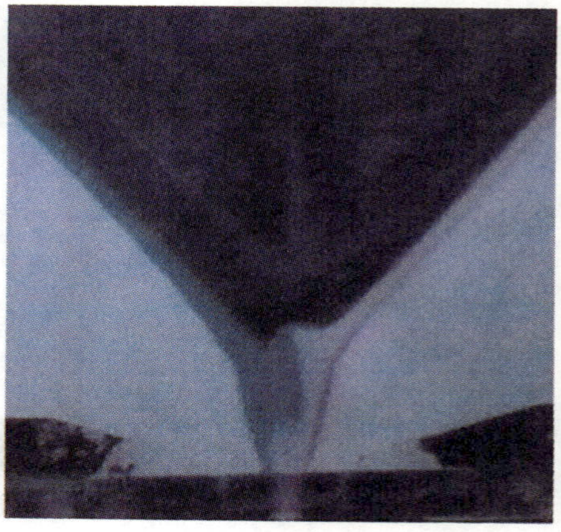

CONVERGENCE PROPERTY AND SUPERHARMONIC FUNCTIONS ON BALAYAGE SPACES

Tadashi Murazawa
Department of Mathematics
Kyoto Perfectural University
Shimogamo-Sakyo, Kyoto 606,
Japan

Abstract. Let us consider relationships between superharmonic functions and functions u of W which satisfy conditions that $[u < \infty]$ is finely dense on a balayage space (X, W). We construct an inequality on (X, W), which gives the convergence property under suitable conditions.

1. Definitions and preliminaries

Let (X, W) be a balayage space with countable base in the sense of Bliedtner and Hansen [2]. We assume that there exists a function $k(x, y) : X \times X \to [0, \infty]$ such that $k(x, y)$ is lower semi-continuous and finite continuous if $x \neq y$.

We use the following notations:

$k(x, y) = k_y(x) = k_x^*(y) = k^*(y, x)$,
$\mathcal{S}^+$: the convex cone of all non-negative superharmonic functions on X,
$\mathcal{P}$: the convex cone of finite, continuous potentials on X,
$\mathcal{P}_c$: $\{p \in \mathcal{P} \mid$ the superharmonic support, s-supp (p), is compact$\}$,
S : $\{s \in W \mid [s < \infty]$ is finely dense$\}$,
S^* : the dual of S in the sense of H-cone [3],
S^{**} : $(S^*)^*$,
S_0 : $\{s \in S \mid s$ is universally continuous on $X\}$.

Conditions:

[C] $1 \in W$,

[T_1] for every point $y \in X$, k_y is a potential whose superharmonic support, s-supp(k_y), is $\{y\}$.

[T_2] for every point $y \in X$ and for every open set U, there exists a neighborhood V of y such that

$$\sup\{k(x, y') \mid x \in X,\ y' \in V\} < \infty ,$$

K. GowriSankaran et al. (eds.), Classical and Modern Potential Theory and Applications, 353–357.

[T_3] k_x is finite, continuous if x is a finely isolated point,

[R_{cp}] for every potential $p \in \mathcal{P}$, there exists a non-negative measure μ such that $p = k\mu$.

By using results of H-cone, we get the following

Proposition 1. S is standard H-cone of functions on the set X and X is nearly saturated.

Corollary 2. S and S^{**} are isomorphic.

2. Properties of $\mu \in M_K^+(X) \cap S_0^*$

Under some conditions, let us make non-negative measures with compact support supp (μ) belonging to S_0^*. From now on in this section, let us assume that conditions [C], [T_1], [T_2], [T_3], and [R_{cp}] are satisfied on the balayage space (X, W).

Now let us recall the following measure [8].

Let ψ be a finite continuous functions with compact support on X and let x be an element of X not in $S(\psi)$, the support of ψ. For each $v \in W$, we define

$$v\psi(x) := \frac{1}{\sup\psi} \int_0^{\sup\psi} \hat{R}_v^{(\psi>\alpha)}(x) d\alpha \ ,$$

where $\{\psi > \alpha\} := \{x \in X \mid \psi(x) > \alpha\}$. Then we have $v\psi \in W$ and $v\psi \leq v$, and also that there exists a non-negative measure $\epsilon_x^\psi \in M^+(X)$ such that for all $v \in W$

$$\epsilon_x^\psi(v) = \int v d\epsilon_x^\psi = v\psi(x)$$

and supp $(\epsilon_x^\psi) \subset S(\psi)$, which is called a Mokobodzki measure.

Here if we set $v = k_y$, then we get

$$\epsilon_x^\psi(k_y) = \int k_y(z) d\epsilon_x^\psi(z) = \int k_z^*(y) d\epsilon_x^\psi(z) = k^*\epsilon_x^\psi(y).$$

We get the following result.

Lemma 3. There exists a non-negative measure λ such that

(a) $\lambda(X) < \infty$,

(b) $k^*\lambda$ is continuous and $k^*\lambda > 0$ on X.

where $k^*\lambda(y) = \int k^*(x, y) d\lambda(x)$.

Remark 4. On the above proposition we may regard the measure λ as a weak unit in S^* in the sense of H-cone. In fact, it is known that $\{\lambda_n\}_{n\in\mathbf{N}}$ separates the set E_V of V-excessive functions with respect to the resolvent V with $E_V = W$ on (X, W), i.e., for every V-excessive functions u_1 and u_2, $u_1 \neq u_2$, of E_V there exists a measure λ_n such that $\lambda_n(u_1) \neq \lambda_n(u_2)$ (see [7] Lemma 1). Hence, we have that if $\lambda_n(u) = 0$ for all $n \in \mathbf{N}$ implies $u = 0$ for each $u \in S$. Thus from the definition of λ, λ is a weak unit of S^*.

Proposition 5. For the standard H-cone S of functions on the set X, there exists a sequence $\{\mu_n\}$ of positive measures μ_n belonging to $M_K^+(X) \cap S_0^*$ such that

(a) $k^*\mu_n \in C(X)$,

(b) $\{\mu_n\}$ is increasingly dense in S^*.

Corollary 6. Let $u \in W$. Then the following are equivalent:

(1) $\mu_n(u) < \infty$ for all $n \in N$, where $\{\mu_n\}$ are the measures in Proposition 5,

(2) the set $\{x \in X \mid u(x) < \infty\}$ is finely dense,

(3) the set $\{x \in X \mid u(x) < \infty\}$ is dense.

3. Relationship between S and $\mathcal{S}^+$

In the following, let us discuss on the relationships between S and $\mathcal{S}^+$ under the following conditions on the balayage space (X, W).

Theorem 7. Let (X, W) be the balayage space without isolated points. Let conditions $[C]$ and $[T_2]$ be satisfied on (X, W). Then the following is obtained:

$$\mathcal{S}^+ \subset \{s \in W \mid [s < \infty] \text{ is dense}\} .$$

On the other hand, in general, we cannot get the inverse result of this theorem. But if we add some hypotheses, it is possible to do so. Let us assume that conditions $[C]$, $[T_1]$, $[T_2]$, $[T_3]$ and $[R_{cp}]$ are satisfied on the balayage space (X, W) with the kernel function $k(x, y) : X \times X \to [0, \infty]$ such that $(x, y) \to k(x, y)$ is lower semi-continuous, and continuous if $x \neq y$. Then it is known that there exists the dual balayage space (X, W^*) of (X, W) with the dual relation of balayage

$$\hat{R}^E_{k_y}(x) = {}^*\hat{R}^E_{k^*_x}(y)$$

for every subset $E \subset X$ and $x, y \in X$, where ${}^*\hat{R}^E_{k^*_x}(y)$ denotes a balayage function with respect to E and k^*_x [5].

Lemma 8. Let A be a subset of X with compact closure $\overline{A}$ and $u \in S$. Then $R^A u$ is continuous on $X \backslash \overline{A}$.

Theorem 9. Let (X, W) be a balayage space with the above kernel function $k(x, y)$ satisfying condition $[C]$, $[T_1]$, $[T_2]$, $[T_3]$ and $[R_{cp}]$. Then $s \subset \mathcal{S}^+$.

4. An inequality $< I >$ on S

We get the following inequality on the balayage space (X, W) with the kernel function $k(x, y) : X \times X \to [0, \infty]$ under conditions $[C]$, $[T_1]$, $[T_2]$, $[T_3]$ and $[R_{cp}]$.
Theorem 10. Let ϕ be a functional on S with properties;

(a) $0 \leq \phi(s) < \infty$ for all $s \in S$,

(b) $\phi(cs) = c\phi(s)$ for all positive real numbers c and for all $s \in S$

(c) for every s, $s' \in S$ with $s \leq s'$, $\phi(s) \leq \phi(s')$.

Then there exists a natural number $n_0 \in N$ and a positive real number M such that

$$\phi(u) \leq M \sum_{i=1}^{n_0} \mu_i(u)$$

for all u in S, where $\{\mu_i\}$ is as defined in Proposition 5.

Corollary 11. Let U be a relatively compact open set and K be a compact set with $K \subset U$. Then there exists a natural number i_0 and a positive real number M such that

$$< I > \qquad \sup_{x \in K} \hat{R}_u^{cu}(x) \leq M \sum_{i=1}^{i_0} \mu_i(u)$$

for all $u \in S$.

Now, let us consider Theorem 10 and Corollary 11 on a $\mathcal{P}$-harmonic space with countable base. Then we have

Theorem 12. Let $(X, \mathcal{U})$ be a $\mathcal{P}$-harmonic space with countable base, and the function $k(x, y) : X \times X \to [0, \infty]$ be lower semi-continuous and continuous if $x \neq y$ Then, under conditions $[C]$, $[T_1]$ and $[R_{cp}]$, the inequality $< I >$ is equivalent to the Doob convergence property.

REFERENCES

[1] H. BAUER: Harmonische Raume und ihre Potentialtheorie, Lecture Note in Math., Springer, Berlin, 1966.

[2] J. BLIEDTNER and W. HANSEN: Potential Theory. An Analytic and Probabilistic Approach to Balayage, Springer, Berlin, Berlin, 1986

[3] N. BOBOC, G. BUCUR and A. CORNEA: Order and Convexity in Potential Theory: H-Cones, Lecture Note in Math. 853 Springer, Berlin, 1981.

[4] C. CONSTANTINESCU and A. CORNES: Potential Theory on Harmonic Spaces, Springer, Berlin, 1972.

[5] T. IKEGAMI: Duality on Balayage Spaces, (To appear).

[6] K. JANSSEN: On the Existence of a Green Function for Harmonic Spaces, Math. Ann. 208 (1974) 295-303.

[7] H. MAAGLI: Representation integrale des potentiels, Sem. de Theorie du potentiel, no. 8. Lecture Notes in Math. 1235, Springer 1987, 114-119.

[8] G. MOKOBODZKI: Representation integrale des fonctions surharmoniques au moyen des reduites, Ann. Inst. Foureier 15 (1965) 103-112.

MEAN VALUE PROPERTY AND HARMONIC FUNCTIONS

Ivan Netuka and Jiří Veselý
Charles University
Mathematical Institute
Sokolovská 83
186 00 Praha 8
Czech Republic

ABSTRACT. The present state as well as the historical development of problems related to the mean value property of harmonic and caloric functions are discussed. The article is based on contributions of the authors at the meeting; it is a substantially extended update of [Net2].

KEY WORDS. Mean values, mean value property, harmonic functions, restricted mean value property, minimum principle, two-radius theorems, Laplace equation, heat equation, potential.

1. Introduction and notation

The unifying theme of the results that we survey in this article is *the mean value property of harmonic functions.* Results related to this theme are very numerous and diverse, so it is necessary to group them under various headings, according to subject matter.

CONTENTS:

In what follows, D is an open subset of $\mathbb{R}^m$, the m-dimensional Euclidean space, $m \in \mathbb{N}$. When necessary, other special properties of D are specified. Recall that a function h is said

K. GowriSankaran et al. (eds.), Classical and Modern Potential Theory and Applications, 359–398.

to be *harmonic* on D provided $h \in \mathcal{C}^2(D)$ and satisfies the Laplace equation $\Delta h = 0$ on D. Of course, Δ is the Laplace operator

$$\Delta := \sum_{j=1}^{m} \frac{\partial^2}{\partial x_j^2} .$$

For example, if $D \subset \mathbb{R}^1$, then h is harmonic on D if and only if it is affine on any interval contained in D; in this case the Laplace equation reduces to $h'' = 0$.

Let λ and σ stand for the Lebesgue measure in $\mathbb{R}^m$ and the $(m-1)$-dimensional Hausdorff measure in $\mathbb{R}^m$, respectively. For $x \in \mathbb{R}^m$ and $r > 0$ put

$$B(x,r) := \{y \in \mathbb{R}^m : |x-y| < r\}, \quad S(x,r) := \{y \in \mathbb{R}^m : |x-y| = r\} .$$

A very important role in the study of harmonic functions is played by the Poisson integral; see e.g., [Hel, p. 17], [ABR, p. 12]:

1.1 THEOREM. *Suppose that $m \geq 2$, h is harmonic on D and $B(y,r)$ is contained with its closure in D. Then*

$$h(x) = \frac{r^{m-2}}{\sigma(S(y,r))} \int_{S(y,r)} \frac{r^2 - |y-x|^2}{|z-x|^m} h(z)\, d\sigma(z) \tag{1.1}$$

whenever $x \in B(y,r)$.

2. Mean value property

For a ball $B(x,r)$ and a function f integrable on $B(x,r)$ with respect to λ put

$$A(f,x,r) := \frac{1}{\lambda(B(x,r))} \int_{B(x,r)} f(y)\, d\lambda(y) . \tag{2.1}$$

Similarly, for a sphere $S(x,r)$ and a function f integrable on $S(x,r)$ with respect to σ define

$$L(f,x,r) := \frac{1}{\sigma(S(x,r))} \int_{S(x,r)} f(y)\, d\sigma(y) . \tag{2.2}$$

The numbers $A(f,x,r)$ and $L(f,x,r)$ are called the *volume mean* (of f on $B(x,r)$) and the *spherical mean* (of f on $S(x,r)$), respectively. For $m = 1$, the value $L(f,x,r)$ is defined as the arithmetic mean value of $f(x-r)$ and $f(x+r)$. If $B(x,r)$ is contained with its closure in D, then $B(x,r)$ and $S(x,r)$ will be called *admissible* with respect to D.

2.1 THEOREM. *If h is harmonic on D and $S(x,r)$ is admissible with respect to D, then*

$$h(x) = L(h,x,r) . \tag{S}$$

(It follows immediately from Theorem 1.1 by setting $x = y$ in (1.1).) This is known as the *spherical mean value property* of harmonic functions.

The statement was called by C. Neumann in [Neu2, p. 24], cf. [Neu1], [Bac, p. 26], [BM, p. 480], [Kel1, p. 223], "der Gauss'sche Satz des arithmetischen Mittels" in view of the following theorem proved by C. F. Gauss (1840); let us quote from Article 20, [Gau, p. 222]:

LEHRSATZ. *Bedeutet V das Potential einer wie immer vertheilten Masse in dem Elemente einer mit dem Halbmesser R beschriebenen Kugelfläche ds, so wird, durch die ganze Kugelfläche integrirt,*

$$\int V ds = 4\pi(RM^0 + RRV^0) \tag{2.3}$$

wenn man mit M^0 die ganze im Innern der Kugel befindliche Masse, mit V^0 das Potential der ausserhalb befindlichen Masse in Mittelpunkt der Kugel bezeichnet, und dabei die Massen, die etwa auf der Oberfläche der Kugel stetig vertheilt sein mögen, nach Belieben den äussern oder innern Massen zuordnet.

If no "Masse" lies in the ball (i.e., $M^0 = 0$ in (2.3)) we obtain

$$V^0 = \frac{1}{4\pi R^2} \int V ds \ , \tag{2.4}$$

which is nothing else than the *arithmetic mean of the function V on the sphere.* (The footnote in [Kel1, p. 82] explains the terminology.) We remark that Theorem 2.1 can be also derived from the theorem on three potentials; see e.g., [Kel1, p. 219, 223]. For another proof, see e.g., [ABR, p. 5], [Dem]. Sometimes harmonic functions are *defined* by the mean value property; cf. [Bre, p. 166], [Doo2, p. 4], [PS, p. 85], [Usp] and Section 3 below.

Integration of (S) with respect to radii gives us a similar property involving balls:

2.2 THEOREM. *If h is harmonic on D and $B(x, r)$ is admissible with respect to D, then*

$$h(x) = A(h, x, r) \ . \tag{B}$$

This is called the *volume mean value property* of harmonic functions.

3. Converse of the mean value property

Let us start with the following quotation from [Kel1, p. 224]: "The property of harmonic functions given by Gauss' theorem is so simple and striking, that it is of interest to inquire what properties functions have which are, as we shall express it, their own arithmetic means on the surface of spheres." Certain one-dimensional results, which have a long history inspire some important observations (see [Acz, p. 5, 35], also for numerous references): the so-called Cauchy functional equation

$$f(x+y) = f(x) + f(y) \ , \qquad x, y \in \mathbb{R}^1 \ , \tag{3.1}$$

(the solutions of (3.1) obviously satisfy the condition $f(x) = L(f, x, r)$ whenever $x \in \mathbb{R}^1$ and $r > 0$) was known already to A. M. Legendre (1791) and to Gauss (1809). In an article published in 1821, A. L. Cauchy studied this functional equation and proved that any continuous solution of (3.1) is given by $f(x) = f(1) \cdot x$, $x \in \mathbb{R}^1$. Results obtained later by G. Darboux (1875, 1880), W. Sierpiński (1920) and A. Ostrowski (1929) considerably weakened the continuity assumption up to measurability of f or its boundedness on a set of positive measure while a complementary result of G. Hamel (1905) shows the existence of rather "bad" solutions (non-measurable, having dense graph in $\mathbb{R}^2$) of (3.1). What can be said about the problem in $\mathbb{R}^m$ for $m > 1$? Now it is well known that, under mild regularity assumptions on h, condition (S) for all admissible spheres with respect to D is equivalent to the harmonicity of h. We start with the following

3.1 THEOREM. *Suppose that h is a continuous function on D. Then the following statements are equivalent:*

(i) *the function h is harmonic on D;*
(ii) *if $S(x,r)$ is admissible with respect to D, then $h(x) = L(h,x,r)$;*
(iii) *if $B(x,r)$ is admissible with respect to D, then $h(x) = A(h,x,r)$;*
(iv) *if $B(x,r)$ is admissible with respect to D, then $L(h,x,r) = A(h,x,r)$.*

It is useful to observe that, in the case where D is *connected*, property (ii) (or (iii)) implies the strong maximum (or minimum) principle; see [Gau, p. 223] and [Hel, p. 18]: if h attains its maximum or minimum at a point $x \in D$, then it is constant. To see this, let us denote by M the set $\{y \in D : h(x) = h(y)\}$. Continuity of h implies that M is closed while from (ii) (or (iii)) we get that M is open. Hence $M = \emptyset$ or $M = D$.

Now we can easily prove (ii) $\Longrightarrow$ (i) (or (iii) $\Longrightarrow$ (i)) as follows. Note, that it is enough to choose an admissible ball $B(y,r) \subset D$ and prove that h is harmonic in $B(y,r)$. Define the function v by replacing h on $B(y,r)$ by its Poisson integral; cf. (1.1). Now since v is continuous on $\overline{B(y,r)}$ and harmonic on $B(y,r)$ (the Poisson integral gives an explicit formula for solution of the Dirichlet problem for a ball; this was first proved in [Sch]), the function $w = v - h$ is continuous on $\overline{B(y,r)}$ and the above mentioned maximum principle leads to $w = 0$. Hence h is harmonic on $B(y,r)$. The proof of the remaining implication (iv) $\Longrightarrow$ (i) in Theorem 3.1 can be found e.g., in [BR1].

Implications (ii) $\Longrightarrow$ (i) (or (iii) $\Longrightarrow$ (i)) are usually called the converse of theorem(s) on the mean value property. The first converse seems to be due to P. Koebe (1906). He proved in [Koe] the following theorem:

SATZ. *Ist u eine in der Ebene oder im Raume erklärte stetige reelle Funktion, welche inbezug auf jede ganz im Innern des Definitionsbereiches liegende Kreisfläche bzw. Kugel die erwähnte Gaußsche Mittelwerteigenschaft besitzt, so ist u eine Potentialfunction.*

Despite the different terminology (the term "*harmonic function*" was used for the first time in the work of W. Thompson (Lord Kelvin) and P. G. Tait in the year 1873 and it took some time until it was generally accepted; see [TT], [ABR, p. 25], [BM, p. 468]), we recognize in the quoted Satz the theorem which is nowadays known to almost every mathematician.

A more general theorem was proved in [Lev] by E. Levi (1909), apparently independently of Koebe. Following ideas of his brother Beppo Levi, he proved a theorem which can be expressed in modern terms as follows: *If D is a bounded domain, f is a bounded measurable function on D, and* (S) *holds for all admissible spheres with respect to D, then f is harmonic.* He applied Fubini's theorem to prove that (B) holds for all admissible balls with respect to D; from this he derived the continuity and differentiability of f; in fact, $f \in C^\infty(D)$. Then he verified directly that f satisfies the Laplace equation.

Thus the way to define harmonic functions without a priori differentiability requirements using Theorem 3.1 was opened. Levi also mentioned a further possible generalization: provided that Fubini's theorem is applicable, one needs only integrability over all admissible spheres with respect to D in order to get (B) for all admissible balls with respect to D.

Soon afterwards, a much more important result appeared: it is closely related to the material studied in Section 5. In the note [Vot, p. 264], V. Volterra (1909) discovered the importance of the maximum principle for such considerations. He proved the following *uniqueness* theorem:

TEOREMA I. *La funzione u assolutamente continua e finita nel campo σ è determinata quando: 1° in ogni punto A interno al campo si conosce*

$$\frac{1}{M(A)}\,P[\,u, M(A)\,] - u(A);$$

2° si conoscono i valori della funzione u al contorno del campo; 3° tutti i punti interni al campo sono connessi col contorno.

Let us give a modernized version of the statement; we need first some notation. Let D be a bounded domain. We suppose that to every point $x \in D$ a probability measure μ_x with support $E(x)$ is associated in such a way that $\mu_x(\mathbb{R}^m \setminus D) = 0$. We say that a point $x \in D$ can be joined with the boundary ∂D of D provided that for every $\varepsilon > 0$ there exists a finite sequence $\{x_1, \dots, x_n\}$ of points of D such that

$$x_1 \in E(x),\ x_2 \in E(x_1) \dots, x_n \in E(x_{n-1})\ , \quad \text{and} \quad \operatorname{dist}(x_n, \partial D) < \varepsilon\ .$$

(Note that a similar concept also appeares in modern potential theory, see [Doo1, p. 53].) Now the assertion can be formulated as follows:

3.2 THEOREM. *A function u continuous on $\overline{D}$ is uniquely determined provided that*

1° *the difference $\int u d\mu_x - u(x)$ is given for every $x \in D$;*
2° *the function u is given on ∂D;*
3° *every $x \in D$ can be joined with the boundary ∂D.*

Observe that Volterra considered also discrete measures. As an application of his Teorema I, he proved the following theorem inspired by [Lev]: *Assume that D is a bounded domain, f is a continuous function on $\overline{D}$ satisfying condition* (S) *at every point $x \in D$ for* one radius $r(x)$, $0 < r(x) \le \operatorname{dist}(x, \partial D)$. *Suppose that the above condition* 3° *(where now μ_x is the normalized surface measure on $S(x,r)$) is fulfilled. If there is a function h continuous on $\overline{D}$ and harmonic on D such that $f = h$ on ∂D, then f is harmonic on D (and $f = h$).*

To get the function with which f is compared with the help of a minimum principle he needed to solve the Dirichlet problem for the boundary condition $f|_{\partial D}$ and hence he supposed that " ... *il teorema di esistenza delle funzioni armoniche valga pel campo σ* ... ". Of course, the same reasoning can be applied for the mean values (B); Volterra carefully described differences between his theorem and the result of Levi (see [Vot, p. 266]) and showed by an example the importance of the continuity up to boundary ∂D.

A further weakening of the hypothesis in the theorem proved in [Lev] was provided by L. Tonelli in [Ton] (1909). He proved: *A function f integrable on D and having the property* (S) *at every $x \in D$ for every admissible sphere for which the corresponding mean value exists is harmonic.* In fact he proved it by the method of Levi, showing first that the function f must be continuous.

Later G. Vitali (1912) proved in [Vit] that Volterra's condition 3° can be replaced by a weaker condition which is automatically fulfilled in case of admissible circles. He filled a gap in [Ton] where Tonelli did not verify condition 3° when applying the above mentioned result of Volterra. In Vitali's work we can already find the "nearest point" tool that was later used many times; see e.g., [Cim], [CH, p. 279] and [Bur]. In [Ton] Tonelli also took a step in the direction towards proving part (b) of Theorem 3.5 below which provides an answer to one of the following quite natural problems:

3.3 PROBLEM. *Let $C \subset D$ and let* (S) *hold for a continuous function h and all admissible $S(x,r)$ with $x \in C$. Under what conditions imposed on C may one conclude that h is harmonic?*

3.4 PROBLEM. *Let* (S) *hold for a continuous function h and for any admissible sphere $S(x,r)$ with respect to D with radii r from a set $R_x \subset]0,\infty[$. How "big" must R_x be to ensure that f is harmonic?*

(Both problems have been known implicitly for a long time; they were explicitly formulated e.g., in [Fla3].)

3.5 THEOREM. *A function f is harmonic on D provided one of the following conditions holds:*

(a) *The function f is locally integrable on D and* (B) *holds for every admissible ball with respect to D.*
(b) *The function f is continuous on D and there is a dense subset C of D such that* (B) *holds for every admissible ball with respect to D with centre in C.*
(c) *The function f is continuous on D and for every $x \in D$ there is a sequence of positive numbers $\{r_n(x)\}$ tending to 0 such that* (B) *holds for all balls $B(x, r_n(x))$ with $x \in D$ and $n \in \mathbb{N}$.*

These assertions are now relatively well known and can be easily proved. For the proof of part (a), see [ABR, p. 18], part (b) follows by the Lebesgue convergence theorem and for the proof of (c), see e.g., [Lit1] or [ABR, p. 16]. Note that (a), (b) and (c) hold with (S) instead of (B). (Let us mention that in [Rie] and [Lit1] the mean value property plays a key role in the introduction of non-smooth subharmonic functions.)

We remark that there are characterizations of harmonic functions expressed in terms of zero flux through regular surfaces or spheres; see e.g., [Sev], [Bec], [Koe], [Kel1, p. 227], [Boc], [Ger], [Sas1] [Eva1], [Eva2].

Similar problems like 3.3 and 3.4 can be studied for (B). In fact, results on spherical and volume mean values are related: see comments from [Lit2] quoted in Section 5. Part (b) of Theorem 3.5 provides answer to Problem 3.3 while part (c) gives information related to Problem 3.4.

A more interesting and somewhat surprising result concerning Problem 3.3 was proved by L. Flatto in [Fla3]:

3.6 THEOREM. *Suppose that h is a continuous function on $\mathbb{R}^m$, $m \geq 2$ and* (S) *holds for every admissible sphere with the centre in C. If C is not nowhere dense in $\mathbb{R}^m$ (i.e., the interior of $\overline{C}$ is not empty), then h is harmonic.*

The same article contains deeper assertions connected with Problem 3.4. Now there is a whole family of similar theorems: they concern both spherical and volume means and in all of them R_x contains for each $x \in D$ two radii or one radius. We shall deal with these theorems in the following three sections.

4. Two radius theorems

Now our starting position has dramatically changed: *arbitrarily small radii* for which (S) or (B) hold are not available. Since harmonicity is a local property, while (S) and (B) for a finite number of r's is of a non-local character, the interplay between those two things is complicated; cf. [Zal3].

First we shall present some examples on an interval $D \subset \mathbb{R}^1$: the restriction f to D of a *discontinuous* solution of the Cauchy functional equation shows that mere (S) for *all* admissible spheres does not imply the harmonicity of f. On the other hand, if we add continuity, the only solutions are locally affine and hence harmonic. If continuity on D is supposed and (S) holds for only *one* admissible sphere at each point, then again harmonicity is not guaranteed. Moreover, nothing can be saved even by the imposition of additional boundedness hypotheses on the set D and/or the function f; see [CH, p. 281] for a pleasing "zigzag" example of M. Schiffman or [Huc]. The following example is more important:

4.1 EXAMPLE. Consider the function $\sin(\pi x)$ on $\mathbb{R}^1$. It satisfies the spherical mean value property with respect to all spheres $B(x,n)$, $x \in \mathbb{R}^1$, $n \in \mathbb{N}$. (Observe that the set of ratios n_1/n_2, $n_1, n_2 \in \mathbb{N}$, of radii of these spheres is dense in $]\,0,\infty\,[$.)

In view of this example, a theorem saying that if D is the whole space "two radii suffice", is somewhat astonishing. The result is due to J. Delsarte (1958) and appeared in [Del1]; see also [Del2], [DL]. Let us quote (with a small change in notation) from the paper [DL] in which the result was extended to cover also other differential equations:

Dans une note récente [Del1], *l'un d'entre-nous a donné le théorème suivant: Si f est une fonction indéfiniment dérivable dans $\mathbb{R}^m$, désignons par $L(f,x,r)$ la moyenne de f sur la sphère de centre x et de rayon r. Soient alors a et b deux nombres positifs fixes, distincts. Si l'on a, pour tout $x \in \mathbb{R}^m$:*

$$L(f,x,a) = L(f,x,b) = f(x) \ ,$$

alors la fonction f est harmonique (sauf peut-être pour des valeurs exceptionnelles de a et b en nombre fini).

(The authors tacitly suppose that $m \geq 2$.) For the description of exceptional sets we need Bessel functions; cf. [Zal3], [Wat]. For n integer, let $J_n(z)$ be the n-th Bessel function of the first kind defined by

$$J_n(z) := \sum_{k=0}^{\infty} \frac{(-1)^k}{k!(k+n)!} \left(\frac{z}{2}\right)^{2k+n} , \qquad n \geq 0 , \tag{4.1}$$

$$J_n(z) := (-1)^n J_{-n}(z) , \qquad n < 0 . \tag{4.2}$$

There is a natural way of extending the definition to Bessel function J_α for α non-integer. It is known that, for $\alpha > -1$, all zeros of J_α are real. If α is an odd multiple of $1/2$, then J_α can be expressed in terms of trigonometric functions; in particular,

$$J_{-1/2}(z) = \left(\frac{2}{\pi z}\right)^{1/2} \cos z \ , \qquad J_{1/2}(z) = \left(\frac{2}{\pi z}\right)^{1/2} \sin z \ .$$

The exceptional sets will be described by means of the function

$$j_p(z) := 2^p \Gamma(p+1) J_p(z)\, z^{-p} \ . \tag{4.3}$$

(We shall need the case of p equal to integer multiples of $1/2$ only.) The above mentioned theorem of Delsarte was announced in [Del1] (1958). For more than ten years it was considered to be a rather special result on harmonicity; see [Zal3]. The following theorem is contained in [Fla3] (1965):

4.2 THEOREM. *Suppose that* $m \geq 2$ *and* f *is a continuous function on* $\mathbb{R}^m$. *If for two radii* $r_1, r_2 > 0$

$$f(x) = L(f, x, r_1) = L(f, x, r_2) \tag{4.4}$$

for all $x \in \mathbb{R}^m$, *then the function* f *is harmonic unless* $r_1/r_2 = z_1/z_2$ *where* $z_k \neq 0$, $j_p(z_k) = 1$ *for* $p = m/2 - 1$ *and* $k = 1, 2$.

Even though the theorem is essentially the same as that quoted above, it describes the exceptional set in detail; of course, the description is not simple and the finiteness of the exceptional set is by no means clear. A similar theorem holds for *volume means*; cf. [Zal1]:

4.3 THEOREM. *Suppose that* $m \geq 2$ *and* f *is a locally integrable function defined on* $\mathbb{R}^m$. *If for two radii* $r_1, r_2 > 0$

$$f(x) = A(f, x, r_1) = A(f, x, r_2) \tag{4.5}$$

for all $x \in \mathbb{R}^m$, *then the function* f *is harmonic unless* $r_1/r_2 = z_1/z_2$ *where* $z_k \neq 0$, $j_p(z_k) = 1$ *for* $p = m/2$ *and* $k = 1, 2$.

For $m = 2$ both of these properties are rather close to analogous theorems on functions of a complex variable. We do not want to go into details; we recommend nicely written articles of Zalcman [Zal1], [Zal2], [Zal3], [Zal4] and [Zal6]. Just to have a chance to compare Theorems 4.2 and 4.3 with a "complex Delsarte type theorem" let us give another result from [Zal1]:

4.4 THEOREM. *Suppose that* f *is a locally integrable (complex) function in the complex plane* $\mathbb{C}$. *Denote by* $C_r(z)$ *the circle of radius* $r > 0$ *about* $z \in \mathbb{C}$. *If for two different radii* $r_1, r_2 > 0$

$$\int_{C_{r_1}(z)} f(\zeta)\, d\zeta = \int_{C_{r_2}(z)} f(\zeta)\, d\zeta = 0 \tag{4.6}$$

for λ*-almost all* $z \in \mathbb{C}$, *then* f *is equal almost everywhere to an entire function unless* $r_1/r_2 = z_1/z_2$ *where* $z_k > 0$, $J_1(z_k) = 0$ *for* $k = 1, 2$.

4.5 EXCEPTIONAL SETS. Denote the exceptional sets of ratios r_1/r_2 in the previous three theorems by $S[m]$, $B[m]$ and C, respectively. Then for $m = 2$ all three sets are described in terms of J_0 and J_1, but in a different manner.

Delsarte proved in [Del2] that (for all $m \geq 2$) the exceptional sets $S[m]$ *are finite*. He succeeded in proving that $S[3]$ *is empty* and conjectured that the same is true of $S[m]$ for all other $m \geq 2$. This seems to be still an open problem.

Theorem 4.2 holds also for $m = 1$. The set $S[1]$ *is infinite and dense in* $]0, \infty[$. Indeed, for $p = 1/2 - 1 = -1/2$ the equation $j_p(z) = 1$ reduces to $\cos z = 1$ and $S[1]$ is the set of quotients of all non-zero $z \in \{2n\pi : n \in \mathbb{N}\}$, i.e., the set of rational numbers; cf. Example 4.1.

It should be mentioned here that analogous results can be obtained for $B[m]$; see [Zal3].

4.6 REMARKS ON METHODS. The proof from [DL] is based on the following fact, already known to Poisson; $u(x,r) := L(f,x,r)$ is a solution of the Cauchy problem for a singular hyperbolic equation called in [Zal1] the Euler-Poisson-Darboux equation:

$$\frac{\partial^2 u}{\partial r^2} + \frac{m-1}{r}\frac{\partial u}{\partial r} = \Delta_x u \ ,$$
$$u(x,0) = f(x) \ , \qquad \frac{\partial}{\partial r}u(x,0) = 0 \ .$$

The theories of mean periodic functions and transmutation operators were used. Results in [DL] cover not only the Laplace equation.

Flatto in [Fla3] worked with iterated spherical means and also with the theory of mean periodic functions; the second tool is a standard one and it is used in all papers. While Flatto's approach is more accessible than that of [Del3], his result covers only the spherical case.

Zalcman proved Delsarte type theorems in [Zal1] (1972) independently, also for volume means. His proof is based on tools related to the Fourier transform.

Another relevant paper has its starting point also in problems concerning complex functions. In [BST] (1973) L. Brown, B. M. Schreiber and B. A. Taylor attacked the problem of the mutual relation of *Pompeiu and Morera problems.* They produced results which enabled them to get Theorem 4.2 for a function $f \in \mathcal{C}(\mathbb{R}^2)$ as a by-product. The main tool which they used was spectral synthesis.

4.7 LOCALIZATION. For a relatively long time it was not clear how to generalize two-radius theorems to the case of *bounded* domains in $\mathbb{R}^m$. A solution was provided independently in articles by J. D. Smith (1973) in [Smi], by C. A. Bernstein and R. Gay (1986) in [BG] and by V. V. Volchkov (1991) [Vol1], [Vol2]. Note that in the next theorem part (a) is from [BG], parts (b) and (c) are taken from [Vol1]. Volchkov mentioned at the end of [Vol1] that "*similar theorems are valid for circles $C(z,r)$ and spheres.*"

4.8 THEOREM. *Let f be continuous on $B(0,R) \subset \mathbb{R}^m$, $m \geq 2$ and $r_1, r_2 > 0$. Then the following hold:*

(a) *If $f(x) = L(f,x,r_1)$ for all x, $|x| < R - r_1$, $f(x) = L(f,x,r_2)$ for all x, $|x| < R - r_2$, $r_1 + r_2 < R$ and r_1/r_2 does not belong to $S[m]$, then f is harmonic.*
(b) *If $f(x) = A(f,x,r_1)$ for all $|x| < R - r_1$, $f(x) = A(f,x,r_2)$ for all $|x| < R - r_2$, $r_1 + r_2 < R$ and r_1/r_2 does not belong to $B[m]$, then f is harmonic.*
(c) *If $r_1 + r_2 > R$ or $r_1/r_2 \in B[m]$, then there is a non-harmonic function f satisfying conditions imposed on $A(f,x,r_1)$ and $A(f,x,r_2)$ in* (b).

In fact, as shown very recently in [Vol3], part (b) can be extended to locally integrable functions (and the equalities required for almost all x only). Also, part (c) is improved and completed: if $r_1 + r_2 > R$, a non-harmonic $\mathcal{C}^\infty$-function exist whereas f can be obtained real analytic non-harmonic provided that $r_1/r_2 \in B[m]$. Finally, the case $r_1 + r_2 = R$ and $r_1/r_2 \notin B[m]$ is discussed: f is harmonic, if f is a $\mathcal{C}^\infty$-function; on the other hand, for every $k \in \mathbb{N}$ there are non-harmonic $\mathcal{C}^k$-functions with $f(x) = A(f,x,r_j)$ for almost all x satistying $|x| < R - r_j$, $j = 1, 2$.

It should be mentioned that among a number of theorems which Volchkov announced in [Vol1] there are also Delsarte type theorems for the Laplace, Poisson and Helmholz equations. For the Laplace equation he proved in [Vol2] a theorem on two functions (Delsarte type theorem for *volume means* can be obtained from it by setting $f = g$); some notation is needed.

Let us denote, for all $m \geq 1$,

$$\kappa_m := \inf\{|\operatorname{Im} z| : j_p(z) = 1, z \neq 0, p = m/2\} , \tag{4.7}$$

$$h_m(x,r) := |x|^{(1-m)/2} \cdot \exp\left(\frac{\kappa_m |x|}{r}\right) , \qquad x \in \mathbb{R}^m, \quad r > 0 . \tag{4.8}$$

4.9 THEOREM. *Let f, g be continuous on $\mathbb{R}^m$, $0 < r_1 < r_2$ and*

$$f(x) = A(f,x,r_1) , \qquad g(x) = A(g,x,r_2) , \qquad x \in \mathbb{R}^m . \tag{4.9}$$

(a) *If $r_1/r_2 \notin B[m]$ and $f(x) - g(x) = o(h_m(x,r_2))$ as $|x| \to \infty$, then f and g are harmonic functions.*
(b) *If $r_1/r_2 \notin B[m]$ and $f(x) - g(x) = O(h_m(x,r_2))$ as $|x| \to \infty$, then f is harmonic, but g can be a non-harmonic function.*
(c) *In all remaining cases there exist continuous non-harmonic functions f and g such that (4.9) holds.*

4.10 ON BESSEL FUNCTIONS. Delsarte type theorems on harmonic functions can be proved in several ways, all of which ultimately involve Bessel functions. We shall close this section by a quotation from [Zal3, p. 164]: *People ask, " Where do the Bessel functions come from ?" I used to think I knew the answer to that question; now I am not so sure. There are by now a number of different approaches to the theorem, and in each approach the Bessel functions arise in a slightly different way: as Fourier transforms, as eigenfunctions of the Laplacian, as spherical functions, or as solutions to certain ordinary differential equations. On one level, of course, these are all the same. What is mildly troubling is that the theorem extends in a number of different directions (some of which we explore below) and that now one, now another (but not all) of the techniques available provides the key to the appropriate extension. No doubt there is a final synthesis; since we have not yet attained it, it seems best to answer humbly that, like Kronecker's integers, the Bessel functions have been granted us by Providence: they are simply a fact of nature.*

5. Restricted mean value property: volume means

Examples showing that the one-radius volume mean value property does not in general guarantee harmonicity seem to be more complicated than their spherical analogues (see Littlewood's problems below). Inspired by Theorem 4.3 and formulae for Bessel functions let us present such an example for $\mathbb{R}^1$:

5.1 EXAMPLE. Given $r_0 > 0$ consider $\theta = \alpha + i\beta$ satisfying

$$\frac{\sin \theta r_0}{\theta r_0} = 1 .$$

Clearly, such a solution exists; define $f(t) := e^{-\beta t}\cos\alpha t$. Now it is easily seen that

$$f(x) := \frac{1}{2r_0}\int_{x-r_0}^{x+r_0} f(t)\,dt\ , \qquad x \in \mathbb{R}^1\ ,$$

and f is *not affine* on $\mathbb{R}^1$. For $\mathbb{R}^m$ with $m \geq 2$ put $g(x) = g(x_1,\ldots,x_m) := f(x_1)$ to get a function for which $g(x) = A(g,x,r_0)$, $x \in \mathbb{R}^m$; cf. [CH, p. 281]. For another example, see [Net1] where a problem posed in [Mar1] was solved; the function is constructed in $\mathbb{R}^m$, with radii depending on x. Another example for $B(0,1) \subset \mathbb{R}^2$ can be found in [Huc, p. 492]; see also [Lit2, p. 22]).

5.2 EXAMPLE. A simple example in $\mathbb{R}^m$ (appearing e.g., in [Vee2], [Hea]) is given by $f(x) = f(x_1,\ldots,x_m) := \operatorname{sgn}(x_1)$. Put $M = \{x \in \mathbb{R}^m : x_1 = 0\}$ and for any $x \in \mathbb{R}^m$ define $r = r(x) := \operatorname{dist}(x,M)$ provided $x \notin M$, while for $x \in M$ choose $r = r(x)$ arbitrarily, e.g., $r = 1$. Obviously, $f(x) = A(f,x,r(x)) = L(f,x,r(x))$ and f is surely not harmonic since it is discontinuous.

The above examples show that even with volume means the "one-radius" mean value property does not guarantee the harmonicity of the function and hence some additional assumptions are necessary.

Here and in the next section we shall be interested in the so called restricted mean value property for volume and spherical means. In this case, the set R_x from Problem 3.4 contains only *one* radius $r = r(x)$.

Recall again that one-radius theorems go back to [Vot]. Before confining ourselves to the main topic of this section, volume means, we emphasize the distinction between restricted mean value properties with respect to spheres and with respect to balls. This important distinction was soon recognized: we quote from a collection of problems which J. E. Littlewood published in [Lit2, p. 22]; see also [Huc].

5.3 LITTLEWOOD'S PROBLEMS. *If w is continuous in the closure $\overline{D}$ of a bounded domain D, and "1-circle," that is, equal at each point to its average over some one circumference about the point and lying (with its interior) in D, then w is harmonic on D (Kellogg, [Kel3]). There can be a finite number of exceptional boundary points provided w is bounded. It is easy to see that even when D is a circle it is not enough for w to be merely continuous in the open D, but the obvious counter-example, a function, of r only, $(a_n \log r + b_n)$'s pieced together is strongly unbounded. Hence the question: If w is continuous on D, and bounded on D, and 1-circle, is it necessarily harmonic? I expect the answer "No," even when D is a circle. Difficult, because (a) a function of r only won't do, and (b) a discontinuity at 1 point (only) of the boundary won't do (and this rules out one natural approach). (...)*

Note that the positive result is weaker when the average is taken over one circular area (though a paper has been published, proving the weaker case, and with a "smooth" boundary, oblivious to this fact). The negative results are correspondingly stronger.

The final remark is a simple consequence of the mean value theorem of elementary calculus: if for a continuous function h and a fixed point x the equality $h(x) = A(h,x,r_0)$ holds for a certain $r_0 > 0$, then there is an $r \in\ ]0,r_0[$ with $h(x) = L(h,x,r)$.

Let us introduce the following terminology. A function δ defined on D is said to be *admissible*, if $0 < \delta(x) \leq \operatorname{dist}(x,\partial D)$, $x \in D$.

Let δ be an admissible function on D, f a function defined on D and $x \in D$. Put

$$A_\delta f(x) := A(f, x, \delta(x)) \ ,$$

provided f is integrable on $B(x, \delta(x))$.

Functions f satisfying $f = A_\delta f$ are sometimes said to have *the restricted mean value property on balls*. They are also called *δ-harmonic* functions.

This restricted mean value property is not just a curiosity without any application. Soon after [Vot] appeared, H. Lebesgue (1912) in [Leb3] showed how it is related to the Dirichlet problem; cf. [Leb4]: he put $\delta(x) := \operatorname{dist}(x, \partial D)$, $x \in D$, and given a boundary condition $g \in \mathcal{C}(\partial D)$ he considered a continuous extension G of g to $\overline{D}$. (It is worth mentioning that he proved the existence of such an extension earlier in [Leb1].) He defined

$$AG : x \mapsto A(G, x, \delta(x)) \ , \quad x \in D \ ,$$

and put $A^0G := G$, $A^nG := A(A^{n-1}G)$ for $n \in \mathbb{N}$. Then he proved that for any regular domain D, the sequence $\{A^nG\}$ converges to the solution of the Dirichlet problem for D and the boundary condition g. (We remark that later the assumption of the regularity of D was removed by F. W. Perkins (1927) in [Per]; for a generalization, see [Car], [Ves1].)

At the beginning of the development of theorems on the restricted mean value property there was no substantial difference between the use of spherical and volume means. Recall (cf. Theorem 3.2) the role of the minimum principle in [Vot] and the freedom in the choice of non-trivial measures μ_x associated to $x \in D$. Proofs of harmonicity are in fact quite easy, given continuity up to the boundary.

Good behaviour of the studied function f near the boundary gives (for example for a regular D) a nice harmonic function with which f can be compared: it is the corresponding solution of the Dirichlet problem. This reasoning is applicable in more general situations, if enough information on the boundary behaviour of f and a sufficiently powerful minimum principle are available; see [Ves2], [OS].

The following theorem was proved by W. Veech (1973) in [Vee2], cf. also [Vee1]; it shows that we can limit ourselves to Borel measurable functions δ and f.

5.4 THEOREM. *Let δ be an admissible function on D and f* a positive *function satisfying $A_\delta f = f$ on D. Then there exist Borel measurable functions f_0 and δ_0 on D such that $f_0 \geq 0$, $0 < \delta_0 \leq \delta$, $f = f_0$ λ-almost everywhere and $A_{\delta_0} f_0 = f_0$ on D.*

The first positive non-trivial result which did not depend on continuity up to the boundary is due to F. Huckemann (1954); see [Huc]. Note that the limits are *not* assumed to be finite.

5.5 THEOREM. *Suppose that f is continuous on $]-1,1[\, \subset \mathbb{R}^1$ and has the restricted mean value property on balls. If for the primitive function F of f the limits $F(-1+)$, $F(1-)$ exist, then f is affine (i.e., harmonic).*

The first step away from a "good boundary behaviour" via abstract boundaries was made with the help of ergodic theory methods by M. A. Akcoglu and R. W. Sharpe (1968) in [AS, p. 458]. They used results proved by W. Feller in [Fel] and obtained as an illustrative example the following assertion:

5.6 THEOREM. *Suppose that* $D = B(0,1) \subset \mathbb{R}^2$ *and* $\delta(x) := \text{dist}(x, \partial D)$, $x \in D$. *Then a bounded measurable function* f *satisfies the condition* $A_\delta f(x) = f(x)$ *for all* $x \in D$ *if and only if* f *is harmonic on* D.

Note that in this case D is a very special bounded domain having smooth boundary and that the function f is supposed to be bounded. (W. A. Veech in [Vee2] attributes the assertion of Theorem 5.6 directly to [Fel].)

The further development consisted of a search for an optimal mixture of less restrictive assumptions. J. R. Baxter in [Bax1] (1972) improved the result from Theorem 5.6 using similar methods:

5.7 THEOREM. *Let* f *be a bounded measurable function on* D *and* δ *an admissible function on* D. *Suppose that* $\overline{D}$ *is a compact* C^1 *manifold with boundary and that there is a positive constant* α *such that*

$$\alpha \, \text{dist}(x, \partial D) \leq \delta(x) , \qquad x \in D .$$

Then $A_\delta f = f$ *on* D *implies that* f *is harmonic on* D.

In the article, the harmonicity of f was derived via the mean value property (i.e., for all admissible balls).

Similar results were obtained by S. Alinhac (1972); see [Ali1], [Ali2]. He assumed smoothness of the boundary ∂D, δ to be a Lipschitz function and worked in certain Sobolev spaces; his mean values were more general than (B).

Other new results appeared soon after: the articles of Veech and of D. Heath were prepared influencing each other. Results and methods differ but both apply probabilistic tools. The following result was proved in [Hea]:

5.8 THEOREM. *Let* D *be a proper open subset of* $\mathbb{R}^m$, $\alpha > 0$ *and* δ *an admissible function on* D *such that*

$$\alpha \, \text{dist}(x, \partial D) < \delta(x) < (1 - \alpha) \, \text{dist}(x, \partial D) , \qquad x \in D .$$

If f *is a function satisfying* $A_\delta f = f$ *on* D, *then* f *is harmonic on* D.

Thus the assumption of smoothness of the boundary was completely removed but certain control of δ from above was added. As remarked in [Vee2], Theorem 5.8 was improved by Heath and S. Orey; cf. also [Bru].

To formulate the next result, let us introduce a useful notion: if there is a harmonic function v on D such that $|f| \leq v$, then the function f is said to be *harmonically dominated* (on D). The article [Vee2] contains the following theorem:

5.9 THEOREM. *Suppose that* $D \subset \mathbb{R}^m$ *is a bounded Lipschitz domain and that* δ *is an admissible function locally bounded away from zero, i.e., for any compact* $K \subset D$ *we have* $\inf \delta(K) > 0$. *Let* f *be a function on* D *which is harmonically dominated. If* $A_\delta f = f$ *on* D, *then* f *is harmonic on* D.

In the next result taken from [Vee3], d is a function satisfying

$$0 < d(x) \leq \text{dist}(x, \partial D) \quad \text{and} \quad |d(x) - d(y)| \leq |x - y| , \qquad x, y \in D . \tag{5.1}$$

5.10 THEOREM. *Let $D \subset \mathbb{R}^m$ be a bounded Lipschitz domain, $\alpha > 0$ such that*

$$\alpha d(x) < \delta(x) < (1-\alpha)d(x) , \qquad x \in D .$$

If f is a positive measurable function on D satisfying $A_\delta f = f$, then f is harmonic on D.

The result of R. A. Hunt and R. L. Wheeden (1970) from [HuW] on the identity of the Euclidean and Martin boundaries for Lipschitz domains is used. Veech also conjectured that, for a bounded domain, any positive solution of $A_\delta f = f$ is harmonic provided that δ is locally bounded away from zero. This conjecture seems to be open.

In [Bax2] (1978), a theorem similar to Theorem 5.9 of Veech *removing any restriction on boundary ∂D* is proved. Also more general mean values are considered.

Proofs of the above results heavily depend on probability theory; see also [Sus]. Only recently two purely analytic proofs of analogous results appeared. In a series of articles, W. Hansen and N. Nadirashvili attacked successfully Littlewood's problems. The next theorem was proved in [HNa5]; see also [HNa1], [HNa2]:

5.11 THEOREM. *Let D be a domain in $\mathbb{R}^m$ (with non-polar complement when $m \leq 2$). Suppose that δ is admissible and f is a measurable and harmonically dominated function satisfying $A_\delta f = f$ on D. If f is continuous or δ is locally bounded away from zero, then f is harmonic on D.*

This very nice result solves one of the problems of Littlewood mentioned above. The authors generate PWB-like solutions and use the Schrödinger equation and Martin boundary. As a by-product, they obtain the assertion of Theorem 5.9 *without any restriction on ∂D*. We remark that in [HNa1], [HNa2] and [HNa4] more general equations than the Laplace equation are investigated.

With the use of completely different methods and also working in a more general context, A. Cornea and the second author obtained in [CV] a theorem which in classical potential theory leads to the following result from [Bax2]:

5.12 THEOREM. *Suppose that $D \subset \mathbb{R}^m$ is a Green domain. Let d satisfy (5.1), δ be a measurable function on D and $\alpha > 0$ be such that*

$$\alpha d(x) < \delta(x) < (1-\alpha)d(x) , \qquad x \in D .$$

Then any positive measurable f satisfying $A_\delta f = f$ on D is harmonic.

We remark that in the case of $m > 2$ every domain is a Green domain and recall that in $\mathbb{R}^2$ the Green domains are precisely those with non-polar complement.

The main ingredients used in [CV] are a deep study of a Martin-like kernel generated from a certain function (already used by Veech), some Harnack type inequalities and a Martin representation.

For $D = \mathbb{R}^m$, the equation $A_\delta f = f$ does not imply the harmonicity of a continuous function f, even if δ is constant; cf. Example 5.2. Two-radius theorems are useful in this situation. As a consequence of it we can get one-radius theorems with a growth condition. This is illustrated by the next result of Volchkov; see [Vol2] and cf. Theorem 4.9 for $g = 0$:

5.13 THEOREM. *Suppose that f is continuous on $\mathbb{R}^m$ and satisfies $A_\delta f = f$ for a constant function δ. If κ_m and h_m are defined by (4.7) and (4.8) and $f(x) = o(h_m(x,\delta))$ for $|x| \to \infty$, then f is harmonic. There is a non-harmonic continuous f satisfying $A_\delta f = f$ such that $f(x) = O(h_m(x,\delta))$ for $|x| \to \infty$.*

Another one-radius theorem involving a growth condition has very recently been proved in [HNa4].

5.14 THEOREM. *Let $m > 2$ and $D \subset \mathbb{R}^m$ be a domain. Suppose that δ is an admissible function on D such that $\delta(x) \le |x| + M_0$ for some $M_0 > 0$ and any $x \in D$. Let f be a Lebesgue measurable function on D which is harmonically dominated and let $A_\delta f = f$ on D. If f is continuous or δ is locally bounded away from zero, then f is harmonic.*

Observe that δ automatically satisfies the growth condition provided that $D \neq \mathbb{R}^m$. On the other hand, for every $m \ge 1$, a non-constant rotation invariant bounded $\mathcal{C}^\infty$ function f may be constructed such that $A_\delta f = f$ for a function δ satisfying $|x| + 1 \le \delta(x)$ for every $x \in \mathbb{R}^m$ and $\delta(x) = o(|x| \log |x|^{3/m} \log\log |x|)$ for $|x| \to \infty$.

For a result of this kind, see [HNa4]; for a similar and older result for spherical means, see below.

6. Restricted mean value property: spherical means

We shall briefly repeat some facts from the previous sections: Examples 4.1, 5.1 (together with the note in Section 5 on Littlewood's problems) and 5.2 show that there are non-harmonic functions with the restricted spherical mean value property.

For further exposition we need some notation; we shall follow the schema of the previous section. Recall that a function δ defined on D was said to be admissible if $0 < \delta(x) \le \operatorname{dist}(x, \partial D)$, $x \in D$.

Let δ be an admissible function on D, f a function defined on D and $x \in D$. Put

$$L_\delta f(x) := L(f, x, \delta(x)) ,$$

provided f is integrable on $S(x, \delta(x))$.

Functions f satisfying $f = L_\delta f$ on D are sometimes also said to have *the restricted mean value property on spheres*.

We have already described Volterra's result and the following development of assertions based directly on a minimum principle. It is worth mentioning that quite an elementary approach to theorems of this kind together with a brief description of their history is contained in an article written by R. B. Burckel; see [Bur].

Let us start with a non-trivial result obtained by Heath in the above mentioned article [Hea] (1973):

6.1 THEOREM. *Suppose that D is a proper open subset of $\mathbb{R}^m$, $m \ge 2$ and δ is a Lipschitz function satisfying for some $0 < \alpha < 1$*

$$|\delta(x) - \delta(y)| < \alpha |x - y| , \qquad x, y \in D .$$

If f is a bounded continuous function on D and $L_\delta f = f$, then f is harmonic on D.

This result (like its volume analogue from [Hea]) was obtained by a heavy use of probabilistic tools. Other contributions to the restricted mean value property on spheres are due to P. C. Fenton. In [Fen1] (1976) he proved:

6.2 THEOREM. *Suppose that $D = \mathbb{R}^2$, δ is continuous and that there exists a point x_0 such that the set $\{x \in \mathbb{R}^2 : |x - x_0| < \delta(x)\}$ is bounded. If f is a continuous function satisfying $L_\delta f = f$ on D and majorized by a harmonic function h on D, then f is harmonic on D. In fact, $f = h + K$ for a constant K.*

In [Fen3] Fenton proved a theorem covering the case of the unit disc. The result reads as follows:

6.3 THEOREM. *Let us suppose that $D = B(0,1) \subset \mathbb{R}^2$ and δ is a function which satisfies $0 < \delta(x) < 1 - |x|$, $x \in D$. Let f be a continuous function having a harmonic minorant g_1 and a harmonic majorant g_2 on D such that the following conditions hold:*

$$\lim_{x\to\theta} f(x) \quad \text{exists for } \sigma\text{-almost every } \theta \in \partial D \text{ and}$$

$$\limsup_{r\to 1-} \big(g_2(r\theta) - g_1(r\theta)\big) < \infty \quad \text{for every } \theta \in \partial D\ .$$

If f satisfies $L_\delta f = f$ on D, then f is harmonic on D.

There are other theorems of this kind proved by Fenton in case where $D = B(0,1) \subset \mathbb{R}^2$; they use assumptions like the existence of radial limits of f almost everywhere on ∂D; cf. [Fen2].

The next result was obtained by X. Gong in [Gon] (1983). (We thank to Xun Zhang for his assistance in deciphering the Chinese text.) In order to formulate the assertion, we introduce the following notation. Let $D \subset \mathbb{R}^m$, f be a continuous function on D and δ be a function on D satisfying

$$0 < \delta(x) < \operatorname{dist}(x, \partial D)\ , \qquad x \in D\ . \tag{6.1}$$

Set $B_y := B(y, \delta(y))$ and denote by P_y the solution of the Dirichlet problem for B_y and the boundary condition f on ∂B_y. Finally for $x \in D$ put

$$M(x) := \sup\{P_y(x) : x \in B_y\}\ , \qquad N(x) := \inf\{P_y(x) : x \in B_y\}\ .$$

6.4 THEOREM. *Let D be a bounded Lipschitz domain in $\mathbb{R}^m$, $m > 2$, let f be continuous on D and let δ satisfy (6.1). If f has a finite limit σ-almost everywhere on ∂D and the greatest harmonic minorant of $M - N$ is a BLD function, then f is harmonic on D.*

Further results involving boundary behaviour of f were proved in [Fol] and [OS]. The next assertion is taken from [OS].

6.5 THEOREM. *Let $D \subset \mathbb{R}^m$ be a domain such that the harmonic measure of ∂D is strictly positive. Let δ be an admissible function on D which is locally bounded away from zero. If f is a function on $\overline{D}$ which is continuous at almost every point of ∂D (with respect to harmonic measure) and $L_\delta f = f$ on D, then f is harmonic on D.*

To solve the second of Littlewood's problems, Hansen and Nadirashvili proved an "annulus theorem"; see [HNa3]. For $0 < s < t$ put

$$R(x, s, t) := \{y \in \mathbb{R}^2 : s < |x - y| < t\}\ .$$

6.6 THEOREM. *Let $D = B(0,1) \subset \mathbb{R}^2$, $0 < \alpha < 1$. Then there exist continuous functions s, t and f on D such that $0 < s(x) < t(x) \le \alpha \operatorname{dist}(x, \partial D)$ and*

$$f(x) = \frac{1}{\lambda(R(x, s(x), t(x)))} \int_{R(x,s(x),t(x))} f \, d\lambda \,, \qquad x \in D$$

but f is not harmonic *on D.*

Clearly, easy reasoning based on the mean value theorem (cf. Part 5.3 above) leads to the following

6.7 COROLLARY. *Let $0 < \alpha < 1$. Then there exists a continuous non-harmonic function f, $0 \le f \le 1$, such that for some δ satisfying $0 < \delta(x) < \alpha \operatorname{dist}(x, \partial D)$, $x \in D$ the equality $L_\delta f(x) = f(x)$ holds for every $x \in D$.*

Corollary 6.7 gives an affirmative answer to a long-standing conjecture of Littlewood formulated in the second part of 5.3. It is of interest that the solution was found by a probabilistic technique and that the method does not directly generalize to $\mathbb{R}^m$, $m > 2$.

We shall close this section with a theorem of Flatto [Fla3]; it is probably the first theorem on the restricted mean value property on spheres obtained via a two-radius theorem. We need again some notation. Similarly as in (4.7) we set

$$\eta_m := \inf\{|\operatorname{Im} z| : j_p(z) = 1, z \ne 0, p = m/2 - 1\} \,. \tag{6.2}$$

6.8 THEOREM. *Suppose that f is continuous on $\mathbb{R}^m$ and satisfies $L_\delta f = f$ for a constant function δ. If η_m is defined by (6.2) and $f(x) = O(\exp k|x|)$ for $|x| \to \infty$ with $0 \le k < \eta_m/\delta$, then f is harmonic.*

Note that Flatto also shows that the result is the best possible.

7. Inverse mean value property: volume means

The following statement is an immediate consequence of Theorem 2.2.

7.1 THEOREM. *Let $r > 0$ and $A = B(0,r)$. Then*

$$h(0) = \frac{1}{\lambda(A)} \int_A h \, d\lambda \tag{7.1}$$

whenever h is an integrable harmonic function on A.

The following problem is quite natural:

7.2 PROBLEM. *Characterize open sets A in $\mathbb{R}^m$ with $0 < \lambda(A) < \infty$ and $0 \in A$ such that (7.1) holds for every integrable harmonic function on A.*

To the best of our knowledge, the first result in this direction can be traced to [Bro]. Brödel's result reads as follows:

Es sei Γ ein nicht kreisförmiges, konvexes Gebiet; $\rho = k(\varphi)$ sei die Polargleichung der Randkurve bezogen auf einen inneren Punkt Π als Pol. Es sei n_0 die größte natürliche Zahl derart, daß $I_n^2 + J_n^2 = 0$ für $1 \le n \le n_0$ gilt. Dann ist die Gesamtheit der zu (Γ, Π) gehörigen Funktionen mit Flächenmittelwerteigenschaft identisch mit der Gesamtheit derjenigen Polynome höchstens n_0-ten Grades in x und y, die der Laplaceschen Differentialgleichung genügen.

Here I_n and J_n are defined by

$$I_n := \int_0^{2\pi} k^{n+2}(\varphi) \cos n\varphi \, d\varphi \ , \qquad J_n := \int_0^{2\pi} k^{n+2}(\varphi) \sin n\varphi \, d\varphi \ ,$$

and a continuous function h is in the class of "zu (Γ, Π) gehörigen Funktionen mit Flächenmittelwerteigenschaft", if

$$h(\varphi(\Pi)) = \frac{1}{\lambda(\varphi(\Gamma))} \int_{\varphi(\Gamma)} h \, d\lambda$$

holds for every similarity φ (i.e., every composition of a dilation with centre Π and a translation). So the following theorem is an easy consequence of Brödel's result stated in [Bro, p. 15].

7.3 THEOREM. *Let $A \subset \mathbb{R}^2$ be a bounded open convex set, $0 \in A$ and let A not coincide with a disc. Then there exists a harmonic polynomial h such that*

$$h(0) \neq \frac{1}{\lambda(A)} \int_A h \, d\lambda \ .$$

A higher dimensional analogue can be found as a special case of results obtained in [FL1, pp. 156, 158]. Before stating the theorem, we shall introduce the following notation: For a measurable set $A \subset \mathbb{R}^m$ with $0 < \lambda(A) < \infty$, $B(A)$ stands for the open ball centered at the origin for which $\lambda(B(A)) = \lambda(A)$.

7.4 THEOREM. *Let $A \subset \mathbb{R}^m$, $m \ge 2$ be a bounded domain with C^2 boundary, let $\mathbb{R}^m \setminus \overline{A}$ be connected and $0 \in A$. If (7.1) holds for all harmonic polynomials, then $A = B(A)$.*

(For more general results for globally defined harmonic functions, see [GHR1].)

Methods of complex function theory (reproducing kernels and conformal mapping) were used by B. Epstein in the proof of the following result; see [Eps].

7.5 THEOREM. *Let $A \subset \mathbb{R}^2$ be a simply connected domain with $\lambda(A) < \infty$ and $0 \in A$. If (7.1) holds for every integrable harmonic function on A, then $A = B(A)$.*

The same conclusion holds for more general averages, see [Bos]. The following assertion valid also for higher dimensions was proved by B. Epstein and M. M. Schiffer in [ES].

7.6 THEOREM. *Let $A \subset \mathbb{R}^m$, $m \ge 2$ be a domain with $\lambda(A) < \infty$ and $0 \in A$. If $\mathbb{R}^m \setminus A$ has non-empty interior and (7.1) holds for every integrable harmonic function, then $A = B(A)$.*

(For a proof based on the notion of a modified Schwarz potential, see [Gus, p. 203] and [Shp2, p. 30].)

M. Goldstein and W. H. Ow showed in [GO] that the hypothesis on $\mathbb{R}^m \setminus A$ can be replaced in the case of $m = 2$ by the assumption that at least one component of the boundary of A is a continuum. They also proved that simple connectedness of plane regions is a necessary condition for the mean value property of type (7.1).

A simple and elegant proof of the following general result is due to Ü. Kuran; see [Kur1], cf. also [Kur2].

7.7 THEOREM. *Let $A \subset \mathbb{R}^m$, $m \geq 2$ be an open set with $\lambda(A) < \infty$ and $0 \in A$. If* (7.1) *holds for every integrable harmonic function h on A, then $A = B(A)$.*

Actually, connectedness of A is supposed in [Kur1]. The proof given below which is a slight modification of the original proof shows that the connectedness assumption is superfluous. Our proof uses a simplification suggested by V. Souček and is copied from [Net2]. An extension to open sets is also mentioned in [GHJ].

In order to prove Theorem 7.7, let $y \in \mathbb{R}^m \setminus A$ be the nearest point to the origin and $B := B(0, |y|)$. Observe that $\int_{A\setminus\overline{B}} h\, d\lambda = 0$ whenever h is an integrable harmonic function on A with $h(0) = 0$. Indeed,

$$0 = h(0) \cdot \lambda(A) = \int_A h\, d\lambda = \int_{A\setminus\overline{B}} h\, d\lambda + \int_B h\, d\lambda =$$

$$\int_{A\setminus\overline{B}} h\, d\lambda + h(0) \cdot \lambda(B) = \int_{A\setminus\overline{B}} h\, d\lambda\ .$$

Putting $K(x) := (|x|^2 - |y|^2)/|x-y|^m$, $x \in \mathbb{R}^m \setminus \{y\}$ and $h := K - K(0)$ on A, we arrive at an integrable harmonic function (cf. (1.1)) on A with $h(0) = 0$ and strictly positive on $A \setminus \overline{B}$. Consequently, $\lambda(A \setminus \overline{B}) = 0$, which implies $A \subset \overline{B}$ and $A = B$.

There are nice applications of the above stated inverse mean value property. Theorem 7.7 can be used to answer affirmatively the following question: Let P be a solid, homogeneous body in $\mathbb{R}^3$ which induces on its exterior a gravitational field identical to that of a point mass. Must P be a ball? For proofs, generalizations and a further discussion, see [ASZ], [Zal5].

The inverse mean value property turned out to be important in investigations of the best harmonic L^1 approximation to subharmonic functions; see [GHJ], [GHR1], [AG4], [Hau].

It was observed in [GHR1] that in order to have $A = B(A)$ it suffices that (7.1) holds for harmonic functions belonging to $L^p(A)$ with a fixed p satisfying $1 \leq p < m/(m-1)$. Using a different approach, the following result (in which $m/(m-2)$ is replaced by ∞ for $m = 2$) was proved in [AG2].

7.8 THEOREM. *Let $A \subset \mathbb{R}^m$, $m \geq 2$ be an open set with $\lambda(A) < \infty$ and $0 \in A$. If $0 < p < m/(m-2)$ and* (7.1) *holds for every positive harmonic $h \in L^p(A)$, then $A = B(A)$.*

The conclusion $A = B(A)$, however, fails, if $p = m/(m-2)$. The following two theorems are taken from [AG2]; see also [AG4].

7.9 THEOREM. *Let A be as in Theorem 7.8. If* (7.1) *holds for every bounded harmonic function on A, then $B(A) \setminus A$ is polar.*

7.10 THEOREM. *Let A be as in Theorem 7.8. Let $m > 2$, $0 < p < m-2$ and let $M_p(A)$ be the class of all integrable positive harmonic functions h satisfying*

$$\sup\{(1/r^{m-p}) \int_{B(y,r)\cap A} h\, d\lambda : y \in \partial A, r > 0\} < \infty\ .$$

If (7.1) *holds for every $h \in M_p(A)$, then $B(A) \setminus A$ is a null set for the Hausdorff measure of dimension $m - 2 - p$.*

Results of [Kur1], [AG2], [AG4] are included in a more general approach introduced in [HNe2]. All remaining results in this section are taken from this paper.

First we introduce the following notation. With every positive measure μ on $\mathbb{R}^m$, $m \geq 2$, we associate its potential

$$G^{\mu} : x \mapsto \int G(x-y)\, d\mu(y) \ , \quad x \in \mathbb{R}^m \ ,$$

corresponding to the kernel G defined on $\mathbb{R}^m$ by

$$G(z) := \begin{cases} -\log|z| & \text{when } m = 2 \ , \\ |z|^{2-m} & \text{when } m > 2 \ . \end{cases}$$

Recall that, for a compact set $E \subset \mathbb{R}^m$, the potential $G^{1_E\lambda}$ (of the restriction of λ to E) is a C^1-function on $\mathbb{R}^m$, bounded on $\mathbb{R}^m$ if $m > 2$, behaving like a multiple of G near ∞ if $m = 2$ and harmonic on $\complement E$, the complement of E.

7.11 THEOREM. *Let A be a Lebesgue measurable set in $\mathbb{R}^m$, $m \geq 2$, such that $0 < \lambda(A) < \infty$. Then* (7.1) *holds for every bounded $h = G^{1_E\lambda} - G^{1_F\lambda}$, E, F compact sets contained in $\complement A$ if and only if $\lambda(B(A) \setminus A) = 0$.*

In order to discuss the role of exceptional sets, we introduce one more notation. Let $\mathcal{M}$ be a family of finite positive measures with compact support in $\mathbb{R}^m$ such that $1_E\lambda \in \mathcal{M}$ and $1_E\mu \in \mathcal{M}$ for every compact set $E \subset \mathbb{R}^m$ and every $\mu \in \mathcal{M}$. For $P \subset \mathbb{R}^m$, let $\mathcal{M}_P$ stand for all $\mu \in \mathcal{M}$ with support in $\complement P$.

A set $M \subset \mathbb{R}^m$ is said to be an $\mathcal{M}$-*null set* provided that $\mu(E) = 0$ whenever $E \subset M$ is compact and $\mu \in \mathcal{M}$.

7.12 THEOREM. *Let A be as in Theorem 7.11. The equality* (7.1) *holds for every $h = G^{\mu} - G^{\nu}$ bounded near ∞, $\mu, \nu \in \mathcal{M}_A$, if and only if $B(A) \setminus A$ is an $\mathcal{M}$-null set.*

We conclude this section by stating several results illustrating the generality of Theorem 7.11 and Theorem 7.12.

7.13 COROLLARY. *The equality* (7.1) *holds for every bounded $h = G^{\mu} - G^{\nu}$, μ, ν finite positive measures on $\mathbb{R}^m$ supported by a compact subset of $\complement A$ with G^{μ}, G^{ν} continuous if and only if $B(A) \setminus A$ is inner polar (i.e., every compact subset of $B(A) \setminus A$ is polar).*

7.14 COROLLARY. *Let $0 < \alpha < 1$. Then* (7.1) *holds whenever $h \in \operatorname{Lip}\alpha$ is bounded and harmonic on a neighbourhood of A if and only if $B(A) \setminus A$ has zero inner measure with respect to the Hausdorff measure of dimension $m - 2 + \alpha$.*

(Here $\operatorname{Lip}\alpha$ denotes the set of all $f : \mathbb{R}^m \longrightarrow \mathbb{R}$ satisfying locally a Hölder condition of order α.) Finally let us give an application to fine potential theory; cf. [Fugl].

7.15 COROLLARY. *Let A be a finely open subset of $\mathbb{R}^m$ such that $0 \in A$ and $\lambda(A) < \infty$. Then* (7.1) *holds for every bounded finely harmonic function h on A if and only if $B(A) \setminus A$ is polar.*

8. Inverse mean value property: spherical means

An analogue of Theorem 7.1 for spherical means follows easily from Theorem 2.1:

8.1 THEOREM. *Let $r > 0$ and $D = B(0,r)$. Then*

$$h(0) = \frac{1}{\sigma(\partial D)} \int_{\partial D} h \, d\sigma \tag{8.1}$$

whenever h is continuous on $\overline{D}$ and harmonic on D.

Recall that σ stands for the $(m-1)$-dimensional Hausdorff measure in $\mathbb{R}^m$. We shall ask the following question, which is similar to Problem 7.2.

8.2 PROBLEM. *Suppose that $D \subset \mathbb{R}^m$ is a bounded domain with $0 \in D$ and $\sigma(\partial D) < \infty$. Is it true that D is a ball provided that (8.1) holds for every function h continuous on $\overline{D}$ and harmonic on D?*

We recall that D satisfying the above mentioned requirements and having the boundary homeomorphic to the unit sphere in $\mathbb{R}^m$ is called a *pseudosphere* in [LV1]. A good question (posed e.g., in [Shp1]) is whether there exists a pseudosphere in $\mathbb{R}^m$ which is not a sphere.

A construction of plane pseudospheres different from circles was first achieved in [KL] by M. V. Keldysch and M. A. Lavrentieff in order to show the existence of so-called non-Smirnov domains; cf. [Pri5], [Dur]. (For another construction, see [DSS], [Dur] or [Pir].)

In fact, neither a mean value property nor a relation between length and harmonic measure was behind investigations from [KL]. Let us quote from [KL, p. 1]:

Soit D un domaine simplement connexe, univalent, limité par une courbe simple et rectifiable, situé dans le plan de la variable complexe. Désignons par $w = f(z)$ la fonction qui réalise la représentation conforme du domaine D sur le cercle $|w| < 1$, et soit $z = \varphi(w)$ la fonction inverse. Quelques problèmes aux limites de la théorie des fonctions se réduisent à la question suivante: la fonction harmonique $\log|\varphi'(w)|$ est-elle représentable dans le cercle $|w| < 1$ par l'intégrale de Poisson de ses valeurs limites sur la circonférence $|w| = 1$?

Par example, M. V. Smirnoff a démontré que la condition nécessaire et suffisante pour que le système des polynomes orthogonaux sur le contour Γ du domaine D soit complet, est que la fonction $\log|\varphi'(w)|$ *soit représentable par l'intégrale de Poisson. M. Smirnoff a démontré que cette représentation est, en même temps, une condition nécessaire et suffisante pour que chaque fonction définie dans D et représentable par l'intégrale de Cauchy satisfasse au principe du maximum.*

The result from [KL, p. 15] on the existence of a "wild" plane pseudosphere reads as follows:

THÉORÈME. *Quel que soit le nombre positif ρ, il existe un domaine D jouissant des propriétés suivantes:*

1° *Le domaine D est simplement connexe, univalent et limité par une courbe simple et rectifiable Γ, renferment l'origine $z = 0$.*
2° *Le domaine D ne coïncide pas avec le cercle $|z| < 1$ et est contenu dans le cercle $|z| < \rho$.*
3° *Dans la représentation conforme $z = \varphi(w), \varphi(0) = 0$ du domaine D sur le cercle $|w| < 1$, il correspond à chaque arc γ de la courbe Γ un arc de même longueur sur la circonférence $|w| = 1$.*

Quite recently, the following result was established in [LV1].

8.3 THEOREM. *For $m \geq 2$, there exists a pseudosphere in $\mathbb{R}^m$ which is not a sphere.*

The construction shows that D can be chosen so that there exists a homeomorphism f on $\mathbb{R}^m$ with $f(\partial B(0,1)) = \partial D$ and

$$(c(\beta))^{-1}|x-y|^{1/\beta} \leq |f(x)-f(y)| \leq c(\beta)|x-y|^{\beta} ,$$

whenever $\beta \in \,]0,1[$ and $|x-y| \leq 1/2$.

There are also positive results: By [Gre], a convex pseudosphere in $\mathbb{R}^2$ is a disc; cf. [Bro]. Further, a pseudosphere in $\mathbb{R}^2$ which is a Smirnov domain is necessarily a disc; see [Spr]. For $m \geq 2$, G. Fichera showed in [Fic] that every pseudosphere with $\mathcal{C}^1$ boundary is a sphere. Under a more restrictive smoothness assumptions, this also follows from [FL1]. From the review of [KaS] in **Zbl 692**: 60058 we learned that a ball is the only bounded domain such that ∂D is a finite union of regular surfaces and (8.1) holds for every h continuous on $\overline{D}$ and harmonic on D. For another proof (with the assumption that D has a $\mathcal{C}^2$ boundary) and a relation to quadrature surfaces, see [Shg2]; cf. also [Shg1].

The most general result seems to be established in [LV2, pp. 349, 363].

8.4 THEOREM. *In $\mathbb{R}^m$, $m \geq 2$, every pseudosphere with a Lipschitz boundary is a sphere.*

Let us conclude Sections 7 and 8 with the following result from [Kos] including both volume and spherical means:

8.5 THEOREM. *Let $D \subset \mathbb{R}^2$ be a convex domain with $\mathcal{C}^2$ boundary. If*

$$\frac{1}{\lambda(D)}\int_D h\,d\lambda = \frac{1}{\sigma(\partial D)}\int_{\partial D} h\,d\sigma$$

holds for every h continuous on $\overline{D}$ and harmonic on D, then D is a disc.

9. Asymptotic mean value property

The asymptotic behaviour of the mean values can be used as a test of harmonicity. The following result was proved by W. Blaschke:

Eine im Kreise J

$$(x-x_0)^2 + (y-y_0)^2 \leq r^2$$

stetige Funktion $f(x,y)$ habe die Eigenschaft, daß ihr Differentialparameter Δf an jeder inneren Stelle x,y von J existiert und gleich Null ist, d. h. ausführlich geschrieben, es sei

$$\lim_{\rho\to 0}\frac{4}{\rho^2}\left\{\frac{1}{2\pi}\int_{-\pi}^{+\pi} f(x+\rho\cos\vartheta, y+\rho\sin\vartheta)d\vartheta - f(x,y)\right\} = 0$$

für alle x,y die der Bedingung genügen

$$(x-x_0)^2 + (y-y_0)^2 < r^2 .$$

Dann ist $f(x,y)$ ein logarithmisches Potential im gewöhnlichen Sinne des Wortes, hat also auch notwendig stetige Ableitungen beliebiger Ordnung.

For the proof, see [Bla1, p. 6] or [Bla2]. The proof of its slightly modernized following version (for $m \geq 2$) can be found e.g., in [Bre, p. 21].

9.1 THEOREM. *Let h be a function on an open set $D \subset \mathbb{R}^m$. Then h is harmonic on D if and only if h is continuous and*

$$\lim_{r\to 0+} 2mr^{-2}(L(h,x,r) - h(x)) = 0 \ , \quad x \in D \ . \tag{9.1}$$

It is interesting to note that a characterization of this type was proved as early as 1914 by M. Plancherel in [Pla, p. 245] for harmonic functions in spherical domains. Theorem 9.1 follows, however, from the result stated in a modernized version below and essentially going back to the paper [Zar] of S. Zaremba from 1905. The statement of the theorem is taken from [Hop] where a proof can be found. E. Hopf says in [Hop] "*Einer Anregung von Herrn Saks folgend, möchte ich auf einen der Sätze der schönen, mit Unrecht in Vergessenheit geratenen Arbeit von Zaremba hier kurz eingehen.*"

9.2 THEOREM. *Let $D \subset \mathbb{R}^m$ be an open set and let S be a vector space of continuous functions containing $C^2(D)$. Let Φ be a linear operator associating with every $f \in S$ a real-valued function Φf on D in such a way that the following two conditions hold:*

(i) *The operator Φ is an extension of the Laplace operator in the sense that $\Phi f = \Delta f$ on D, whenever $f \in C^2(D)$;*

(ii) *If $f \in S$ and f attains its maximum on D at a point $x_0 \in D$, then $(\Phi f)(x_0) \le 0$.*

If $h \in S$ and $\Phi h = 0$ on D, then h is harmonic on D.

Observe that Theorem 9.1 follows immediately from Theorem 9.2. Indeed, let S be the set of all continuous functions h on D for which the limit in (9.1) exists for every $x \in D$. Denote this limit by $(\Phi h)(x)$. Then the conditions (i) and (ii) of Theorem 9.2 are fulfilled, see [Bre, p. 19].

Similarly, instead of (9.1) one can consider the operator Φ_1 given by

$$\Phi_1 h(x) := \lim_{r\to 0+} 2(m+2)r^{-2}(A(h,x,r) - h(x)) \ , \quad x \in D \ , \tag{9.2}$$

on continuous functions h on D for which the limit in (9.2) exists.

A direct proof of the characterization of harmonic functions based on the operator Φ_1 can be found in [Bre, p. 21, 30], [OFa] and [Pri1]. It is of interest to note that Theorem 9.2 gives easily the following assertion: *If f is continuous on D, $\frac{\partial^2 f}{\partial x_j{}^2}$ exists at each point of D for $j = 1, \ldots, m$ and $\Delta f = 0$ on D, then f is* actually *harmonic*; see [Hop].

Note that operators from (9.1) and (9.2) and their generalizations obtained by replacing "lim" by "lim sup" and "lim inf" play an important role in characterizations of subharmonic and superharmonic functions; see [Bre, p. 20], [Rad], [Cim], [Sas2], [Rea], [Pri2], [Pri3], [Pri4], [Sas1], [Sas3], [Szp], [Koz], [Ars1], [Ars2], [Rud1], [Pot], [Smy2]; for a generalization to other PDE's, see [HJ2].

It should be noted that, in Theorem 9.1, equality (9.1) actually need not be hypothesized to hold everywhere; cf. [Rud1], [Ars2], [Pri3], [Pri4]. Here we only state the following sample result from [Ars2].

9.3 THEOREM. *Let h be a function on an open set $D \subset \mathbb{R}^2$. Then h is harmonic if and only if h is continuous, there exists an F_σ-set Z of capacity zero such that*

$$\limsup_{r\to 0+} 4\,r^{-2}|L(h,x,r) - h(x)| < \infty \ , \quad x \in D \setminus Z \ ,$$

and

$$\lim_{r\to 0+} 4\,r^{-2}(L(h,x,r)-h(x))=0$$

for λ-almost every $x\in D$.

Until now we were interested in the asymptotic behaviour of averages for $r\to 0+$. Less attention seems to have been paid to the case when the asymptotic behaviour for large r is considered.

Motivated by problems of the stabilization of temperature, M. Plancherel and G. Pólya proved in 1931 the following result:

9.4 THEOREM. *Suppose that $f\in L^1_{loc}(\mathbb{R})$ and*

$$\varphi(x) := \lim_{r\to\infty} (2r)^{-1}\int_{x-r}^{x+r} f$$

exists for every $x\in\mathbb{R}$. Then φ is an affine function.

For the proof, see [PP], or [Boa, p. 129]. The idea is to show that $\varphi-\varphi(0)$ satisfies the Cauchy functional equation and, as a Baire 1 function, has to be affine.

In [PP], a two-dimensional analogue is also investigated (which apparently remained unknown to the authors of [BW1]), and the following statement is proved:

9.5 THEOREM. *Suppose that $f\in L^1_{loc}(\mathbb{R}^2)$ and*

$$\varphi(x) := \lim_{r\to\infty} A(f,x,r)$$

exists for every $x\in\mathbb{R}^2$. Then φ is harmonic provided that there exists a function $\psi\in L^1_{loc}(\mathbb{R}^2)$ and a positive locally bounded function r_0 on $\mathbb{R}^2$ such that $|A(f,x,r)|\le\psi(x)$ whenever $r\ge r_0(x)$.

Surprisingly enough, any control of the growth of f is redundant, as is shown in [BW1]:

9.6 THEOREM. *Let f be a continuous function on $\mathbb{R}^m$. If*

$$\lim_{r\to\infty} A(f,\,\cdot\,,r)=f \tag{9.3}$$

uniformly on compact subsets of $\mathbb{R}^m$, then f is harmonic.

On the other hand, there exists a non-harmonic continuous function on $\mathbb{R}^m$ and a sequence $\{r_j\}$ tending to ∞ such that $A(f,\cdot,r_j)\to f$ uniformly on compact sets.

The conclusion of Theorem 9.6 still holds if r in (9.3) tends to ∞ through a sufficiently "thick" set, e.g., through an infinite union of disjoint intervals in $]\,0,\infty\,[$ with lengths bounded away from zero; see [BW1].

It should be mentioned that in [BW1] more general radial measures than the Lebesgue measure are considered.

10. Invariant mean value property

In this section we summarize several results concerning harmonic functions on the unit disc $D=B(0,1)\subset\mathbb{R}^2$.

Recall that, for every $y \in D$, the function

$$\omega_y : x \mapsto \frac{x-y}{1-\bar{y}x}$$

maps D conformally onto itself. (Here we are using the usual notation for complex numbers.) Conversely, every conformal mapping of D onto itself has the form $x \mapsto \omega_y(\varepsilon x)$ for a suitable $y \in D$ and a complex unit ε. Such mappings are usually called *Möbius transformations.* Since conformal mappings preserve harmonicity, the following assertion is a consequence of Theorem 2.2 :

10.1 THEOREM. *If h is an integrable harmonic function on D, then*

$$(h \circ \omega)(0) = \frac{1}{\lambda(D)} \int_D h \circ \omega \, d\lambda \tag{10.1}$$

whenever ω is a Möbius transformation.

Is the property stated above characteristic for harmonic functions on D? For the class of functions uniformly continuous on D, the answer is affirmative; see [Rud2, p. 286] and [AFR]. The same answer was obtained in [AC] for functions with uniformly continuous radialization. The following result is proved in [Eng]:

10.2 THEOREM. *Let $h \in L^\infty(D)$ and let (10.1) hold whenever ω is a Möbius transformation. Then h is harmonic on D.*

The result was extended in [AFR] in two directions: Theorem 10.2 holds for $h \in L^1(D)$ and, further, a natural generalization of (10.1) to functions on the unit ball in $\mathbb{C}^n$ still characterizes harmonicity, but if and only if $n < 12$.

For a "spherical" invariant mean value property, we refer to [AC], [Rud2, p. 51], [GlR], [Glo1], [Glo2] and [Izu] where a "one-radius theorem" is discussed.

11. Mean value property and caloric functions

We shall begin with a translation of a part of the introduction from [Kup4]: "*Mean value theorems related to the Laplace operator are well known, see* [Krt], [Mik]. *They play an important role in the theory of harmonic functions: the strong maximum principle, theorems on compactness of families of harmonic functions* [Mik, p. 238], *Harnack's theorems, analyticity of harmonic functions* [Mik, p. 239] *etc. are their consequences.*

Similar mean value theorems for the heat conduction operator are less known. The paper [Pin] *of B. Pini seems to be the first one including such theorems for the one-dimensional case, though there is no doubt that many mathematicians already knew about these theorems.*"

In fact, in [Pin] we find the following theorem:

Una soluzione regolare di $\mathfrak{M}(u) = 0$ in un campo A, verifica la

$$u(P) = \mu_0(u, P, r) \tag{11.1}$$

in ogni punto P di A e per ogni r ammissibile (tale cioè che $\mathcal{D}(P,r) \subset A$). Reciprocamente: una funzione $u(P)$ continua in A è ivi soluzione regolare di $\mathfrak{M}(u) = 0$ se per ogni punto P di A e per ogni r ammissibile vale la (11.1).

Here $\mathfrak{M}$ is the one dimensional (adjoint) heat conduction operator and $\mu_0(u, P, r)$ a special mean expressed in polar coordinates over the level set of the fundamental solution.

Pini's paper apparently remained unknown to the author of [Ful2] who writes:

"*The Gauss mean value theorem and its converse, due to Koebe, characterize solutions of Laplace's equation* [Kel1]. *In view of the strong analogy between Laplace's equation and the heat equation it seems reasonable to expect an analogue of the Gauss-Koebe result to hold for the heat equation. The purpose of this paper is to present such a result. For simplicity we work in two dimensions, though it is clear from calculations that the result is independent of dimensions.*"

We shall give below the precise statement of Fulks result which is relatively simple and pleasant from the point of view of analytical presentation. For another form we refer to papers [Smy1] and [Kup4] written apparently without knowledge of [Ful2].

First we recall the notion of a caloric function. Let $G \subset \mathbb{R}^{m+1}$ be open and u be a function on G. The function u is said to be *caloric* if $\partial^2 u/\partial x_j^2$ and $\partial u/\partial x_{m+1}$ are continuous on G for $j = 1, ..., m$ and u satisfies on G the heat equation, i.e.,

$$\sum_{j=1}^{m} \frac{\partial^2 u}{\partial x_j^2} - \frac{\partial u}{\partial x_{m+1}} = 0 . \tag{11.2}$$

Now we shall introduce the Fulks measure (or the Pini-Fulks measure); see [Bau1], [Bau2]. We follow the notation from [Bau1].

Let us denote by $\widetilde{W}$ the fundamental solution of (11.2) with the pole at the origin, i.e., for $(y,t) \in \mathbb{R}^m \times \mathbb{R}^1$ put

$$\widetilde{W}(x,t) := \begin{cases} \left(\frac{1}{4\pi t}\right)^{m/2} e^{-|x|^2/4t} & \text{when } t > 0 , \\ 0 & \text{when } t \le 0 , \end{cases}$$

and for $z_0 = (x_0, t_0) \in \mathbb{R}^{m+1}$ define

$$W_{z_0}(x,t) := \widetilde{W}(x_0 - x, t_0 - t) .$$

For $z_{0,} \in \mathbb{R}^{m+1}$ and $c > 0$ put

$$\Omega(z_0, c) := \{(x,t) \in \mathbb{R}^{m+1} : W_{z_0}(x,t) > (1/4\,\pi c)^{m/2}\}$$

(*the heat ball*) and $B(z_0, c) := \partial\Omega(z_0, c)$ (*the heat sphere*).

We define on $(\mathbb{R}^m \times\ ]0, \infty[\,) \cup \{0\}$

$$Q(x,t) := \begin{cases} |x|^2\,[4|x|^2 t^2 + (|x|^2 - 2mt)^2]^{-1/2} & \text{when } t > 0 , \\ 1 & \text{when } (x,t) = 0 . \end{cases}$$

Furthermore, for a point $z_0 = (x_0, t_0)$ in $\mathbb{R}^{m+1}$, we define the density function q_{z_0} on $(\mathbb{R}^m \times\ ]-\infty, t_0\,[\,) \cup \{z_0\}$ by $q_{z_0}(x,t) := Q(x_0 - x, t_0 - t)$ and denote by $q_{z_0,c}$ its restriction to $B(z_0, c)$. Finally, let $\mu_{z_0,c}$ is the measure on $B(z_0, c)$ having the density $(1/4\pi c)^{m/2} q_{z_0,c}$ with respect to the m-dimensional Hausdorff measure on $B(z_0, c)$ (*the Fulks measure*).

11.1 THEOREM. *Let $G \subset \mathbb{R}^{m+1}$ be open and u be a function on G. Then u is caloric on G if and only if u is continuous on G and*

$$h(z_0) = \int h \, d\mu_{z_0,c}$$

whenever $z_0 \in G$ and $\overline{\Omega(z_0,c)} \subset G$.

Theorem 11.1 was proved in [Pin] for $m = 1$ and in [Ful2]. For assertions of a similar nature, see [Smy1], [Kup4].

Volume means can be obtained from spherical means by a suitable integration with respect to "radii", see e.g., [Pin], [Kup4], [Smy1].

Finally, a Blaschke-Privaloff type theorem can be found in [Pin] and [Ful1]. It should be mentioned that averages over rectangles or parallelepipeds are also used for the study of caloric (or supercaloric) functions; cf. e.g., [CC, p. 89], [Was1], [Was2], [Wid], [Doo2, p. 275] [Fri2], [Fri3, p. 130], [Geh]. The corresponding densities are, however, analytically more complicated.

For further results on mean values in parabolic potential theory, see [Was3], [Was4], [Was5], [Was6], [Was7], [Was8], [Kup5].

Let us remark that Fulks measures are used in [Was1] and [Was2] for the definition and study of the so-called supertemperatures. From the result of H. Bauer [Bau2] stated below it follows that supertemperatures are exactly supercaloric functions which in parabolic potential theory correspond to superharmonic functions.

11.2 THEOREM. *For $z_0 \in \mathbb{R}^{m+1}$ and $c > 0$, $\mu_{z_0,c}$ coincides with the balayage of the Dirac measure at z_0 on the complement of $\Omega(z_0,c)$.*

Note that, for classical harmonic functions, the analogous result for balls is obvious. For an alternative proof of Theorem 11.2, see [Net3]. As shown in [Bau1], Theorem 11.1 can be deduced from Theorem 11.2. There is also an interesting capacitary interpretation of the Fulks measure: with respect to the adjoint heat equation, the Fulks measure is the equilibrium measure for the heat ball; see [Brz].

12. A look at related topics

A. Mean value property with respect to general measures.

Until now, mainly averages with respect to the normalized Lebesgue measure on a ball or to the normalized surface measure on a sphere have been considered. There is an extensive literature on functions satisfying a mean value property with respect to more general measures; see e.g., [Bax2], [BW1], [BW2], [Bos], [CD], [CV], [DL], [Den], [Dok], [Fla1], [Fla2], [FJ1], [FJ2], [FL1], [FL2], [Fur1], [Fur2], [Gar], [Gre], [HNe1], [Hir], [Joh], [Log], [Por1], [Por2], [RW], [Smy2], [Vot], [Wei1], [Wei2], [Zal1], [Zal2]. For the mean value property for α-harmonic functions introduced in [Ito] and related to the Riesz potentials, see e.g., [Lan], [BH].

B. Mean value property for solutions of partial differential equations.

We have limited ourselves exclusively to classical harmonic or caloric functions, i.e., to solutions of the Laplace equation or the heat equation. There are many papers and books investigating analogous mean value theorems for more general partial differential equations (also references concerning the heat equation are included); see e.g., [ACL], [Asg], [Bax], [Bos], [BP], [Chu], [CD], [CGL], [CV], [CH], [DL], [FG], [Fla1], [Fla2], [FL2], [Fri1], [Ful1], [Ful2], [FL1], [GL1], [GL2], [GL3], [Gar], [Gav], [HNa1], [HNa2], [HW1], [HW2], [HJ1], [HJ2], [HJ3], [Hor], [JC], [Joh], [Kou1], [Kou2], [Kup1], [Kup2], [Kup3], [Kup4], [Kup5], [Les], [Mih1], [Mih2], [Min], [Nau1], [Nau2], [Nic1], [Nic2], [Nic3], [Nic4], [Pin], [Ros], [Sco], [Smy1], [Smy3], [Vol1], [Vol2], [Was1], [Was2], [Was3], [Was4], [Was5], [Zal2].

C. Mean value property and (harmonic) polynomials.
Here we cite only a few papers on the subject; see e.g., [ACL], [AHM], [BR1], [BR2], [BW1], [Bro], [CD], [Fla3], [Fri1], [Har1], [Har2], [Har3], [Har4], [Har5], [Har6], [RW], [Wal].

D. Mean value property of harmonic functions on more general spaces.
For harmonic functions on manifolds, see e.g., [BZ], [Fur2], [Gra], [GV], [GW], [Kow1], [Kow2], [KO], [KS], [Los], [Mal], [Sak1], [Tal], [Var], [Wil1], [Wil2], [Wil3], [Wil4]. For the mean value property of harmonic functions on a Hilbert space, see e.g., [Bel], [MW]. For harmonic functions on groups etc., see e.g., [Aze], [BW2], [DG], [Dez], [Dok], [Fur1], [Hir], [Lew], [Oba1], [Oba2].

E. Other related topics.
The following references are related to other aspects of the mean value property: [AG1], [AG3], [AN], [Avc], [Bar], [CS], [Eva3], [Fug2], [GR], [GHR2], [GHR3], [God], [HG], [Kel2], [Leb2], [Mak], [Maz], [Met], [Par], [PW1], [PW2], [Ran], [Sak2], [SO], [SW1], [SW2], [Tho], [Sto], [Wei1], [Wei2].

Acknowledgement

This work was in part supported by grant No. 354 of Charles University and by grant No. 201/93/2174 of Czech Grant Agency.

We take this opportunity to thank the following colleagues and friends who in various ways improved our article: David Armitage, Bob Burckel, Josef Král and Larry Zalcman.

References

[Acz] Aczél J., *Lectures on functional equations and their applications*, (Original German edition: Birkhäuser, Basel, Stuttgart 1961), New York, London, 1966.

[AHM] Aczél J., Haruki H., McKiernan M. A., Sakovič G. N., *General and regular solutions of functional equations characterizing harmonic polynomials*, Aequationes Math. **1** (1968), 37–53.

[ASZ] Aharonov D., Schiffer M. M., Zalcman L., *Potato kugel*, Israel J. Math. **40** (1981), 331–339.

[AFR] Ahern P., Flores M., Rudin W., *An invariant volume-mean-property*, J. Funct. Anal. **111** (1993), 380–397.

[AS] Akcoglu M. A., Sharpe R. W., *Ergodic theory and boundaries*, Trans. Amer. Math. Soc. **132** (1968), 447–460.

[Ali1] Alinhac S., *Une charactérisation des fonctions harmoniques dans un ouvert borné par des propriétés de moyenne sur certains boules*, C. R. Acad. Sci. Paris Sér. A–B **275** (1972), 29–31.

[Ali2] Alinhac S., *Une charactérisation des fonctions harmoniques dans un ouvert par des propriétés de moyenne*, Rev. Roumaine Math. Pures Appl. **18** (1973), 1465–1472.

[AFP] Arazy J., Fisher S. D., Peetre J., *Hankel operators on weighted Bergman spaces*, Amer. J. Math. **110** (1988), 989–1053.

[AG1] Armitage D. H., Goldstein M., *Quadrature and harmonic L^1-approximation in annuli*, Trans. Amer. Math. Soc. **312** (1989), 141–154.

[AG2] Armitage D. H., Goldstein M., *The volume mean-value property of harmonic functions*, Complex Variables Theory Appl. **13** (1990), 185–193.

[AG3] Armitage D. H., Goldstein M., *Quadrature and harmonic approximation of subharmonic functions in strips*, J. London Math. Soc. (2) **46** (1992), 171–179.

[AG4] Armitage D. H., Goldstein M., *Characterizations of balls and strips via harmonic functions*, Approximation by solutions of partial differential equations (Fuglede B. et al., eds.), Kluwer, London, New York, 1992, pp. 1–9.

[AN] Armitage D. H., Nelson C. S., *A harmonic quadrature formula characterizing open strips*, Math. Proc. Cambridge Philos. Soc. **113** (1993), 147–151.

[ACL] Aronszajn N., Creese T. M., Lipkin J. L., *Polyharmonic functions*, Oxford Univ. Press, New York, 1983.

[Ars1] Arsove M., *Functions representable as differences of subharmonic functions*, Trans. Amer. Math. Soc. **75** (1953), 327–365.

[Ars2] Arsove M., *The Looman-Menchoff theorem and some subharmonic functions analogues*, Proc. Amer. Math. Soc. **6** (1955), 94–105.

[Asg] Ásgeirsson L., *Über eine Mittelwertseigenschaft von Lösungen homogener linearer partieller Differentialgleichungen 2. Ordnung mit konstanten Koeffizienten*, Math. Ann. **113** (1937), 321–346.

[Avc] Avci Y., *Characterization of shell domains by quadrature identities*, J. London Math. Soc. (2) **23** (1981), 123–128.

[AC] Axler J., Čučković Ž., *Commuting Toeplitz operators with harmonic symbols*, Integral Equations Operator Theory **14** (1991), 1–11.

[ABR] Axler S., Bourdon P., Ramey W., *Harmonic function theory*, Springer, New York, 1992.

[Aze] Azencott R., *Espace de Poisson des groupes locallement compact*, Lecture Notes in Math. 148, Springer, Berlin, 1970.

[Bac] Bacharach M., *Abriss der Geschichte der Potentialtheorie*, Vandenkoeck & Ruprecht's Ver., Göttingen, 1883.

[Bar] Barta J., *Some mean value theorems in the potential theory*, Acta Tech. Acad. Sci. Hungar. **75** (1973), 3–11.

[Bau1] Bauer H., *Heat balls and Fulks measures*, Ann. Acad. Sci. Fenn. Ser. A. I Math. **10** (1985), 67–82.

[Bau2] Bauer H., *Dirichlet problem and irregular boundary points*, Arkhimedes **4** (1991), 290–294.

[Bax1] Baxter J. R., *Restricted mean values and harmonic functions*, Trans. Amer. Math. Soc. **167** (1972), 451–463.

[Bax2] Baxter J. R., *Harmonic functions and mass cancellation*, Trans. Amer. Math. Soc. **245** (1978), 375–384.

[Bec] Beckenbach E. F., *Concerning the definition of harmonic functions*, Bull. Amer. Math. Soc. **51** (1945), 240–245.

[BR1] Beckenbach E. F., Reade M., *Mean values and harmonic polynomials*, Trans. Amer. Math. Soc. **53** (1943), 230–238.

[BR2] Beckenbach E. F., Reade M., *Regular solids and harmonic polynomials*, Duke Math. J. **12** (1945), 629–644.

[Bel] Belyaev A. A., *The mean value theorem for harmonic functions in a domain of Hilbert space*, Vestnik Moskov. Univ. Ser. I Mat. Mekh. **5** (1982), 32–35. (Russian)

[BW1] Benyamini Y., Weit Y., *Functions satisfying the mean value property in the limit*, J. Analyse Math. **52** (1989), 167–198.

[BW2] Benyamini Y., Weit Y., *Harmonic analysis of spherical functions on $SU(1,1)$*, Ann. Inst. Fourier (Grenoble) **42** (1992), 671–694.

[BG] Berenstein C. A., Gay R., *A local version of the two-circles theorem*, Israel J. Math. **55** (1986), 267–288.

[BZ] Berenstein C. A., Zalcman L., *Pompeiu's problem on spaces of constant curvature*, Israel J. Math. **30** (1976), 113–130.

[Bla1] Blaschke W., *Ein Mittelwertsatz und eine kennzeichnende Eigenschaft des logaritmischen Potentials*, Ber. Ver. Sächs. Akad. Wiss. Leipzig **68** (1916), 3–7.

[Bla2] Blaschke W., *Mittelwertsatz der Potentialtheorie*, Jahresber. Deutsch. Math.-Verein. **27** (1918), 157–160.

[BH] Bliedtner J., Hansen W., *Potential theory: An analytic and probabilistic approach to balayage*, Springer, Berlin, 1986.

[Boa] Boas R. P., *A primer of real functions*, The Mathematical Association of America, Washington D.C., 1981.

[Boc] Bôcher M., *On harmonic functions in two dimensions*, Proc. Nat. Acad. Sci. U.S.A. **41** (1906), 557–583.

[Bos] Bose A. K., *Functions satisfying a weighted average property*, Trans. Amer. Math. Soc. **118** (1965), 472–487.

[BP] Bramble J., Payne L. E., *Mean value theorems for polyharmonic functions*, Amer. Math. Monthly **73** (1966), 124–127.

[Bre] Brelot M., *Éléments de la théorie classique du potentiel*, 4e éd., Centre de documentation Universitaire, Paris, 1969.

[Bro] Brödel W., *Funktionen mit Gaussischen Mittelwerteigenschaften für konvexe Kurven und Bereiche*, Deutsche Math. **4** (1939), 3–15.

[BST] Brown L., Schreiber B. M., Taylor B. A., *Spectral synthesis and the Pompeiu problem*, Ann. Inst. Fourier (Grenoble) **23** (1973), 125–154.

[Bru] Brunel A., *Propriété restreinte de valeur moyenne caractérisant les fonctions harmoniques bornées sur un ouvert $\mathbb{R}^n$*, (selon D. Heath et L. Orey), Seminaire Goulaouic-Schwartz, Paris, Exposé No XIV, 1971–72.

[Brz] Brzezina M., *Capacitary interpretation of the Fulks measure*, (preprint).

[BM] Burkhardt H., Meyer W. F., *Potentialtheorie*, Encyklopädie der Mathematischen Wissenschaften, II. A 7b, Teubner-Verlag, Leipzig, 1899–1916, pp. 464–503.

[Bur] Burckel R. B., *A strong converse to Gauss's mean-value theorem*, Amer. Math. Monthly **87** (1980), 819–820.

[Car] Carathéodory C., *On Dirichlet's problem*, Amer. J. Math. **59** (1937), 709–731.

[CS] Ciesielski Z., Semadeni Z., *Survey of some new methods in potential theory III*, Prace matematyczne **11** (1967), 99–128. (Polish)

[CD] Choquet G., Deny J., *Sur quelques propriétés de moyenne caractéristiques des fonctions harmoniques et polyharmoniques*, Bull. Soc. Math. France **72** (1944), 118–140.

[Chu] Chu S. C., *On a mean value property for solutions of a wave equation*, Amer. Math. Monthly **74** (1967), 711–713.

[Cim] Cimmino G., *Formole di maggiorazione nel problema di Dirichlet per le funzioni armoniche*, Rend. Sem. Mat. Uni. Padova **3** (1932), 46–66.

[CGL] Citti G., Garofalo N., Lanconelli N., *Harnack's inequality for sum of squares of vector fields plus a potential*, Amer. J. Math. **115** (1993), 699–734.

[CC] Constantinescu C., Cornea A., *Potential theory on harmonic spaces*, Springer, Berlin, 1972.

[CV] Cornea A., Veselý J., *Martin compactification for discrete potential theory*, Potential Analysis, (to appear).

[CH] Courant R., Hilbert D., *Mathematical methods of physics, vol. II*, Interscience Pub., New York, 1966.

[DG] Debiard A., Gaveau B., *Une formule de Green locale pour certains espaces symmetriques de rang* > 1, Bull. Sci. Math. (2) **105** (1981), 29–47.

[Del1] Delsarte J., *Note sur une propriété nouvelle des fonctions harmoniques*, C. R. Acad. Sci. Paris Sér. A–B **246** (1958), 1358–1360.

[Del2] Delsarte J., *Lectures on topics in mean periodic functions and the two-radius theorem*, Tata Inst. Fund. Research, Bombay, 1961.

[DL] Delsarte J., Lions J. L., *Moyennes géneralisées*, Comment. Math. Helv. **33** (1959), 59–69.

[Dez] Delzant A., *Fonctions harmoniques sur les groupes semisimples*, Sém. Théorie du potentiel (Brelot – Choquet – Deny), Paris, Exposé No. X, 1962-63.

[Dem] Demidovic B. P., *A simple proof of the mean-value theorem for harmonic functions*, Uspekhi Mat. Nauk **9** (1954), vyp.3 (61), 213–214. (Russian)

[Den] Deny J., *Familles fondamentales. Noyaux associés*, Ann. Inst. Fourier (Grenoble) **3** (1951), 73–101.

[Dok] Dokken D. P., *μ-harmonic functions on locally compact groups*, J. Analyse Math. **52** (1989), 1–25.

[Doo1] Doob J.L., *Probability methods applied to the first boundary value problem*, Proceedings of the third Berkeley symposium on mathematical statistics and probability, Vol. 2 (Neyman J., ed.), Berkeley, 1956, pp. 49–80.

[Doo2] Doob J.L., *Classical potential theory and its probabilistic counterpart*, Springer, New York, 1984.

[Dur] Duren P., *Theory of H^p spaces*, Academic Press, New York, 1970.

[DSS] Duren P., Shapiro H., Shields A., *Singular measures and domains not of Smirnov type*, Duke Math. J. **33** (1966), 247–254.

[Eng] Engliš M., *Functions invariant under the Berezin transform*, (preprint).

[Eps] Epstein B., *On the mean-value property of harmonic functions*, Proc. Amer. Math. Soc. **13** (1962), 830.

[ES] Epstein B., Schiffer M. M., *On the mean-value property of harmonic functions*, J. Analyse Math. **14** (1965), 109–111.

[Eva1] Evans G. C., *Problems of potential theory*, Nat. Acad. Proc. **7** (1921), 89–98.

[Eva2] Evans G. C., *Note on a theorem of Bôcher*, Amer. J. Math. **50** (1928), 123–126.

[Eva3] Evans G. C., *On potentials of positive mass*, Trans. Amer. Math. Soc. **37** (1935), 226–253.

[FG] Fabes E. B., Garofalo N., *Mean value properties of solutions of parabolic equations with variable coefficients*, J. Math. Anal. Appl. **121** (1987), 305–316.

[Fel] Feller W., *Boundaries induced by non-negative matrices*, Trans. Amer. Math. Soc. **83** (1956), 19–54.

[Fen1] Fenton P. C., *Functions having the restricted mean value property*, J. London Math. Soc. (2) **14** (1976), 451–458.

[Fen2] Fenton P. C., *On sufficient conditions for harmonicity*, Trans. Amer. Math. Soc. **253** (1979), 139–147.

[Fen3] Fenton P. C., *On the restricted mean value property*, Proc. Amer. Math. Soc. **100** (1987), 477–481.

[Fic] Fichera G., *A characteristic property of the sphere in $\mathbb{R}^n$*, Atti Acad. Sci. Istit. Bologna Cl. Sci. Fis. Rend. **14 (2)** (1984/85), 235–244.

[Fla1] Flatto L., *Functions with a mean value property*, J. Math. Mech. **10** (1961), 11–18.

[Fla2] Flatto L., *Functions with a mean value property II*, Amer. J. Math. **85** (1963), 248–270.

[Fla3] Flatto L., *The converse of Gauss's theorem for harmonic functions*, J. Differential Equations **1** (1965), 483–490.

[FJ1] Flatto L., Jacobson D., *Functions satisfying the mean value property for product measures*, Amer. J. Math. **97** (1975), 1049–1059.

[FJ2] Flatto L., Jacobson D., *Functions satisfying a discrete mean value property*, Aequationes Math. **22** (1981), 173–193.

[Fol] Föllmer H., *Ein Littlewood Kriterium für gleichmässig integrable Martingale und insbesondere für Dirichlet-Lösungen*, Elliptische Differentialgleichungen, Bd. 2, Akademie Verlag, Berlin, 1970, pp. 113–119.

[Fri1] Friedman A., *Mean-values and polyharmonic polynomials*, Michigan Math. J. **4** (1957), 67–74.

[Fri2] Friedman A., *Parabolic equations of the second order*, Trans. Amer. Math. Soc. **93** (1959), 509–530.

[Fri3] Friedman A., *Partial differential equations of parabolic type*, Prentice-Hall, INC., New York, 1964.

[FL1] Friedman A., Littman W., *Bodies for which harmonic functions satisfy the mean value property*, Trans. Amer. Math. Soc. **102** (1962), 147–166.

[FL2] Friedman A., Littman W., *Functions satisfying the mean value property*, Trans. Amer. Math. Soc. **102** (1962), 167–180.

[Fug1] Fuglede B., *Finely harmonic functions*, Lecture Notes in Math. 289, Springer, Berlin, 1972.

[Fug2] Fuglede B., *On the mean value property of finely harmonic and finely hyperharmonic functions*, Aequationes Math. **39** (1990), 198–203.

[Ful1] Fulks W., *An approximate Gauss mean value theorem*, Pacific J. Math. **14** (1964), 513–516.

[Ful2] Fulks W., *A mean value theorem for the heat equation*, Proc. Amer. Math. Soc. **17** (1966), 6–11.

[Fur1] Furstenberg H., *A Poisson formula for semi-simple groups*, Ann. of Math. **77** (1963), 335–386.

[Fur2] Furstenberg H., *Boundaries of Riemannian symmetric spaces*, Symmetric spaces (Boothby W. M., Weiss G. L., eds.), Marcel Dekker Inc., New York, 1972, pp. 359–377.

[GL1] Garofalo N., Lanconelli E., *Wiener's criterion for parabolic equations with variable coefficients and its consequences*, Trans. Amer. Math. Soc. **308** (1988), 811–836.

[GL1] Garofalo N., Lanconelli E., *Asymptotic behaviour of fundamental solutions and potential theory of parabolic operators with variable coefficients*, Math. Ann. **283** (1989), 211–239.

[GL2] Garofalo N., Lanconelli E., *Level sets of the fundamental solution and Harnack inequality for degenerate equations of Kolmogorov type*, Trans. Amer. Math. Soc. **321** (1990), 775–792.

[Gar] Garsia A. M., *A note on the mean value property*, Trans. Amer. Math. Soc. **102** (1962), 181–186.

[GR] Garsia A. M., Rodemich E., *On functions satisfying the mean value property with respect to a product measure*, Proc. Amer. Math. Soc. **17** (1966), 592–594.

[Gau] Gauss C. F., *Algemeine Lehrsätze in Beziehung auf die im verkehrtem Verhältnisse des Quadrats der Entfernung Wirkenden Anziehungs- und Abstossungs-Kräfte, 1840, Werke, 5. Band*, Göttingen, 1877.

[Gav] Gaveau B., *Principle of moindre action, propagation de la chaleur et éstimées sous élliptiques sur certain groupes nilpotents*, Acta Math. **139** (1977), 95–153.

[Geh] Gehring F. W., *On solutions of the equation of heat conduction*, Michigan Math. J. **5** (1958), 191–202.

[Ger] Gergen J. J., *Note on a theorem of Bôcher and Koebe*, Bull. Amer. Math. Soc. **37** (1931), 591–596.

[Glo1] Globevnik J., *Integrals over circles passing through the origin and a characterization of analytic functions*, J. Analyse Math. **52** (1989), 199–209.

[Glo2] Globevnik J., *Zero integrals on circles and characterizations of harmonic and analytic functions*, Trans. Amer. Math. Soc. **317** (1990), 313–330.

[GlR] Globevnik J., Rudin W., *A characterization of harmonic functions*, Indag. Math. (N.S.) **91** (1988), 419–426.

[God] Godement R., *Une généralisation du théorème de la moyenne pour les fonctions harmoniques*, C. R. Acad. Sci. Paris Sér. A – B **234** (1952), 2137–2139.

[GHJ] Goldstein M., Haussmann W., Jetter K., *Best harmonic L^1 approximation to subharmonic functions*, J. London Math. Soc. (2) **30** (1984), 257–264.

[GHR1] Goldstein M., Haussmann W., Rogge L., *On the mean value property of harmonic functions and best harmonic L^1-approximation*, Trans. Amer. Math. Soc. **305** (1988), 505–515.

[GHR2] Goldstein M., Haussmann W., Rogge L., *Characterizations of open strips by harmonic quadrature*, Approximation by solutions of partial differential equations (Fuglede B. et al., eds.), Kluwer, London, New York, 1992, pp. 87–92.

[GHR3] Goldstein M., Haussmann W., Rogge L., *On the inverse mean value property of harmonic functions on strips*, Bull. London Math. Soc. **24** (1992), 87–92.

[GO] Goldstein M., Ow W. H., *On the mean-value property of harmonic functions*, Proc. Amer. Math. Soc. **29** (1971), 341–344.

[Gon] Gong X., *Functions with the restricted mean value property*, Acta Sci. Natur. Univ. Amoien **27** (1988), 611–615. (Chinese)

[Gra] Gray A., *Mean values in Riemannian manifolds*, Proceedings of the IV. International Colloquium on Differential Geometry, Univ. Santiago de Compostela, Santiago de Compostela, 1978, Cursos Congr. Univ. Santiago de Compostela 15, Univ. Santiago de Compostela, Santiago de Compostela, 1979, pp. 140–146.

[GV] Gray A., Vanhecke L., *Power series expansions, differential geometry of geodesic spheres and tubes, and mean-value theorems*, Geometry and differential geometry (Proc. Conf., Univ. Haifa, Haifa, 1979), Lecture Notes in Math. 792, Springer, Berlin, 1980, pp. 252–259.

[GW] Gray A., Willmore T. J., *Mean-value theorems for Riemannian manifolds*, Proc. Roy. Soc. Edinburgh Sect. A **92** (1982), 343–364.

[Gre] Green J. W., *Mean values of harmonic functions on homothetic curves*, Pacific J. Math. **6** (1956), 279–282.

[Gus] Gustafsson B., *On quadrature domains and an inverse problem in potential theory*, J. Analyse Math. **55** (1990), 172–216.

[HNa1] Hansen W., Nadirashvili N., *A converse to the mean value theorem for harmonic functions*, Acta Math., (to appear).

[HNa2] Hansen W., Nadirashvili N., *Mean values and harmonic functions*, Math. Ann. **297** (1993), 157–170.

[HNa3] Hansen W., Nadirashvili N., *Littlewood's one circle problem*, J. London Math. Soc. (2), (to appear).

[HNa4] Hansen W., Nadirashvili N., *Liouville's theorem and the restricted mean value property*, (preprint).

[HNa5] Hansen W., Nadirashvili N., *On the restricted mean value property for measurable functions*, (preprint).

[HNe1] Hansen W., Netuka I., *Volume densities with the mean value property for harmonic functions*, Proc. Amer. Math. Soc., (to appear).

[HNe2] Hansen W., Netuka I., *Inverse mean value property of harmonic functions*, Math. Ann. **297** (1993), 147–156.

[HW1] Hartman P., Wintner A., *On the inverse of the parabolic differential operator* $\frac{\partial^2}{\partial x^2} - \frac{\partial}{\partial t}$, Amer. J. Math. **75** (1953), 598–610.

[HW2] Hartman P., Wintner A., *Mean value theorems and linear operators*, Amer. Math. Monthly **62** (1955), 217–222.

[Har1] Haruki S., *On two functional equations connected with a mean-value property of polynomials*, Aequationes Math. **6** (1971), 275–277.

[Har2] Haruki S., *On a relation between the "square" functional equation and the "square" mean-value property*, Canad. Math. Bull. **14** (1971), 161–165.

[Har3] Haruki S., *Four different unknown functions satisfying the triangle mean value property for harmonic polynomials*, Ann. Polon. Math. **33** (1976/77), 219–221.

[Har4] Haruki S., *Four different unknown functions satisfying the triangle mean value property for harmonic polynomials II*, Ann. Polon. Math. **44** (1984), 239–243.

[Har5] Haruki S., *On the theorem of S. Kakutani, M. Nagumo and J. L. Walsh for the mean value property of harmonic and complex polynomials*, Pacific J. Math. **94** (1981), 113–123.

[Har6] Haruki S., *On the mean value property of harmonic and complex polynomials*, Proc. Japan Acad. Ser. A Math. Sci. **57** (1981), 216–218.

[Hau] Haussmann W., *Approximation by harmonic functions*, Topics in multivariate approximation (Santiago, 1986), Academic Press, Boston (MA), 1987, pp. 111–124.

[Hea] Heath D., *Functions possessing restricted mean value properties*, Proc. Amer. Math. Soc. **41** (1973), 588–595.

[Hel] Helms L. L., *Introduction to potential theory*, Wiley-Interscience, New York, 1969.

[HG] Hersch R., Griego R. J., *Brownian motion and potential theory*, Scientific American (1967), no. 3 (March), 67–74.

[Hir] Hirschfeld M. R., *Sur les fonctions μ-harmoniques dans un espace localement compact mesuré*, C. R. Acad. Sci. Paris Sér. A - B **262** (1966), 174–176.

[HJ1] Hoh W., Jacob N., *Quelques remarques sur la théorie du potentiel de l'equation de A. N. Kolmogorov*, C. R. Acad. Sci. Paris Ser. I Math. **310** (1990), 657–659.

[HJ2] Hoh W., Jacob N., *Remarks on mean value properties of solutions of second order differential operators*, Exposition. Math. **9** (1991), 357–387.

[HJ3] Hoh W., Jacob N., *On the potential theory of the Kolmogorov equation*, Math. Nachr. **154** (1991), 51–66.

[Hop] Hopf E., *Bemerkungen zur Aufgabe 49*, Jahresber. Deutsch. Math.-Verein., 2. Teil **39** (1930), 4–6.

[Hor] Horiuchi T., *On the potential theory for certain degenerated elliptic operators*, Bull. Fac. Sci. Ibaraki Univ. Ser. A **20** (1988), 27–45.

[Huc] Huckemann F., *On the 'one circle' problem for harmonic functions*, J. London Math. Soc. (2) **29** (1954), 491–497.

[HuW] Hunt R. A., Wheeden R. L., *Positive harmonic functions on Lipschitz domains*, Trans. Amer. Math. Soc. **147** (1970), 507–527.

[Ito] Itô M., *On α-harmonic functions*, Nagoya Math. J. **26** (1966), 205–221.

[Izu] Izuchi K., *The one-radius theorem is not true for bounded real-analytic functions*, Proc. Amer. Math. Soc. **103** (1988), 823–830.

[JC] Ji X. H., Chen D. Q., *Fundamental properties of solutions of a degenerate elliptic equation in extended space*, J. Differential Equations **58** (1985), 192–211.

[Joh] Johnston E. H., *The mean value theorem and analytic functions*, Proc. Edinburgh Math. Soc. (2) **26** (1983), 289–295.

[KaS] Kaneko H., Sakai A., *Mean value of harmonic functions and Brownian motion*, Sûgaku **41** (1989), 182–184. (Japanese)

[KL] Keldysch M. V., Lavrentieff M. A., *Sur la représentation conforme des domaines limites par des courbes rectifiables*, Ann. Sci. École Norm. Sup. **54** (1937), 1–38.

[Kel1] Kellogg O. D., *Foundations of potential theory*, Springer, Berlin, 1929.

[Kel2] Kellogg O. D., *Les moyennes arithmétiques dans la théorie du potentiel*, Enseign. Math. **27** (1928), 14–26.

[Kel3] Kellogg O. D., *Converses of Gauss's theorem on the arithmetic mean*, Trans. Amer. Math. Soc. **36** (1934), 227–242.

[Koe] Koebe P., *Herleitung der partiellen Differentialgleichungen der Potentialfunktion aus deren Integraleigenschaft*, Sitzungsber. Berlin. Math. Gessellschaft **5** (1906), 39–42.

[Kos] Kosmodemjanskij A. A. Jr., *Converse of the mean value theorem for harmonic functions*, Uspekhi Mat. Nauk **36** (1981), 175–176. (Russian)

[Kou1] Kounchev O. I., *Mean value theorems for domains other than balls and L^1-approximation by harmonic functions*, C. R. Acad. Bulgare Sci. **41** (1988), 17–20.

[Kou2] Kounchev O. I., *Mean value property for biharmonic functions characterizing the annulus*, C. R. Acad. Bulgare Sci. **42** (1989), 33–35.

[Kow1] Kowalski O., *The second mean-value operator on Riemannian manifolds*, Proceedings of the Conference on Differential Geometry and its Applications (Nové Město na Moravě, 1980), Univ. Karlova, Prague, 1982, pp. 33–45.

[Kow2] Kowalski O., *A comparison theorem for spherical mean-value operators in Riemannian manifolds*, Proc. London Math. Soc. (3) **47** (1983), 1–14.

[KO] Kozaki M., Ogura Y., *On geometric and stochastic mean values for small geodesic spheres in Riemannian manifolds*, Tsukuba J. Math. **11** (1987), 131-145.

[KS] Kozaki M., Sumi H., *On normal forms of Laplacian and its iterations in harmonic spaces*, Comment. Math. Univ. Carolin. **30** (1989), 795–802.

[Koz] Kozakiewicz W., *Un théorème sur les opérations et son application à la théorie des laplaciens généralisés*, C. R. Soc. Sci. Varsovie **26** (1933), 1–7.

[Kup1] Kuptsov L. P., *Mean value theorem and the maximum principle for Kolmogorov's equation*, Mat. Zametki **15** (1974), 479–489 (Russian); English transl. in Math. Notes **15** (1974), 280–286.

[Kup2] Kuptsov L. P., *The mean property and the maximum principle for the parabolic equation of second order*, Dokl. Akad. Nauk SSR **242** (1978), 529–532 (Russian); English transl. in Soviet. Math. Dokl. **19** (1978), 1140–1144.

[Kup3] Kuptsov L. P., *On parabolic means*, Dokl. Akad. Nauk SSSR **252** (1980), 296–301 (Russian); English transl. in Soviet Math. Dokl. **21** (1980), 741–746.

[Kup4] Kuptsov L. P., *Mean property for the heat-conduction equation*, Mat. Zametki **29** (1981), 211–223, 317 (Russian); English transl. in Math. Notes **29** (1981), 110–116.

[Kup5] Kuptsov L. P., *A class of means for solutions of the heat-conduction equation*, Mat. Zametki **35** (1984), 201–220 (Russian); English transl. in Math. Notes **35** (1984), 105–115.

[Kur1] Kuran Ü., *On the mean-value property of harmonic functions*, Bull. London Math. Soc. **4** (1972), 311–312.

[Kur2] Kuran Ü., *Two results on means of harmonic functions*, Proc. Symp. Complex Analysis, Canterbury 1973 (Clunie J., Hayman W. K., eds.), Lond. Math. Soc. Lecture Notes, vol. 12, Univ. Press, Cambridge, 1974, pp. 101–104.

[Krt] Kurant R., *Partial differential equations*, Mir, Moskva, 1964. (Russian)

[Lan] Landkof N. S., *Foundations of modern potential theory*, (translated from 1966 Russian edition), Springer, Berlin, 1972.

[Leb1] Lebesgue H., *Sur le problème de Dirichlet*, Rend. Circ. Mat. Palermo **24** (1907), 371–402.

[Leb2] Lebesgue H., *Sur le théorème de la moyenne de Gauss*, Bull. Soc. Math. France **40** (1912), 16–17.

[Leb3] Lebesgue H., *Sur le problème de Dirichlet*, C. R. Acad. Sci. Paris **154** (1912), 335–337.

[Leb4] Lebesgue H., *Sur le théorème de la moyenne et le problème de Dirichlet*, C. R. de Séances de la Soc. Math. de France **48** (1920), 26.

[Les] Lesky P. Jr., *Zur Charakterisierung von Loesungen polyharmonischer Gleichungen im Ganzraum*, Z. Angew. Math. Mech. **69** (1989), 162-163.

[Lev] Levi E., *Sopra una proprietà caratteristica delle funzione armoniche*, Rend. Accad. d. Lincei Roma (5) **18** (1909), 10–15.

[LV1] Lewis J. L., Vogel A., *On pseudospheres*, Rev. Mat. Iberoamericana **7** (1991), 25–54.

[LV2] Lewis J. L., Vogel A., *On some almost everywhere symmetry theorems*, Progress in nonlinear differential equations (Lloyd N. G. et al., eds.), Nonlinear diffusion equations and their equilibrium states 3, Birkhäuser, Boston, Basel, Berlin, pp. 347–374.

[Lew] Lewkowicz R. E., *Harmonic functions on homogeneous spaces of commutator induced extensions of compact groups*, Colloq. Math. **55** (1988), 285–304.

[Lit1] Littlewood J. E., *On the definition of a subharmonic function*, J.London Math. Soc. (2) **2** (1927), 189–192.

[Lit2] Littlewood J. E., *Some problems in real and complex analysis*, Hath. Math. Monographs, Massachusetts, 1968.

[Log] Loges F., *Rational determination of a special kind of volume integrals based upon generalization of an average theorem of potential theory*, Wiss. Z. Martin-Luther-Univ. Halle-Wittenberg, Math.-Natur. Reihe **35** (1986), 140–149.

[Los] Losev A. G., *Harmonic functions on manifolds of negative curvature*, Mat. Zametki **40** (1986), 738–742 (Russian); English transl. in Math. Notes **40** (1986), 915–917.

[Mak] Makai E., *A property of mean of harmonic functions*, Rend. Circ. Mat. Palermo **63** (1942), 33–40.

[Mal] Malliavin P. (notes prepared by Dehen D., Michel D.), *Geometrie differentielle stochastique*, Seminaire de Mathématiques Superieures 64, Presses de l'Université de Montréal, Montréal (Que.), 1978.

[Mar1] Mařík J., *Problem No. 10*, Časopis Pěst. Mat. **81** (1956), 470. (Czech)

[Maz] Mazaika P. K., *A lattice summation using the mean value theorem for harmonic functions*, SIAM Rev. **26** (1984), 113–115.

[Met] Metz V., *Potentialtheorie auf dem Sierpinski gasket*, Math. Ann. **289** (1991), 207–237.

[Mik] Mikhailov V. P., *Partial differential equations*, Nauka, Moskva, 1976. (Russian)

[Mih1] Mikhlin S. G., *The problem of minimum of a quadratic functional*, Holden-Day, Inc., San Francisco, London, Amsterdam, 1965.

[Mih2] Mikhlin S. G., *Linear partial differential equations*, Izdat. "Vyss. Skola", Moscow, 1977. (Russian)

[Min] Min-Teh Ch., *On a theorem of Nicolesco and degenerated Laplace operators*, Proc. Amer. Math. Soc. **2** (1951), 77–86.

[MW] Mingareli A. B., Wang S., *A Laplace operator in infinite dimensional Hilbert space*, (preprint).

[Nau1] Naumov V. V., *General solution of an equation of the harmonic oscillations of an elastic body*, Dinamika Sploshn. Sredy **80** (1987), 102–112, 168. (Russian)

[Nau2] Naumov V. V., *Mean value theorems for the equation of harmonic oscillations of an elastic body*, Dinamika Sploshn. Sredy **82** (1987), 147–153, 168. (Russian)

[Net1] Netuka I., *Solution of problem No. 10*, Časopis Pěst. Mat. **94** (1969), 223–225. (Czech)

[Net2] Netuka I., *Harmonic functions and mean value theorems*, Časopis Pěst. Mat. **100** (1975), 391–409. (Czech)

[Net3] Netuka I., *Fine behaviour of solutions of the Dirichlet problem near an irregular point*, Bull. Sci. Math. **114** (1990), 1–22.

[Neu1] Neumann C., *Revision einiger allgemeiner Sätze aus der Theorie des Newtonschen Potentials*, Math. Ann. **3** (1871), 424–434.

[Neu2] Neumann C., *Untersuchungen über das logarithmische und Newtonsche Potential*, Teubner, Leipzig, 1877.

[Nic1] Nicolescu M., *Une propriété caractéristique de moyenne des solutions régulieres de l'equation de la chaleur*, Com. Acad. R. P. Romåne **2** (1952), 677–679.

[Nic2] Nicolescu M., *Sur une propriété caractéristique de moyenne des fonctions polycaloriques*, Com. Acad. R. P. Romåne **4** (1954), 551–554.

[Nic3] Nicolescu M., *Propriété de moyenne des fonctions harmoniques bornées dans un demi-plan ou dans un angle droit*, Rev. Roumaine Math. Pures Appl. **1** (1956), 43–50.

[Nic4] Nicolescu M., *Sur les moyennes généralisées successives d'une fonction*, Rev. Roumaine Math. Pures Appl. **6** (1961), 429–441, (see also Mathematica (Cluj) 27 (1962), 107–121).

[Oba1] Obata N., *Analysis of the Levy Laplacian*, Soochow J. Math. **14** (1988), 105–109.

[Oba2] Obata N., *The Levy Laplacian and mean value theorem*, Probability measures on groups IX, Lecture Notes in Math. 1379, Springer, Berlin, 1989, pp. 242–253.

[OFa] O'Farrell A. G., *A characterization of harmonic functions*, Proc. Roy. Irish Acad. Sect. A. **78** (1978), 33–36.

[OS] Øksendal B., Strook D. W., *A characterization of harmonic measure and Markov processes*, Ann. Inst. Fourier (Grenoble) **32** (1982), 221–232.

[Par] Parreau M., *Sur les moyennes des fonctions harmoniques et analytiques et la classification des surfaces de Riemann*, Ann. Inst. Fourier (Grenoble) **3** (1951), 103–197.

[Per] Perkins F. W., *Sur la résolution du problème de Dirichlet par des médiations réitérées*, C. R. Acad. Sci. Paris **184** (1927), 182–183.

[PW1] Picardello M. A., Woess W., *A converse to the mean value property on homogeneous trees*, Trans. Amer. Math. Soc. **311** (1989), 209–225.

[PW2] Picardello M. A., Woess W., *Ends of infinite graphs, potential theory and electrical networks*, Cycles and Rays (Hahn G. et al., eds.), Kluwer, London, New York, 1990, pp. 181–196.

[Pin] Pini B., *Maggioranti e minoranti delle soluzioni delle equazioni paraboliche*, Ann. Mat. Pura Appl. **(4) 37** (1954), 249–264.

[Pir] Piranian G., *Two monotonic, singular uniformly almost smooth functions*, Duke Math. J. **33** (1966), 255–262.

[Pla] Plancherel M., *Les problèmes de Cantor et de du Bois-Reymond*, Ann. Sci. École Norm. Sup. **31** (1914), 223–262.

[PP] Plancherel M., Pólya G., *Sur les valeurs moyennes des fonctions réelles définies pour toutes valeurs de la variable*, Comment. Math. Helv. **3** (1931), 114–121.

[Por1] Poritsky H., *On operations permutable with the Laplacian*, Amer. J. Math. **54** (1932), 667–691.

[Por2] Poritsky H., *Generalizations of the Gauss law of spherical mean*, Trans. Amer. Math. Soc. **43** (1938), 199–225.

[PS] Port S. C., Stone C. J., *Brownian motion and classical potential theory*, Academic Press, New York, 1978.

[Pot] Potts D. H., *A note on the operator of Blaschke and Privaloff*, Bull. Amer. Math. Soc. **54** (1948), 782–787.

[Pri1] Privaloff I., *Sur les fonctions harmoniques*, Mat. Sb. **32** (1924/25), 464–469.

[Pri2] Privaloff I., *On a theorem of S. Saks*, Mat. Sb. **51** (1941), 457–460.

[Pri3] Privaloff I., *Sur la definition d'une fonction harmonique*, Dokl. Akad. Nauk SSSR **31** (1941), 102–103.

[Pri4] Privaloff I., *Quelques applications de l'opérateur généralisé de Laplace*, Dokl. Akad. Nauk SSSR **31** (1941), 104–105.

[Pri5] Privalow I., *Randeigenschaften analytischer Funktionen*, Deutscher Verlag, Berlin, 1956.

[Rad] Radó T., *Subharmonic functions*, Springer, Berlin, 1937.

[RW] Ramsey T., Weit Y., *Mean values and classes of harmonic functions*, Math. Proc. Cambridge Philos. Soc. **96** (1984), 501–505.

[Ran] Randol B., *Some remarks on the stability of a property related to the mean value theorem for harmonic functions*, Proc. Amer. Math. Soc. **114** (1992), 175–179.

[Rea] Reade M. O., *Some remarks on subharmonic functions*, Duke Math. J. **10** (1943), 531–536.

[Rie] Riesz F., *Über die subharmonischen Funktionen und ihre Role in der Funktionentheorie und in der Potentialtheorie*, Acta Sci. Math. (Szeged) **2** (1925), 87–100.

[Ros] Rosculet M. N., *Differential equations and applications*, Editura Academiei Republicii Socialiste Romania, Bucharest, 1984. (Roumanian)

[Rud1] Rudin W., *A theorem on subharmonic functions*, Proc. Amer. Math. Soc. **2** (1951), 209–212.

[Rud2] Rudin W., *Function theory in the unit ball of* $\mathbb{C}^n$, Springer, New York, 1980.

[Sak1] Sakai M., *The sub-mean-value property of subharmonic functions and its application to the estimation of the Gaussian curvature of the span metric*, Hiroshima Math. J. **9** (1979), 555–593.

[Sak2] Sakai M., *Quadrature domains*, Lecture Notes in Math. 934, Springer, Berlin, 1982.

[Sas1] Saks S., *Note on defining properties of harmonic functions*, Bull. Amer. Math. Soc. **38** (1932), 380–382.

[Sas2] Saks S., *On subharmonic functions*, Acta Litt. Sci. Univ. Szeged **5** (1930-32), 187–193.

[Sas3] Saks S., *On the operators of Blaschke and Privaloff for subharmonic functions*, Mat. Sb. **(9) 51** (1941), 451–456.

[SW1] Schiff J. L., Walker W. J., *Finite harmonic and geometric interpolation*, Rocky Mountain J. Math. **13** (1983), 651–657.

[SW2] Schiff J. L., Walker W. J., *Finite harmonic interpolation II*, J. Math. Anal. Appl. **87** (1982), 1–8.

[Sch] Schwarz H. A., *Über die Integration der partiellen Differentialgleichungen* $\frac{d^2u}{dx^2} + \frac{d^2u}{dy^2} = 0$ *für die Fläche eines Kreises*, Vierteljschr. Naturforsch. Ges. Zürich **15** (1870), 113–128.

[Sco] Scornazzani V., *Sul problema di Dirichlet per l'operatore di Kolmogorov*, Boll. Un. Mat. Ital. A (5) **18** (1981), 43–62.

[Sev] Severini G., *Sopra una proprietà caratteristica delle funzioni armoniche*, Boll. Accad. Gioenia (2), fasc. 2.

[Shg1] Shahgholian H., *A characterization of the sphere in terms of single-layer potentials*, Proc. Amer. Math. Soc. **115** (1992), 1167–1168.

[Shg2] Shahgholian H., *Quadrature surfaces as free boundaries*, (preprint).

[Shp1] Shapiro H. S., *Remarks concerning domains of Smirnov type*, Michigan Math. J. **13** (1966), 341–348.

[Shp2] Shapiro H. S., *The Schwarz function and its generalization to higher dimensions*, Wiley-Interscience Publication, John Wiley & Sons, INC., New York, 1992.

[SO] Sibagaki W., Ono A., *On the mean-value theorem of harmonic functions*, Mem. Fac. Sci. Kyushu Univ. Ser. A **7** (1952), 41–50.

[Smi] Smith J. D., *Harmonic analysis of scalar and fields in* $\mathbb{R}^n$, Proc. Camb. Phil. Soc. **72** (1973), 403–416.

[Smy1] Smyrnelis E. P., *Sur les moyennes des fonctions paraboliques*, Bull. Sci. Math. (2) **93** (1969), 163–173.

[Smy2] Smyrnelis E. P., *Mesures normales et fonctions harmoniques*, Bull. Sci. Math. (2) **95** (1971), 197–207.

[Smy3] Smyrnelis E. P., *Une proprieté de moyenne des fonctions biharmoniques*, Bull. Sci. Math. (2) **109** (1985), 103–111.

[Spr] Spraker J. S., *Note on arc-length and harmonic measure*, Proc. Amer. Math. Soc. **105** (1989), 664–665.

[Sto] Stoll M., *Mean value theorems for harmonic and holomorphic functions on bounded symmetric domains*, J. Reine Angew. Math. **290** (1977), 191–198.

[Sus] Susanka L., *Potential theoretic and stopping time methods for converse mean value property*, (Thesis) (1991), University of Minnesota, Minneapolis, 1–71.

[Szp] Szpilrajn E., *Remarques sur les fonctions sousharmoniques*, Ann. of Math. **34** (1933), 588–594.

[Tal] Talenti G., *A note on the Gauss curvature of harmonic and minimal surfaces*, Pacific J. Math. **101** (1982), 477–492.

[Tho] Thompson J. M., *Distribution of mass for averages of Newtonian potential functions*, Bull. Amer. Math. Soc. **41** (1935), 744–752.

[TT] Thompson W., Tait P. G., *Treatise on Natural Philosophy*, Cambridge University Press, Cambridge, 1879.

[Ton] Tonelli L., *Sopra una proprietà caratteristica delle funzione armoniche*, Rend. Accad. d. Lincei Roma (5) **18** (1909), 557–582.

[Usp] Uspenskij V. A., *Geometric derivation of basic properties of harmonic functions*, Uspekhi Mat. Nauk **4** (1949), vyp.2 (30), 201–205. (Russian)

[Var] Varopulos N. T., *Fonctions harmoniques et diffusion sur une varieté riemannienne*, C. R. Acad. Sci. Paris Sér. I. Math. **294** (1982), 277–280.

[Vee1] Veech W. A., *A converse to Gauss' theorem*, Bull. Amer. Math. Soc. **78** (1971), 444–446.

[Vee2] Veech W. A., *A zero-one law for a class of random walks and a converse to Gauss' mean value theorem*, Ann. of Math. (2) **97** (1973), 189–216.

[Vee3] Veech W. A., *A converse to the mean value theorem for harmonic functions*, Amer. J. Math. **97** (1975), 1007–1027.

[Ves1] Veselý J., *Sequence solutions of the Dirichlet problem*, Časopis Pěst. Mat. **106** (1981), 84–93.

[Ves2] Veselý J., *Restricted mean value property in axiomatic potential theory*, Comment. Math. Univ. Carolin. **23** (1982), 613–628.

[Vit] Vitali G., *Sopra una proprietà caratteristica delle funzioni armoniche*, Rend. Accad. d. Lincei Roma (5) **21** (1912), 315–320.

[Vol1] Volchkov V. V., *Mean-value theorems for some differential equations*, Dokl. Akad. Nauk Ukrain. SSR **1991**, no. 6, 8–11. (Russian)

[Vol2] Volchkov V. V., *Theorems on spherical averages for some differential equations*, Dokl. Akad. Nauk Ukrain. SSR **1992**, no. 5, 8–11. (Russian)

[Vol3] Volchkov V. V., *On some questions related to the Pompeiu problem*, (preprint). (Russian)

[Vot] Volterra V., *Alcune osservazioni sopra proprietà atte ad individuare una funzione*, Rend. Accad. d. Lincei Roma (5) **18** (1909), 263–266.

[Wal] Walsh J. L., *A mean value theorem for polynomials and harmonic polynomials*, Bull. Amer. Math. Soc. **42** (1936), 923–930.

[Wat] Watson G. N., *A treatise on the theory of Bessel functions*, Cambridge University Press, Cambridge, 1962.

[Was1] Watson N. A., *A theory of subtemperatures in several variables*, Proc. London Math. Soc. **(3) 26** (1973), 385–417.

[Was2] Watson N. A., *Green functions, potentials, and the Dirichlet problem for the heat equation*, Proc. London Math. Soc. **(3) 33** (1976), 251–298, (Corrigendum: ibid. 37 (1978), 32–34).

[Was3] Watson N. A., *A convexity theorem for local mean values of subtemperatures*, Bull. London Math. Soc. **22** (1990), 245–252.

[Was4] Watson N. A., *A thermic analogue of the fundamental function of the distance*, Bull. London Math. Soc. **22** (1990), 253–257.

[Was5] Watson N. A., *Mean values and thermic majorization of subtemperatures*, Ann. Acad. Sci. Fenn. Ser. A. I Math. **16** (1991), 113–124.

[Was6] Watson N. A., *Mean values of subharmonic functions over Green spheres*, Math. Scand. **69** (1991), 307–319.

[Was7] Watson N. A., *Generalization of the spherical mean convexity theorem on subharmonic functions*, Ann. Acad. Sci. Fenn. Ser. A. I Math. **17** (1992), 241–255.

[Was8] Watson N. A., *Mean values of subtemperatures over level surfaces of Green functions*, Ark. Mat. **30** (1992), 165–185.

[Wei1] Weit, Y., *An asymptotic characterization of harmonic functions.*, Contemp. Math. **91** (1989), 295–297.

[Wei2] Weit Y., *On a generalized asymptotic mean value property*, Aequationes Math. **41** (1991), 242–247.

[Wid] Widder D. V., *The heat equation*, Academic Press, New York, 1975.

[Wil1] Willmore T. J., *Mean-value theorems in Riemannian geometry*, Contributions to geometry, Proc. Geom. Sympos., Siegen, 1978, Birkhauser, Basel, Boston (MA), 1979, pp. 281–287.

[Wil2] Willmore T. J., *Mean-value theorems in riemannian geometry*, Contributions to geometry (Tölke J., Wills J. M., eds.), Birkhäuser, Basel, 1979, pp. 281–287.

[Wil3] Willmore T. J., *An extension of Pizzetti's formula to Riemannian manifolds*, Analysis on manifolds (Conf., Univ. Metz, Metz, 1979), Asterisque, 80, Soc. Math. France, Paris, 1980, pp. 53–56.

[Wil4] Willmore T. J., *Harmonic spaces*, Med. Konink. Acad. Wetensch. België **43** (1981), 1–11.

[Zal1] Zalcman L., *Analyticity and the Pompeiu problem*, Arch. Rational Mech. Anal. **47** (1972), 237–254.

[Zal2] Zalcman L., *Mean values and differential equations*, Israel J. Math. **14** (1973), 339–352.

[Zal3] Zalcman L., *Offbeat integral geometry*, Amer. Math. Monthly **87** (1980), 161–175.

[Zal4] Zalcman L., *Modern perspectives in classical function theory*, Rocky Mountain J. Math. **12** (1982), 75–92.

[Zal5] Zalcman L., *Some inverse problems of potential theory*, Contemp. Math. **63** (1987), 337–350.

[Zal6] Zalcman L., *A bibliographic survey of the Pompeiu problem*, Approximation by solutions of partial differential equations (Fuglede B. et al., eds.), Kluwer, London, New York, 1992, pp. 185–194.

[Zar] Zaremba S., *Contributions à la théorie d'une équation fonctionelle de la physique*, Rend. Circ. Mat. Palermo **19** (1905), 140–150.

FARRELL AND MERGELYAN SETS FOR THE SPACE OF BOUNDED HARMONIC FUNCTIONS

Fernando PEREZ-GONZALEZ and Rodrigo TRUJILLO-GONZALEZ

Departamento de Análisis Matemático, Universidad de La Laguna
38271 La Laguna, Tenerife, Spain.

1. Introduction

Let $\mathbb{D}$ be the unit disc, F a proper relatively closed subset of $\mathbb{D}$, and A a linear space of analytic functions equipped with a topology τ such that the analytic polynomials P are τ-dense in A. Two problems of joint approximations in (A,τ) have been considered:

a) To describe those subsets F such that every function f in A whose restriction to F is bounded can be approximated (simultaneously) by polynomials in the topology τ, and a such a way that the approximating polynomials satisfy the same control as the absolute value of the function f on F. (Farrell sets for (A,τ)).

b) To describe those subsets F such that every function f in A whose restriction to F is uniformly continuous can be approximated (simultaneously) by polynomials in the topology τ, and uniformly on F. (Mergelyan sets for (A,τ)).

These approximation problems was introduced to A. Stray by L.A. Rubel for 1970 (See **[R]** and **[St3]**). A. Stray gave in **[St1]** a complete geometric characterization of Farrell and Mergelyan sets for $A = H(\mathbb{D})$ endowed with the topology of uniform convergence in the compact subsets of $\mathbb{D}$, and in **[St2]**, a description of Farrell sets for the space $H^\infty(\mathbb{D})$ endowed with the topology of pointwise bounded convergence. The problem of describing the same sets for the Hardy spaces $H^p(\mathbb{D})$ was solved by L.A. Rubel and A. Stray in **[RS]** when $1 \le p < \infty$ and τ is the weak topology for $H^p(\mathbb{D})$; by the first author in **[PG]** when τ is the norm topology for $H^p(\mathbb{D})$; and by A. Stray and F. Pérez-González for any $p>0$ in **[PGS]** (See also **[St3]**). Finally, A.G. O'Farrell and the first author have studied in **[OFPG]** Farrell sets for bounded analytic functions in bounded open subsets of the plane.

K. GowriSankaran et al. (eds.), Classical and Modern Potential Theory and Applications, 399–412.

We are interested in this type of approximation problems in spaces of harmonic functions in the unit ball of $\mathbb{R}^n$ equipped with a topology such that harmonic polynomials are dense. In this paper we give a (geometric) necessary and suficient condition for a relatively closed subset of the unit ball B in $\mathbb{R}^{n+1}$ to be a Farrell set for the space of bounded harmonic functions in B (Theorem 3.2), and we show that this condition is necessary for Mergelyan sets in the same class (Theorem 4.1).

2. Definitions and Preliminar Results

Let B be the unit ball of $\mathbb{R}^{n+1}$ and $S = S^n$ its boundary, i. e., the unit sphere of $\mathbb{R}^{n+1}$; σ denotes the normalized surface-area measure on S so that $\sigma(S) = 1$. The points in $\mathbb{R}^{n+1}$ will represented by $z = (x,y)$, where $x \in \mathbb{R}^n$ and $y \in \mathbb{R}$, and the upper half-space will denoted by $\mathbb{R}^{n+1}_+ = \{(x,y) : x \in \mathbb{R}^n, y>0\}$.

Let $h^\infty(B)$ be the space of the complex-valued bounded harmonic functions in B endowed with the weak-* topology so that harmonic polynomials are weak-* dense in $h^\infty(B)$. Actually, (see **[SW,** p. 141**]**), we can assume that for every $u \in h^\infty(B)$ there exists a sequence $\{p_n\}$ of harmonic polynomials such that $p_n(z) \longrightarrow u(z)$ for all $z \in B$, and $\|p_n\|_\infty \longrightarrow \|u\|_\infty$ as $n \to \infty$.

DEFINITION 2.1 A relatively closed subset F of B is called a *Farrell set for* $h^\infty(B)$ if for every $f \in h^\infty(B)$ there exists a sequence of harmonic polynomials $\{p_n\}$ such that

(i) $p_n(z) \longrightarrow f(z)$, for all $z \in B$,

(ii) $\|p_n\|_\infty \longrightarrow \|f\|_\infty$, and

(iii) $\|p_n\|_F \longrightarrow \|f\|_F$

as $n \to \infty$, where as usual $\|f\|_\infty = \sup\{|f(z)| : z \in B\}$ and $\|f\|_F = \sup\{|f(z)| : z \in F\}$.

Similarly,

DEFINITION 2.2 F is said to be a *Mergelyan set for* for $h^\infty(B)$ if for every function $f \in h^\infty(B)$ uniformly continuous on F, there exists a sequence $\{p_n\}$ of harmonic polynomials verifying:

(i) $p_n(z) \longrightarrow f(z)$, for all $z \in B$,

(ii) $\|p_n\|_\infty \longrightarrow \|f\|_\infty$, and

(iii) $\|p_n - f\|_F \longrightarrow 0$

as $n \to \infty$.

F will always represent a relatively closed subset of B.

DEFINITION 2.3 We shall say that F satisfies the non-tangential condition (in short, NTC) if almost every point of $\overline{F}\cap S$ with respect to the measure σ on S is a nontangential limit of points of F.

The *tangential boundary* F_t of F is the set of all points of $\overline{F}\cap S$ such that *only* can be reached tangentially from F.

We will use standard notations and well known results of the theory of harmonic functions of several variables that the reader can find in **[SW]** or in the recent book **[ABR]** by S. Axler, P. Bourdon and W. Ramey. We firstly remind the relationship between B and $\mathbb{R}^{n+1}_+$. The inversion $z \longrightarrow z^*$ of $\mathbb{R}^{n+1}\cup\{\infty\}$ relative to the unit sphere, where $z^* = z/|z|^2$ if $z \neq 0, \infty$; $z^* = 0$ if $z = \infty$; and $z^* = \infty$ if $z = 0$, is a conformal map on $\mathbb{R}^n\setminus\{0\}$. This map takes spheres containing the origin onto hyperplanes, and takes balls onto open half-spaces. Compositions of the inversion map with appropriate translations and dilations will give us a one-to-one map from B to $\mathbb{R}^{n+1}_+$. Of course there are many such maps, but the one we choose below has the advantage of being its own inverse under composition. Let $\mathbf{N}=(0,1)$ and $\mathbf{S}=(0,-1)$ where 0 denotes the origin in $\mathbb{R}^n$. We can think of $\mathbf{N}$ and $\mathbf{S}$ as the north and south poles, respectively, of the unit sphere S. Define $\Phi : \mathbb{R}^{n+1}\setminus\{\mathbf{S}\} \longrightarrow \mathbb{R}^{n+1}\setminus\{\mathbf{S}\}$ by $\Phi(z) = 2(z - \mathbf{S})^* + \mathbf{S}$. It is easy to see that Φ is a one-to-one map of $\mathbb{R}^{n+1}\setminus\{\mathbf{S}\}$ on itself. We can regard Φ as a homeomorphism of $\mathbb{R}^{n+1}\setminus\{\infty\}$ onto itself by defining $\Phi(\infty) = \mathbf{S}$ and $\Phi(\mathbf{S}) = \infty$. The next result summarizes the basic properties of Φ that we need to work as in B as in $\mathbb{R}^{n+1}_+$ at our convenience without producing any confusion.

PROPOSITION 2.4 The map Φ has the following properties:

(a) $\Phi(\Phi(z)) = z$ for all $z \in \mathbb{R}^{n+1}\cup\{\infty\}$.

(b) Φ is a conformal, one-to-one map of $\mathbb{R}^{n+1}\setminus\{\mathbf{S}\}$ onto $\mathbb{R}^{n+1}\setminus\{\mathbf{S}\}$.

(c) Φ maps B onto $\mathbb{R}^{n+1}_+$ and $\mathbb{R}^{n+1}_+$ onto B.

(d) Φ maps $S\setminus\{\mathbf{S}\}$ onto $\mathbb{R}^n$ and $\mathbb{R}^n$ onto $S\setminus\{\mathbf{S}\}$.

PROOF: See ([**ABR,** p. 134]).

The property of the tangential boundary F_t pointed out below is probably known. However, since we have not been able to find explicitly it in the literature we have prefered to include its details in order to make the paper selfcontained and understand better the geometry of the F_t.

PROPOSITION 2.5 If F is a relative closed subset of the unit ball B of $\mathbb{R}^{n+1}$, then its tangential boundary F_t has no interior.

PROOF: Via the transformation Φ, we can consider that F is a relatively closed subset of $\mathbb{R}^{n+1}_+$, and we will assume F_t has non-empty interior in $\mathbb{R}^n$. So, if I is an open set in $\mathbb{R}^n$ and $I \subset F_t$, without loss of generality we can suppose that $I = \{(x_1,\dots,x_n,y) \in \mathbb{R}^{n+1}_+ : y=0,\ 0<x_i<1 \quad i=1,..,n\}$. We take the sequence $\{Q^j_m\}$ of dyadic cubes in $\mathbb{R}^{n+1}$ defined by

$$Q^j_m = \{ (x,y) : \frac{j-1}{2^m} \le x_i \le \frac{j}{2^m} \quad i=1,..,n \ ; \ \frac{1}{2^m} \le y \le \frac{1}{2^{m-1}} \},$$

where $m = 1,2,3,\dots,$ and j is a positive integer ($1 \le j \le 2^m$). Select a cube $Q^{j_1}_{m_1}$ containing a point of F (a such cube exists because of $I \subset \overline{F} \cap \mathbb{R}^n$). Next, we look at those cubes (placed under $Q^{j_1}_{m_1}$) belonging to the m-generation ($m>m_1$) whose projections on $\mathbb{R}^n$ are included in $I_1 = \{(x,0) : \frac{j_1-1}{2^{m_1}} \le x_i \le \frac{j_1}{2^{m_1}}, i = 1,..,n\}$, and choose a cube $Q^{j_2}_{m_2}$ ($m_2>m_1$) such that $Q^{j_2}_{m_2} \cap F \neq \emptyset$. If $I_2 = \{(x,0) : \frac{j_2-1}{2^{m_2}} \le x_i \le \frac{j_2}{2^{m_2}},\ i = 1,..,n\}$ denotes the projection of $Q^{j_2}_{m_2}$ on $\mathbb{R}^n$, clearly $I_2 \subset I_1$. In the same way, we then look at those dyadic cubes of a posterior generation which are placed under $Q^{j_2}_{m_2}$ and proceed to pick a $Q^{j_3}_{m_3}$ intersecting to F, and so on. By this procedure, we construct a sequence $\{Q_k\}$ of dyadic cubes in $\mathbb{R}^{n+1}_+$, each one intersecting F, and a decreasing sequence $\{I_k\}$ of closed n-cubes in $\mathbb{R}^n$ with $\bigcap_{k=1}^{\infty} I_k = (x_0,0) \in I$.

Now, we are going to find a cone in $\mathbb{R}^{n+1}_+$ with vertex at this $(x_0,0)$ that contains a sequence of points in F converging to $(x_0,0)$, so that $(x_0,0)$ would not belong to F_t and we get a contradiction. Indeed, for any cube Q_k and any point $(a,b) \in \mathbb{R}^{n+1}_+$ ($a \in \mathbb{R}^n$, $b>0$) such that $(a,b) \in F \cap Q_k$, we have

$$|a-x_0| \le \frac{1}{2^{m_k}} \sqrt{n},$$

and

$$\frac{1}{2^{m_k}} \le b \le \frac{1}{2^{m_k - 1}},$$

so that $|a-x_0| \le b\sqrt{n}$. But then, the cone in $\mathbb{R}^{n+1}_+$ with vertex at $(x_0,0)$ defined by

$$\Delta((x_0,0)) = \{ (x,y) \in \mathbb{R}^{n+1}_+ : |x - x_0| \le \sqrt{n}\, y \}$$

contains all cubes Q_k for k large enough, and hence a sequence of points in $F \cap \Delta((x_0,0))$ converging to $(x_0,0)$. The contradiction is clear.

COROLLARY 2.6 If $\bar{F} \cap S$ has non-empty interior, then every open subset of $\bar{F} \cap S$ contains a dense set of points of nontangential approach from F.

3. Farrell Sets

We shall need to estimate from below the harmonic measure of some "spherical caps" for proving the main result of this Section (Theorem 3.2). Fixed $\alpha > 1$, we define for every $z \in B$ the set

$$J_z = \{ \zeta \in S : |\zeta - z| < \alpha(1-|z|) \}. \tag{3.1}$$

The harmonic measure $d\omega_z$ in S for a point z of B is defined by $d\omega_z(\xi) = P(z,\xi)\, d\sigma(\xi) = \dfrac{1-|z|^2}{|z-\xi|^{n+1}}\, d\sigma(\xi)$, and it will be useful record some simple facts:

(a) For any $\zeta \in S$, $J_{b\zeta} \subset J_{a\zeta}$, whenever $a<b<1$.

(b) If τ is a rotation in $\mathbb{R}^{n+1}$, then $\tau(J_z) = J_{\tau z}$ and $\omega_{\tau z}(J_{\tau z}) = \omega_z(J_z)$.

(c) Furthermore, it is well-known that if T_z is the spherical sector in B whose base is J_z, i.e., $T_z = \{ r\zeta : 0 \le r \le 1, \zeta \in J_z \}$, then $\sigma(J_z) = m(T_z)$ where m denotes the normalized Lebesgue measure in $\mathbb{R}^{n+1}$.

The next result is the n-dimensional version of a well known property of the harmonic measure for subsets in the unit circle. We think it can have an independent interest.

LEMMA 3.1 For every $\alpha > 1$ there exists a constant $c \in (0,1)$, depending only on the dimension and on α, such that $\omega_z(J_z) \ge c$ for every point z in B.

PROOF: For any point z in B, we have

$$\omega_z(J_z) = \int_{J_z} \frac{1-|z|^2}{|z-\xi|^{n+1}}\, d\sigma(\xi) \geq \frac{\sigma(J_z)}{\alpha^{n+1}(1-|z|)^n} \tag{3.2}$$

Since σ is rotation-invariant, by (b) above it is enough to estimate the right hand side in (3.2) for points z $(0<|z|<1)$ in any fixed direction, say the direction of $e_1 = (1,0,..,0)$; any lower bound for such points will also hold for all other z in B. By (c) we only need to find a lower bound for $m(T_z)$, and we consider the right cone C_z inside T_z generated by revolving round the axis of e_1 the isosceles triangle of height $h = 1 - \frac{(\alpha^2-1)(1-|z|)^2}{2|z|}$ and semi-base $r = (1-h^2)^{1/2}$. Let $k(\alpha)$ be a constant depending only on α such that $h \geq 1/4$ whenever $|z| \geq k(\alpha)$. Then $m(C_z) = r^n h$.

Hence, for $|z| \geq k(\alpha)$ we have

$$\omega_z(J_z) \geq \frac{m(C_z)}{\alpha^{n+1}(1-|z|)^n}$$

$$\geq \frac{h\left(\frac{(\alpha^2-1)(1-|z|)^2}{2|z|}\right)^{n/2}}{\alpha^{n+1}(1-|z|)^n}$$

$$> \frac{1}{2\alpha^{n+1}}\left(\frac{(\alpha^2-1)}{2}\right)^{n/2}.$$

On the other hand, for $|z|< k(\alpha)$, the property (a) above implies that

$$\omega_z(J_z) \geq \frac{\sigma(J_z)}{\alpha^{n+1}(1-k(\alpha))^n} \geq \frac{\sigma(J_{k(\alpha)e_1})}{\alpha^{n+1}(1-k(\alpha))^n}.$$

Therefore, if we do $c = \min\left\{ \frac{1}{2\alpha^{n+1}}\left(\frac{(\alpha^2-1)}{2}\right)^{n/2}, \frac{\sigma(J_{k(\alpha)e_1})}{\alpha^{n+1}(1-k(\alpha))^n} \right\}$ we are done.

We already are in position to establish the following result.

THEOREM 3.2 F is a Farrell set for $h^\infty(B)$ if and only if F satisfies the NTC.

PROOF: We shall follow ideas in **[St2]**. Suppose that $\sigma(F_t) > 0$ and we construct a bounded harmonic function u in B which cannot be pointwise boundedly approximated by harmonic polynomials with the same control on F than u, so that F would not be a Farrell set for $h^{\infty}(B)$.

Fix $\alpha > 1$. For $s > 0$ and $\zeta \in S$ we define the sets

$$\Gamma(\zeta,s) = \{w \in B : 0< 1-|w| \le s \ , \ |\zeta-w| < \alpha(1-|w|) \}.$$

and

$$E_s = \{\zeta \in \overline{F} \cap S : \Gamma(\zeta,s) \cap F = \emptyset \}$$

Note that the E_s's are closed subsets of S and $F_t = \bigcup_s E_s$. Consequently, there is a $E = E_{s_0}$ with $\sigma(E) > 0$. Let u be the Poisson integral of the boundary function $u(\zeta) = \chi_E(\zeta) + k\chi_{S\setminus E}(\zeta)$ of $L^{\infty}(S)$, where k is any fixed constant, $0 < k < 1$. It is clear that $\|u\|_{\infty} = 1$, and we claim that $\|u\|_F < 1$. Indeed, take any $z \in F$ with $|z| > 1-s_0$ so that $J_z \cap E = \emptyset$, where J_z is as in (3.1). By Lemma 3.1, there exists a constant $c \in (0,1)$ such that $\omega_z(J_z) \ge c$. Thus, we have the inequalities $\omega_z(E) \le 1-c$ and $\omega_z(S\setminus E) \ge \omega_z(J_z) \ge c$, and then

$$|u(z)| \le \int_E |u| \, d\omega_z + \int_{S\setminus E} |u| \, d\omega_z \le \omega_z(E) + k\, \omega_z(S\setminus E)$$
$$\le k + (1-k)(1-c) < 1.$$

for any $z \in F$ with $|z| > 1-s_0$. Since $\sup\{|u(z)| : |z| \le 1-s_0\} < 1$, we conclude that $\|u\|_F < 1$.

Suppose now that $\{p_n\}$ is a sequence of harmonic polynomials converging pointwise to u in B and $\|p_n\|_{\infty}$ tends to $\|u\|_{\infty}$ as $n \to \infty$. It remains to be shown that $\|p_n\|_F \longrightarrow 1$. Given $\varepsilon > 0$ arbitrary, by Fatou's theorem there exists a $z \in B$ such that

$$|u(z)| > 1 - \varepsilon, \tag{3.3}$$

and

$$\omega_z(E) > 1 - \varepsilon. \tag{3.4}$$

For such z, there exists $n_0 \in \mathbb{N}$ such that $|p_n(z) - u(z)| < \varepsilon$, whenever $n \ge n_0$. Hence

$$|p_n(z)| \ge |u(z)| - \varepsilon > 1 - 2\varepsilon, \tag{3.5}$$

whenever $n \ge n_0$ by (3.3). On the other hand

$$|p_n(z)| \le \int_E |p_n|\, d\omega_z + \int_{S\setminus E} |p_n|\, d\omega_z$$

$$\le \|p_n\|_E\, \omega_z(E) + \|p_n\|_\infty\, \omega_z(S\setminus E)$$

$$\le \|p_n\|_E + \varepsilon$$

by (3.4), and we get that $1 \ge \|p_n\|_F \ge \|p_n\|_E \ge 1-3\varepsilon$ for any $n \ge n_o$, by (3.5) and because of $E \subset \bar{F}$. This means that $\|p_n\|_F \longrightarrow 1$ as $n \to \infty$. Reciprocally, let $u \in h^\infty(B)$ and suppose that $\|u\|_\infty = 1$ and $\|u\|_F = \eta$. Let $\varepsilon > 0$ be arbitrary (but fixed) and consider the set

$$N_\varepsilon = \{\ g \in C(\bar{B}) : \|g\|_\infty \le 1+\varepsilon\ ,\ \|g\|_F \le \eta+\varepsilon\ \}.$$

It is enough to prove the following

Claim: Given a compact $J \subset B$, u can be uniformly approximated in J by harmonic polynomials belonging to N_ε.

In fact, taking the claim for granted, if $(J_m)_{m=1}^\infty$ is an increasing sequence of compact subsets of B such that $B = \bigcup_{m=1}^\infty J_m$ and given $\varepsilon_n = 1/n$ ($n = 1,2,3,\ldots$), for each J_m there is a sequence of harmonic polynomial $\{p_k^{n,m}\}_{k=1}^\infty$ that verify

$$\|p_k^{n,m}\|_\infty \le 1 + 1/n,$$

$$\|p_k^{n,m}\|_F \le \eta + 1/n, \qquad \text{and}$$

$$\|p_k^{n,m} - u\|_{J_m} < 1/k,$$

for k large enough. Then, for any $\varepsilon > 0$ and any compact subset $J \subset B$ the diagonal sequence $\{q_n\}_{n=1}^\infty = \{p_{n,n}^n\}_{n=1}^\infty$ satisfies that

1) $\|q_n\|_\infty < 1 + \varepsilon$
2) $\|q_n\|_F < \eta + \varepsilon$
3) $\|q_n - u\|_J < \varepsilon$

for n large enough and F must be a Farrell set for $h^\infty(B)$.

Proof of the claim:

In the space $C(\bar{B})$ of all those continuos functions in the closed unit ball we define the norm

$$\|g\|_* = \max\ \{\ (1+\varepsilon)^{-1}\|g\|_\infty\ ,\ (\eta+\varepsilon)^{-1}\|g\|_F\}$$

which is equivalent to the uniform norm. We denote by $\mathcal{P}_h$ the closure of the harmonic polynomials in $(C(\bar{B}), \|\cdot\|_*)$ norm, and we have to prove that u is in the closure of $\mathcal{P}_h \cap N_\varepsilon$. Since $\mathcal{P}_h \cap N_\varepsilon$ is a convex set of $C(\overline{B})$, by the Hahn–Banach theorem we have only to to show that $|\mu(u)| \le 1$ for every complex regular Borel measure μ with compact support in B such that $|\mu(g)| \le 1$ for any $g \in \mathcal{P}_h \cap N_\varepsilon$, (as usual, we identify measures with linear functionals).

Since $\mathcal{P}_h$ is a subspace of $C(\overline{B})$, the functional μ admits an preserving-norm continuous extension to $C(\overline{B})$ which can be represented by a complex measure ν with compact support in $\overline{B}$. Now, $\|\mu\|_{\mathcal{P}_h} \le 1$, ($\|\mu\|_{\mathcal{P}_h}$ is, of course, the norm of μ on $\mathcal{P}_h$) because N_ε is the unit ball of $(C(\overline{B}), \|\cdot\|_*)$; hence, $\|\nu\| \le 1$. We observe that $\|\nu\| = |\nu|(B) = 1$, and therefore

$$|\nu|(\overline{B}\backslash\overline{F}) + \eta|\nu|(\overline{F}) \le 1. \tag{3.6}$$

We assert that the restriction of ν to S is absolutely continuous with respect to the Lebesgue measure in S. Indeed, if K is any compact subset of S and $\sigma(K) = 0$, then there exists a sequence $\{a_n\} \subset \mathcal{P}_h$ such that $\|a_n\| = 1$, $a_n = 0$ in K and $a_n \dashrightarrow 1$ in $\overline{B}\backslash K$, so that $(\nu-\mu)\ (a_n) = 0$ for all n; by the dominated convergence theorem, as $n \to \infty$ we have that

$$0 = (\nu-\mu)\ (a_n) \longrightarrow (\nu-\mu)\ (K) = \nu(K)$$

because supp $\mu \subset B$ and $K \subset S$.

Therefore, u is defined almost everywhere, modulus ν, and $|u| \le 1$ ν-a.e. of $\overline{B}\backslash\overline{F}$, and $|u| \le \eta$ ν-a.e. of $\overline{F}$, (here we have used the hypothesis of NTC over F). All these conditions guarantee that $\nu(u)$ is defined.

If $\{p_n\} \subset \mathcal{P}_h$ is a uniformly bounded sequence that converges pointwise to u in B and a.e. in S, then, by an argument of dominated convergence we have

$$\mu(u) = \lim \mu(p_n) = \lim \nu(p_n) = \nu(u).$$

This equality, together with (3.6) and the estimates of u (with respect to ν) in $\overline{B}$ and $\overline{F}$, give

$$|\mu(u)| = |\nu(u)| \le |\nu|(\overline{B}\backslash\overline{F}) + \eta|\nu|(\overline{F}) \le 1$$

and the proof is complete.

4. Mergelyan Sets

When F_t has positive measure, a careful reading of the proof of Proposition 3.1 in (**[De,** p. 332**]**) lead us to make up a certain subset H of F_t with $\sigma(H) > 0$ that is crucial for showing below that NTC is also a necessary condition for F to be a Mergelyan set for $h^\infty(B)$. The construction of such set H runs as follows:

Suppose $\sigma(F_t) > 0$. For simplicity, via the transformation Φ in Section 2, we place F_t on the boundary of the $\mathbb{R}^{n+1}_+$ and we develop our construction there. Fix a sequence $\{\varepsilon_j\}$, $0 < \varepsilon_j < 1$, verifying

$$\prod_{j=1}^{\infty}(1-\varepsilon_j) = 1/2. \tag{4.1}$$

For each $z = (z,0) \in F_t$, $\Gamma(z,h(z),\theta)$ denotes the truncated cone with vertex at z, height h(z) and amplitude θ, i. e.

$$\Gamma(z,h(z),\theta) = \{(x,y) \in \mathbb{R}^{n+1}_+ : |z-x| \le (\mathrm{tg}\theta)\ y,\ 0 \le y \le h(z)\}.$$

Take, for instance, the sequence of angles $\{\theta_k\} = \{\pi/2^{k+2}\}$. For $\theta_1 = \pi/8$ and for each $z \in F_t$, we fix $h_1(z) > 0$ satisfying $\Gamma(z,h_1(z),\theta_1) \cap F = \emptyset$. Then, for every $j = 1,2,\ldots$, we consider the sets

$$E_{1,j} = \{\ z \in F_t : h_1(z) \ge \frac{1}{j}\ \}.$$

It is clear that

$$\ldots \subset E_{1,j} \subset E_{1,j+1} \subset \ldots$$

and

$$F_t = \bigcup_{j=1}^{\infty} E_{1,j}.$$

Now, we select $j_1>0$ and a compact subset $E_1 \subset E_{1,j_1}$ such that

$$\sigma(E_{1,j_1}) > \sigma(E_1) > (1-\varepsilon_1)\sigma(F_t)$$

and

$$\Gamma(z,1/j_1,\theta_1) \cap F = \emptyset,$$

for any $z \in E_1$. Next, for $\theta_2 = \pi/16$ and for each $z \in E_1$, by repeating the process, we pick $j_2 > j_1$ and a compact set $E_2 \subset E_1$ satisfying:

$$\sigma(E_2) > (1-\varepsilon_2)\sigma(E_1) > (1-\varepsilon_2)(1-\varepsilon_1)\sigma(E)$$

and

$$\Gamma(z,\ 1/j_2,\ \theta_2) \cap F = \emptyset,$$

for any $z \in E_2$. By this procedure we generate a sequence $\{E_k\}$ of compacts sets decreasing to the closed set $H = \bigcap E_k$. Note that $\sigma(H) > \frac{1}{2}\sigma(F_t) > 0$ by (4.1), and fixed an arbitrary angle θ_k, there exists a j_k so large we please such that

$$\Gamma(z,1/j_k,\theta_k) \cap F = \emptyset \tag{4.2}$$

for all z in H.

THEOREM 4.1 If F is a Mergelyan set for $h^{\infty}(B)$ then F satisfies the NTC.

PROOF: Assume that $\sigma(F_t) > 0$, and let H' be a subset of F_t as obtained above with $\sigma(H') > 0$. Let $u \in h^{\infty}(B)$ be the Poisson integral of the function defined as u = 1 on H' and u = k on S\H', where k is constant, $0<k<1$.

It is clear that $\|u\|_{\infty} = 1$ and take for granted that u is uniformly continuous on F. Just as in proving Theorem 3.2, we can check that $\|u\|_F < 1$. However, if for this function u there exists a sequence of harmonic polynomials $\{p_n\}$ satisfying that

$$\text{i) } p_n(z) \longrightarrow u(z) \quad \text{for all } z \in B,$$
$$\text{ii) } \|p_n\|_{\infty} \longrightarrow \|u\|_{\infty}, \quad \text{and}$$
$$\text{iii) } \|p_n - u\|_F \longrightarrow 0,$$

as $n \to \infty$; (iii) would imply that $\|p_n\|_F \longrightarrow \|u\|_F$, so that the sequence $\{\|p_n\|_F\}$ should converge to $\|u\|_F$. But it can be shown with the same arguments that were used in the proof of Theorem 3.2 that $\|p_n\|_F \longrightarrow 1$, and it would mean that F cannot be a Mergelyan set for $h^{\infty}(B)$. Thus we will be ready as soon as it was shown that, in fact, u is uniformly continuous on F.

Since H' is a compact subset of S and $\sigma(H') > 0$, we have that S\H' can be represented as a countable union of disjoint open subsets of S, say S\H' = $\bigcup I'_n$, with $\sum \sigma(I'_n) < 1$. Furthermore, u(z) extends continuously from F to each point on $\overline{F} \cap (\bigcup I'_n)$ because $u \equiv k$ on every I'_n, and we will show that u even extends with continuity from F to *every* point on H', so that u is a continuous function in $\overline{F}$ and hence u is uniformly continuous on F. In other words, we have to prove that $\lim_{m\to\infty} u(z_m) = k$ for every sequence $\{z_m\} \subset F$ convergent to any point in H'.

Put $H = \Phi(H')$, and $\Phi(z_m) = (x_m,y_m)$ with $x_m \in \mathbb{R}^n$ and $y_m > 0$, for any m. Let M be a sufficiently large constant satisfying

$$\text{arctag } M > \left(1 - \frac{\varepsilon}{2}\right)\frac{\pi}{2}\,, \tag{4.3}$$

and fix an $k_o \in \mathbb{N}$ so that $\text{ctg } \theta_k > M$ for all $k \geq k_0$, where $\theta_k = \pi/2^{k+2}$. Then, by (4.2), it follows that $(x_m,y_m) \notin \bigcup_{z\in H} \Gamma(z,1/j_{k_o},\theta_{k_o})$, for all m bigger than some m_0. We are now in position of drawing a cone $\mathbb{R}^{n+1}_+$ with vertex at (x_m,y_m) and whose base is the ball $J_m = B(x_m,r_m) \in \mathbb{R}^n$ centered at x_m and radius r_m satisfying

$$\frac{r_m}{y_m} = \operatorname{ctg} \theta_{k_o} > M, \tag{4.4}$$

for $m \geq m$. If we call $J' = \Phi(J\ )$ then $u \equiv k$ on J', and we can write

$$u(z_m) = k \int_{J'_m} P(z_m,t)\, d\sigma(t) + \int_{S\setminus J'_m} P(z_m,t)\, u(t)\, d\sigma(t).$$

It will be enough to show that given $\varepsilon > 0$ there is a positive integer m_0 such that

$$1-\varepsilon < \int_{J'_m} P(z_m,t)\, d\sigma(t) < 1, \tag{4.5}$$

for $m \geq m_0$. Indeed, since u is bounded, it would imply that $\lim_{m\to\infty} \int_{S\setminus J'_m} P(z_m,t)u(t)d\sigma(t) = 0$ and then $\lim_{m\to\infty} u(z_m) = k$.

After this preparation, we will estimate the integral in (4.5) by using, in essence, the change of variables formula between B and $\mathbb{R}^{n+1}_+$ and the relationship between their respective Poisson kernels (see **[ABR,** p. 148 and ch. VII]).

$$\int_{J'_m} P(z_m,t)\, d\sigma(t) = c_{n+1}(|x_m|^2+(y_m+1)^2)^{\frac{n-1}{2}} \int_{J_m} \frac{y_m}{(1+|t|^2)^{\frac{n-1}{2}} (|x_m-t|^2+y_m^2)^{\frac{n+1}{2}}}\, dt$$

$$\geq c_{n+1}(|x_m|^2+(y_m+1)^2)^{\frac{n-1}{2}} \int_B d\sigma \int_0^{r_m} \frac{r^{n-1}\, y_m}{(1+(|x_m|+r)^2)^{\frac{n-1}{2}} (r^2+y_m^2)^{\frac{n+1}{2}}}\, dr$$

$$\geq \frac{(|x_m|^2+(y_m+1)^2)^{\frac{n-1}{2}}}{(1+(|x_m|+r_m)^2)^{\frac{n-1}{2}}}\, c_{n+1}\omega_n \int_0^{r_m} \frac{r^{n-1} y_m}{(r^2+y_m^2)^{\frac{n+1}{2}}}\, dr, \tag{4.6}$$

where c_{n+1} and ω_n are the normalizing constants of the Poisson kernel for the upper half-space and ω_n is the area surface of the unit ball of $\mathbb{R}^n$, respectively.

The first factor in (4.6) goes to one as $m \to \infty$; hence we must show that the second one is bigger than $1-\varepsilon$ for m large enough. Now, if we call

$$P_n(r) = \int \frac{r^{n-1}}{(r^2+y_m^2)^{\frac{n+1}{2}}}\, dr,$$

a simple integration by parts produces the recurrence formula

$$P_n(r) = \left(\frac{n-2}{n-1}\right) P_{n-2}(r) - \frac{1}{n-1} \frac{r^{n-2}}{(r^2+y_m^2)^{\frac{n-1}{2}}} \tag{4.7}$$

for $n \geq 3$, and it is easy to compute that

$$P_1(r) = \frac{1}{y_m} \operatorname{arctg}(r/y_m), \qquad \text{and} \qquad P_2(r) = \frac{-1}{(r^2+y_m^2)^{\frac{1}{2}}}.$$

Therefore, the factor in (4.6) to be estimated can be written as

$$c_{n+1}\omega_n \int_0^{r_m} \frac{r^{n-1} y_m}{(r^2+y_m^2)^{\frac{n+1}{2}}}\, dr = c_{n+1}\omega_n\, y_m \Big(P_n(r_m) - P_n(0)\Big). \tag{4.8}$$

For analyzing the behaviour of $P_n(r_m)$ and $P_n(0)$ when m is large, we use (4.7) and proceed by induction over n separately according to n be even or odd. First, if n is odd, we observe that $P_1(0) = P_3(0) = 0$, and follows from (4.7) that $P_n(0) = 0$. On the other hand, by using (4.3) it can be shown that

$$c_{n+1}\omega_n\, y_m\, P_n(r_m) \geq 1 - \varepsilon \tag{4.9}$$

for m sufficiently large. The identity $c_{n+1}\omega_n y_m P_n(0) = -1$ holds for any n even; and a straightforward computation and (4.4) yield $-\varepsilon \leq c_{n+1}\omega_n y_m P_n(r_m) \leq 0$, when m is large enough. Hence

$$c_{n+1}\omega_n\, y_m \Big(P_n(r_m) - P_n(0)\Big) = 1 + c_{n+1}\omega_n\, y_m\, P_n(r_m) \geq 1 - \varepsilon \tag{4.10}$$

for n even. (4.9) and (4.10) show that the inequality (4.5) holds and the proof is finished.

ACKNOWLEGMENTS. We are very grateful to José L. Fernández, Manuel Flores and Arne Stray for several interesting discussions hold with them while doing this paper. Both authors were partially supported by the grant PS89-0135 from DGICYT of the Ministerio de Educación y Ciencia of Spain.

References

[ABR] S. AXLER, P. BOURDON and W. RAMEY, *Harmonic Function Theory*, Springer Verlag, New York, 1992.

[De] J. DETRAZ, Algèbres de fonctions analytiques dans le disque, *Ann. Scient. Ecole Norm Sup.* 4e serie (1970), 313–352.

[OFPG] A.G. O'FARRELL and F. PEREZ-GONZALEZ, Pointwise bounded approximation by polnomials, *Math. Proc. Camb. Phil. Soc.* **112**, (1992), 147–155.

[PG] F. PEREZ-GONZALEZ, H^p joint approximation, *Proc. Amer. Math. Soc.* **102,** (1988), 577–580.

[PGS] F. PEREZ-GONZALEZ and A. STRAY, Farrell and Mergelyan sets for H^p spaces ($0<p<1$), *Michigan Math. J.*, **36** (1989), 379–386.

[R] L.A. RUBEL, Joint approximation in the complex domain, *Symposium on Complex Analysis Canterbury 1973,* (J. Clunie and W.K. Hayman, eds.), Cambridge University Press, London Mathematical Society, Lect. Notes Series **12**, (1974), 115–118.

[RS] L.A. RUBEL and A. STRAY, Joint approximation in the unit disc, *J. Approx. Theory* **37** (1983), 44–50.

[SW] E. M. STEIN and G. WEISS, *Fourier Analysis on euclidean spaces.* Princeton University Press, 1971.**[St1]** A. STRAY, Characterization of Mergelyan sets, *Proc. Amer. Math. Soc.* **44** (1974), 347–352.

[St2] A. STRAY, Pointwise bounded approximation by functions satisfying a side condition, *Pacific J. Math.* **51** (1974), 301–305.

[St3] A. STRAY, Vitushkin's method and some problems in approximation theory, *Aspects of Contemporary Complex Analysis* (D.A. Brannan, J. Clunie, eds.) Academic Press, London, (1980), 351–365.

Méthodes Analytiques en dimension infinie

ARNAUD DE LA PRADELLE
Equipe d'analyse
Université Paris VI
75252 Paris 5
France

1. Introduction. De nombreux auteurs ([1],[2],[3],[9],[10],[11],[13]) ont étudié l'EDS du type de Langevin

$$dX_t = \sqrt{2a} dW_t - AX_t dt \tag{1}$$

où W_t est un mouvement brownien à valeurs dans un espace de Banach E muni d'une mesure gaussienne centrée ξ. Une méthode probabiliste classique consiste d'abord à montrer l'existence d'un processus cylindrique X_t à valeurs dans E, et à montrer ensuite la continuité des trajectoires.

La méthode analytique que nous utiliserons ici (*cf.* [7],[8]) consiste à résoudre le problème de Cauchy ordinaire

$$v'(t) = \sqrt{2a} u'(t) - Av(t)dt \tag{2}$$

relativement à l'espace de Sobolev $\mathcal{H} = W_0^2(\mathbf{R}^+, H)$ où $H \subset E$ désigne l'espace de Cameron-Martin associé à ξ.

La solution de (1) apparaît alors comme le prolongement de (2) à $\mathcal{C}(\mathbf{R}^+, E)$, selon un résultat général sur les prolongements d'opérateurs linéaires que nous rappelons.

2. Hypothèses et rappels sur les prolongements d'opérateurs linéaires.

E est un espace localement convexe lusinien et ξ est une mesure gaussienne centrée sur E chargeant tous les ouverts de E.

On note H' l'adhérence de E' dans $L^2(\xi)$ et H son dual qui est l'espace de Cameron-Martin et est compactement inclus dans E (*cf.* [4]).

Exemples. 1) On prend $E = \mathbf{R}^{\mathbf{N}}$ et ξ la mesure gaussienne canonique. On a $H = \ell^2$.
2) On prend $E = \mathcal{C}_0(\mathbf{R}^+)$ et ξ la mesure de Wiener. On a $H = W_0^2(\mathbf{R}^+)$.
3) On prend $E = \mathcal{S}'(\mathbf{R})$ et ξ la mesure du bruit blanc. On a $H = L^2(\mathbf{R}, dx)$.
4) Si E est fortement convexe (i.e. tel que l'enveloppe convexe fermée de tout compact est compacte), alors il existe une mesure gaussienne centrée μ sur un sous-espace Ω de type K_σ de $\mathcal{C}(\mathbf{R}^+, E)$ tel que pour toute $f \in E'$ le processus $f(W_t)$ soit

K. GowriSankaran et al. (eds.), Classical and Modern Potential Theory and Applications, 413–417.

un brownien de variance $t|f|^2_{H'}$, où les W_t sont les coordonées sur Ω. L'espace de Cameron-Martin est alors l'espace $W^2_0(\mathbf{R}^+, H)$ (*cf.* [7],[8]).

Si $u \in H$, on note k_u la fonction définie par $k_u(x) = \exp(\tilde{u}(x) - |u|^2/2)$ où $\tilde{u}$ est l'élément de H' correspondant à u.

1. PROPOSITION: *si $f \in H'$ il existe $\tilde{f}$ linéaire ξ-mesurable sur E, essentiellement unique et tel que*

$$\tilde{f}(u) = \int f k_u d\xi = \langle f, u\rangle_{H',H}.$$

Réciproquement, si f est linéaire ξ-mesurable, alors f appartient à H'.

2. PROPOSITION: *soit T un opérateur linéaire de E et scalairement ξ-mesurable. On a les propriétés suivantes: a) T est mesurable au sens de Lusin, c'est à dire que pour tout $\varepsilon > 0$, il existe un compact K tel que $\xi(E \backslash K) < \varepsilon$, et que la restriction de T à K soit continue. b) il existe un sous-espace borélien E_0 de E portant ξ et tel que la restriction de T à E_0 soit borélienne.*

Le résultat fondamental que nous utiliserons est le suivant

3. THÉORÈME: *soit A un opérateur borné de H. Il existe un opérateur $\tilde{A}$ linéaire et ξ-mesurable de E essentiellement unique et prolongeant A. De plus tout sous-espace portant ξ porte aussi $\tilde{A}(\xi)$. On dit que $\tilde{A}$ est le prolongement essentiel de A.*

Réciproquement, tout opérateur linéaire ξ-mesurable A de E tel que tout sous-espace portant ξ porte aussi $A(\xi)$ induit par restriction un opérateur linéaire continu de H.

4. COROLLAIRE: *soient A_1 et A_2 deux opérateurs bornés de H. Alors $\tilde{A}_2 \circ \tilde{A}_1$ est ξ-mesurable et vaut $A_2 \circ A_1$ presque partout. En particulier, si A est bijectif, $\tilde{A}$ est "presque bijectif".*

Citons une première application:

5. THÉORÈME: (quantification) *soit A un opérateur de H, de norme ≤ 1. Posons $S = \sqrt{I - A^*A}$. On définit $\Gamma(A)$ par la "formule de Mehler"*

$$\Gamma(A)f(x) = \int f(A^*x + Sy)d\xi(y)$$

pour toute fonction $f \geq 0$ et borélienne. Dans cette formule, A^* et S désignent les prolongements essentiels. Alors aux ensembles négligeables près, $\Gamma(A)f$ ne dépend *pas du prolongement essentiel particulier de A. De plus pour tout* $p \geq 1, \Gamma(A)$ *est l'unique opérateur sous-markovien et contractant de* $L^p(E, \xi)$ *vérifiant*

$$\Gamma(A)k_u = k_{Au}$$

pour tout $u \in H$. *On a de plus* $\Gamma(AB) = \Gamma(A)\Gamma(B)$.

3. Processus d'Ornstein-Uhlenbeck.

Nous nous plaçons dans le cadre de l'exemple 4). On va d'abord résoudre le problème de Langevin

$$dX_t = dW_t - AX_t dt \tag{3}$$

où A est le générateur d'un semi-groupe hermitien C_t de type $\mathcal{C}_0$ sur H.

6. THÉORÈME: *l'équation (3) possède la solution faible*

$$X_t(x, \omega) = C_t x + \int_0^t C_{t-s} dW_s(\omega) \tag{4}$$

où l'intégrale du second membre converge au sens de Wiener faible. Pour $\xi \otimes \mu$ *tout* $(x, \omega) \in E \times \Omega$ *la trajectoire* X *est continue à valeurs dans* E. *Il existe plus précisément un sous-espace borélien* E_1 *de* E *portant* ξ, *et un sous-espace borélien* Ω_1 *de* Ω *portant* μ, *tels que l'application* $(x, \omega, t) \to X_t(x, \omega)$ *soit borélienne sur* $E_1 \times \Omega_1 \times \mathbf{R}^+$ *et continue en* t.

Démonstration: on considère le problème de Cauchy classique

$$v'(t) = u'(t) - Av(t)dt \tag{5}$$

avec la condition initiale $v(0) = 0$. Il est classique que cela définit un opérateur $u \to [t \to v(t) = \int_0^t C_{t-s} u'(s) ds]$ qui est borné sur $\mathcal{H}$. Notons X^0 son extension linéaire μ-mesurable en opérateur de Ω. Il est clair que l'on a au sens faible

$$X_t^0(\omega) = \int_0^t C_{t-s} dW_s(\omega). \tag{6}$$

D'autre part, pour tout $\varepsilon > 0$ l'opérateur $u \to [t \to C_{t+\varepsilon} u]$ est un opérateur borné de H dans $\mathcal{H}$. Il a donc un prolongement linéaire ξ-mesurable soit $x \to [t \to C_{t+\varepsilon} x]$ de E dans Ω. On en déduit un processus $C_t x$ à trajectoires continues sur $]0, +\infty[$. On peut en fait montrer ([7],[8]) la continuité jusqu'à l'origine en utilisant convenablement un théorème ergodique de Stein ([14]).

On peut alors énoncer

7. THÉORÈME: *soit de plus a un opérateur hermitien positif (non nécessairement borné), et qui commute avec A. On suppose que* E' *est inclus dans le domaine de*

l'opérateur $\sqrt{a}$, et que $\sqrt{a}$ est prolongeable en opérateur linéaire ξ-mesurable de E dans E. Alors le problème de Cauchy stochastique

$$dX_t = \sqrt{2a}dW_t - AX_tdt \tag{1}$$

a la solution faible

$$X_t(x,\omega) = C_tx + \sqrt{2a}\int_0^t C_{t-s}dW_s \tag{7}$$

avec les mêmes propriétés de régularité qu'au théorème 6.

Si l'on a de plus $a = A$, alors le processus X, admet le semi-groupe d'Ornstein-Uhlenbeck généralisé

$$\Gamma(C_t)f(x) = \int f(C_tx + S_ty)d\xi(y) \tag{8}$$

comme semi-groupe de transition. ($S_t = \sqrt{I - C_{2t}}$). En ce cas, la forme de Dirichlet associée vaut

$$\int\int f'(x,\sqrt{A}u)g'(x,\sqrt{A}u)d\xi(x)d\xi(u). \tag{9}$$

BIBLIOGRAPHIE

1. S. Albeverio, M. Röckner, *Stochastic differential equations in infinite dimensions: solutions via Dirichlet forms*, Probability theory and related fields, Springer-Verlag, 1991.

2. A. Antoniadis, R. Carmona, *Eigenfunctions Expansions for Infinite Dimensional Ornstein-Uhlenbeck Processes*, Probability Theory Rel. Fields, 74, (1987), p.31-54.

3. G. Da Prato, J. Zabczyck, *Stochastic equations in infinite dimensions*, Cambridge University Press, no. 44, 1992.

4. D. Feyel, A. de La Pradelle, *Capacités gaussiennes*, Ann. Inst. Fourier, t.41, f.1, p.49-76, 1991.

5. D. Feyel, A. de La Pradelle, *Opérateurs linéaires et espaces de Sobolev sur l'espace de Wiener*, CRAS, t.316, série I, p.227-229, 1991.

6. D. Feyel, A. de La Pradelle, *Opérateurs linéaires gaussiens*, à paraître in Potential Analysis, 1993.

7. D. Feyel, A. de La Pradelle, *Processus browniens de dimension infinie*, Preprint de l'Université d'Evry-Val d'Essonne, 1993.

8. D. Feyel, A. de La Pradelle, *Brownian Processes in Infinite Dimension* à paraître in Potential Analysis, 1994.

9. X. Fernique, *La régularité des fonctions aléatoires d'Ornstein-Uhlenbeck à valeurs dans ℓ^2; le cas diagonal*, CRAS, Paris, 309, série I, 1989, p.59-62.

10. X. Fernique, *Fonctions aléatoires dans les espaces lusiniens*, Expositiones Math., 8, no.4, 1990, p.289-364.

11. I. Iscoe, D. McDonald, M. Talagrand, J. Zinn, *Continuity of ℓ^2-valued Ornstein-Uhlenbeck processes*, Ann. Prob. 18(1990), p.68-84.

12. S. Kusuoka, *Dirichlet forms and Diffusion processes on Banach spaces*, J. Fac. Sci. Univ. Tokyo, sec IA, 29 (1982).

13. A. Millet, W. Smoleński, *On the continuity of Ornstein-Uhlenbeck processes infinite dimension*, à paraître, 1992.

14. E.M. Stein, *Singular Integrals and differentiability properties of functions*, Princeton University Press, 1970.

Construction d'un processus à deux paramètres à partir d'un semigroupe à un paramètre

Shiqi SONG, Université Evry Val d'Essonne, Boulevard des Coquibus, 91025 EVRY Cedex, FRANCE

Résumé. Supposons que J est soit R, soit une demi-droite ouverte. Supposons que μ est une mesure σ-finie sur J. Pour un semigroupe classique P_t sur J donné (voir §2) qui est μ-symétrique, on prouve que la propriété de c-semigroupe (voir §3) est une condition nécessaire et suffisante pour qu'un processus Z à deux paramètres à valeurs dans J existe, qui satisfait les conditions suivantes :

a) La filtration $F_{s,t} = \sigma\{Z_{u,v}, u \in [0,s], v \in [0,t]\}$ satisfait la condition (F4).

b') : Le processus $t \to (Z_{s,t}, s \geq 0)$ est un processus de Markov symétrique.

c') : Pour s fixé, $t \to Z_{s,t}$ est un processus de Markov de semigroupe de transition P_t.

d) Pour tout couple (s,t), la loi de $Z_{s,t}$ est μ.

On montre ensuite que la capacité $c_{2,2}$ associé avec un c-semigroupe P_t lipchitzien (voir §8) est équivalente à la probabilité de rencontre du processus Z avec a-b'-c'-d.

Mots clés. Forme de Dirichlet sur un espace vectoriel, capacité, processus de Markov symétrique, processus à deux paramètres, forme de Dirichlet image, mouvement brownien, processus de Bessel.

AMS classification des sujets. 31C15, 60J25, 60J30, 60J45.

K. GowriSankaran et al. (eds.), Classical and Modern Potential Theory and Applications, 419–451.

§1 Introduction

Dans ce papier, certains problèmes concernant les capacités des espaces de Sobolev d'indices (2,2) vont être étudiés. Les capacités des espaces de Sobolev d'indices (r,p) ont fait l'objet des nombreux travaux, cf. par exemple, Fukushima-Kaneko [6], Malliavin [11], Ren [12], DelaPradelle-Feyel [5], Sugita [16], Kazumi-Shigekawa [10], Hirsch [8], Song [14], etc.

Soit J un espace topologique. Soit μ une mesure σ-finie sur J. Soit P_t un semigroupe de Markov sur J symétrique par rapport à μ. Nous définissons l'opérateur :

$$Uf(x) = \frac{1}{\sqrt{\pi}} \int_0^\infty t^{-1/2} e^{-t} P_t f(x)\, dt, \quad f \in L^2(J,\mu),$$

et l'espace de Sobolev $W^{2,2} = U^2(L^2(J,\mu))$. A partir de l'espace de Sobolev $W^{2,2}$, nous pouvons définir une capacité $c_{2,2}$ sur J.

Dans Hirsch [8] et Song [14], on s'est intéressé au problème de donner une interprétation probabiliste à la capacité $c_{2,2}$. Alors, lorsque le semigroupe P_t est un semigroupe brownien ou un semigroupe d'Ornstein-Uhlenbeck, un processus continu Z à deux paramètres à valeurs dans J a été construit tel que les conditions suivantes sont satisfaites :

a) La filtration (à deux indices) $F_{s,t} = \sigma\{Z_{u,v}, u \in [0,s], v \in [0,t]\}$ satisfait la condition (F4) (voir [18]).

b)
b' : Par rapport à la filtration $F_{\infty,t}$, l'application $t \to (Z_{s,t}, s \geq 0)$ est un processus de Markov, symétrique.
b" : Par rapport à la filtration $F_{s,\infty}$, l'application $s \to (Z_{s,t}, t \geq 0)$ est un processus de Markov, symétrique.

c)
c' : Pour s fixé, par rapport à la filtration $F_{\infty,t}$, l'application $t \to Z_{s,t}$ est un processus de Markov de semigroupe de transition P_t.
c" : Pour t fixé, par rapport à la filtration $F_{s,\infty}$, l'application $s \to Z_{s,t}$ est un processus de Markov de semigroupe de transition P_t.

d) Pour tout couple (s,t), la loi de $Z_{s,t}$ est μ.

Et le théorème suivant a été démontré :

Théorème 1 : Soit Z un processus qui vérifie les conditione a-b-c-d en relation avec le semigroupe P_t. Alors, les trois inégalités suivantes sont valables :

$$|| \sup_{(s,t)\in R_+^2} e^{-s-t} |u|(Z_{s,t}) ||_2 \leq C\, c_{2,2}(u),$$

$$|| \sup_{(s,t)\in R_+^2} \left| \int_{u\in[0,s],v\in[0,t]} e^{-u-v} f(Z_{u,v}) du dv \right| ||_2 \leq C ||U^2 f||_2,$$

$$c_{2,2}(A) \leq P[\exists (s,t) \in R_+^2, Z_{s,t} \in A],$$

où C est une constante, et A est un borélien dans J. □

Maintenant nous nous intéressons à un semigroupe markovien P_t quelconque. Est-ce qu'il est toujours possible de construire un processus Z qui satisfait les conditions a-b-c-d afin d'obtenir les trois inégalités précédentes ? La réponse dépend certainement des propriétés de P_t. Mais, quelles sont-elles ? Ces questions méritent bien une explication. Alors, quelle sera la théorie appropriée pour en donner des réponses ?

Dans une situation extrêmement simple où l'espace J est soit R, soit une demi-droite, nous allons donner des réponses aux questions posées ci-dessus. Nous utilisons la théorie des espaces de Dirichlet.

Dans le §2, la notion de semigroupe classique est introduite.

Dans le §3, la notion de c-semigroupe est introduite. Il sera ensuite démontré que la propriété de c-semigroupe est une condition nécessaire pour qu'un semigroupe classique P_t puisse donner lieu à un processus à deux paramètres avec les conditions a-b'-c'-d.

Dans les sections §4 à §7, on démontrera que la propriété de c-semigroupe est aussi une condition suffisante.

Dans les sections §8 et §9, sous une condition technique, on démontrera que les conditions a-b-c-d sont équivalentes aux conditions a-b'-c'-d.

Dans le §10, on démontrera que les conditions a-b-c-d entrainent les trois inégalités du Théorème 1.

Dans notre étude, certains problèmes de base de la théorie de l'espace de Dirichlet seront concernés. Notamment, interviendront dans notre texte la fermabilité des pré-formes de Dirichlet sur un espace de Banach et l'existence du processus de Markov associé à une forme de Dirichlet sur un espace de Banach. Les formes de Dirichlet considérées dans notre travail n'ont pas été traitées ailleur. Nos résultats sur la fermabilité (cf. §4) et l'existence du processus de Markov (cf. §5) sont nouveaux dans la matière.

Techniquement, la notion de la forme de Dirichlet image introduite par Bouleau-Hirsch [4] joue un rôle central dans notre étude.

Terminons cette section par la remarque suivante : les semigroupes de Bessel de dimensions entières (qui sont, d'après Yor [17], symétriques par rapport à ses mesures de vitesse) sont des c-semigroupes, ainsi que les solutions des équations différentielles stochastiques $dX_t = dB_t + b(X_t)\,dt$, si $b \in C_b^\infty(R)$ avec $b' \leq 0$.

§2 Quelques notations pour commencer

Pour faciliter l'exposé de notre travail, on commence par introduire quelques notations de base utilisées dans le texte. On considère des différents ensembles :

J est soit R, soit une demi-droite;
$E = C(R_+,J)$ muni de la topologie de Fréchet standard;
$\Xi = C(R_+,E)$;
D est l'ensemble des nombres diadiques positifs ou nul;
$F = \bar{J}^D$ muni de la topologie produite, où $\bar{J}$ est la fermeture de J dans $[-\infty,\infty]$.

On va identifier Ξ avec $C(R_+^2,J)$. Ainsi, les éléments de Ξ pourront être considérés comme des fonctions de deux variables : la première variable désigne la coordonnée de E (coordonnée horizontale), la deuxième variable désigne la coordonnée de Ξ (coordonnée verticale). Les points de J ou de F seront notés par x, y, z, etc. Lorsque $x \in F$, x_t, $t \in D$, désignera la t^e-coordonnée de x. Un élément de E sera désignés par ξ. Les notations ξ_t, $t \in R_+$, désigneront les coordonnées de ξ. Quand G désigne un sous-ensemble de R_+, ξ_G désigne la famille des coordonnées $\{\xi_t, t \in G\}$. Pour $k \in N$, $C_c^k(J)$ désigne l'ensemble des fonction k-fois continument dérivables à support compact dans J. On introduit aussi des espaces de fonctions sur l'espace E de la façon suivante : Pour un sous-ensemble I de D, $FC_c^k(I)$, $k \geq 0$, désigne l'ensemble des fonctions cylindriques $u(\xi) = f(\xi_{t_1},\dots, \xi_{t_n})$ sur E, dont la base $(t_1,\dots, t_n)$ est contenue dans I, telle que f est à support compact dans J^n, et est k-fois continument dérivable. L'ensemble des coordonnées $(t_1,\dots, t_n)$ sera appelé une base de la fonction u. Il est clair que l'intersection de deux bases est encore une base de u. On appellera la base (minimale) de u l'intersection de toutes bases de u. Les espaces $FC_c^k(I)$ ne sont pas des espaces vectoriels. On désigne par $LFC_c^k(I)$ l'espace vectoriel engendré par $FC_c^k(I)$. Notons que l'application :

$$\iota : \xi \in E \to (\xi_t, t \in D) \in F$$

est une injection continue de E dans F. Par l'injection ι, on va considérer E comme un sous-ensemble de F. Ainsi, les espaces fonctionnels $FC_c^k(I)$ définis sur E seront également considérés comme des espaces fonctionnels sur F. Pour une fonction $u \in FC_c^1(D)$, pour $t \in D$, $\partial_t u$ désignera la dérivée de u par rapport à la t^e-coordonnée de F ou de E. En particulier, si $u(\xi) = f(\xi_s,\xi_t)$ avec $f(x,y) \in C_c^1(J^2)$, $\partial_t u(\xi) = \frac{\partial f}{\partial y}(\xi_s,\xi_t)$.

Posons $\Pi(D)$ l'ensemble des fonctions sur E de la forme $\prod_{i \in D} f_i(\xi_i)$, où les fonctions f_i sont des constantes 1, sauf pour un nombre fini d'entre elles qui sont des fonctions de $C_c^1(J)$. Posons $L\Pi(D)$ l'espace vectoriel engendré par $\Pi(D)$. Certes, $\Pi(D)$ et $L\Pi(D)$ sont des sous-ensembles de $LFC_c^1(D)$.

Soit μ une mesure de Radon sur J et $P_t(x,dy)$, $x \in J$, $dy \subset J$, un semigroupe de transition Markovien sur J. Le semigroupe P_t sera appelé un semigroupe classique par rapport à μ, s'il satisfait les conditions suivantes :

i) $P_t(x,dy)$ est symétrique par rapport à le mesure μ. Dans ce cas, $P_t(x,dy)$ induit un unique semigroupe d'opérateurs linéaires continus dans $L^2(J,\mu)$, que l'on dénote par (P_t).

ii) Soit $\mathcal{E}$ la forme de Dirichlet sur $L^2(J,\mu)$ associée avec P_t. Alors, $D(\mathcal{E})$ contient $C_c^1(J)$ et $\mathcal{E}$ est régulière sur $C_c^1(J)$ (cf. Fukushima [7] pour la notion).

iii) Le domaine du générateur A de $\mathcal{E}$ contient $C_c^2(J)$. De plus, il existe une fonction b dans $L_{loc}^2(J,\mu)$ telle que pour toute fonction $u \in C_c^2(J)$, $Au = u'' + b\,u'$.

iv) P_t est conservatif, c'est-à-dire, $P_t 1 = 1$.

Lorsque P_t est un semigroupe classique par rapport à μ, il possède de nombreuses bonnes propriétés (Voir, par exemple, Albeverio-Kusuoka-Röckner [1], Bouleau-Hirsch [4]).

α : La mesure μ est nécessairement absolument continue par rapport à la mesure de Lebesgue dx sur J. Si p(x) désigne la fonction de densité, alors p a une version continue qui est strictement positive sur J, et la dérivée de p(x) au sens des distributions est b(x)p(x) qui est une fonction dans $L^2_{loc}(J,dx)$.

β : Toute fonction dans $D(\mathcal{E})$ a une version continue, dont la dérivée au sens des distributions est dans $L^2(J,\mu)$, c'est-à-dire, $D(\mathcal{E}) \subset \{h; h \in L^2(J,\mu), h' \in L^2(J,\mu)\}$. On a également $\{h; h$ est à support compact dans J, $h' \in L^2(J,\mu)\} \subset D(\mathcal{E})$.

χ : Toute fonction dans D(A) a une version dérivable dont la dérivée est continue sur J.

A partir de maintenant on fixe une mesure μ de Radon sur J. Tout les semigroupes classiques seront par rapport à la mesure μ.

Supposons que P_t est un semigroupe classique. Alors, il existe un unique processus de Hunt continu X à valeurs dans J dont le semigroupe de transition est $P_t(x,dy)$. On désigne par P_x la loi sur E du processus X issu de $x \in J$, et $m = P_\mu$. Le processus X possède aussi des bonnes propriétés. Premièrement, sous m, X_t satisfait l'équation stochastique suivante :

$$dX_t = \sqrt{2}\, dB_t + b(X_t)\, dt,$$

où B est un mouvement brownien réel standardpar rapport à la filtration $\{\sigma(X_s, s \le t)\}_{t\ge 0}$. Sans perdre de la généralité, on va supposer que cette équation est valable sous P_x pour tout $x \in J$. Ensuite, le temps de vie de X est infini, ce qui est conséquence de la propriété iv). Enfin, pour tout $t \ge 0$, la loi de X_t est m.

Soit P_t un semigroupe classique. On dit qu'il génère un processus Z à deux paramètres, si les conditions suivantes sont vérifiées : (comparer ces conditions avec celles présentées dans le §1)

a) La filtration (à deux indices) $F_{s,t} = \sigma\{Z_{u,v}, u \in [0,s], v \in [0,t]\}$ satisfait la condition (F4) (voir [18]).

b') Par rapport à la filtration $F_{\infty,t}$, l'application $t \to (Z_{s,t}, s \ge 0)$ est un processus de Markov à valeurs dans E, symétrique, de loi initiale m. (Signalons que dans notre travail, le mot "loi" peut désigner une mesure σ-finie.)

c') Pour s fixé, par rapport à la filtration $F_{\infty,t}$, l'application $t \to Z_{s,t}$ est un processus de Markov de semigroupe de transition P_t.

d) Pour tout couple (s,t), la loi de $Z_{s,t}$ est μ.

Remarquons que les conditions ici sont moins fortes que celles présentées dans le §1. On verra plus tard qu'en fait, elles sont équivalentes, quitte à supposer quelques hypothèses techniques.

Pour terminer la section, on énonce deux lemmes qui seront utilisés plus tard. On s'épargne les preuves qui sont standards.

Lemme 1 : Soit $d \in N$. Soit $\Pi(J^d)$ l'ensemble des fonctions sur J^d de la forme $\prod_{1\leq i\leq d} f_i(x_i)$, où f_i sont des fonctions de $C_c^1(J)$. Posons $L\Pi(J^d)$ l'espace vectoriel engendré par $\Pi(J^d)$. Alors, pour toute fonction v de $C_c^1(J^d)$, il existe une suite des fonctions v_n de $C_c^1(J^d)$ telles que :

1) $\|v - v_n\|_\infty + \sum_{1\leq i\leq d} \|\partial_i v - \partial_i v_n\|_\infty \to 0$, lorsque n tend vers l'infini.

2) Les supports des fonctions v_n sont contenus dans un même compact de J^n.

3) Les fonctions v_n appartiennet à $L\Pi(J^d)$. □

Lemme 2 : (Formule d'intégration par parties) Pour toute fonction $v \in D(\mathcal{E})$ à support compact dans J, la formule suivante est valable :

$$\int v'(x)\, \mu(dx) = - \int v(x)\, b(x)\, \mu(dx). \square$$

§3 Propriété de c-semigroupe est une condition nécessaire

Dans cette section on cherche une condition nécessaire pour qu'un semigroupe classique P_t puisse générer un processus continu à deux paramètres Z. Notre condition nécessaire est décrite par la définition suivante :

Définition 1 : Un semigroupe classique P_t sur J est appelé un c-semigroupe, si la condition suivante est satisfaite : Pour tout $t \geq 0$, il existe un noyau $c_t(x,y)$ sur J^2, telle que $|c_t(x,y)| \leq 1$, pour tout $(t,x,y) \in R_+ \times J^2$, et que

$$\frac{d}{dx} P_t v(x) = \int v'(y)\, c_t(x,y)\, P_t(x,dy) = P_t[v' c_t(x,\bullet)](x), \quad \mu\text{-p.p.},$$

pour toute fonction $v \in D(\mathcal{E})$. □

Théorème 2 : Si un semigroupe classique P_t génère un processus continu Z à deux paramètres, il est nécessairement un c-semigroupe.

Preuve : Le processus $t \to Z_t = (Z_{s,t}, s \geq 0)$ est un processus de Markov m-symétrique. On désigne par L son générateur. Soient $0 \leq s < s'$ deux nombres dyadiques. Les conditions sur le processus Z impliquent que les deux processus en $t \geq 0$ suivants :

$$\frac{1}{\sqrt{2}}\left(Z_{s,t} - Z_{s,0} - \int_0^t b(Z_{s,u})\, du \right) \text{ et } \frac{1}{\sqrt{2}}\left(Z_{s',t} - Z_{s',0} - \int_0^t b(Z_{s',u})\, du \right)$$

sont deux mouvements browniens réels standards pour la filtration $F_{\infty,t}$. Il existe alors une fonction $cov_{s,s'}(\xi)$, $\xi \in E$, telle que $\langle Z_{s,\bullet}, Z_{s',\bullet} \rangle(du) = 2\, cov_{s,s'}(Z_u)\, du$, où $\langle\ \rangle$ désigne le crochet oblique des deux processus. Notons que, d'après Iketa-Watanabe [9], la fonction $cov_{s,s'}(\xi)$ est de valeur absolue au plus 1.

Maintenant il est facile de voir que $FC_c^2(D) \subset D(L)$ et que

$$L(uv) = uLv + vLu + 2\, f'(\xi_s)\, g'(\xi_{s'})\, cov_{s,s'}$$

si $u(\xi) = f(\xi_s)$ et $v(\xi) = g(\xi_{s'})$ avec $f, g \in C_c^2(J)$. D'autre part, d'après la condition §2.c', on a $Lu(\xi) = (Af)(\xi_s)$. A l'aide de ces deux égalités, on obtient

$$\int h(\xi_s)\, f'(\xi_s)\, g'(\xi_{s'})\, cov_{s,s'}(\xi)\, m(d\xi) = \int h(x)\, f'(x)\, (P_{s'-s}g)'\, \mu(dx),$$

avec $h \in C_c^2(J)$. D'où, $P_{s'-s}[g'c_{s'-s}(x,\bullet)](x) = (P_{s'-s}g)'(x)$, μ-p.p., avec $c_{s'-s}(x,y)$ l'espérance conditionnelle de $cov_{s,s'}(\xi)$ sachant $\xi_s = x$ et $\xi_{s'} = y$ sous la mesure m. On s'épargne le reste de la démonstration qui est facile. □

§4 Propriété de c-semigroupe et espace de Dirichlet sur $C(R_+,J)$

A partir de maintenant on fixe un c-semigroupe P_t. Nous utiliserons les notations associées avec P_t introduites dans §2 et §3. Ainsi, $c_t(x,y)$ est le noyau de P_t, $\mathcal{E}$ est la forme de Dirichlet associée avec P_t, et X est le processus de Hunt continu dont le semigroupe de transition est P_t, et $m = P_\mu$.

Dans cette section et les trois sections suivantes, on va démontrer que la propriété de c-semigroupe est aussi une condition suffisante pour que P_t génère un processus continu Z à deux paramètres. Notre démonstration est divisée en quatre parties. Dans la présente section, on va construire une forme de Dirichlet ϑ sur l'espace $L^2(E,m)$. Dans le §5, on va démontrer que cette forme de Dirichlet ϑ est associée avec un processus de Markov Z continu m-symétrique à valeurs dans E. Dans le §6, on va démontrer que Z possède une propriété de Markov particulière que l'on appellera propriété de Markov locale. Enfin, dans le §7 on va démontrer que Z satisfait les conditions a-b'-c'-d du §2.

On commence par étudier le noyau $c_t(x,y)$.

Lemme 1 : On a : $c_t(y, x) = c_t(x, y)$, $P_t(x,dy)$-p.p. y, μ-p.p. x.

Preuve : C'est une conséquence de la symétrie du semigroupe P_t par rapport à m. □

Lemme 2 : Soient 0 < s, t. Alors, pour tout x, $c_{s+t}(x,z)$ est l'espérance conditionnelle de $c_s(x,y)c_t(y,z)$ sachant z sous la mesure $P_s(x,dy)P_t(y,dz)$. Plus généralement, $c_{s_1+\dots+s_n}(x_1,x_n)$ est l'espérance conditionnelle de

$$c_{s_1}(\xi_{t_1},\xi_{t_2})c_{s_2}(\xi_{t_2},\xi_{t_3})\dots c_{s_n}(\xi_{t_n},\xi_{t_{n+1}})$$

sachant $\xi_{t_{n+1}} = x_n$ sous la probabilité P_{x_1}, où $t_1 = 0$ et $t_{i+1} = s_1 + \dots s_i$.

Preuve : C'est une conséquence de la propriété de semigroupe de P_t. □

Soit $n \in \mathbb{N}$. Soit $I_n = \{\frac{k}{2^n}; k = 0, 1, 2, 3, \dots\}$ l'ensemble des nombres dyadiques positifs ou nuls de base 2^n. Définissons sur $LFC_c^1(I_n)$,

$$\vartheta_n(u, v) = \int \sum_{i,j} \partial_i u\, \partial_j v\, c_{i,j}^n\, dm,$$

où ∂_i est la dérivée sur la coordonnée $\frac{i}{2^n}$, et les fonctions $c_{i,j}^n$ sont définies comme suit : pour $i \neq j$, $c_{i,j}^n = c_{j,i}^n$, et si $i < j$ on a $c_{i,j}^n(\xi) = \prod_{i \le k < j} c_a(\xi_k,\xi_{k+1})$, où $a = \frac{1}{2^n}$ et ξ_k est la coordonnée $\frac{k}{2^n}$; pour $i = j$, $c_{i,i}^n = 1$.

Lemme 3 : Pour toute fonction $u, v \in LFC_c^1(I_n)$, $\vartheta_{n+1}(u,v) = \vartheta_n(u,v)$.

Preuve : Cela résulte du Lemme 2. □

Lemme 4 : La matrice $(c^n_{i,j})$ est définie positive ou nulle. En particulier, $\vartheta_n(u,u) \geq 0$ pour toute $u \in LFC^1_c(I_n)$.

Preuve : Nous démontrons que la matrice $(c^n_{i,j})_{i,j\leq k}$ est une matrice définie positive ou nulle pour tout $k \geq 0$. Nous le faison par récurrence. Lorsque $k = 0$, c'est évident. Pour $k = 1$, c'est encore facile, car $\left|c_t(x,y)\right| \leq 1$. Supposons que $(c^n_{i,j})_{i,j\leq k}$ est une matrice définie positive ou nulle. Nous constatons que la matrice $(c^n_{i,j})_{i,j\leq k+1}$ admet la factorisation suivante :

$$(c^n_{i,j})_{i,j\leq k+1} = Q^t \begin{pmatrix} (c^n_{i,j})_{i,j\leq k} & , & 0 \\ 0 & , & 1-(c^n_{k,k+1})^2 \end{pmatrix} Q,$$

où Q est une matrice inversible. Maintenant l'hypothèse de récurrence nous permet de conclure que $(c^n_{i,j})_{i,j\leq k+1}$ est aussi définie positive ou nulle. □

Lemme 5 : Pour chaque $n \in N$, ϑ_n satisfait la propriété de contraction au sens de Fukushima [7].

Preuve : Cela se découle de la définition de ϑ_n et de la positivité de $(c^n_{i,j})$. □

Lemme 6 : Pour tout $n \in N$, pour tout $0 = t_0 < t_1 <...< t_n$, pour toute fonction $v \in C^1_c(J^{n+1})$, considérons la fonction :

$$x \to \int P_{t_1}(x,dx_1)...P_{t_n}(x_{n-1},dx_n)\, v(x, x_1,...,x_n),\ x \in J.$$

Alors, la dérivée au sens des distributions de cette fonction est donnée par :

$$\sum_{0\leq i\leq n} \int P_{t_1}(x,dx_1)...P_{t_n}(x_{n-1},dx_n)\, \partial_i v(x, x_1,...,x_n) \prod_{1\leq k\leq i} c_{t_k}(x_{k-1},x_k),$$

avec $x_0 = x$ et $\prod_{1\leq k\leq 0} c_{t_k}(x_{k-1},x_k) = 1$, et cette fonction appartient à $D(\mathcal{E})$.

Preuve : Prendre les dérivées et utiliser la propriété de c-semigroupe. □

Lemme 7 : Soit $n \in N$ et $i \in I_n$. Soit $f \in D(\mathcal{E})$ à support compact dans J, soit $g \in FC_c^1(I_n)$. Considérons la fonction $v(\xi) = f(\xi_i)g(\xi)$ et la fonction $f(x)\, E_m[g \mid \xi_i = x]$. Alors, la dérivée au sens des distributions de la fonction conditionnelle est donnée par :

$$\frac{d}{dx}\{f(x)E_m[g \mid \xi_i = x]\} = E_m[\sum_j \partial_j v\, c^n_{i,j} \mid \xi_i = x].$$

En particulier, elle appartient à $D(\mathcal{E})$.

Preuve : C'est une conséquence du Lemme 6 et de la symétrie de m par rapport au retournement du temps. □

Lemme 8 : Pour $i \in I_n$ fixé, pour $u \in LFC_c^2(I_n)$, $v \in LFC_c^1(I_n)$, on a :

$$\int \sum_j \partial_i u\, \partial_j v\, c^n_{i,j}\, dm = - \int v\, [b(\xi_i)\, \partial_i u + \sum_j \partial_{i,j} u\ c^n_{i,j}]\, dm.$$

Preuve : C'est une conséquence du Lemme §2.2 et du Lemme 7. □

Définissons un opérateur L_n sur $LFC_c^2(I_n)$ comme suit :

$$L_n u = \sum_{i,j} c^n_{i,j}\, \partial_{i,j} u + \sum_i b(\xi_i)\, \partial_i u, \quad u \in LFC_c^2(I_n).$$

Lemme 9 : L_n est un opérateur symétrique par rapport à la mesure m. Plus précisément, l'opérateur L_n et la forme ϑ_n sont liés par l'équation suivante :

$$\vartheta_n(u, v) = - \int v\, L_n u\, dm, \quad u, v \in LFC_c^2(I_n).$$

Preuve : Il suffit de sommer sur i les égalités du Lemme 8. □

Soit $0 \le s < t$ deux nombres dyadiques. On introduit une filtration $\{F_n\}$ sur E :

$$F_n = \sigma\{\xi_u;\, u \in I_n,\, s < u < t\} \vee \sigma\{\xi_s, \xi_t\}.$$

Alors, il est clair que les fonctions $c^n_{s,t}$ sont F_n-mesurables. Mais on peut démontrer plus.

Lemme 10 : Quelque soit $0 \le a \le s$, sous la probabilité conditionnelle $E_m[\bullet \mid \xi_a]$, la suite des fonctions $\{c^n_{s,t}\}_n$ forment une martingale bornée par rapport à la filtration $\{F_n\}$. Soit $cov_{s,t}$ la limite de cette martingale. Alors, $t \in D \cap [s,\infty[\to \{cov_{s,t}\}$ est une fonctionnelle multiplicative du processus de Markov X. Plus précisément, on a $cov_{s,t}\, cov_{t,u} = cov_{s,u}$, p.p. pour tout $0 \le s < t < u \in D$.

Preuve : La première partie du lemme résulte du Lemme 2. Le reste est évident. □

Définissons maintenant une forme sur $LFC^1_c(D)$:

$$\vartheta(u, v) = \int \sum_{i,j} \partial_i u\, \partial_j v\, cov_{i,j}\, dm, \quad u, v \in LFC^1_c(I_n),\ n \in N,$$

où $cov_{i,j} = cov_{j,i} = cov_{t_i,t_j}$, si $t_i < t_j$, $t_i, t_j \in I_n$.

Théorème 11 : La forme ϑ sur $LFC^1_c(D)$ est fermable dans $L^2(E,m)$. Désignons encore par ϑ sa fermeture dans $L^2(E,m)$. Alors, ϑ est une forme de Dirichlet locale. Soit L le générateur de ϑ. Alors, D(L) contient $LFC^2_c(D)$. De plus, ϑ possède un opérateur carré du champs Γ (cf. Bouleau-Hirsch [4]) dont l'action sur $LFC^1_c(D)$ est $\Gamma(u,v) = \sum_{i,j} \partial_i u\, \partial_j v\, cov_{i,j}$, où $v, u \in LFC^1_c(D)$, et les indices i et j parcourent respectivement la base de u et la base de v.

Preuve : On remarque que la restriction de ϑ sur chaque $LFC^1_c(I_n)$ est exactement ϑ_n. La formule d'intégration par parties de ϑ_n démontrée dans le Lemme 9 permet de dire que ϑ est fermable dans $L^2(E,\mu)$ (cf. Fukushima [7]). La même formule d'intégration par partie implique aussi que l'espace $FC^2_c(D)$ est contenu dans D(L). Pour déterminer l'opérateur carré du champs de ϑ, on considère l'équation suivante (cf. Bouleau-Hirsch [4]) : f, g, h $\in LFC^2_c(I_n)$, où $n \in N$ est fixé,

$$\vartheta(fh, g) + \vartheta(gh, f) - \vartheta(h, fg) = \int h\, [L_n(fg) - gL_n f - fL_n g]\, dm$$

$$= 2 \int h\, \Gamma_n(f,g)\, dm = 2 \int h\, \Gamma(f,g)\, dm.$$

Cette dernière égalité résulte du Lemme 10. La localité de la forme de Dirichlet ϑ est claire. Le théorème est prouvé. □

Les deux lemmes suivants seront utilisés plus tard. Les preuves sont épargnées.

Lemme 12 : $L\Pi(D)$ est dense dans $D(\vartheta)$ pour la norme ϑ_1. □

Lemme 13 : Pour toutes fonctions $v \in FC_c^1(D)$, $u \in FC_b(D) \cap D(\vartheta)$, où $FC_b^1(D)$ désigne l'ensemble des fonctions cylindriques sur E de bases contenues dans D continûment dérivables, on a la formule suivante :

$$\Gamma(u, v) = \sum_{i,j} \partial_i u \, \partial_j v \, \mathrm{cov}_{i,j},$$

où les indices i et j parcourent respectivement la base de u et la base de v. □

§5 Existence du processus associé à la forme de Dirichlet ϑ

Dans ce paragraphe on démontre qu'il existe un processus de Markov Z m-symétrique continu à valeurs dans E dont la forme de Dirichlet associée est ϑ. La méthode utilisée est une généralisation de celle introduite dans Schmuland [13]. Elle consiste à construire d'abord le processus de Markov Z dans F, continu pour la topologie de F, et ensuite à montrer que $P[\exists t < \zeta, Z_t \notin E] = 0$, où ζ désigne le temps de vie du processus Z, et que Z est aussi continu pour la topologie de E. Rappelons que par l'injection ι (voir le §2), E est considéré comme un sous-ensemble de F.

On aura besoin des notations et des terminologies suivants: Sur l'espace E, il y a deux topologies : la topologie induite par F et la topologie originale. On les appellera respectivement la topologie faible et la topologie forte. Posons m_D la mesure image de m par l'application ι. Notons que la mesure m_D est portée par E (on verra plus tard que E est un sous-ensemble mesurable de F). Posons ϑ_D la forme de Dirichlet sur $L^2(F,m_D)$ qui est l'image de la forme de Dirichlet ϑ par l'application ι. Pour $n \in N$, $a \geq 0$, pour $y \in F$ et $z \in E$, posons

$$\left|y - z\right|_{n,a} = \sup_{s\in I_n,\ s\leq a} \left|y_s - z_s\right|, \text{ et}$$

$$\left|y - z\right|_{a} = \sup_{s\in D,\ s\leq a} \left|y_s - z_s\right|.$$

Ces quantités peuvent prendre des valeurs infinies. Soit $\{z_k\}$ une suite d'éléments dense dans E pour la topologie forte. Soit φ une fonction continue sur $[0,\infty]$ nulle en zéro strictement croissante lipchitzienne sur $[0,\infty[$. Posons $w_a(y) = \inf\{\varphi(|y - z|_a); z \in (z_k)\}$.

Le point clé de la méthode de Schmuland [13] est de démontrer que les fonctions :

$$y \to w_a(y) \text{ et } y \to \varphi(|y - z|_a),$$

pour $z \in (z_k)$ et $a > 0$, sont des fonctions ϑ_D-quasi-continues.

On commence par quelques lemmes techniques.

Lemme 1 : Pour tout $x \in J$, le noyau de probabilité $P_t(x,dy)$ est absolument continu par rapport à la mesure de Lebsegue sur J. De même, pour tout $n \in N$, pour tout $0 < t_1 <...< t_n$, la mesure :

$$\mu(dx_0)P_{t_1}(x_0,dx_1)...P_{t_n}(x_{n-1},dx_n)$$

est absolument continue par rapport à la mesure de Lebesgue sur J^{n+1}.

Preuve : Il suffit de comparer $P_t(x,dy)$ avec la loi du mouvement brownien à l'aide du théorème de Girsanov. □

Lemme 2 : a) Soit $\{u_n\}$ une suite des fonctions ϑ_D-quasi-continues bornées dans $D(\vartheta_D)$ pour la norme $\vartheta_{D,1}$. Supposons que u est une fonction telle que $\lim_n u_n(y) = u(y)$ pour tout $y \in F$. Alors, u est également une fonction de $D(\vartheta_D)$, qui est ϑ_D-quasi-continue.

b) Soit Cap la capacité de la forme de Dirichlet ϑ_D. Alors,

$$\mathrm{Cap}(\cup_{i\in D}\{y_i \in \bar{J} - J\}) = 0.$$

Preuve : Routine (voir Schmuland [13] pour une idée de la démonstration). □

On va étudier les fonctions $w_a(y)$ et $\varphi(|y - z|_a)$. Pour cela, désignons par f n'importe quelle fonction Lipchitzienne positive ou nulle à support compact dans J. Désigons par ϕ n'importe quelle fonction de $C_c^\infty(R)$ telle que $0 \le \phi \le 1$, $|\phi'| \le 1$. La fonction ϕ sera considérée comme une fonction continue sur $[0,\infty]$ si on définit $\phi(\infty) = 0$. Posons

$$g(y) = f(y_0)\,(\phi\varphi)(\left|y - z\right|_{n,a}),\ y \in F.$$

Il est clair que $g \in D(\vartheta_D)$. Utilisant Lemme 1, on peut voir que la dérivée $\partial_i g(y)$, $i \in I_n$, $y \in E$, est m-presque surement égale à la fonction :

$$f'(y_0)(\phi\varphi)(\left|y - z\right|_{n,a})\,1_{\{i=0\}} + f(y_0)\,(\phi\varphi)'(\left|y - z\right|_{n,a})\,\mathrm{sign}(y_i - z_i)\,1_{A_i},$$

où $A_i = \{\left|y_i - z_i\right| > \left|y_j - z_j\right|,\ \forall j \in I_n,\ j \neq i\}$. D'où :

$$\Gamma(g,g)(y) = f'(y_0)^2\,(\phi\varphi)(\left|y - z\right|_{n,a})^2$$

$$+\,2\,f'(y_0)\,(\phi\varphi)(\left|y - z\right|_{n,a})\,f(y_0)\,(\phi\varphi)'(\left|y - z\right|_{n,a})\sum_i \mathrm{sign}(y_i - z_i)\,1_{A_i}\,c_{0,i}$$

$$+\,f(y_0)^2\,(\phi\varphi)'(\left|y - z\right|_{n,a})^2\sum_i 1_{A_i}$$

$$\leq 2C^2(f'(y_0) + f(y_0))^2 \leq 8\,C^2B^2,$$

où $y \in E$, C est une constante majorant φ et φ', et B est une constante majorant f et f '. Comme $\left|f'(y_0)\right|$ et $f(y_0)$ appartiennent à $L^2(E,m)$, les fonctions g sont uniformément bornées dans $D(\vartheta_D)$ pour la norme $\vartheta_{D,1}$. A partir de cette remarque et le Lemme 2, il est routine de montrer successivement que les fonctions :

$$y \to f(y_0)\,\varphi(\left|y - z\right|_{n,a}),\ y \in F,$$
$$y \to f(y_0)\varphi(\left|y - z\right|_a),\ y \in F,$$

sont des fonctions ϑ_D-quasi-continues dans $D(\vartheta_D)$ (Voir, par exemple, Schmuland [13], p.87-88). Notons aussi que d'après le Lemme 2, $1_{F'} = 1$ ϑ_D-quasi-partout, où $F' = F - \cup_{i\in D}\{y_i \in \bar{J} - J\})$. Par conséquent, on a le lemme suivant :

Lemme 3 : La fonction : $y \to \varphi(\left|y - z\right|_a)$ est ϑ_D-quasi-continue pour tout $a \geq 0$, $z \in E$.

□

Remarquons que les inégalités précédentes sur les fonctions g s'étendent pour les fonctions $v_{z,a}(y) = f(y_0)\varphi(\left|y - z\right|_a)$:

$$\Gamma(v_{z,a}, v_{z,a}) \leq 2C^2(f'(y_0) + f(y_0))^2 \leq 8\, C^2B^2.$$

Posons maintenant

$$v_{k,a}(y) = v_{z_k,a}(y),$$

$$V_{d,a}(y) = \inf\{v_{k,a};\, k = 1, ..., d\} = f(y_0)\inf\{\varphi(|y - z|_a);\, z = z_1,...,z_d\},$$

$$V_a(y) = \inf_d V_{d,a}(y) = f(y_0)w(y).$$

On va démontrer que V_a est une fonction ϑ_D-quasi-continue. Pour cela, on étudie d'abord la fonction $\inf(x) = x_1\wedge...\wedge x_d$, $x \in R^d$. Cette fonction est visiblement une fonction lipchitzienne. Posons maintenant $A_i = \{x;\, x_i < x_j,\ 1\leq j\leq n,\ j \neq i\}$, $1 \leq i \leq n$. Posons $h(x) = \sum_{1\leq i\leq n} 1_{\partial A_i}(x)$. D'après Schmuland [13], $\partial_i \inf(x) = 1_{A_i}(x) + 1_{\partial A_i}(x)\, h(x)^{-1}$. Soit (a_{ij}) une famille des nombres bornés par $A \geq 0$. On a :

$$\left|\sum_{i,j} \partial_i \inf(x)\, \partial_j \inf(x)\, a_{ij}\right|$$

$$\leq A \sum_{i,j} (1_{A_i}(x) + 1_{\partial A_i}(x)\, h(x)^{-1})\, (1_{A_j}(x) + 1_{\partial A_j}(x)\, h(x)^{-1})$$

$$= A \sum_{i,j} (1_{A_i}(x)\, 1_{A_j}(x) + 1_{\partial A_i}(x)\, h(x)^{-1}\, 1_{A_j}(x)$$

$$+ 1_{A_i}(x) 1_{\partial A_j}(x)\, h(x)^{-1} + 1_{\partial A_i}(x)\, h(x)^{-1}\, 1_{\partial A_j}(x)\, h(x)^{-1})$$

$$\leq 4\, A.$$

Considérons les fonctions $V_{d,a}$. Comme $\left|\Gamma(v_{i,a}, v_{j,a})\right| \leq 8C^2B^2$, on a :

$$\sum_{i,j\leq d} \partial_i \inf(\{v_{k,a}\})\, \partial_j \inf(\{v_{k,a}\})\, \Gamma(v_{i,t}, v_{j,t}) \leq 32\, C^2B^2.$$

Les fonctions $V_{d,a}$ sont alors uniformément bornées dans $D(\vartheta_D)$ pour la norme $\vartheta_{D,1}$. Maintenant, on applique le Lemme 2, et on obtient :

Lemme 4 : V_a est une fonction ϑ_D-quasi-continue de $D(\vartheta_D)$. Par conséquent, w_a est ϑ_D-quasi-continue. □

Les quasi-continuités des fonctions w_a et $v_{z,a}$ étant démontrées, on peut énoncer le théorème suivant dont la démonstration est la même que celle de Schmuland [13].

Théorème 5 : Il existe un processus de Markov Z à valeurs dans E, continu pour la topologie forte de E, symétrique par rapport à la mesure m, dont la forme de Dirichlet associée est ϑ. □

§6 Propriété de Markov locale

On exhibe une propriété intéréssante du processus de Markov Z construit dans la section précédente : une propriété que l'on appellera propriété de Markov locale. Voici la définition : Soit Q_t le semigroupe de transition du processus Z sur E.

Définition 1 : On dit que le semigroupe Q_t possède la propriété de Markov locale, si pour tout $0 \le a \le b < \infty$, le semigroupe Q_t laisse invariante la tribu $\sigma(\xi_{[a,b]})$, c'est-à-dire, pour toute fonction $f \in \sigma(\xi_{[a,b]})$, $Q_t f \in \sigma(\xi_{[a,b]})$. □

Pour étudier cette propriété pour le semigroupe Q_t, on introduit les opérateurs d'espérance conditionnelle. Soient $0 \le a \le b$ deux points dyadiques. Posons $K = [a,b] \cap D$, et $E^K = E[\bullet \mid \xi_K]$, où l'espérance conditionnelle est prise par rapport à la mesure m.

Lemme 2 : Posons $D_K(\vartheta) = D(\vartheta) \cap \sigma\{\xi_K\}$. Alors, quelque soit une fonction $f \in L\Pi(D)$, $E^K[f]$ est limite des fonctions de $LFC_c^1(D) \cap D_K(\vartheta)$ pour la norme ϑ_1.

Preuve : Il suffit de remarquer que, pour $f \in \Pi(D)$, $E^K[f]$ s'écrit sous la forme $E^K[f] = P_{\xi_a}[g_1]\, g_2\, P_{\xi_b}[g_3]$. □

Lemme 3 : L'opérateur E^K est un opérateur continu de $D(\vartheta)$ dans $D(\vartheta)$ pour la norme ϑ_1. Plus précisément, pour toute fonction $f \in D(\vartheta)$, on a $\Gamma(E^K[f], E^K[f]) \le E^K[\Gamma(f,f)]$.

Preuve : En fait, appliquant les résultats du §4, on obtient

$$E^K[\Gamma(v,f)] = \Gamma(v, E^K[f]),$$

pour toute $f \in \Pi(D)$. Le lemme résulte aisément de cette égalité. □

Lemme 4 : $LFC_c^1(D) \cap D_K(\vartheta)$ est dense dans $D_K(\vartheta)$ pour la norme ϑ_1.

Preuve : Pour toute $u \in D_K(\vartheta)$, il existe une suite des fonction $u_n \in L\Pi(D)$ qui converge vers u pour la norme ϑ_1. D'après le Lemme 3, $E^K[u_n]$ converge vers $E^K[u] = u$ pour la norme ϑ_1. Alors, le lemme est prouvé, car d'après le Lemme 2, chaque fonction $E^K[u_n]$ est limite des fonctions de $LFC_c^1(D) \cap D_K(\vartheta)$ pour la norme ϑ_1. □

Lemme 5 : Soit P^K la projection orthogonale de $D(\vartheta)$ sur $D_K(\vartheta)$ par rapport au produit scalaire ϑ_1. Alors, P^K coincide avec l'opérateur E^K.

Preuve : C'est une conséquence de la formule $\Gamma(v, E^K[f]) = E^K[\Gamma(v,f)]$, $f \in D(\vartheta)$, $v \in LFC_c^1(D) \cap D_K(\vartheta)$. □

Lemme 6 : Soit m_K la mesure image de m par l'application $\xi \to \xi_K$. Soit ϑ_K la forme de Dirichlet sur $L^2(m_K)$ image de ϑ par l'application ξ_K. Soit L_K son générateur. Alors, $D(L_K) \circ \xi_K = D(L) \cap \sigma(\xi_K)$. De plus, pour tout $u \in D(L_K)$, $L(u \circ \xi_K) = (L_K u) \circ \xi_K$.

Preuve : D'après Bouleau-Hirsch [4], on a déja $D(L_K) \circ \xi_K \supset D(L) \cap \sigma(\xi_K)$. Pour démontrer l'inclusion inverse, soit $f \in D(\vartheta)$ et soit $g \in D(\vartheta_K)$ telle que $P^K(f)(\xi) = g(\xi_K)$. Soit $u \in D(L_K)$. On a :

$$\vartheta(f, u \circ \xi_K) = \vartheta(P^K(f), u \circ \xi_K) = \vartheta_K(g,u) = -\int g \, L_K u \, dm_K$$
$$= -\int P^K(f)(\xi) \, (L_K u)(\xi_K) \, dm = -\int f(\xi) \, (L_K u)(\xi_K) \, dm,$$

car $P^K = E^K$. Cette égalité implique $u \circ \xi_K \in D(L)$ et $L(u \circ \xi_K) = (L_K u)(\xi_K)$. Le lemme est démontré. □

Soient maintenant R_λ et $\tilde{R}_\lambda$, $\lambda \geq 0$, les résolventes de ϑ et respectivement de ϑ_K. Alors, pour toute fonction $u \in LFC_c^2(K)$, on a :

$$(\lambda - L)\,[(\tilde{R}_\lambda u)\circ\xi_K] = [(\lambda - L_K)\,\tilde{R}_\lambda u]\circ\xi_K = u\circ\xi_K = (\lambda - L)\,R_\lambda(u\circ\xi_K).$$

Comme $\lambda - L$ est injectif, on a en fait démontré $(\tilde{R}_\lambda u)\circ\xi_K = R_\lambda(u\circ\xi_K)$. Autrement dit, si $\tilde{Q}_t$ est le semigroupe de ϑ_K, on a l'égalité : $Q_t(u\circ\xi_K) = (\tilde{Q}_t u)\circ\xi_K$. En particulier, $Q_t(u\circ\xi_K) \in \sigma(\xi_K)$.

Théorème 7 : Le semigroupe Q_t possède de la propriété de Markov locale. □

Utilisant la propriété de Markov du processus Z, appliquant le Théorème 7, on peut facilement démontrer le corollaire suivant :

Corollaire 8 : Pour toute fonction bornée $G \in \sigma(Z_{u,v},\ u \in [a,b],\ v \geq 0)$, on a $Q_\xi G \in \sigma(\xi_{[a,b]})$, où Q_ξ désigne la loi du processus Z_t issu de $Z_0 = \xi \in E$. (Rappelons qu'un processus à valeurs dans E est considéré comme un processus à deux paramètres à valeurs dans J.) Par conséquent, le processus $t \to (Z_{u,t},\ u \in [a,b])$ est un processus de Markov pour la filtration naturelle de Z. □

§7 Propriété a-b'-c'-d

Comme annoncé, on va démontrer les propriétés a-b'-c'-d pour le processus de Markov Z. On commence par la propriété (F4). Soit $F_{s,t} = \sigma\{Z_{u,v};\ 0\leq u\leq s,\ 0\leq v\leq t\}$, $s, t \in R_+$, la filtration naturelle à deux indices du processus Z. Rappelons que la condition (F4) signifie que les opérateurs d'espérance conditionnelle sachant respectivement $F_{\infty,t}$ ou $F_{s,\infty}$ commutent. L'importance de la condition (F4) est qu'elle entraine l'inégalité de martingale à deux paramètres (cf. [18]).

Quand on considère un processus de Markov à valeurs dans E, on peut le considérer comme un seul processus, ou une famille de processus issus des points de E. On précise que, dans cette section, Z désigne le seul processus de Markov symétrique de loi initiale m, dont la forme de Dirichlet associée est ϑ. Les calculs des espérances, avec conditionnement ou non, sont prises sous la loi du processus Z.

Théorème 1 : La filtration naturelle $F_{s,t}$ du processus à deux paramètres $Z_{s,t}$ satisfait la condition (F4).

Preuve : Soit $s \geq 0$, $t \geq 0$. Soit G une fonction bornée sur l'espace Ξ. On démontre d'abord l'égalité suivante $E[G \mid F_{s,\infty} \mid F_{\infty,t}] = E[G \mid F_{s,t}]$. Pour cela, il suffit de montrer $E[G \mid F_{\infty,t}] \in F_{s,t}$, pour toute fonction bornée $G \in F_{s,\infty}$. Soit (θ_t) le groupe des translations du processus de Markov Z. D'après le théorème de classe monotone, il suffit de considérer une fonction $G = G_1\, G_2 \circ \theta_t$, où $G_2 \in F_{s,\infty}$, et $G_1 \in F_{s,t}$. On a $E[G \mid F_{\infty,t}] = G_1\, Q_{Z_t}[G_2] \in F_{s,t}$, d'après le Corollaire §6.8.

On démontre maintenant l'autre égalité $E[G \mid F_{\infty,t} \mid F_{s,\infty}] = E[G \mid F_{s,t}]$. On n'a besoin que de démontrer que, pour toute fonction H qui est $F_{\infty,t}$-mesurable, $E[H \mid F_{s,\infty}] = E[H \mid F_{s,t}]$. Pour cela, considérons deux variables aléatoires : $G_1 \in F_{s,t}$, $G_2 \in F_{s,\infty}$. Posons $G = G_1\, G_2 \circ \theta_t$. On a :

$$E[G\, H] = E[\, G_1\, H\, E(G_2 \circ \theta_t \mid F_{\infty,t})\,] = E[G_1\, H\, Q_{Z_t}(G_2)\,]$$

$$= E[G_1\, E(H \mid F_{s,t})\, Q_{Z_t}(G_2)\,],$$

car, d'après le Corollaire §6.8, $Q_{Z_t}(G_2) \in F_{s,t}$. En utilisant le théorème de classe monotone, on conclut $E[H \mid F_{s,\infty}] = E[H \mid F_{s,t}]$. Le théorème est démontré. □

Une conséquence du Théorème 1 est que le processus Z satisfait la condition a) du §2. Il satisfait évidemment la condition b'. Pour la propriété c', on a :

Théorème 2 : Le processus Z saitisfait la propriété c') du §2.

Preuve : Prenons d'abord un $s \in D$. D'après les résultats du §6 appliqués pour $a = s = b$, on voit déjà que $t \rightarrow Z_{s,t}$ est un processus de Markov pour la filtraiton $F_{\infty,t}$. Il est en plus un processus de Markov symétrique dont la forme de Dirichlet est $\vartheta_{[s]}$. Pour identifier sa loi, on remarque que $\vartheta_{[s]}$ est régulière sur $C_c^1(J)$ comme démontré dans le Lemme §6.4. Elle coincide visiblement avec $\mathcal{E}$ sur $C_c^1(J)$. Donc, elle est égale à $\mathcal{E}$. Par conséquent, la loi du processus $t \rightarrow Z_{s,t}$ est égale à celle du processus X, c'est-à-dire, m.

Maintenant considérons un nombre positif $s \notin D$. Prenons une suite d'intervalles $[a_i,b_i]$, a_i et $b_i \in D$, tels que $\cap_i [a_i,b_i] = [s]$. Par la loi 0 ou 1 d'un processus de Hunt, on voit que $\cap_i \sigma(\xi_{[a_i,b_j]}) = \sigma(\xi_s)$, m-a.s.. Comme $Q_t[f \circ \xi_{[s]}] \in \cap_i \sigma(\xi_{[a_i,b_j]})$ d'après la propriété de Markov locale, $Q_t[f \circ \xi_{[s]}]$ est une fonction de ξ_s. Cette remarque implique que le processus $t \to Z_{s,t}$ est un processus de Markov pour la filtration $F_{\infty,t}$. Quant à sa loi, elle est limite des lois de $t \to Z_{a_i,t}$ qui sont toutes égales à m. Le théorème est démontré. □

Exhibons quelques conséquences du Théorème 2 sans démonstrations :

Corollaire 3 : Pour s, t, u trois nombres positifs, pour f une fonction bornée sur J, on a :

$$E[f(Z_{s,t+u}) \mid F_{\infty,t}] = P_u f(Z_{s,t}),$$

$$E[f(Z_{s+u,t}) \mid F_{s,\infty}] = P_u f(Z_{s,t}). \square$$

Corollaire 4 : Soient $a \geq 0$, $b \geq 0$. Alors, le processus $(Z_{a,b+t}, t \geq 0)$ et le processus $(Z_{a+t,b}, t \geq 0)$ sont deux processus de Markov de même loi P_x, conditionnellement indépendants sachant $Z_{a,b} = x$. □

Corollaire 5 : Pour quatre nombres positifs ou nuls, pour une fonction f bornée sur J, on a :

$$E[f(Z_{s+a,t+b}) \mid F_{s,t}] = P_{a+b} f(Z_{s,t}). \square$$

Revenons maintenant à la propriété d.

Lemme 6 : Le temps de vie ζ du processus Z est égale à infini presque sûrement. En particulier, la loi de Z_t est invarainte par translation du temps.

Preuve : On doit montrer que $Q_t 1_E = 1_E$ m-p.p. pour tout $t \geq 0$. Soit $\{f_n\}$ une suite croissante des fonctions de $C_c^1(J)$ qui convergent vers 1. Posons $v_n(\xi) = f_n(\xi_0)$. Alors, d'après ce qu'on a démontré, on a $Q_t[1_E v_n](\xi) = P_t[f_n](\xi_0)$. Faisant tendre n vers l'infini, on démontre le lemme. □

Théorème 7 : Z vérifie la condition d).

Preuve : Il résulte de la chaine suivante :

$$\text{loi}(Z_{s,t}) = \text{loi}(Z_{s,0}) = \text{loi}(X_s) = \text{loi}(X_0) = \mu. \square$$

§8 Propriété de Markov horizontale I

Dans cette section et la section suivante, on va démontrer que les conditions a-b'-c'-d, sont en fait équivalentes aux conditions a-b-c-d sous l'hypothèse suivante :

* : Quelque soit $t > 0$ fixé, le noyau $c_t(x,y)$, $x,y \in J$, est localement lipchitzienne sur $J \times J$, et la valeur absolue $\left|c_t(x,y)\right|$ est strictement inférieure à 1 sur $J \times J$. La fonction b est localement lipchitzienne sur J.

Nous devons démontrer les propriétés b" et c". La propriété c" est déja démontrée par le Corollaire §7.4. Il ne nous reste plus qu'à démontrer la propriété b". Notre démonstration est fait en deux temps. Dans cette section, on démontre que, quelque soit $s > 0$, $s' > 0$, la loi conditionnelle de $(Z_{s+s',t}, t \in R_+)$ sachant $F_{s,\infty}$ est $\sigma\{Z_{s,t}, t \in R_+)$ mesurable. Dans la section suivante, on démontre en plus que Z est horizontalement symétrique et homogène.

Démontrons maintenant la prorpiété de Markov horizontale du processus Z. Sans perdre de généralité, on va seulement démontrer que la loi conditionnelle de $(Z_{2,t}, t \in R_+)$ sachant $F_{1,\infty}$ est $\sigma\{Z_{1,t}, t \in R_+)$ mesurable. La démonstration est faite en deux étapes. Dans la première étape, on discrètise le processus Z. On démontre que les processus discrètisés ont la propriété de Markov horizontale. Ensuite, on prend la limite sur les processus discrètisés. On démontre que le processus Z est la limite des processus discretisés et que la propriété de Markov horizontale passe au processus Z.

Les processus discrètisés correspondent à des discrètisations de la forme de Dirichlet ϑ. Pour un nombre entier positif $n \in N$, on pose $K_n = [0,1] \cup (I_n \cap [1,2])$. On définit ε_n comme la forme de Dirichlet image de ϑ par l'application $\xi \to \xi_{K_n}$. Posons ensuite $H_n = I_n \cap [1,2]$, et η_n la forme de Dirichlet image de ε_n par l'application $\xi \to \xi_{H_n}$.

Lemme 1 : La forme de Dirichlet ε_n est associée avec un processus de Markov continu symétrique $Z^{(n)}$ à valeurs dans $C(R_+, C(K_n, J))$. Le processus $Z^{(n)}$ possède la propriété de Markov locale. En particulier, l'application $t \to \xi_{H_n}(Z_t^{(n)})$ est un processus de Markov symétrique qui est de plus le processus associé avec la forme de Dirichlet η_n. Par ailleur, pour tout $s \in K_n$, le processus $t \to (Z_{s,t}^{(n)}, t \geq 0)$ est égale en loi à X.

Preuve : En fait, la forme de Dirichlet ε_n possède le même genre de propriétés que la forme de Dirichlet ϑ. Par conséquent, les arguments des sections §5 et §6 sont applicables à ε_n. □

Le processus $Z^{(n)}$ est notre processus discrètisé. Pour l'instant on fixe un processus $Z^{(n)}$. On utilisera les notations suivantes : Pour un sous ensemble G de K_n, $Z^{(n)}_G$ désigne la famille des variables aléatoires $(Z^{(n)}_{s,t}, s \in G, t \in R_+)$. On va démontrer que $Z^{(n)}_{[0,1]}$ et $Z^{(n)}_{H_n}$ sont conditionnellement indépendantes sachant $Z^{(n)}_{[1]}$. L'idée essentielle est de montrer que $Z^{(n)}_{H_n}$ est fonction d'une famille des mouvements browniens indépendants, que nous définissons comme suit :

$$\gamma_s(t) = \xi_s(Z^{(n)}_t) - \xi_s(Z^{(n)}_0) - \int_0^t b(\xi_s(Z^{(n)}_u))\, du, \quad s \in H_n.$$

$$C_1(t) = \frac{1}{\sqrt{2}}\, \gamma_1(t), \text{ et}$$

$$C_s(t) = \int_0^t \frac{1}{\sqrt{1 - c^2_{s,s-}(u)}} \left(\frac{1}{\sqrt{2}} \gamma_s(dt) - c_{s,s-}(u) \frac{1}{\sqrt{2}} \gamma_{s-}(dt) \right), \quad s \in H_n, \text{ et } s \neq 1,$$

où $s- = \sup\{u \in H_n, u < s\}$, et $c_{i,j}(t) = c^n_{i,j}(Z^{(n)}_t)$ (voir §4).

Pour voir que les processus C_s, $s \in H_n$, sont des mouvements browniens indépendants, il suffit de calculer leurs crochets obliques. Cela sera possible à l'aide de l'expression de la forme de Dirichlet η_n :

$$\eta_n(u,v) = \int \sum_{i,j} \partial_i u\, \partial_j v\, c^n_{i,j}\, dm_n,$$

où $u, v \in FC^1_c(H_n)$ et m_n est la mesure image de m par l'application $\xi \to \xi_{H_n}$.

On a besoin aussi de savoir la relation entre les processus C_s, $s \in H_n$, et $Z^{(n)}_{[0,1]}$. Les trois lemmes suivants préparent le terrain. Les deux premiers lemmes sont vrais pour tout processus de Markov symétrique conservatifs avec un opérateur carré du champs. On va les énoncer (sans démonstrations) pour le processus Z au lieu du processus $Z^{(n)}$ pour éviter la complexité des notations.

Lemme 2 : Soit $0 \leq t < t' \leq b$. Soit H une fonction $F_{\infty,t}$-mesurable de carré intégrable. Soit f et φ deux fonctions de $FC_c^2(D)$. Alors,

$$E[H\, Q_{b-t'}f(Z_{t'})\, \varphi(Z_{t'})] - E[H\, Q_{b-t}f(Z_t)\, \varphi(Z_t)]$$

$$= \int_t^{t'} du\, \{E[H\, \Gamma(Q_{b-u}f, \varphi)(Z_u)] + E[H\, Q_{b-u}f(Z_u)\, L\varphi(Z_u)]\}. \square$$

Lemme 3 : Soit γ(t) la partie martingale du processus $t \to \varphi(Z_t)$. On la formule suivante :

$$E[H\, Q_{b-t'}f(Z_{t'})\, \gamma(t')] - E[H\, Q_{b-t}f(Z_t)\, \gamma(t)]$$

$$= \int_t^{t'} du\, \{E[H\, \Gamma(Q_{b-u}f, \varphi)(Z_u)].$$

Par conséquent, le crochet oblique de la martingale $M_t = Q_{b-t}f(Z_t)$ avec γ(t) est donné par $\langle M, \gamma\rangle(dt) = \Gamma(Q_{b-t}f, \varphi)(Z_t)\, dt.$ □

Revenons à notre processus $Z^{(n)}$.

Lemme 4 : Pour toute martingale M de carré intégrable dans la filtration $F_{1,t}^{(n)} = \sigma\{Z_{u,v}^{(n)}, u \in [0,1], v \in [0,t]\}$, $\langle M, C_s\rangle = 0$ quelque soit $s > 1$.

Preuve : Désignons par $Q_{b-t}^{(n)}$ le semigroupe de $Z^{(n)}$. Utilisant la propriété de Markov locale du processus $Z_t^{(n)}$, on voit que l'ensemble des martingales qui sont de la forme

$$\sum_i G_i Q_{t_{i+1}-t}^{(n)} f_i(Z_t^{(n)})\, 1_{[t_i,t_{i+1}[}(t),$$

où (t_i) est une suite des points de R croissant vers l'infini, G_i est $F^{(n)}_{1,t_i}$-mesurable, $f_i \in FC^2_c(D\cap[0,1])$, est totale dans l'espace des $F^{(n)}_{1,t}$ -martingales de carré intégrable. Pour démontrer le lemme il suffit de considérer des martingales de la forme $M_t = Q^{(n)}_{b-t}f(Z^{(n)}_t)$.

Remarquons que la fonction $\xi \to \xi_s$ est localement une fonction de $D(\vartheta)$. Par conséquent, $\Gamma(Q_{b-t}f, \xi_s)$ est bien définie. Utilisant un argument de localisation et le Lemme 3, on voit que

$$\langle M, \gamma_s\rangle(dt) = \Gamma(Q_{b-t}f, \xi_s)(Z_t)dt.$$

D'où :

$$\langle M, C_s\rangle(dt) = \frac{1}{\sqrt{2}}\,\frac{1}{\sqrt{1-c^2_{s,s-}(t)}}(\langle M, \gamma_s\rangle(dt) - c_{s,s-}(t)\,\langle M, \gamma_{s-}\rangle(dt))$$

$$= \frac{1}{\sqrt{2}}\,\frac{1}{\sqrt{1-c^2_{s,s-}(t)}}(\Gamma(Q_{b-t}f, \xi_s)(Z_t) - c_{s,s-}(t)\,\Gamma(Q_{b-t}f, \xi_{s-})(Z_t))\,dt$$

$$= 0,$$

car, pour toute fonction v de $D(\varepsilon_n)\cap\sigma(\xi_{[0,1]})$,

$$\Gamma(v, \xi_s) = c_{s,s-}(t)\,\Gamma(v, \xi_{s-}),\quad s \in H_n.$$

(Vérifier cela d'abord pour $v \in FC^1_c(D\cap[0,1])$, et puis passer à la limite.) □

Lemme 5 : La famille $(C_s, s \in H_n)$ est une famille de mouvements browniens indépendants. De plus, les mouvements browniens C_s, $s > 1$, $s \in H_n$, sont encore indépendants de $Z^{(n)}_{[0,1]}$.

Preuve : Il suffit de calculer les crochet obliques des C_s pour voir qu'ils sont des mouvements browniens indépendants (notons que les processus $\frac{1}{\sqrt{2}}\gamma_s$ sont des mouvements browniens qui ne sont pas indépendants). La deuxième partie du lemme résulte de Lemme 4 et de la formule des martingales exponentielles. □

Lemme 6 : Le processus $t \to (Z^{(n)}_{s,t}, s \in H_n)$ est solution de l'équation différentielle stochastique suivante :

$$\frac{1}{\sqrt{2}} Y_s(dt) = \sum_{u \leq s, u \in H_n} c_{s,u}(t) \sqrt{1 - c^2_{u,u-}(t)}\, C_u(dt) + \frac{1}{\sqrt{2}} b(Y_s(t))\, dt,$$

pour tout $s \in H_n$. Par conséquent, $Z^{(n)}_{H_n}$ est fonction des processus C_s, $s \in H_n$.

Preuve : L'équation du lemme est une traduction des identités suivantes qui sont évidentes :

$$\frac{1}{\sqrt{2}} \xi_s(Z^{(n)}_t) = \frac{1}{\sqrt{2}} \xi_s(Z^{(n)}_0) + \frac{1}{\sqrt{2}} \gamma_s(t) + \int_0^t \frac{1}{\sqrt{2}} b(\xi_s(Z^{(n)}_u))\, dt.$$

L'équation stochastique du lemme a des coefficients localement lipchitziens sur J^{H_n}. Elle a donc l'unicité trajectorielle de la solution jusqu'à $\sup_k \tau_k(Y)$ avec

$$\tau_k(Y) = \inf\{t; Y(t) \notin (K_k)^{H_n}\},$$

où K_k est une suite strictement croissante de sous intervalles compacts de J dont la réunion est J. Comme le temps de vie du processus de Hunt X est infini, comme $Z^{(n)}_s$, $s \in H_n$, ont les mêmes loi que X, $\sup_k \tau_k(Y)$ est infini presque partout. Il en résulte que presque toutes les trajectoires de $Z^{(n)}_{H_n}$ sont fonctions de C_s, $s \in H_n$. Le lemme est démontré. □

Corollaire 7 : $E[f(Z^{(n)}_{H_n}) \mid Z^{(n)}_{[0,1]}] \in \sigma(Z^{(n)}_{[1]})$.

Preuve : D'après le lemme précédent, il existe une application F telle que $Z^{(n)}_{H_n} = F(C_s, s \in H_n)$. Comme $C_1 \in \sigma(Z^{(n)}_{[1]})$, et comme les C_s sont indépendants de $Z^{(n)}_{[0,1]}$ pour $s \in H_n \cap]1,2]$, le lemme en découle. □

Ce corollaire montre que, pour chaque n fixé, $Z^{(n)}$ possède (partiellement) la propriété de Markov horizontale. On va maintenant étudier les limites de $Z^{(n)}$. Fixons un nombre positif λ. Posons p_n la projection orthogonale de $D(\vartheta)$ sur $D(\vartheta)\cap\sigma(\xi_{K_n})$ par rapport au produit scalaire ϑ_λ. Soient respectivement R_λ et $R_\lambda^{(n)}$ les résolvents de Z et de $Z^{(n)}$.

Lemme 8 : On a : $p_n R_\lambda v = (R_\lambda^{(n)} v)(\xi_{K_n})$ pour tout $v \in \sigma(\xi_{K_n})$ bornée, et $\lim_n R_\lambda^{(n)} v(\xi_{K_n}) = R_\lambda v(\xi)$ pour toute v bornée de base finie contenue dans D.

Preuve : La première partie est évidente. Pour voir la deuxième partie, il suffit de remarquer que $\cup_n D(\vartheta)\cap\sigma(\xi_{K_n})$ est dense dans $D(\vartheta)$ muni du produit scalaire ϑ_λ (cf. les arguments dans Albeverio-Song [2]). □

Dans la suite, on va écrire simplement $R_\lambda^{(n)} v$ pour $(R_\lambda^{(n)} v)(\xi_{K_n})$. Soit $k \in N$. Soient u_1, ..., u_k des fonctions bornées de base finie dans K_n. On introduit les notations suivantes :

$$V_{n,1} = R_\lambda^{(n)} u_1,$$
$$V_{n,k} = R_\lambda^{(n)} (u_k V_{n,k-1}),\ k > 1.$$

On définit V_k de la même façon en remplaçant $R_\lambda^{(n)}$ par R_λ.

Lemme 9 : Pour tout $k \geq 1$, on a : $\lim_n V_{n,k} = V_k$ dans $L^2(E,m)$.

Preuve : Démontrer cela par récurrence en k. □

Les processus $Z^{(n)}$ peuvent être considérés comme des processus à valeurs dans E, si on interpolle les coordonnées $s \in H_n$ linéairement. Avec cette interprétation, on a le lemme suivant :

Lemme 10 : $\lim_n Z_{[0,2]}^{(n)} = Z_{[0,2]}$ en loi.

Preuve : Considérons $Z^{(n)}$ comme des variables aléatoires dans $F^{[0,2]}$. Soit K un intervalle compact de J. La famille des lois de $Z^{(n)}$ sur $F^{[0,2]}\cap\{Z_{0,0}^{(n)} \in K\}$ est clairement tendue (voir Billinsley [3]). Pour identifier les points d'adhérent de cette famille avec la loi de Z, il suffit de remarquer que la loi de $Z_{0,0}^{(n)}$ est μ pour tout n, et que les égalités suivantes sont vraies pour tout compact K:

$$\lim_n \int_0^\infty e^{-\lambda t_1 - \ldots - \lambda t_k} dt_1 \ldots dt_k \; E[1_K(Z^{(n)}_{0,0}) u_1(Z^{(n)}_{t_1}) u_2(Z^{(n)}_{t_1+t_2}) \ldots u_k(Z^{(n)}_{t_1+\ldots+t_k})]$$

$$= \lim_n E[1_K(Z^{(n)}_{0,0}) V_{n,k}] = E[1_K(Z_{0,0}) V_k]$$

$$= \int_0^\infty e^{-\lambda t_1 - \ldots - \lambda t_k} dt_1 \ldots dt_k \; E[1_K(Z_{0,0}) u_1(Z_{t_1}) u_2(Z_{t_1+t_2}) \ldots u_k(Z_{t_1+\ldots+t_k})],$$

où u_i sont des fonctions de $FC^1_c(D)$. □

Enfin, on peut conclure par le théorème suivant :

Théorème 11 : Pour tout $k \in N$, $E[f(Z_{H_k}) \mid Z_{[0,1]}] \in \sigma(Z_{[1]})$.

Preuve : Prenons des fonctions continues et bornées f et u. D'après le Corollaire 4, il existe une suite de fonctions uniformément bornées F_n telles que

$$E[u(Z^{(n)}_{[0,1]}) \, f(Z^{(n)}_{H_k})] = E[u(Z^{(n)}_{[0,1]}) \, F_n(Z^{(n)}_{[1]})].$$

On a alors :

$$E[u(Z_{[0,1]}) \, f(Z_{H_k})] = \lim_n \, E[u(Z^{(n)}_{[0,1]}) \, f(Z^{(n)}_{H_k})]$$

$$= \lim_n \, E[u(Z^{(n)}_{[0,1]}) \, F_n(Z^{(n)}_{[1]})] = \lim_n \, E[u(Z_{[0,1]}) \, F_n(Z_{[1]})],$$

car $Z^{(n)}_{[0,1]}$ a la même loi que $Z_{[0,1]}$ (regarder leurs espaces de Dirichlet associés). Comme F_n est uniformément bornée, $F_n(Z_{[1]})$ converge faiblement dans L^2 (quitte à extraire une sous suite) vers une fonction de $\sigma(Z_{[1]})$. Le théorème est prouvé. □

§9 Propriété de Markov horizontale II

Dans cette section, l'hypothèse * de la section précédente est toujours en vigueur. En plus de la propriété de Markov du processus horizontal $s \to (Z_{s,t}, t \in R_+)$, il est en fait un processus de Markov homogène et symétrique.

Théorème 1 : Soit $0 \le a < b$. Le processus horizontal est un processus stationnaire, c'est-à-dire, $Z_{[a,b]}$ est égal en loi à $Z_{[a+d,b+d]}$ pour tout $d > 0$. En particulier, le processus horizontal est un processus de Markov homogène.

Preuve : C'est simplement parce que la mesure m est invariante par la translation. □

Soit $a > 0$. Définissons l'opérateur de l'inversion du temps sur l'intervalle $[0,a]$: $\theta_a(\xi)_t = \xi_{a-t}$ pour $t \in [0,a]$. Le théorème suivant est facile à démontrer, car m est invariante par l'inversion du temps.

Théorème 2 : Soi $\vartheta \circ \theta_a$ la forme de Dirichlet image de $\vartheta_{[0,a]}$ par l'application θ_a, où $\vartheta_{[0,a]}$ est la restriction de ϑ sur $D_{[0,a]}(\vartheta)$. Alors, $\vartheta \circ \theta_a = \vartheta_{[0,a]}$. On a donc $\theta(Z_{[0,1]}) = Z_{[0,1]}$ en loi. Par conséquent, le processus horizontal est un processus de Markov symétrique. □

§10 Inégalités relatives à la capacité $c_{2,2}$.

On rappelle que le noyau U est défini par l'égalité suivante :

$$Uf(x) = \frac{1}{\sqrt{\pi}} \int_0^\infty t^{-1/2} e^{-t} P_t f(x)\, dt, \quad f \in L^2(J,\mu).$$

On désigne par $W^{2,2}$ l'ensemble des fonctions de la forme $U^2 f$ avec $f \in L^2(E,\mu)$ muni de la norme $N_{2,2}(U^2 f) = N_2(f)$, où $N_2(f)$ indique la norme usuelle d'un L^2-espace. On définit la capacité $c_{2,2}$: Pour une fonction f s.c.i. positive ou nulle,

$$c_{2,2}(f) = \inf\{\, N_{2,2}(h) \,;\, h \ge f \text{ presque partout, } h \in W^{2,2}) \,\},$$

et pour une fonction g quelconque,

$$c_{2,2}(g) = \inf\{\, c_{2,2}(f) \,;\, f \ge |g| \text{ partout, f fonction s.c.i.} \}.$$

On va démontrer les trois inégalités du Théorème §1.1 pour le c-semigroupe P_t qui satisfait l'hypothèse du §8. En fait, on a besoin de démontrer seulement les deux premières :

$$N_2[(\sup_{s,t \ge 0} e^{-s-t} |u|(Z_{s,t}))] \le C\, c_{2,2}(u),$$

$$N_2[(\sup_{s,t\geq 0} \left| \int_{0\leq u\leq s, 0\leq v\leq t} e^{-u-v} f(Z_{u,v}) dudv \right|) \leq C\, N_2(U^2 f).$$

La troisième en découle par la méthode de Hirsch [8].

Les résultats du §7 et la méthode de Song [14] permet de démontrer le théorème suivant.

Théorème 1 : Formule de Dynkin généralisée : Pour toute fonction u dans $L^2(J,\mu)$, pour tout a > 0, b > 0, on a :

$$E[\int_0^\infty e^{-s-t} u(Z_{s,t})\, dsdt \mid F_{a,b}]$$

$$= \int e^{-s-t} u(Z_{s,t}) 1_{\{0\leq s\leq a, 0\leq t\leq b\}}\, dsdt + \int e^{-s-b} U^2 u(Z_{s,b}) 1_{\{0\leq s\leq a\}}\, ds$$

$$+ \int e^{-a-t} U^2 u(Z_{a,t}) 1_{\{0\leq t\leq b\}}\, ds + e^{-a-b} U^4 u(Z_{a,b}). \ \square$$

Théorème 2 : On a l'égalité :

$$E[(\int_0^\infty e^{-s-t} u(Z_{s,t})\, dsdt)^2] = E[(U^2 u)^2].$$

Preuve : Calcul direct. $\square$

Corollaire 3 : Pour toute fonction u, on a l'inégalité suivante :

$$N_2[(\sup_{s,t\geq 0} e^{-s-t} |u|(Z_{s,t}))] \leq C \ c_{2,2}(u),$$

où C est une constante.

Preuve : Il suffit de reprendre la démonstration de Song [14]. Notons qu'on a la propriété (F4) qui assure l'inégalité de martingale à deux paramètres. $\square$

Théorème 4 : On a l'inégalité suivante :

$$N_2[(\sup_{a,b\geq 0} \left| \int_{0\leq s\leq a, 0\leq t\leq b} e^{-s-t} u(Z_{s,t})\, dsdt \right|) \leq C\, N_2(U^2 u),$$

où C est une constante.

Preuve : On réécrit la formule de Dynkin généralisée :

$$\int e^{-s-t}\, u(Z_{s,t})1_{\{0\leq s\leq a, 0\leq t\leq b\}}\, dsdt$$

$$= E[\int_0^\infty e^{-s-t}\, u(Z_{s,t})\, dsdt \mid F_{a,b}]$$

$$- \int e^{-s-b}\, U^2u(Z_{s,b})1_{\{0\leq s\leq a\}}\, ds$$

$$- \int e^{-a-t}\, U^2u(Z_{a,t})1_{\{0\leq t\leq b\}}\, ds$$

$$- e^{-a-b}\, U^4u(Z_{a,b}).$$

On considère les termes à droite. Le premier terme est une martingale à deux paramètres. Avec la condition (F4) et le Théorème 2, on montre facilement

$$N_2[(\sup_{a,b\geq 0} \left| E[\int_0^\infty e^{-s-t}\, u(Z_{s,t})\, dsdt \mid F_{a,b}] \right|)] \leq C\, N_2(U^2u).$$

Pour estimer le deuxième et le troisième termes, on a besoin d'introduire quelques notations. Définissons les opérateurs V et R comme suit :

$$Vf(\xi) = \frac{1}{\sqrt{\pi}}\, E[\int_0^\infty s^{-1/2}\, e^{-s}\, f(Z_{s,\bullet})\, ds \mid Z_{0,\bullet} = \xi],$$

$$Rf(\xi) = \frac{1}{\sqrt{\pi}}\, E[\int_0^\infty t^{-1/2}\, e^{-t}\, f(Z_{\bullet,t})\, dt \mid Z_{\bullet,0} = \xi],\ \ \xi \in E.$$

D'après les résultats du §7, on a les relations suivantes :

$$U^2u(Z_{s,t}) = R(Uu\circ\xi_s)(Z_{\bullet,t}) = V(Uu\circ\xi_t)(Z_{s,\bullet}).$$

On dénote par respectivement c_V et c_R les $c_{1,2}$-capacités définies à partir de V et de R. Maintenant considérons le deuxième terme. On a :

$$N_2[(\sup_{a,b\geq 0} \left| \int e^{-s-b}\, U^2u(Z_{s,b})1_{\{s\leq a\}}\, ds \right|)]$$

$$\leq N_2[(\sup_{b\geq 0} e^{-b} R[\sup_{a\geq 0}\left| \int e^{-s}\, U u\circ\xi_s\, 1_{\{s\leq a\}}\, ds\right|](Z_{\bullet,b}))]$$

$$\leq C\, c_R[R[\sup_{a\geq 0}\left| \int e^{-s}\, U u\circ\xi_s\, 1_{\{s\leq a\}}\, ds\right|],\ \text{d'après Song [14]},$$

$$\leq C\, N_2[\sup_{a\geq 0}\left| \int e^{-s}\, U u\circ\xi_s\, 1_{\{s\leq a\}}\, ds\right|],\ \text{d'après Song [14]},$$

$$\leq C\, N_2(U^2 u),\ \text{d'après Song [14]}.$$

Le troisième terme peut être estimé de la même manière, quitte à remplacer l'opérateur R par l'opérateur V.

Le dernier terme peut être estimé en utilisant le résultat du Corollaire 3. Le present théorème est démontré. □

Rappelons qu'une conséquence des résultats de la présente section est qu'on a une théorie du potentiel pour le processus Z à deux paramètres, comme elle est faite dans Song [14] et dans Hirsch [8].

Reférences

[1] Albeverio, S., Kusuoka, S., Röckner, M. On partial integration in infinite dimensional space and applications to Dirichlet forms, J. London Math. Sc. 42, 122-136, 1990.

[2] Albeverio, S., Song, S. Markovian uniqueness for logarithmically concave measures, Preprint 1993.

[3] Billingsley, P. Convergence of probability measures, John Wiley and Sons, New York, 1968.

[4] Bouleau, N., Hirsch, F. Dirichlet forms and analysis on Wiener space, Berlin - New York, de Gruyter, 1991.

[5] De La Pradelle, A., Feyel, D. Capacités Gaussiennes. Ann. Inst. Fourier, Grenoble , 41, 1, 49-76, 1991.

[6] Fukushima, M., Kaneko, H. On (r,p)-capacities for general Markov semigroups, Infinite dimensional analysis and stochastic processes, Edited by Albeverio-Boston-Pitman, Research notes in mathematics, 124, p.41-47, 1985.

[7] Fukushima, M. Dirichlet forms and Markov processes. North-Holland/Kodansha. 1980.

[8] Hirsch, F. Capacité associée à une somme de processus indépendants à accroissements indépendants, à paraître dans CRAS, 1993.

[9] Ikeda, N., Watanabe, S. Stochastic differential equations and diffusion processes, North-Holland/Kodansha, 1981.

[10] Kazumi, T., Shigekawa, I. Measures of finite (r,p)-energy and potentials on a separable metric space. Séminaire de Probabilités XXVI, Lecture Notes in Mathematics, 1526, Springer-Verlag, Berlin-Heidelberg-New York, 1992.

[11] Malliavin, P. Implicit functions in finite corank on the Wiener space, Stochastic analysis, Proceedings of the Taniguichi international symposium, Katata and Kyoto, Edited by K. Itô, North-Holland mathematical Library 32, p.369-386, Amsterdam, North-Holland, 1984.

[12] Ren, J. Topologie p-fine sur l'espace de Wiener et théorème des fonctions implicites. Bull.Sc.Math.,2e série ,114, 99-114, 1990.

[13] Schmuland, B. An alternative compactification for classical Dirichlet forms on topological vector spaces. Stochastics 33, 75-90, 1990.

[14] Song, S. Inégalités relatives aux processus d'Ornstein-Uhlenbeck à n-paramètres et capacité gaussienne $c_{n,2}$, à paraître dans Séminaire de Probabilités, 1993.

[15] Song, S. A study on Markovian maximality, change of probability and regularity, à paraître dans Potential Analysis, 1992.

[16] Sugita, H. Positive generalized Wiener functions and potential theory over abstract Wiener spaces. Osaka J. Math. 25, p.665-696, 1988.

[17] Yor, M. Intertwings of Bessel processes, Preprint 1991.

[18] Proceedings, Paris 1980 : Processus aléatoires à deux indices. Lecture Notes in Mathematics. 863. Springer-Verlag Berlin Heidelberg New York.

CAPACITIES AND HARMONIC MEASURES FOR UNIFORMLY ELLIPTIC OPERATORS OF NONDIVERGENCE FORM

Jang-Mei Wu
Department of Mathematics
University of Illinois
1409 West Green Street
Urbana, Illinois 61801 USA

In this note, we consider uniformly elliptic operators of nondivergence form

$$L = \sum A_{ij}(x) \frac{\partial^2}{\partial x_i \partial x_j} \tag{1}$$

where A is symmetric, bounded and continuous in a domain $\Omega \subseteq \mathbf{R}^n$, and satisfies

$$\Lambda^{-1}|\xi|^2 \leq \sum A_{ij}(x)\xi_i\xi_j \leq \Lambda|\xi|^2 \tag{2}$$

for all $x \in \Omega$, $\xi \in \mathbf{R}^n$, and some constant $\Lambda \geq 1$.

A function $u \in W^{2,n}_{\text{loc}}(\Omega)$ is a solution of $L = 0$ if $Lu = 0$ a.e., in Ω. The maximum principle, the existence and uniqueness of the solution to the Dirichlet problem, and the Harnack principle are well known [GT]. A lower semicontinuous function v is called an L-supersolution on Ω, if for any closed ball $D \subseteq \Omega$ and any L-solution u in D with u continuous on $\overline{D}$, the inequality $v \geq u$ on ∂D implies that $v \geq u$ in D. A function $v \in C^2(\Omega)$ is an L-supersolution if and only if $Lv \leq 0$.

We shall discuss sets of capacity zero and sets of harmonic measure zero for these operators, and compare them with known results for divergence operators.

1. Capacities.

Let B be the unit ball in $\mathbf{R}^n$, $\overline{x} = \left(\frac{1}{2}, 0, 0, \cdots, 0\right)$ and S be a compact set in $\left\{|x| \leq \frac{1}{4}\right\}$. The L-capacity of S is defined as

$$\inf\{v(\overline{x}) : \ v \text{ is a positive } L-\text{supersolution on } B \text{ and } v \geq 1 \text{ on } S\}.$$

When A_{ij} are Dini continuous, sets of L-capacity zero are precisely those of capacity zero for the Laplacian Δ; this follows from the growth of the Green function for these operators ([A] and [K]). In general, a result of Gilbarg and Serrin

K. GowriSankaran et al. (eds.), Classical and Modern Potential Theory and Applications, 453–459.

states that there are operators L with continuous coefficients when $n = 2$, bounded coefficients when $n \geq 3$, for which a single point has positive L-capacity; again this reflects the behavior of the Green function ([Bau1], [Bau2], [GS] and [M]).

Recall that for the Laplacian Δ, a set has positive capacity if it has positive h-Hausdorff measure Λ_h for some $h > 0$ satisfying $\int_0^1 h(r)/r^{n-1} dr < \infty$; and a set has zero capacity if it has finite $(n-2)$-dimensional Hausdorff measure when $n \geq 3$, or finite logarithmic measure when $n = 2$.

In general, sets of L-capacity zero do not coincide with those of vanishing capacity for the Laplacian. The following theorems are proved in [W1].

Theorem 1. *Let $n \geq 2$, $a > 0$ and $h(r) = r^{n-2} \left(\log \frac{1}{r}\right)^{-1-a}$. There exist a set $S \subseteq \frac{1}{4}B \subseteq \mathbf{R}^n$ of dimension $n-2$, positive Hausdorff h-measure (thus $\Delta-$cap $S > 0$), and a nondivergence operator L with continuous coefficents such that L-cap $S = 0$.*

Let $k(r) = r^{n-2} \left(\log \frac{1}{r}\right)^{a}$. There exist a set $S_0 \subseteq \frac{1}{4}B \subseteq \mathbf{R}^n$ of dimension $n-2$, and an operator L_0 in $\mathbf{R}^n$ with continuous coefficients so that the Hausdorff measure $\Lambda_k(S_0) = 0$ (thus $\Delta - \text{cap } S_0 = 0$) but $L_0 - \text{cap } S_0 > 0$.

Moreover, the modulus of continuity for the coefficients of both operators can be chosen to be $\left(\log \frac{1}{t}\right)^{-1}$, and the ellipticity constants Λ can be chosen to be bounded above by $1 + 0(1) \left(\log \frac{1}{\text{dist}(x,S)}\right)^{-1}$.

The definition of L-capacity zero can be extended to operators whose coefficients are only bounded and continuous on $D \backslash S$. In this case, we say a set $S \subseteq \frac{1}{4}B$ has L-capacity zero if

$$\inf\{v(\overline{x}) \;:\; v \text{ is a positive } L-\text{supersolution on } D\backslash S, \text{ and } \lim_{x \to x_0} \inf v(x) \geq 1 \text{ at each } x_0 \in S\} = 0;$$

otherwise we say S has positive L-capacity. The following is proved in [W1].

Theorem 2. *Let $0 < \alpha < n-2 < \beta < n$. Then there exist compact sets S and S_0 in $\frac{1}{4}B$ and operators L and L_0 with coefficients bounded continuous in $B\backslash S$ and $B\backslash S_0$ respectively, so that $\dim S = \alpha$ but L-cap $S > 0$, and $\dim S_0 = \beta$ but L_0-cap $S_0 = 0$. Moreover the ellipticity constants can be chosen to be $1 + 0(1)(n-2-\alpha)$ as $\alpha \to n-2$, or $1 + 0(1)(\beta - n + 2)$ as $\beta \to n-2$ respectively.*

Since a point may have positive L-capacity, the only new part about S and L in Theorem 2 is the relation between the ellipticity constant and the dimension.

Write $\Delta = \frac{\partial^2}{\partial r^2} + (n-1)r^{-1}\frac{\partial}{\partial r} + r^{-2}\delta$, where $r = |x|$ and δ is the spherical part of the Laplacian, and let

$$L_1 = \frac{\partial^2}{\partial r^2} + b(x)(n-1)r^{-1}\frac{\partial}{\partial r} + b(x)r^{-2}\delta$$
$$= \sum \left[b(x)\delta_{ij} - (b(x)-1)x_i x_j r^{-2}\right] \frac{\partial^2}{\partial x_i \partial x_j}$$

for some $b(x) > 0$. It is observed in [GS] that when $b(x)$ is a radial function, the function

$$u(x) = \int_{|x|}^{1/2} \exp\left[\int_s^{1/2} (n-1)b(t)t^{-1}\,dt\right] ds$$

is a solution of $L_1 = 0$ in $\mathbf{R}^n \backslash \{0\}$ and it plays the role of Green function for L_1. By choosing the function $b(t)$ with various moduli of continuity at 0, one may obtain $u(x)$ of different growth at 0. In particular, if $n = 2$ and $b(t) = 1 + 2(\log t)^{-1}$, then $u(x) = c_1 + c_2(\log|x|)^{-1}$, which is continuous in $\mathbf{R}^n$ and has a nonremovable singularity at 0. Thus $\{0\}$ has positive L_1-capacity.

In the theorems above, the sets constructed are Cantor sets in the form $\bigcap_k \bigcup_j B_{k,j}$, where $\{B_{k,j}\}_j$ are mutually disjoint balls of the same size, situated very close to the centers of $\{B_{k-1,i}\}_i$. The operators in question resemble L_1 in a sequence of rings $\{R_{k,j}\}$; each $R_{k,j}$ is a ring of a large modulus in $B_{k,j} \setminus \bigcup_i B_{k+1,i}$. After the operator and the set are properly chosen, one may produce an L-supersolution having the desired growth near each point in the Cantor set. This is the main idea for the proof; see [W1] for detail.

2. Harmonic Measures.

Let L be the operator of the nondivergence form (1) defined in $\mathbf{R}^n_+$. In view of the maximum principle, and the existence and the uniqueness of the solution for the Dirichlet problem, there exists a unique positive Borel measure ω_L^x on $\partial\mathbf{R}^n_+$ so that for any $g \in C_0(\partial\mathbf{R}^n_+)$,

$$u(x) = \int_{\partial\mathbf{R}^n_+} g(y)\,d\omega_L^x(y)$$

is the solution of $Lu = 0$ in $\mathbf{R}^n_+$, which is continuous on $\overline{\mathbf{R}^n_+}$ and satisfies $u|\partial\mathbf{R}^n_+ = g$. The measure ω_L^x is called the L-harmonic measure at x.

Similarly, one may define the L-harmonic measure ω_L for a uniformly elliptic operator L of divergence form in $\mathbf{R}^n_+$:

$$L = \sum \frac{\partial}{\partial x_i}\left(A_{ij}(x)\frac{\partial}{\partial x_j}\right), \tag{3}$$

where A_{ij} are bounded measurable and satisfy the ellipticity condition (2).

In a sequence of papers ([CFK], [MM] and [MMS]), Modica, Mortola, Salsa, Caffarelli, Fabes and Kenig have proved the following.

Theorem 3. *There are uniformly elliptic operators L of nondivergence form* (1) *or of divergence form* (3), *with coefficients continuous in $\overline{\mathbf{R}^n_+}$, so that the corresponding L-harmonic measures on $\partial\mathbf{R}^n_+$ are singular with respect to the surface measure σ.*

In these examples, first an operator L_{x_{n-1},x_n} is constructed on the half plane $\{(x_{n-1},x_n) : x_n > 0\}$, so that the L_{x_{n-1},x_n}-harmonic measure is singular with respect to length; then the operator is extended to $\mathbf{R}^n_+$ by adding the Laplacian of the other $n-2$ variables. Therefore the operators and the supports of the corresponding harmonic measures have the forms

$$\text{(A)}\quad \sum_1^{n-2} \frac{\partial^2}{\partial x_j^2} + L_{x_{n-1},x_n} \qquad \text{and} \qquad \text{(B)}\quad \mathbf{R}^{n-2}\times E\times\{0\} \ \text{ with } \ E\subseteq \mathbf{R}^1\ ,$$

respectively.

The method in [CFK] is based on well-known theorems of Beurling and Ahlfors [BA] on quasiconformal extensions of quasisymmetric functions on $\mathbf{R}^1$, and on the existence of singular quasisymmetric functions on $\mathbf{R}^1$. The operator L_{x_{n-1},x_n} is the pullback of the Laplacian under quasiconformal mappings in $\mathbf{R}^2$ and is in the divergence form. This method breaks down in $\mathbf{R}^n (n \geq 3)$, because the pullback of the Laplacian under quasiconformal mappings need not be uniformly elliptic when $n \geq 3$.

In [MM] and [MMS], the authors constructed L_{x_{n-1},x_n} in the divergence form and the nondivergence form by considering Riesz products.

There are some natural questions. How small can the Hausdorff dimension of the support of an L-harmonic measure be? Is there a genuinely n-dimensional approach that will produce operators L whose harmonic measures are singular with respect to the surface measure, but which cannot be rebuilt from 2-dimensional operators in any obvious way?

In the *divergence case*, the proposition below follows from the boundary Harnack principle, a comparison theorem between harmonic measure and the Green function and a well-known theorem of Littman, Stampacchia and Weinberger on the Green function.

Proposition 1. *On $\partial\mathbf{R}^n_+$, a set of Hausdorff dimension $\leq n-2$ has zero L-harmonic measure for all divergence operators L in $\mathbf{R}^n_+$.*

The number $n-2$ is sharp. In fact, Tukia [T] recently proved that given any $\epsilon > 0$, there exists a quasisymmetric homeomorphism on $\mathbf{R}^1$ which maps a set of dimension $< \epsilon$ onto a set whose complement in $\mathbf{R}^1$ has dimension $< \epsilon$. Using Tukia's Theorem instead of that of Beurling and Ahlfors in the 2-dimensional quasiconformal mapping approach used by Caffarelli, Fabes and Kenig, one can easily obtain the following.

Proposition 2. *Given* $\alpha \in (n-2, n-1]$, *there exist a divergence operator* L *of the form* (A) *so that the* L*-harmonic measure is supported on a product set* $\mathbf{R}^{n-2} \times E \times \{0\}$ *of dimension at most* α.

In [W2], we produced a large class of porous sets on $\mathbf{R}^{n-1}$, with precisely described porosity condition, which are perfect sets of zero $(n-1)$-dimensional measure but have positive ω_L-measure for some uniformly elliptic operator L in $\mathbf{R}^n_+$. The set and operator are truly higher dimensional, and cannot be reproduced from (A) or (B) in any obvious way. The porosity condition introduced there is sharp. That is, if a set has "larger holes in some sense", it must have zero ω_L-measure for all operators of divergence form or nondivergence form.

Unfortunately, the porous sets constructed there have dimension exactly $n-1$. Therefore this method, although it is n-dimensional, only produces operators whose harmonic measures are singular.

In the *nondivergence case*, the n-dimensional coefficient building technique produces a stronger result [W3].

Theorem 4. *For each* $\alpha \in (n-2, n-1]$, *there exists a nondivergence operator* L *in* $\mathbf{R}^n_+$ *so that the support of* ω_L *has dimension* $\leq \alpha$.

We are unaware of any analogue of Proposition 1 for nondivergence operators, due to the fact that the Green function for a nondivergence operator need not be locally bounded even away from the pole; see the work of Bauman [Bau1]. Since a point may have positive capacity for some L, we conjecture the following.

Conjecture. *For each* $\alpha \in (0, n-1]$, *there exists a nondivergence* L in $\mathbf{R}^n_+$ *so that the support of* ω_L *has dimension* $\leq \alpha$.

The operator constructed in Theorem 4 has diagonal coefficient matrix, with $A_{jj} \equiv 1$ for all $j \neq n$. After subdividing $\mathbf{R}^n_+$ into rectangular cubes, we assign $A_{nn}(x)$ to be a large constant near the lateral boundary of each cube, and $A_{nn}(x)$ to be a small constant practically everywhere else. This enables the lateral boundaries of these cubes to serve as traps and to guide the stochastic process to the small portion of $\partial\mathbf{R}^n_+$ intended. Coefficients are explicitly written down. And the proof is potential theoretic-comparing solutions with carefully constructed subsolutions. Solution of nondivergence operators are in the strong sense, and sometimes smooth subsolutions can be found by inspection. Because of this, we are able to obtain stronger examples for nondivergence operators.

In all our examples, coefficients are not continuous up to the boundary of $\mathbf{R}^n_+$. *If the coefficients are continuous on* $\overline{\mathbf{R}^n_+}$ *and the vertical modulus of continuity of the coefficients satisfies a certain Dini condition, then the harmonic measure* ω_L *and the surface measure are mutually absolutely continuous on* $\partial\mathbf{R}^n_+$. There is considerable literature in this subject due to the contributions of Dahlberg, Fefferman, Kenig, Pipher and Barceló ([B1], [B2], [D], [F] and [FKP]).

References.

[A] A. Ancona, *Principe de Harnack a la frontière et théorème de Fatou pour un opérateur elliptique dans un domaine Lipschitzien*, Ann. Inst. Fourier (Grenoble) **28** (1978), 169-213.

[B1] T. Barceló, *A completely singular harmonic measure for a nondivergence form elliptic operator with a drift*, Communications in P.D.E. **14** (1989), 931-958.

[B2] T. Barceló, *On the harmonic measure for nondivergence elliptic equations*, Communications in P.D.E. **16** (1991), 1367-1422.

[Bau1] P. Bauman, *Equivalence of the Green's functions for diffusion operators in* $\mathbf{R}^n$*: a counter-example*, Proc. Amer. Math. Soc. **91** (1984), 64-68.

[Bau2] P. Bauman, *A Wiener test for nondivergence structure, second-order elliptic equations*, Indiana Univ. Math. J. **34** (1985), 825-844.

[BA] A. Beurling and L. Ahlfors, *The boundary correspondence under quasiconformal mappings*, Acta Math. **96** (1956), 125-142.

[CFK] L. Caffarelli, E. Fabes and C. Kenig, *Completely singular elliptic-harmonic measures*, Indiana Univ. Math. J. **30** (1981), 917-924.

[D] B. Dahlberg, *On the absolute continuity of elliptic measures*, Amer. J. of Math. **108** (1986), 1119-1138.

[F] R. Fefferman, *A criterion for the absolute continuity of the harmonic measure associated with an elliptic operator*, Journal of the AMS **1**:1 (1989), 127-136.

[FKP] R. Fefferman, C. Kenig and J. Pipher, *The theory of weights and the Dirichlet problem for elliptic equations*, Ann. Math. **134** (1991), 65-124.

[GS] D. Gilbarg and J. Serrin, *On isolated singularities of solutions of second order elliptic differential equations*, J. Analyse Math. **4** (1955-56), 309-340.

[GT] D. Gilbarg and N.S. Trudinger, *Elliptic partial differential equations of second order*, 2nd ed., Springer-Verlag, 1983.

[K] N.V. Krylov, *The first boundary-value problem for second-order elliptic equations*, Differentsial'nye Uravneniya **3** (1967), 315-326; English transl., Differntial Equations **3** (1967), 158-164.

[M] K. Miller, *Nonequivalence of regular boundary points for the Laplace and nondivergence equations, even with continuous coefficients*, Ann. Scuola Norm. Sup. Pisa (3) **24** (1970), 159-163.

[MM] L. Modica and S. Mortola, *Construction of a singular elliptic-harmonic measure*, Manuscripta Math. **33** (1980), 81-98.

[MMS] L. Modica, S. Mortola and S. Salsa, *A nonvariational second order elliptic operator with singular elliptic measure*, Proc. Amer. Math. Soc. **84** (1982), 225-230.

[T] P. Tukia, *Hausdorff dimension and quasisymmetric mappings*, Math Scand. **65** (1989), 152-160.

[W1] J.-M. Wu, *Examples of capacity for some elliptic operators*, Trans. Amer. Math. Soc. **333** (1992), 379-395.

[W2] J.-M. Wu, *Porous sets and null sets for elliptic harmonic measures*, Trans. Amer. Math. Soc., to appear.

[W3] J.-M. Wu, *Harmonic measures for elliptic operators of nondivergence form*, preprint.

PROBLEMS

Wiener test for the Dirichlet problem for Δ^2

Find necessary and sufficient conditions on the boundary of a bounded domain $\Omega \subset \mathbb{R}^n$, $n \geq 4$, preferably of a geometric nature, that imply

$$\lim_{\substack{x \to x_0 \\ x \in \Omega}} u(x) = 0$$

for $u \in W_0^{2,2}(\Omega)$ satisfying $\Delta^2 u = f$ in Ω, $f \in C^\infty(\bar{\Omega})$. In [M], Maz'ya gives the sufficient condition

$$(*) \qquad \int_0 r^{4-n}\, cap_2(\Omega^c \cap B\overline{x_0, r}))\frac{dr}{r} = +\infty$$

for $n = 4, 5, 6, 7$. Here $cap_2(K) = \inf \int |\Delta v|^2\, dx$, where $v \in W_0^{2,2}(\Omega)$ and $v \geq 1$ on K a.e.; K = compact set in $\mathbb{R}^n$. Is condition $(*)$ sufficient for $n \geq 8$? Is it necessary? This problem has been posed in [PB], page 507f.

References

[M2] V.G. Maz'ya, *Behavior of solutions to the Dirichlet problem for the biharmonic operator at the boundary point*, Wissenschaftliche Inform. **10**, Tech. Hoch. Karl-Marx-Stadt, 1979.

[M2] V.G. Maz'ya, *Über die regularität eines randpunktes für elliptische differentialgleichungen*, Linear and Complex Analysis Problem Book, ed. V.P. Havin, S.V. Hruščëv, N.K. Nikol'skii, Lecture Notes in Math. **1043**, Springer-Verlag 1984.

David R. Adams (Kentucky)

Estimates for p-harmonic measure

Let $w_p(x) = w_p(x; K)$ be the $W^{1,p}$-solution to

$$\Delta_p w_p = 0 \quad \text{in } B$$

$$w_p = \chi_K, \quad K \subset \partial B,$$

$1 < p < \infty$, where B = unit ball centered at the origin in $\mathbb{R}^n$, K is a compact subset of ∂B, and Δ_p is the usual p-Laplace operator ($\Delta_p u = \operatorname{div}(|\nabla u|^{p-2}\nabla u)$);

K. GowriSankaran et al. (eds.), Classical and Modern Potential Theory and Applications, 461–470.

see [HKM]. Then the question is: how does $w_p(0;K)$ depend on K? When $p=2$, $w_2(0;K)$ is a measure, the harmonic mass on ∂B, and then very precise estimates for w_2 are known; indeed $w_2(0;K)$ is just Hausdorff $(n-1)$ measure H^{n-1} on ∂B. One would like estimates for $w_p(0;K)$, perhaps interms of H^{n-1}. Is $w_p(0;K) \leq \text{const.}\ [H^{n-1}(K)]^{(n-p+1)/(n-1)}$ true; how about simply $w_p(0;B(x_0,r)\cap B) \leq \text{const.}\ r^{n-p+1}$, where $x_0 \in \partial B$? It is not difficult to show

$$(*) \qquad w_p(0;K) \leq \text{const.}\ (H^{n-p}_{(\infty)}(K))^{1/p-1},$$

where $H^s_{(\infty)}(K)$ is the infimum of the sum $\sum_j r_j^s$ over all covers of K by balls $B(x_j,r_j)$. If fact one has $w_p(0;K) \leq \text{const.}\ C_p(K)^{1/p-1}$, where $C_p(K) = \inf\{\int|\nabla\phi|^p\,dx : \phi \in C_0^\infty(\mathbf{R}^n),\ \phi \geq 1 \text{ on } K\}$. The case $p=n$ is discussed in [HM]. The proof of $(*)$ is via the techniques of [KM].

References

[HM] J. Heinonen, O. Martio, *Estimates for F-harmonic measures and Oksendal's theorem for quasi-conformal mapping*, Ind. U. Math. J. **36** (1987), 659-683.

[HKM] J. Heinonen, T. Kilpeläinen, O. Martio, *Nonlinear Potential Theory of Degenerate Elliptic Equations*, Oxford Math. Monographs, Oxford Press 1993.

[KM] T. Kilpeläinen, J. Malý, *The Wiener test and potential estimates for quasilinear elliptic equations*, preprint Jyväskylä U., Feb. 1993.

David R. Adams (Kentucky)

Conditions for Separately Subharmonic Functions to be Subharmonic

Let Ω be a non-empty open subset of $\mathbf{R}^m \times \mathbf{R}^n$ ($m \geq 2$, $n \geq 2$). A function u on Ω is *separately subharmonic* (s.s.h.) on Ω if, for each fixed point $(x,y) \in \Omega$, the functions $u(x,\cdot)$ and $u(\cdot,y)$ are subharmonic (s.h.) on $\{\eta \in \mathbf{R}^n : (x,\eta) \in \Omega\}$ and $\{\xi \in \mathbf{R}^m : (\xi,y) \in \Omega\}$ respectively.

REMARKS. (i) A s.s.h. function is not necessarily s.h. This was proved by Wiegerinck [6] as recently as 1988 (see also Wiegerinck and Zeinstra [7]).

(ii) Let u be s.s.h. on Ω. Riihentaus [5] showed that if u has a majorant which is locally in L^p for some $p > 0$, then u is s.h. on Ω. The cases $p = +\infty$ and $p = 1$ are due to Avanissian [3] and Arsove [2] respectively. Arsove [2] also proved the measurability of s.s.h. functions.

RECENT RESULTS. Using ideas of Domar [4], we prove [1, Theorem 1] a result which substantially improves earlier theorems and allows us to produce examples of functions $\phi : [0, +\infty) \to [0, +\infty)$ with the following property: if u is s.s.h. on Ω and $\phi(\log^+ u^+)$ is locally integrable on Ω, then u is s.h. on Ω. Let us say that such a function ϕ is *admissible.*

EXAMPLE: If $\epsilon > 0$, then

$$\phi(t) = t^{m+n-2+\epsilon},$$

$$\phi(t) = t^{m+n-2}(\log(1+t))^{\max\{m,n\}-1+\epsilon}$$

are admissible. However, if ϕ is increasing and $\phi(t) = o(t^{m+n-2})$ as $t \to +\infty$, then ϕ is not admissible [1, Theorem 2].

PROBLEM. Ideal form: Characterize admissible functions.
Modest form: Determine whether, for example, $\phi(t) = t^{m+n-2}$ is admissible.

References

1. Armitage, D.H. and S.J. Gardiner: Conditions for separately subharmonic functions to be subharmonic, *Potential Analysis* **2**, 255-261 (1993).

2. Arsove, M.G.: On subharmonicity of doubly subharmonic functions, *Proc. Amer. Math. Soc.* **17**, 622-626 (1966).

3. Avanissian, V.: Fonctions plurisousharmoniques et fonctions doublement sous-harmoniques, *Ann. Sci. École Norm. Sup.* **78**, 101-161 (1961).

4. Domar, Y.: On the existence of a largest subharmonic minorant of a given function, *Ark. Mat.* **3**, 429-440 (1957).

5. Riihentaus, J.: On a theorem of Avanissian-Arsove, *Exposition. Math.* **7**, 69-72 (1989).

6. Wiegerinck, J.: Separately subharmonic functions need not be subharmonic, *Proc. Amer. Math. Soc.* **104**, 770-771 (1988).

7. Wiegerinck, J. and R. Zeinstra: Separately subharmonic functions: when are they subharmonic, *Proc. Symp. Pure Math.* **52**, 245-249 (1991).

D.H. Armitage (Queen's University, Belfast)
S.J. Gardiner (University College, Dublin)

Axiom of Domination and Choquet Theory

In harmonic spaces (or even balayage spaces) certain properties as the axiom of polarity (every semipolar set is polar), axiom of thinness (every semipolar set is totally thin), or ellipticity can be characterized by properties suggested by the Choquet theory of the cones

$$S(U) := \{s \in \mathcal{C}(\overline{U}) : s \text{ is superharmonic on } U\},$$

where U is an open (relatively compact) set (see [1,2]).

Are there similar properties of $S(U)$ which characterizes the axiom of domination?

References

1. J. Bliedtner, W. Hansen: *A simplicial characterization of elliptic harmonic spaces.* Math. Ann. **222** (1976), 261-274.

2. J. Bliedtner, W. Hansen: *Potential Theory. An Analytic and Probabilistic Approach to Balayage.* Universitext, Springer-Verlag (1986).

J. Bliedtner (Frankfurt)

Two problems of potential theory arising in the study of the distribution of values of entire functions

For an entire function f define the Nevanlinna *characteristic* $T(r) = T(r,f)$ and the *proximity function* $m(r,a) = m(r,a,f)$ by

$$T(r) = \frac{1}{2\pi}\int_{-\pi}^{\pi} \log^+ |f(re^{i\theta})|\, d\theta,$$

and

$$m(r,a) = \frac{1}{2\pi}\int_{-\pi}^{\pi} \log^+ \frac{1}{|f(re^{i\theta}) - a|}\, d\theta.$$

The characteristic measures the growth of f. For all $a \in \mathbf{C}$ we have $m(r,a) \leq T(r) + O(1)$, $r \to \infty$. The proximity function measures the average approximation of f to a constant a on the circles $\{z : |z| = r\}$. For most values of a the proximity function is small in comparison with $T(r)$, namely $m(r,a) = o(T(r,a))$, $r \to \infty$ for all a except a set of zero logarithmic capacity. The defect of f at the point a is defined as

$$\delta(a,f) = \liminf_{r\to\infty} m(r,a)/T(r).$$

It follows from definitions that $0 \leq \delta(a, f) \leq 1,\ a \in \mathbf{C}$. The Nevanlinna's defect relation in the case of entire functions says that the set of deficient values (i.e. such that $\delta(a, f) > 0$) is at most countable and

$$\sum_{a \in \mathbf{C}} \delta(a, f) < 1. \tag{1}$$

Nothing else can be said on deficiencies for *all* entire functions as examples in [3] show. But in the important class of entire functions of finite order much more is true. The order is defined by

$$\rho = \limsup_{r \to \infty} \log T(r, f) / \log r.$$

Lewis and Wu proved in [5] that for entire functions of finite order

$$\sum_{a \in \mathbf{C}} \delta^{\alpha}(a, f) < \infty$$

with $\alpha = 1/3 - 2^{-264}$. (Similar result for *meromorphic* functions of finite order is true with $\alpha = 1/3$ and this is best possible [6,3]). In the opposite direction there are examples [1] of entire functions f of finite order with $\delta(a_n, f) \geq c^{-n}$ for some sequence a_n and $c < 1$. So what is the best possible estimate of the rate of decrease of deficiencies of entire functions of finite order? The problem is equivalent to the following [2].

Problem 1 *Let $u < 1$ be a subharmonic function in the annulus $A = \{z : 1 < |z| < 2\}$. Let $E_n \subset A$ be some open sets such that $u(z) < 0,\ z \in E_n$ and $u(z) = 0,\ z \in \partial E_n \cap A$. Set*

$$\delta_n = - \inf_{r \in (1,2)} \int_{re^{i\theta} \in E_n} u(re^{i\theta})\, d\theta.$$

What is the best possible estimate of the rate of decrease of δ_n?

Another restriction on deficiencies of entire (and meromorphic) functions of finite order is the following: if we have equality in (1) then the number of deficient values is finite. Actually it is at most ρ in the case of entire functions. There is an old conjecture of Edrei, Fuchs and Hayman [4], that this result is stable in the following sense: there exists a positive function $\epsilon(\rho)$ such that for every entire function of order ρ the condition

$$\sum_{a \in \mathbf{C}} \delta(a, f) > 1 - \epsilon(\rho)$$

implies that the number of deficient values is at most ρ. This is equivalent to the following problem of potential theory.

Problem 2 *Let u_n be a sequence of subharmonic functions in the square $S = \{x + iy : |x| < 1, |y| < 1\}$. Assume that $u_n(z) \to \Re z,\ n \to \infty$ uniformly in S. Consider the sets $E_n = \{z \in S : u_n(z) < 0\}$. Denote by E_n' the connected component* *of E_n which contains the point $-1/2$. Is it possible that $E_n \backslash E'_n$ intersect every horizontal line $\Im z = t$, $-1 < t < 1$?*

This question seems to be related to the estimates of of Cauchy potentials

$$\int (z-t)^{-1}\, d\mu_t$$

outside an exceptional set.

References

1. A. Eremenko. A counterexample to the Arakelyan Conjecture. *Bull. Amer. Math. Soc.*, 27, 1, 1992, 159-164.

2. A. Eremenko. On the set of deficient values of entire function of finite order. *Ukrainian Math. J.*, 39, 1987, 377-390 (English transl.)

3. W. Hayman. *Meromorphic Functions.* Oxford, Clarendon Press, 1964.

4. W. Hayman. *Research Problems in Function Theory.* Athlone Press, London, 1967.

5. J. Lewis, Jang-mei Wu. On Conjectures of Arakelyan and Littlewood. *J. Analyse Math.*, 50, 1988, 259-283.

6. A. Weitsman. A theorem on Nevanlinna deficiencies. *Acta math.*, 128, 1972, 41-52.

A. Eremenko (Purdue University)

Non-tangential Limits in Polydiscs

Let $\mathbf{D}^n = \mathbf{D} \times \mathbf{D} \times \cdots \times \mathbf{D}$ be the n-polydisc consisting of the product of unit discs. It is well known that the non-tangential limit of $u(z_1, z_2, \cdots, z_n)$ as $(z_1, z_2, \cdots, z_n)$ converges to $(e^{it_1}, \cdots, e^{it_n})$ exists for all bounded n-harmonic functions on $\mathbf{D}^n$. It is also known that such a limit may not exist for a n-harmonic (> 0) function u represented by a singular measure on the n-torus (relative to the product of Poisson kernels). [G2, Z]. Here is the problem:

Suppose w is a positive n-harmonic function whose unique (canonical) representing measure on $\mathbf{T}^n$ is absolutely continuous relative to the Lebesgue (= Haar) measure on $\mathbf{T}^n$ [G1]. Do the non-tangential limits of $w(z_1, \cdots, z_n)$ exist (which should then equal a Radon-Nikodym derivative of the measure of w relative to the Lebesgue measure) for almost every element of $\mathbf{T}^n$? See also [G3].

References

[G1] K. GowriSankaran, "Multiply Harmonic functions", Nagoya J. Math. 28(1966), 22-48.

[G2] K. GowriSankaran, "On a problem of Doob concerning multiply superharmonic functions", Nagoya J. Math. 39(1970), 127-132.

[G3] K. GowriSankaran, "Polydiscs and non-tangential limits", Proc. Amer. Math. Soc. 115, 4, (1992), 977-984.

[Z] A. Zygmund, "Trigonometrical Series", vol. 2, 2nd serie ed. Cambridge University Press, New York, 1959.

K. GowriSankaran (McGill University)

Problems
Subharmonicity of doubly subharmonic functions

Let $u : D_1 \times D_2 \mapsto \mathbf{R} \cup \{\infty\}$ be a function defined on the product of two discs $D_1 \subset \mathbf{R}^n$, $D_2 \subset \mathbf{R}^n$ with centres respectively at x_0 and y_0. Suppose for all x in D_1 (resp. all $y \in D_2$) $u(x, \cdot)$ is harmonic (resp. $u(\cdot, y)$ is superharmonic). It has been an open problem to decide if u is superharmonic on $D_1 \times D_2$. This problem reduces to proving that u is locally lower bounded or integrable. The following is a new development.

THEOREM. There exists an open ball D_2' of positive radius centred at some point y_1 in D_2 such that $u(x, y)$ is the sum of a (uniformly and absolutely) convergent power series (or if preferred, convergent series of homogeneous harmonic polynomials in y) with coefficients that are superharmonic in $x \in D_1$.

Does this result make any contribution to solve the original problem?

References

See the list provided by D.H. Armitage and S.J. Gardiner.

K. GowriSankaran (McGill University)

Fractional Order Sobolev Spaces on Wiener Space

For $1 < p < \infty$ and $0 < \alpha \leq 1$, we consider the Sobolev space

$$W^{\alpha,p} = D_{L^p(\lambda)}((I - \Delta)^{\alpha/2}),$$

where λ is the Lebesgue measure on $\mathbf{R}^n$ and Δ the Laplace operator. When $\alpha = 1$, $W^{1,p} = \{f \in L^p : f' \in L^p\}$. We have the following

THEOREM: If F is a Lipschitz function on $\mathbf{R}$ then $F(f) \in W^{\alpha,p}$ for every $f \in W^{\alpha,p}$.

QUESTION. *Replace λ by a Gaussian measure μ and Δ by a self-adjoint operator $A = \Delta - x \cdot \nabla$ with respect to μ. Let*

$$W^{\alpha,p} = D_{L^p(\mu)}((I - A)^{\alpha/2}).$$

When $p = 2$ or $\alpha = 1$, it is known that the above theorem still holds. Is the theorem true for $p \neq 2$ and $\alpha < 1$?

F. Hirsch (Evry-Essone)

A Differentially Algebraic Version of Runge's Theorem

A set of meromorphic functions is said to be uniformly differentially algebraic (UDA) if there is one and the same algebraic differential equation satisfied by all the functions in the set. Given an open region G in the plane and a holomorphic function f on G that satisfies an algebraic differential equation, must f be the limit of a UDA sequence of rational functions with poles in the holes of G? Even in case G is the open unit disc D, the problem is open. Note that there exist differentially algebraic functions in D such that the partial sums of the Taylor series of f around 0 do not form a UDA family. One example is $1/(cos z)$.

Lee A. Rubel (Urbana)

Connected Level Sets

Given two nonconstant entire functions f and g, one would expect that for most complex constants q, the zero set of $f(z) + g(w) + q$ would be a connected set in complex two-space. (It has been shown to be true for all constants q if f and g are polynomials.) Prove that there always exist at least two such constants q. The example $f = g = e^z$ and $q = 0$ shows that it is not always true for all q. (It has been shown that if f, g, h are nonconstant entire functions, then the zero set of $f(x) + g(y) + h(z)$ is always connected.)

Lee A. Rubel (Urbana)

An approximation problem

Let D be a domain in R^n, $n \geq 2$. We ask what conditions on D ensure that every positive harmonic function $u(x)$ in D is the pointwise, and so locally uniform, limit of a sequence of functions $u_n(x)$ positive, (individually) bounded and harmonic in D.

A necessary condition is that the intersection of every open set G with the complement of D is either empty or has positive capacity. For by a Theorem of Brelot (see e.g. Hayman and Kennedy, subharmonic functions I, Theorem 5.18, p. 237) a bounded harmonic function in $G\backslash E$, where E has capacity zero, can be extended to a bounded harmonic function in G. The analogous property for positive harmonic functions clearly fails.

The above condition is not, however, sufficient. Let D be a Denjoy-domain, i.e. a domain whose complement lies in a hyperplane π and has zero $(n-1)$-dimensional measure but positive capacity. Then Armitage, Gardiner and Netuka (Math. Proc. Cambridge Philos. Soc. 113(1993), 561-571) remark that a bounded harmonic function $u(x)$ in D is given on both sides of π by the Poisson integral of the values of u on π. In particular, if x, x' are reflections of each other in π we have

$$u(x) = u(x'). \tag{1}$$

However, the authors note that (1) does not always hold for positive harmonic functions in D.

On the other hand if the boundary of D is sufficiently smooth, then positive harmonic functions u in D have a Riesz-Herglotz representation in terms of a boundary measure μ. (For the case when D is a ball, see e.g. Hayman, Subharmonic functions II, Theorem 6.1, p.286). We can approximate μ by suitably smooth measures μ_n and thus obtain an approximation of u by bounded positive harmonic functions u_n. These conditions are satisfied for instance by Dini-smooth domains. (Dahlberg, Proc. London Math. Soc. 33(1976), 238-250).

H.S. Shapiro, transmitted by W.K. Hayman (York University)

PH-regularity Points for Compact Sets

Let E be a compact subset of $\mathbf{C}^n$. We say that E is of *positive capacity with respect to pluriharmonic (PH) polynomials*, if there exists a positive constant M such that

$$log|u_k(z)| \leq kM + log||u_k||_E + klog(1+|z|), \quad z \in \mathbf{C}^n, \quad k = 1, 2, ...$$

where u_k is any PH polynomial of degree $\leq k$.

We say that E is *PH-regular at* $z_0 \in E$, if for every $b > 1$ there exist a positive constant M and a neighbourhood Ω of z_0 such that

$$||u||_\Omega \leq M b^{deg u} ||u||_E$$

for every PH-polynomial u.

QUESTION. *Let E be a compact set in $\mathbf{C}^n$ of positive capacity with respect to PH-polynomials. Is E then PH-regular at some point $z_0 \in E$?*

J. Siciak (Krakow)

Non-linear Potential Theory and Harmonic Measure

The following is a question of O. Martio.

Consider the p-Laplace operator $\Delta_p = \div(|\nabla|^{p-2}\nabla)$ on $\mathbf{R}^n$, $1 < p < \infty$. Let $\omega_p = \omega_p(x, K)$ be the $W^{1,p}_{loc}$-solution to

$$\begin{cases} \Delta_p u = 0 & \text{on } B \\ u = \chi_k & \text{on } \partial B \end{cases}$$

where B is the unit ball in $\mathbf{R^n}$ and K a compact subset of ∂B. ω_p plays the role of harmonic measure in the non-linear potential theory.

QUESTION. (Martio): Let E and F be compact subsets of ∂B. Does $\omega_p(E) = \omega_p(F) = 0$ imply $\omega_p(E \cup F) = 0$?

Remark: 1. The answer is yes if E and F are disjoint. ([GLM])

2. The answer is yes if $\dim(F)$ is smaller than a nonnegative number which depends only on p and n. [AM]).

References

[GLM] S. Granlund, P. Linqvist and O. Martio, *F*-harmonic measure in space, Ann. Acad. Sci. Fenn. AI Math 7(1982), 233-247.

[AM] P. Avilés and J. Manfredi, On null sets of p-harmonic measures, Partial Differential Equations with Minimal Smoothness and Applications (ed. B. Dahlberg, E. Fabes, R. Fefferman, D. Jerison, C. Kenig, J. Pipher), pp. 33-36. Springer-Verlag, Berlin.

Jang-Mei Wu (Urbana)